HANDBOOK OF
WOOD CHEMISTRY AND WOOD COMPOSITES

HANDBOOK OF WOOD CHEMISTRY AND WOOD COMPOSITES

Edited by
Roger M. Rowell

Taylor & Francis
Taylor & Francis Group

Boca Raton   London   New York   Singapore

A CRC title, part of the Taylor & Francis imprint, a member of the
Taylor & Francis Group, the academic division of T&F Informa plc.

Library of Congress Cataloging-in-Publication Data

Handbook of wood chemistry and wood composites / edited by Roger M. Rowell.
p. cm.
Includes bibliographical references and index.
ISBN 0-8493-1588-3 (alk. paper)
1. Wood—Chemistry—Handbooks, manuals, etc. 2. Engineered wood—Handbooks, manuals, etc.
I. Rowell, Roger M.

TS932.H36 2005
620.1'2—dc22   2004057915

This book contains information obtained from authentic and highly regarded sources. Reprinted material is quoted with permission, and sources are indicated. A wide variety of references are listed. Reasonable efforts have been made to publish reliable data and information, but the author and the publisher cannot assume responsibility for the validity of all materials or for the consequences of their use.

Neither this book nor any part may be reproduced or transmitted in any form or by any means, electronic or mechanical, including photocopying, microfilming, and recording, or by any information storage or retrieval system, without prior permission in writing from the publisher.

All rights reserved. Authorization to photocopy items for internal or personal use, or the personal or internal use of specific clients, may be granted by CRC Press, provided that $1.50 per page photocopied is paid directly to Copyright Clearance Center, 222 Rosewood Drive, Danvers, MA 01923 USA. The fee code for users of the Transactional Reporting Service is ISBN 0-8493-1588-3/05/$0.00+$1.50. The fee is subject to change without notice. For organizations that have been granted a photocopy license by the CCC, a separate system of payment has been arranged.

The consent of CRC Press does not extend to copying for general distribution, for promotion, for creating new works, or for resale. Specific permission must be obtained in writing from CRC Press for such copying.

Direct all inquiries to CRC Press, 2000 N.W. Corporate Blvd., Boca Raton, Florida 33431.

Trademark Notice: Product or corporate names may be trademarks or registered trademarks, and are used only for identification and explanation, without intent to infringe.

Visit the CRC Press Web site at www.crcpress.com

© 2005 by CRC Press

No claim to original U.S. Government works
International Standard Book Number 0-8493-1588-3
Library of Congress Card Number 2004057915
Printed in the United States of America 1 2 3 4 5 6 7 8 9 0
Printed on acid-free paper

3 3001 00907 239 7

For a teacher,
Success is not measured in the number of students
that conform to your way of thinking,
It is measured in the number of students in which
you have confirmed the excitement of the learning process.

Roger Rowell

Preface

Wood has played a major role throughout human history. The earliest humans used wood to make shelters, cook food, construct tools, and make weapons. There are human marks on a climbing pole that were made over 300,000 years ago. We have found wood in the Egyptian pyramids, Chinese temples and tombs, and ancient ships, attesting to the use of wood by past societies. Collectively, society learned very early the great advantages of using a resource that was widely distributed, multifunctional, strong, easy to work, aesthetic, sustainable, and renewable. Wood has been used by people for centuries as a building material; we have accepted its limitations, such as instability toward moisture and degradation due to microorganisms, termites, fire, and ultraviolet radiation, in use. We must accept that wood was designed by nature to perform in a wet environment and that nature is programmed to recycle wood to carbon dioxide and water using the chemistries of biological decay and thermal, ultraviolet, and moisture degradations. By accepting these limitations, however, we also limit our expectations of performance, which, ultimately, limits our ability to accept new concepts for improved materials.

As we start the twenty-first century, we are concerned about issues dealing with the environment, sustainability, recycling, energy, sequestering carbon, and the depletion of our natural resources by a growing world population. In many ways, we are rediscovering wood as a material. We will not, however, be able to realize the full potential of the role that wood and wood products can play in our "modern society" as materials and chemical feed stock until we fully understand their chemistry and material properties. That understanding holds the keys to effective utilization. Wood will not reach its highest use potential until we fully describe it, understand the mechanisms that control its performance properties, and, finally, become able to manipulate those properties to elicit the performance we seek.

The purpose of this book is to present the latest concepts in wood chemistry and wood composites as understood by the various authors who have written the chapters. I thank them for their time and effort in the preparation of this book. The book is an update of an earlier book of this editor, *Chemistry of Solid Wood,* Advances in Chemistry Series No. 207, American Chemical Society, Washington, D.C., 1984, which is long out of print.

Roger M. Rowell

About the Editor

Roger M. Rowell is a project leader and research chemist at the U.S. Department of Agriculture, Forest Service, Forest Products Laboratory, in Madison, Wisconsin. He is also a professor in the Departments of Forest Ecology and Management and Biological Systems Engineering at the Engineering Center for Plasma Research at the University of Wisconsin, Madison, and in the Department of Chemical Engineering, Chalmers University, Gothenburg, Sweden. He has worked on projects for the United Nations and with many universities, institutes, and companies around the world on bio-based composites. He has taught courses in wood chemistry in several countries and has presented many lectures at international and national scientific meetings. His research specialties are in the areas of carbohydrate chemistry, chemical modification of lignocellulosics for property enhancement, water quality, and sustainable materials. He has been a visiting scholar in Japan, Sweden, the United Kingdom, New Zealand, and China. He is a fellow of the International Academy of Wood Science and the American Chemical Society's Division of Cellulose, Paper, and Textiles.

He received his BS degree in chemistry and mathematics from Southwestern College in Winfield, Kansas, his MS in biochemistry from Purdue University in West Lafayette, Indiana, and his PhD in biochemistry from Purdue University. He has edited 9 books and has over 300 publications and 22 patents.

He is married to his wonderful wife, Judith (who has spent many hours helping with this book), and his family includes three grown sons, their wives, and eight grandchildren.

List of Contributors

Lars Bergland
Royal Institute of Technology
Department of Aeronautical
and Vehicle Engineering
Stockholm, Sweden

Von L. Byrd
USDA, Forest Service
Forest Products Laboratory
Madison, WI

Daniel F. Caulfield
USDA, Forest Service
Forest Products Laboratory
Madison, WI

Craig Clemons
USDA, Forest Service
Forest Products Laboratory
Madison, WI

L. Emilio Cruz-Barba
Center for Plasma Aided Manufacturing
University of Wisconsin
Madison, WI

Ferencz S. Denes
Department of Biological Systems Engineering
and Center for Plasma Aided Manufacturing
University of Wisconsin
Madison, WI

W. Dale Ellis
USDA, Forest Service
Forest Products Laboratory
Madison, WI

Charles R. Frihart
USDA, Forest Service
Forest Products Laboratory
Madison, WI

James S. Han
USDA, Forest Service
Forest Products Laboratory
Madison, WI

Rebecca E. Ibach
USDA, Forest Service
Forest Products Laboratory
Madison, WI

Rodney E. Jacobson
AJ Engineering
Madison, WI

Susan L. LeVan-Green
USDA, Forest Service
Forest Products Laboratory
Madison, WI

Sorin Manolache
Center for Plasma Aided Manufacturing
University of Wisconsin
Madison, WI

Regis B. Miller
USDA, Forest Service
Forest Products Laboratory
Madison, WI

Roger Pettersen
USDA, Forest Service
Forest Products Laboratory
Madison, WI

Christopher D. Risbrudt
USDA, Forest Service
Forest Products Laboratory
Madison, WI

Jeffrey S. Rowell
Department of Forest Ecology
and Management
University of Wisconsin
Madison, WI

Roger M. Rowell
USDA, Forest Service
Forest Products Laboratory
Department of Biological Systems Engineering
University of Wisconsin
Madison, WI

Mandla A. Tshabalala
USDA, Forest Service
Forest Products Laboratory
Madison, WI

Alex C. Wiedenhoeft
USDA, Forest Service
Forest Products Laboratory
Madison, WI

R. Sam Williams
USDA, Forest Service
Forest Products Laboratory
Madison, WI

Jerrold E. Winandy
USDA, Forest Service
Forest Products Laboratory
Department of Bio-Based Products
University of Minnesota
St. Paul, MN

Contents

Preface vii

Chapter 1 Wood and Society 1
Christopher D. Risbrudt

PART I Structure and Chemistry

Chapter 2 Structure and Function of Wood 9
Alex C. Wiedenhoeft and Regis B. Miller

Chapter 3 Cell Wall Chemistry 35
Roger M. Rowell, Roger Pettersen, James S. Han, Jeffrey S. Rowell, and Mandla A. Tshabalala

PART II Properties

Chapter 4 Moisture Properties 77
Roger M. Rowell

Chapter 5 Biological Properties 99
Rebecca E. Ibach

Chapter 6 Thermal Properties 121
Roger M. Rowell and Susan L. LeVan-Green

Chapter 7 Weathering of Wood 139
R. Sam Williams

Chapter 8 Surface Characterization 187
Mandla A. Tshabalala

PART III Wood Composites

Chapter 9 Wood Adhesion and Adhesives 215
Charles R. Frihart

Chapter 10 Wood Composites 279

Lars Bergland and Roger M. Rowell

Chapter 11 Chemistry of Wood Strength 303

Jerrold E. Winandy and Roger M. Rowell

Chapter 12 Fiber Webs 349

Roger M. Rowell, James S. Han, and Von L. Byrd

Chapter 13 Wood Thermoplastic Composites 365

*Daniel F. Caulfield, Craig Clemons, Rodney E. Jacobson,
and Roger M. Rowell*

PART IV Property Improvements

Chapter 14 Chemical Modification of Wood 381

Roger M. Rowell

Chapter 15 Lumen Modifications...421

Rebecca E. Ibach and W. Dale Ellis

Chapter 16 Plasma Treatment of Wood 447

Ferencz S. Denes, L. Emilio Cruz-Barba, and Sorin Manolache

Index 475

Hōryū Temple in Nara, Japan. At over 1300 years old, this temple is one of the oldest remaining wooden structures in the world.

1 Wood and Society

Christopher D. Risbrudt
USDA, Forest Service, Forest Products Laboratory, Madison, WI

Forests, and the wood they produce, have played an important role in human activity since before recorded history. Indeed, one of the first major innovations of humankind was utilizing fire, fueled by wood, for cooking and heating. It is very likely that early hominids used wood fires for cooking as long as 1.5 million years ago (Clark and Harris 1985). Clear evidence of this use of wood exists from sites 400,000 years old (Sauer 1962). Since this ancient beginning, the uses of wood, and the value of the forest, have expanded dramatically, as the population of humans and their economies grew. Wood was used in myriad products, including agricultural implements and tools, shelters and houses, bridges, road surfaces, ships and boats, arrows and bows, spears, shoes, wheelbarrows, wagons, ladders, and thousands of others. Other important products that forests provided were food, in the form of berries, nuts, fruits, and wild animals, and, of course, fuel. Wood was the most important material in early human economies, and though other materials have grown in importance, wood used for solid products, fiber, and chemicals is still the largest single type of raw material input by weight—with the one exception of crushed stone, sand, and gravel—into today's economy (Haynes 2003).

Wood is still the major source of cooking and heating fuel for most of the world. In 2002, world consumption of fuelwood and charcoal totaled 1,838,218,860 cubic meters. This represents nearly 54% of the world's consumption of wood. About 43% of this fuelwood consumption occurs in Asia, and Africa consumes 31%. The United States consumes only 4% of the world's total of fuelwood and charcoal (Food and Agriculture Organization of the United Nations 2004). Total world consumption of roundwood, which includes fuelwood, charcoal, and industrial wood, amounted to 3,390,684,310 m^3 in 2002 (FAO of the UN 2004).

Besides producing fuelwood and wood for construction and other uses, forests have always been an important part of the American landscape, playing a key role in the social, economic, and spiritual life of the country. As the American population and economy grew, forests were removed to make way for farms, cities, and roadways. After the first European settlements in North America, forests were often viewed as an obstacle to farming and travel. Huge acreages were cleared in the 19th century to make way for fields, pastures, cities, and industry. In 1800, total cropland area in the United States extended across 20 million acres. By 1850, this had grown to 76 million acres, with pasture and hayland at perhaps twice that amount. Most of this farmland expansion was at the expense of forests (MacCleery 1996). The amount of cropland in the United States peaked in 1932, at about 361 million acres. (USDA National Agricultural Statistical Service 2003) However, although much forestland has been converted to other uses, the net area of forestland has remained relatively stable since the 1920s (Alig et al. 2003). As shown in Figure 1.1, about 70% of the original amount of forested land still remains as forest, although much of it is likely modified from its structure and composition in 1600. Since 1932, however, as farmed land acreage decreased, forest area in the United States has been increasing. Forests have been the beneficiary of the conversion from animal power to mechanical power in farming. An estimated 20 million acres of grain fields and pastures were no longer needed when gasoline tractors replaced horses and mules. As agricultural productivity per acre increased, as a result of plant breeding, fertilizers, and pesticides, forests have reclaimed many acres back from farm fields.

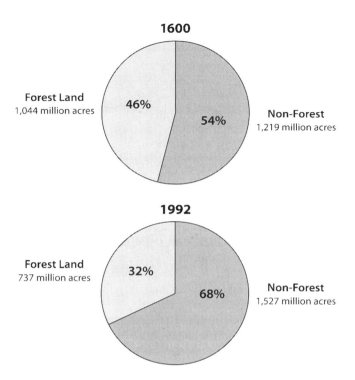

FIGURE 1.1 United States forest area. Forests as a percentage of total U.S. land area, 1600–1992. (From Preliminary data from the 1992 Program Update, USDA Forest Service, 1993.)

Wood has remained an important substance throughout history because of its unique and useful properties. Wood is recyclable, renewable, and biodegradable. Many species are shock resistant, bendable, and stable (although all wood changes dimensions as it loses or gains moisture). Density among species varies greatly; the balsa popular with model-airplane builders can weigh as little as 6 pounds per cubic foot, and some tropical hardwoods weigh more than 70 pounds per cubic foot. Wood and lignin can be converted to many useful industrial chemicals, such as ethanol and plastics. Wood can be treated to resist decay, and with proper construction techniques, and stains or paint, wood buildings can last hundreds of years. The oldest surviving wood structure is an Asian temple, built in the 7th century. Today, wood is used in tools, paper, buildings, bridges, guardrails, railroad ties, posts, poles, mulches, furniture, packaging, and thousands of other products.

Wood's versatility makes many wood products recyclable. Perhaps the earliest and simplest recycling was the burning of used wood for heat, whether in a wood stove, fireplace, or furnace. New technologies are improving the efficiency with which used or scrap wood can generate electricity and heat. The paper and paperboard industry has used recycled paper to augment virgin wood pulp for decades. At first, recycled paper generally found its way into newsprint and other low-grade products, but recent advances in recycling technology permit used paper to go into the manufacture of higher-quality papers, where appearance, texture, and consistency are important. Other products, whether railroad ties or structural timbers from 60-year-old buildings, find second lives as lumber. Affluent consumers, especially in the United States, have long been willing to pay a premium for the aesthetic pleasure of using 100-year-old barn siding as interior paneling. More recently, entrepreneurs have recognized the availability and potential value of millions of board-feet of high-quality lumber in World War II–era buildings sitting on closed (or soon-to-be-closed) military bases.

Other recycling opportunities for wood include the manufacture of wood fiber and plastic composite materials, in which wood fibers improve the strength-to-weight ratio over that of plastic alone—a performance characteristic that has strong appeal in the automotive industry, among others.

The ability of forests to regenerate on abandoned farmland—or after a destructive forest fire—testifies to the renewable nature of the wood resource. Though consumption of wood outpaces growth in some parts of the world, in the United States, trees have been growing and producing wood faster than they have been harvested since the early 1950s. By 1970, U.S. forestlands produced more than 20 billion cubic feet of wood, some 5 billion more than the harvest. (MacCleery 1996) Much wood for U.S. home construction and the nation's paper industry comes from plantations—mostly southern pine—in the southeastern United States. Those plantations depend on the resource's renewability, with trees harvested and new ones planted and then harvested 15 to 30 years later. The major inputs are abundant: rain, sun, airborne carbon, and soil nutrients.

Wood's chemical makeup is largely carbon, hydrogen, and oxygen, arranged as cellulose, hemicellulose, and lignin. As such, wood presents an appetizing feast for a variety of fungal species that can metabolize either the sugar-like celluloses or the more complex lignin. With the help of these fungi in the presence of air and water, wood rots or, in environmental terms, is biologically degraded.

Wood is renewable, recyclable, and biodegradable—characteristics generally accepted as good for the environment. At the beginning of the 21st century, however, the most environmentally friendly aspect of wood may be its role in carbon sequestration. Growing trees soak up great quantities of carbon from atmospheric carbon dioxide (CO_2), widely regarded as a greenhouse gas that traps heat and affects global climates. Indeed, dry wood is roughly half carbon by weight. Each cubic foot of wood contains between 11 and 20 pounds of carbon. A single tree can easily contain a ton or more of carbon. In addition to the carbon in the tree above ground, significant carbon is locked in the roots and surrounding soil. With roughly one third of the U.S. land area, or some 747 million acres, in forestland, the nation's forests hold more than 50 billion tons of carbon out of the atmosphere (U.S. Environmental Protection Agency 2002).

Forests hold unique significance in the environments that Americans value (see Table 1.1). Not only is the United States the world's largest importer *and* exporter of wood products, the forests of America are also highly regarded for their recreational, aesthetic, spiritual, and natural values. Forests are valued for providing wildlife and fish habitat, clean water, and clean air. Forests are further shown to be important sinks of carbon, slowing and ameliorating global warming. These concerns are all melded into the concept of sustainability. This concept, expressed through worldwide concern in the 1992 Rio de Janeiro agreement, has resulted in a multinational effort to measure forest sustainability. The Montreal Process, resulting from the Rio agreement, lists seven criteria and 67 indicators of factors and conditions that can help in the judgment of sustainability:

1. Conservation of biological diversity
2. Maintenance of productive capacity
3. Maintenance of forest ecosystem health and vitality
4. Maintenance of soil and water resources
5. Maintenance of forest contribution to global carbon cycles
6. Maintenance and enhancement of long-term multiple socio-economic benefits to meet the needs of societies
7. Legal, institutional, and economic framework; capacity to measure and monitor changes; and capacity to conduct and apply research and development for forest conservation and sustainable management.

The U.S. has just completed its first assessment of these criteria and indicators (USDA Forest Service 2004).

TABLE 1.1
Indicator Variables for Outputs from FS- and BLM-Administered Lands Suitable for an Ecosystem Market Basket

Indicator Variable

Carbon storage
Ecosystem health
Fire risk to life and property
Fish
Game
Minerals
Passive-use values:
   Existence of salmon
   Existence of other threatened and endangered species
   Existence of unroaded areas
Range
Recreation:
   Access (roads)
   Access to riparian areas
Science integrity
Soil productivity
Special forest and range products
Timber
Visibility
Water quality

Source: Haynes and Horne, 1997.

Not surprisingly, the assessment produces mixed results in almost every category. Although the total area of forest in the United States has remained stable for the past 80 years, the location of forests is changing and the nature of forests and how they are used is changing. Though much forest habitat has remained stable over recent decades, some forest plants, birds, and other animals are at risk of extinction. Net growth on timberland continues to exceed harvest removals. The downside of this is that the net gain can represent overcrowding, which increases risk of wildfire and of susceptibility to insects and disease, serious threats to the forest ecosystems.

Healthy forests function well as sources of water for towns and cities, especially in the Western United States. The first sustainability assessment indicates, however, that at least 10% of the forested counties in each region have areas where forest conditions have deteriorated and water and soil quality have been compromised through reduced oxygen levels, higher sediment, dissolved salts, or acidity.

The assessment of long-term socioeconomic benefits to society reveals the interrelatedness and interdependence of many indicators. Recreational use of the nation's forests is increasing. At the same time, the increased demand for wood and wood products has not led to increased harvest of U.S. trees. It appears that increased wood imports and increased recovery of paper for recycling have enabled Americans to use more wood and paper without cutting more trees from America's forests. The global implications—economic, social, political, and environmental—can only be guessed.

The United States, through government agencies, nongovernmental organizations and institutions, industry, and academia, conducts extensive research regarding forests, wood, recycling, and related topics. Nonetheless, the national debate about the proper care and nurture of forests appears in many cases to be rooted in emotion and politics rather than science. The debate can be shrill at times, which reflects perhaps the intensity with which our culture regards the forests. Perhaps the greatest challenge to ensuring the long-term sustainability of healthy forests that provide our society with valued resources and recreational and aesthetic opportunities, as well as environmentally vital carbon sequestration, lies not in science but in finding a way to sit down at the table together and agree on some goals and the continued need to explore alternative ways to achieve those goals.

REFERENCES

Alig, R.J., Pantinga, A.J., Ahn, S.E., and Kline, J.D. (2003). Land use changes involving forestry in the United States: 1952 to 1997, with projections to 2050. General Technical Report PNW-GTR-587. U.S. Department of Agriculture, Forest Service, Pacific Northwest Research Station, Portland, OR.

Clark, J.D. and Harris, J.W.K. (1985). Fire and its roles in early hominid lifeways. *The African Archaeological Review* 3:3–27.

Food and Agriculture Organization of the United Nations. (2004). FAO statistical data 2004: Forestry data, roundwood, sawnwood, woodbased panels. Available at www.faostat.fao.org (accessed May 14, 2004).

Haynes, R.W. (Tech. Coord.) (2003). An analysis of the timber situation in the United States: 1952–2050. General Technical Report PNW-GTR-560. USDA Forest Service, Portland, OR.

MacCleery, D.W. (1996). *American Forests: A History of Resiliency and Recovery* (rev. ed.). Forest History Society Issues Series. Forest History Society.

Sauer, C.O. (1962). Fire and early man. *Paideuma* 7:399–407.

U.S. Environmental Protection Agency. (2002). Inventory of U.S. Greenhouse Gas Emissions and Sinks: 1990–2000. EPA 430-R-02-003. U.S. Government Printing Office, Washington, DC.

USDA National Agricultural Statistical Service. (2003). *Historical track records*. U.S. Government Printing Office, Washington, DC.

USDA Forest Service. (2004). National Report on Sustainable Forests—2003. Report FS-766. U.S. Government Printing Office, Washington, DC.

Part I

Structure and Chemistry

2 Structure and Function of Wood

Alex C. Wiedenhoeft and Regis B. Miller
USDA, Forest Service, Forest Products Laboratory, Madison, WI

CONTENTS

2.1 The Tree 10
2.2 Softwoods and Hardwoods 11
2.3 Sapwood and Heartwood 12
2.4 Axial and Radial Systems 13
2.5 Planes of Section 13
2.6 Vascular Cambium 14
2.7 Growth Rings 15
2.8 Cells in Wood 17
2.9 Cell Walls 18
2.10 Pits 20
2.11 The Microscopic Structure of Softwoods and Hardwoods 21
    2.11.1 Softwoods 21
        2.11.1.1 Tracheids 21
        2.11.1.2 Axial Parenchyma and Resin Canal Complexes 21
        2.11.1.3 Rays 23
    2.11.2 Hardwoods 24
        2.11.2.1 Vessels 24
        2.11.2.2 Fibers 25
        2.11.2.3 Axial Parenchyma 26
        2.11.2.4 Rays 26
2.12 Wood Technology 27
2.13 Juvenile Wood and Reaction Wood 30
2.14 Wood Identification 31
References 32

Despite the many human uses to which various woods are suited, at a fundamental level wood is a complex biological structure, itself a composite of many chemistries and cell types acting together to serve the needs of the plant. Although humans have striven to understand wood in the context of wood technology, we have often overlooked the key and basic fact that wood evolved over the course of millions of years to serve three main functions in plants: the conduction of water from the roots to the leaves, the mechanical support of the plant body, and the storage of biochemicals. The need for these three functions has driven the evolution of approximately 20,000 different extant species of woody plants, each with unique properties, uses, and capabilities, in both plant and human contexts. Understanding the basic requirements dictated by these three functions and identifying the structures in wood that perform them allows insights into the realm of human wood use

(Hoadley 2000). A scientist with a robust understanding of the interrelationships between form and function can predict the usefulness of a specific wood in a new context.

To begin, it is necessary to define and delimit the component parts of wood at a variety of scales. There is a significant difference in the quality and quantity of wood anatomical expertise necessary for a researcher who is using a solid wood beam compared to the knowledge necessary for an engineer designing a glued-laminated beam, and these are in turn different compared to the knowledge required for making a wood-resin composite with wood flour. In the first case, a large-scale anatomical understanding may help to explain and quantify the mechanical properties of the beam. In the second case, an understanding of anatomical effects on mechanical properties must be coupled with chemical knowledge about the efficacy of various adhesives. In the third case, an understanding of particle size distribution and wood cell wall chemistry will be key pieces of knowledge. The differences in the kinds of knowledge in these three cases are related to the scale at which one intends to interact with wood, and in all three cases the technologically different properties are derived from the biological needs of the living tree. For this reason, the structure of wood will be explained in this chapter at decreasing scales, and in ways that demonstrate the biological rationale for a plant to produce wood with such features. Such background will permit the reader to access primary literature related to wood structure with greater ease.

Although shrubs and many vines form wood, the remainder of this chapter will focus on the wood from trees. As trees are the predominant source of wood for commercial applications and provide examples of virtually all features that merit discussion, this restriction of scope is warranted.

2.1 THE TREE

The general body plan of a tree must be briefly outlined so that all subsequent information can be understood in its proper context within the living organism. A living, growing tree has two main domains, the shoot and the roots. The roots are the subterranean structures responsible for water uptake, mechanical support of the shoot, and storage of biochemicals. The shoot comprises the

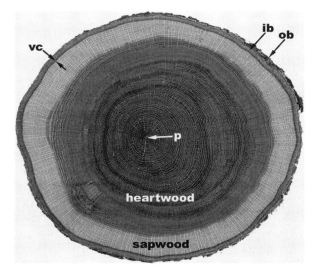

FIGURE 2.1 Macroscopic view of a transverse section of a *Quercus alba* trunk. Beginning at the outside of the tree, there is the outer bark (ob), the inner bark (ib), and then the vascular cambium (vc), which is too narrow to see at this magnification. Interior to the vascular cambium is the sapwood, which is easily differentiated from the heartwood that lies to the interior. At the center of the trunk is the pith (p), which is barely discernible in the center of the heartwood.

trunk or bole of the tree, the branches, and the leaves (Raven et al. 1999). It is with the trunk of the tree that the remainder of the chapter will be concerned.

If one cuts down a tree and looks at the stump, there are several gross observations that can be easily made. The trunk is composed of various materials present in concentric bands. From the outside of the tree to the inside there are six layers: outer bark, inner bark, vascular cambium, sapwood, heartwood, and the pith (Figure 2.1). Outer bark provides mechanical protection to the softer inner bark, and also helps to limit evaporative water loss. Inner bark (phloem) is the tissue through which sugars produced by photosynthesis (photosynthate or "food") are translocated from the leaves to the roots or growing portions of the tree. The vascular cambium is the layer between the bark and the wood that is responsible for producing both these tissues. The sapwood is the active, "living" wood that is responsible for conducting the water (or sap) from the roots to the leaves. It has not yet accumulated the often-colored chemicals that set apart the nonconductive heartwood found as a core of darker-colored wood in the middle of most trees. The pith at the very center of the trunk is the remnants of the early growth of the trunk, before wood was formed.

2.2 SOFTWOODS AND HARDWOODS

To define them botanically, softwoods are those woods that come from gymnosperms (mostly conifers), and hardwoods are woods that come from angiosperms (flowering plants). In the temperate portion of the Northern Hemisphere, softwoods are generally needle-leaved evergreen trees such as pine (*Pinus*) and spruce (*Picea*), whereas hardwoods are typically broadleaf, deciduous trees such as maple (*Acer*) and birch (*Betula*). Not only do softwoods and hardwoods differ in terms of the types of trees from which they are derived, but they also differ in terms of their component cells. The single most important distinction between the two general kinds of wood is that hardwoods have a characteristic type of cell called a vessel element (or pore), whereas softwoods lack these (Figure 2.2). An important cellular similarity between softwoods and hardwoods is that

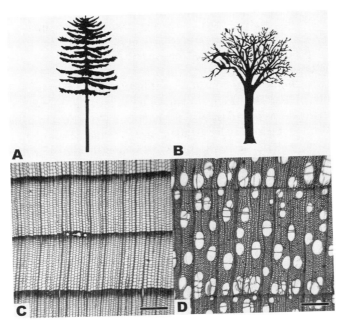

FIGURE 2.2 Softwood and hardwood. (A) The general form of a generic softwood tree. (B) The general form of a generic hardwood tree. (C) Transverse section of *Pseudotsuga mensiezii*, a typical softwood. The three round white spaces are resin canals. (D) Transverse section of *Betula alleghoniensis*, a typical hardwood. The many large, round white structures are vessels or pores, the characteristic feature of a hardwood. Scale bars = 300 μm.

in both kinds of wood, most of the cells are dead at maturity even in the sapwood. The cells that are alive at maturity are known as parenchyma cells, and can be found in both softwoods and hardwoods. Additionally, despite what one might conclude based on the names, not all softwoods have soft, lightweight wood, nor do all hardwoods have hard, heavy wood.

2.3 SAPWOOD AND HEARTWOOD

In both softwoods and hardwoods, the wood in the trunk of the tree is typically divided into two zones, each of which serves an important function distinct from the other. The actively conducting portion of the stem, in which the parenchyma cells are still alive and metabolically active, is referred to as the sapwood. A looser definition that is more broadly applied is that the sapwood is the band of lighter-colored wood adjacent to the bark. The heartwood is the darker-colored wood found to the interior of the sapwood (Figure 2.1).

In the living tree, the sapwood is responsible not only for the conduction of sap, but also for the storage and synthesis of biochemicals. This function is often underappreciated in wood technological discourse. An important storage function is the long-term storage of photosynthate. The carbon that must be expended to form a new flush of leaves or needles must be stored somewhere in the tree, and it is often in the parenchyma cells of the sapwood that this material is stored. The primary storage forms of photosynthate are starch and lipids. Starch grains are stored in the parenchyma cells, and can be easily seen using a microscope. The starch content of sapwood can have important ramifications in the wood industry. For example, in the tropical tree ceiba (*Ceiba pentandra*), an abundance of starch can lead to the growth of anaerobic bacteria that produce ill-smelling compounds that can make the wood unusable (Chudnoff 1984). In the southern yellow pines of the United States, a high starch content encourages the growth of sap-stain fungi that, though they do not effect the strength of the wood, can nonetheless cause a significant decrease in lumber value for aesthetic reasons (Simpson 1991).

The living cells of the sapwood are also the agents of heartwood formation. In order for the tree to accumulate biochemicals, they must be actively synthesized and translocated by living cells. For this reason, living cells at the border between the heartwood and sapwood are responsible for the formation and deposition of heartwood chemicals, one of the important steps leading to heartwood formation (Hillis 1996).

Heartwood functions in the long-term storage of biochemicals of many varieties depending on the species in question. These chemicals are known collectively as extractives. In the past it was thought that the heartwood was a disposal site for harmful by-products of cellular metabolism, the so-called secondary metabolites. This led to the concept of the heartwood as a dumping ground for chemicals that, to a greater or lesser degree, would harm the living cells if not sequestered in a safe place. A more modern understanding of extractives indicates that they are a normal and intentional part of the plant's efforts to protect its wood. Extractives are formed by parenchyma cells at the heartwood-sapwood boundary and are then exuded through pits into adjacent cells (Hillis 1996). In this way it is possible for dead cells to become occluded or infiltrated with extractives despite the fact that these cells lack the ability to synthesize or accumulate these compounds on their own.

Extractives are responsible for imparting several larger-scale characteristics to wood. For example, extractives provide natural durability to timbers that have a resistance to decay fungi. In the case of a wood such as teak (*Tectona grandis*), famed for its stability and water resistance, these properties are conferred by the waxes and oils formed and deposited in the heartwood. Many woods valued for their colors, such as mahogany (*Swietenia mahagoni*), African blackwood (*Diospyros melanoxylon*), Brazilian rosewood (*Dalbergia nigra*), and others, owe their value to the type and quantity of extractives in the heartwood. For these species, the sapwood has little or no value, because the desirable properties are imparted by heartwood extractives. Gharu wood, or eagle wood (*Aquilaria malaccensis*) has been driven to endangered status due to human harvest of the wood to

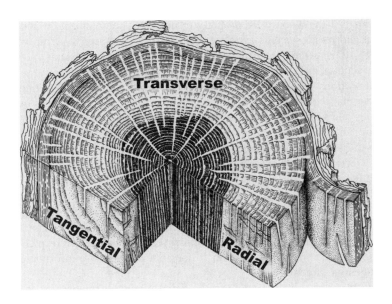

FIGURE 2.3 Illustration of the three planes of section. Note that for the tangential plane of section, only the right-hand portion of the cut is perpendicular to the rays; due to the curvature of the rings, the left portion of the cut is out of plane. From *Biology of Plants*, 4/e, by Peter H. Raven, et. al. © 1971, 1976, 1986 Worth Publishers. Used with permission.

extract valuable resins used in perfume making (Lagenheim 2003). Sandalwood (*Santalum spicatum*), a wood famed for its use in incenses and perfumes, is only valuable if the heartwood is rich with the desired aromatic extractives. The utility of a wood for a technological application can be directly affected by extractives. For example, if a wood high in hydrophobic extractives is used in a composite bonded with a water-based adhesive, weak or incomplete bonding can result.

2.4 AXIAL AND RADIAL SYSTEMS

The distinction between sapwood and heartwood, though important, is a gross feature that is often fairly easily observed. More detailed inquiry into the structure of wood shows that wood is composed of discrete cells that are connected and interconnected in an intricate and predictable fashion to form an integrated system that is continuous from root to twig. The cells of wood are typically many times longer than wide, and are specifically oriented in two separate systems of cells: the axial system and the radial system. The cells of the axial system have their long axes running parallel to the long axis of the organ (e.g., up and down the trunk). The cells of the radial system are elongated perpendicularly to the long axis of the organ, and are oriented like radii in a circle or spokes in a bicycle wheel, from the pith to the bark (Figure 2.3). In the trunk of a tree, the axial system runs up and down, functions in long-distance water movement, and provides the bulk of the mechanical strength of the tree. The radial system runs in a pith-to-bark direction, provides lateral transport for biochemicals, and in many cases performs a large fraction of the storage function in wood. These two systems are interpenetrating and interconnected, and their presence is a defining characteristic of wood as a tissue.

2.5 PLANES OF SECTION

Though one could cut wood in any direction and then look at it, such an approach would, in the vast majority of cases, result in perspectives that can provide only a small proportion of the information that could be gleaned if the wood were properly examined. The organization and interrelationship between the axial and radial systems give rise to three main perspectives from which they can be viewed (Figure 2.3). These three perspectives are the transverse plane of section (the cross-section), the radial plane of section, and the tangential plane of section. The latter planes of section are referred to as longitudinal sections, because they extend parallel to the axial system (along the grain).

The transverse plane of section is the face that is exposed when a tree is cut down; looking down at the stump one sees the transverse section. Cutting a board across the grain exposes the transverse section. The transverse plane of section provides information about features that vary both in the pith-to-bark direction (called the radial direction) and also those that vary in the circumferential direction (call the tangential direction). It does not provide information about variations up and down the trunk.

The radial plane of section runs in a pith-to-bark direction, and it is parallel to the axial system, so it provides information about longitudinal changes in the stem and from the pith to bark along the radial system. To describe it geometrically, it is parallel to the radius of a cylinder, and extending up and down the length of the cylinder. In a practical sense, it is the face or plane that is exposed when a log is split exactly from pith to bark. It does not provide any information about features that vary in a tangential direction.

The tangential plane is at a right angle to the radial plane. Geometrically, it is parallel to any tangent line that would touch the cylinder, and it extends along the length of the cylinder. One way in which the tangential plane would be exposed is if the bark were peeled from a log; the exposed face is the tangential plane. The tangential plane of section does not provide any information about features that vary in the radial direction, but it does provide information about the tangential dimensions of features.

All three planes of section are important to the proper observation of wood, and only by looking at each in turn can a holistic and accurate understanding of the three-dimensional structure of wood be gained. The three planes of section are determined by the structure of wood, and the way in which the cells in wood are arrayed. The cells are laid down in these special arrangements by a special part of the trunk.

2.6 VASCULAR CAMBIUM

The axial and radial systems and their component cells are derived from a special part of the tree called the vascular cambium. The vascular cambium is a thin layer of cells that exists between the inner bark and the wood (Figure 2.1, Figure 2.4A), and is responsible for forming, by means of

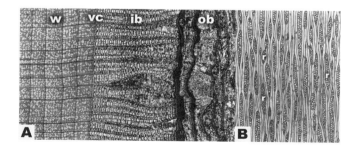

FIGURE 2.4 Light microscopic views of the vascular cambium. (A) Transverse section showing wood (w), vascular cambium (vc), inner bark (ib), and outer bark (ob) in *Tilia americana*. (B) Tangential section through the vascular cambium of *Malus sylvestris*. Ray initials (r) occur in groups that will give rise to the rays. The vertically oriented cells are fusiform initials, which will give rise to the axial system. Scale bars not available. From *Biology of Plants*, 4/e, by Peter H. Raven, et. al. © 1971, 1976, 1986 Worth Publishers. Used with permission.

many cell divisions, wood (or secondary xylem) to the inside, and bark (or secondary phloem) to the outside, both of which are vascular conducting tissues (Larson 1994). As the vascular cambium adds cells to the layers of wood and bark around a tree, the girth of the tree increases, and thus the diameter and total surface area of the vascular cambium itself must increase, and this is accomplished by cell division as well.

The axial and radial systems are generated in the vascular cambium by two component cells: the fusiform initials and the ray initials (Figure 2.4B). The fusiform initials, named to describe their long, slender shape, give rise to the cells of the axial system, and the ray initials give rise to the radial system. For this reason, there is a direct and continuous link between the most recently formed wood, the vascular cambium, and the inner bark. In most cases, the radial system in the wood is continuous into the inner bark, through the vascular cambium. In this way the wood, a water-conducting tissue, stays connected to the photosynthate-conducting tissue, the inner bark. They are interdependent tissues, because the living cells in wood require photosynthate for respiration and cell growth, and the inner bark requires water in which to dissolve and transport the photosynthate. The vascular cambium is an integral feature that not only gives rise to these tissue systems, but also links them so that they may function in the living tree.

In the opening paragraph of this chapter, reference was made to the three functions of wood in the living tree. It is worth reiterating them and their relevance at this point. There is no property of wood, physical, mechanical, chemical, biological, or technological, that is not fundamentally derived from the fact that wood is formed to meet the needs of the living tree. A complementary view is that any anatomical feature of wood can be assessed in the context of the tree's need for water conduction, mechanical support, and storage of biochemicals. To accomplish any of these functions, wood must have cells that are designed and interconnected in ways suitable to perform these functions.

2.7 GROWTH RINGS

Wood is produced by the vascular cambium one layer of cell divisions at a time, but we know from general experience that in many woods there are large cohorts of cells produced more or less together in time, and these cohorts act together to serve the tree. These collections of cells produced together over a discrete time interval are known as growth increments or growth rings. The cells formed at the beginning of the growth increment are called earlywood cells and the cells formed in the latter portion of the growth increment are called latewood cells (Figure 2.5). Springwood and summerwood were terms formerly used to refer to earlywood and latewood, respectively, but their use is anachronistic and not recommended (IAWA Committee 1989).

In the temperate portions of the world and anywhere else where there is a distinct, regular seasonality, trees form their wood in annual growth increments; that is, all the wood produced in one growing season is organized together into a recognizable, functional entity that many sources refer to as annual rings. Such terminology reflects this temperate bias, so a preferred term is growth increment, or growth ring (IAWA Committee 1989). In many woods in the tropics growth rings are not evident. However, continuing research in this area has uncovered several characteristics whereby growth rings can be correlated with seasonality changes (Worbes 1995, Worbes 1999, Callado et al. 2001).

When one looks at woods that form distinct growth rings, and this includes most woods that are likely to be used for wood composites, there are three fundamental patterns within a growth ring: no change in cell pattern across the ring, a gradual reduction of the inner diameter of conducting elements from the earlywood to the latewood, and a sudden and distinct change in the inner diameter of the conducting elements across the ring (Figure 2.6). These patterns appear in both softwoods and hardwoods, but differ in each due to the distinct anatomical structural differences between the two. Many authors use the general term porosity to describe growth rings (recall that vessels and pores are synonymous.)

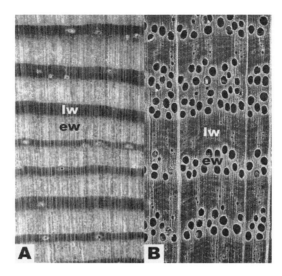

FIGURE 2.5 Hand-lens views (approximately 14x magnification) of the transverse section showing earlywood and latewood. (A) Distinction in a softwood growth ring between earlywood (ew) and latewood (lw) in *Pinus resinosa*. (B) Distinction in a hardwood growth ring between earlywood (ew) and latewood (lw) in *Quercus rubra*.

Nonporous woods (woods without vessels) are softwoods. Softwoods can exhibit any of the three general patterns noted above. Some softwoods such as Western red cedar (*Thuja plicata*), northern white cedar (*Thuja occidentalis*), and species of spruce (*Picea*) and true fir (*Abies*) have growth increments that undergo a gradual transition from the thin-walled wide-lumined earlywood cells to the thicker-walled, narrower-lumined latewood cells (Figure 2.6B). Other woods undergo an abrupt transition from earlywood to latewood, including Southern yellow pine (*Pinus*), larch

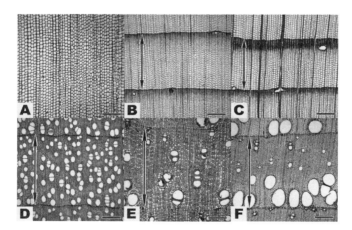

FIGURE 2.6 Transverse sections of woods showing types of growth rings. Arrows delimit growth rings, when present. (A–C) Softwoods: (A) No transition within the growth ring (growth ring absent) in *Podocarpus imbricata*. (B) Gradual transition from earlywood to latewood in *Picea glauca*. (C) Abrupt transition from earlywood to latewood in *Pseudotsuga mensiezii*. (D–F) Hardwoods: (D) Diffuse porous wood (no transition) in *Acer saccharum*. (E) Semi-diffuse porous wood (gradual transition) in *Diospyros virginiana*. (F) Ring porous wood (abrupt transition) in *Fraxinus americana*. Scale bars = 300 μm.

(*Larix*), Douglas fir (*Pseudotsuga menziesii*), bald cypress (*Taxodium disticum*), and redwood (*Sequoia sempervirens*) (Figure 2.6C). Since most softwoods are native to the north temperate regions, growth rings are clearly evident. Only in species such as araucaria (*Araucaria*) and some podocarps (*Podocarpus*) do you find no transition within the growth ring (Figure 2.6A). Many authors have reported this state as growth rings being absent or only barely evident (Phillips 1948, Kukachka 1960).

Porous woods (woods with vessels) are hardwoods, which have two main types of growth rings, and one intermediate form. In diffuse porous woods, the vessels either do not significantly change in size and distribution from the earlywood to the latewood or the change in size and distribution is gradual and no clear distinction between earlywood and latewood can be found (Figure 2.6D). Maple (*Acer*), birch (*Betula*), aspen/cottonwood (*Populus*), and yellow poplar (*Liriodendron tulipifera*) are examples of diffuse porous species.

This pattern is in contrast to ring porous woods in which the transition from earlywood to latewood is abrupt, i.e., the vessels reduce significantly (often by an order or magnitude or more) in diameter and often change their distribution as well. This creates a ring pattern of large, earlywood vessels around the inner portion of the growth increment, alternating with denser, more fibrous tissue in the latewood, as is found in hackberry (*Celtis occidentalis*), white ash (*Fraxinus americana*), shagbark hickory (*Carya ovata*), and northern red oak (*Quercus rubra*) (Figure 2.6F).

Sometimes the vessel size and distribution pattern falls more or less between these two definitions, and this condition is referred to as semi-ring porous (Figure 2.6E). Black walnut (*Juglans nigra*) and black cherry (*Prunus serotina*) are temperate-zone semi-ring porous woods. Most tropical hardwoods are diffuse porous except for Spanish cedar (*Cedrela*) and teak (*Tectona grandis*), which are generally semi-ring porous.

There are no distinctly ring porous species in the tropics and only a very few in the Southern Hemisphere. It is interesting that in genera that span temperate and tropical zones, it is common to have ring porous representatives in the temperate zone and diffuse porous species in the tropics. The oaks (*Quercus*), ashes (*Fraxinus*), and hackberries (*Celtis*) that are native to the tropics are diffuse porous, while their temperate relatives are ring porous. There are numerous detailed texts with more information on growth increments in wood, a few of which are of particular note (Panshin and deZeeuw 1980, Dickison 2000, Carlquist 2001).

2.8 CELLS IN WOOD

To understand a growth ring in greater detail, it is essential to begin with an understanding of the structure, function, and variability of the cells that compose the ring. A single plant cell consists of two primary domains: the protoplast and the cell wall. The protoplast is the sum of the living contents that are bounded by the cell membrane. The cell wall is a non-living, largely carbohydrate matrix extruded by the protoplast to the exterior of the cell membrane. The plant cell wall protects the protoplast from osmotic lysis and can provide significant mechanical support to the plant at large (Esau 1977, Raven et al. 1999, Dickison 2000).

For cells in wood, the situation is somewhat more complicated than this highly generalized case. In many cases in wood, the ultimate function of the cell is borne solely by the cell wall. This means that many mature wood cells not only do not require their protoplasts, but indeed must completely remove their protoplasts prior to achieving functional maturity. For this reason, it is a common convention in wood literature to refer to a cell wall without a protoplast as a cell. Although this is technically incorrect from a cell biological standpoint, it is a convention common in the literature and will be observed throughout the remainder of the chapter.

In the case of a mature cell in wood in which there is no protoplast, the open portion of the cell where the protoplast would have existed is known as the lumen. Thus, in most cells in wood there are two domains: the cell wall and the cell lumen (Figure 2.7). The lumen is a critical component of many cells, whether in the context of the amount of space available for water conduction or in

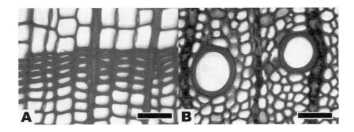

FIGURE 2.7 Transverse sections of wood showing cell walls and lumina. (A) Softwood: All the rectangular cells are of the same type, some with thicker cell walls and narrower lumina, and others with thinner walls and wider lumina in *Pseudotsuga mensiezii*. (B) Hardwood: The large round cells have thick cell walls and very large lumina. Other cells have thinner walls and narrower lumina in *Quercus rubra*. Scale bars = 50 μm.

the context of a ratio between the width of the lumen and the thickness of the cell wall. The lumen has no structure per se, as it is really the void space in the interior of the cell. The relevance of the lumen to the formation of wood composites is the subject of Chapter 15.

2.9 CELL WALLS

The cell walls in wood are important structures. Unlike the lumen, which is a void space, the cell wall itself is a highly regular structure, from one cell type to another, between species, and even when comparing softwoods and hardwoods. The cell wall consists of three main regions: the middle lamella, the primary wall, and the secondary wall (Figure 2.8). In each region, the cell wall has three major components: cellulose microfibrils (with characteristic distributions and organization), hemicelluloses, and a matrix or encrusting material, typically pectin in primary walls and lignin in secondary walls (Panshin and deZeeuw 1980).

To understand these wall layers and their interrelationships, it is necessary to remember that plant cells generally do not exist singly in nature; instead they are adjacent to many other cells, and this association of thousands of cells, taken together, forms an organ such as a leaf. Each of the individual cells must adhere to others in a coherent way to ensure that the cells can act as a unified whole. This means that they must be interconnected with one another to permit the movement of biochemicals (e.g., photosynthate, hormones, cell signaling agents, etc.) and water. This adhesion is provided by the middle lamella, the layer of cell wall material between two or more cells, a part of which is contributed by each of the individual cells (Figure 2.8). This layer is the outermost layer of the cell wall continuum, and in a non-woody organ is pectin rich. In the case of wood, the middle lamella is lignified.

The next layer, formed by the protoplast just interior to the middle lamella, is the primary wall (Figure 2.8). The primary wall is characterized by a largely random orientation of cellulose microfibrils, like thin threads wound round and round a balloon in no particular order, where any microfibril angle from 0 to 90 degrees relative to the long axis of the cell may be present. In cells in wood, the primary wall is very thin, and is generally indistinguishable from the middle lamella. For this reason, the term compound middle lamella is used to denote the primary cell wall of a cell, the middle lamella, and the primary cell wall of the adjacent cell. Even with transmission electron microscopy, the compound middle lamella often cannot be separated unequivocally into its component layers. The compound middle lamella in wood is almost invariably lignified.

The remaining cell wall domain, found in virtually all cells in wood (and in many cells in non-woody plants or plant parts) is the secondary cell wall. The secondary cell wall is composed of three layers (Figure 2.8). As the protoplast lays down the cell wall layers, it progressively reduces the lumen volume. The first-formed secondary cell wall layer is the S_1 layer (Figure 2.8), which is adjacent to compound middle lamella (or technically the primary wall). This layer is a thin layer

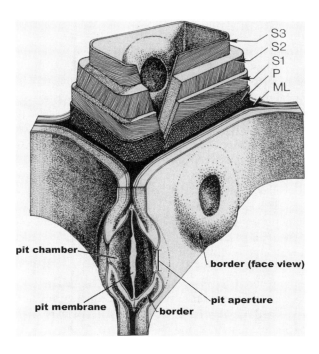

FIGURE 2.8 A cut-away drawing of the cell wall including the structural details of a bordered pit. The various layers of the cell wall are detailed at the top of the drawing, beginning with the middle lamella (ML). The next layer is the primary wall (P), and on the surface of this layer the random orientation of the cellulose microfibrils is detailed. Interior to the primary wall is the secondary wall in its three layers: S1, S2, and S3. The microfibril angle of each layer is illustrated, as well as the relative thickness of the layers. The lower portion of the illustration shows bordered pits in both sectional and face view. The four domains of the pit are illustrated: the pit aperture (pa), the pit chamber (pc), the pit membrane (pm) and the border (b).

and is characterized by a large microfibril angle. That is to say, the cellulose microfibrils are laid down in a helical fashion, and the angle between the mean microfibril direction and the long axis of the cell is large: 50 to 70 degrees.

The next wall layer is arguably the most important in determining the properties of the cell and, thus, the wood properties at a macroscopic level (Panshin and deZeeuw 1980). This layer, formed interior to the S_1 layer, is the S_2 layer (Figure 2.8). This is the thickest secondary cell wall layer and it makes the greatest contribution to the overall properties of the cell wall. It is characterized by a lower lignin percentage and a low microfibril angle: 5 to 30 degrees. There is a strong but not fully understood relationship between the microfibril angle of the S_2 layer of the wall and the wood properties at a macroscopic level (Kretschmann et al. 1998), and this is an area of active research.

Interior to the S_2 layer is the S_3 layer, a relatively thin wall layer (Figure 2.8). The microfibril angle of this layer is relatively high and similar to that of S_1: 70+ degrees. This layer has the lowest percentage of lignin of any of the secondary wall layers. The explanation of this phenomenon is related directly to the physiology of the living tree. In brief, for water to move up the plant (transpiration), there must be a significant adhesion between the water molecules and the cell walls of the water conduits. Lignin is a hydrophobic macromolecule, so it must be in low concentration in the S_3 layer to permit adhesion of water to the cell wall and thus facilitate transpiration. For more detail regarding these wall components and information regarding transpiration and the role of the cell wall, see Chapter 3 or any college-level plant physiology textbook (Kozlowski and Pallardy 1997, Taiz and Zeiger 1991).

2.10 PITS

Any discussion of cell walls in wood must be accompanied by a discussion of the ways in which cell walls are modified to allow communication between the cells in the living plant. These wall modifications, called pit-pairs (or more commonly just pits), are thin areas in the cell walls between two cells, and are a critical aspect of wood structure too often overlooked in wood technological treatments (Figure 2.8). Pits have three domains: the pit membrane, the pit aperture, and the pit chamber. The pit membrane (Figure 2.8) is the thin semiporous remnant of the primary wall; it is a carbohydrate and not a phospholipid membrane. The pit aperture is the opening or hole leading into the open area of the pit, which is called the pit chamber (Figure 2.8). The type, number, size, and relative proportion of pits can be characteristic of certain types of wood, and furthermore can directly affect how wood behaves in a variety of situations, such as how wood interacts with surface coatings (DeMeijer et al. 1998, Rijkaert et al. 2001).

Pits of predictable types occur between different types of cells. In the cell walls of two adjacent cells, pits will form in the wall of each cell separately, but in a coordinated location so that the pitting of one cell will match up with the pitting of the adjacent cell (thus a pit-pair). When this coordination is lacking and a pit is formed only in one of the two cells, it is called a blind pit. Blind pits are fairly rare in wood. Understanding the type of pit can permit one to determine what type of cell is being examined in the absence of other information. It can also allow one to make a prediction about how the cell might behave, particularly in contexts that involve fluid flow. Pits occur in three varieties: bordered, simple, and half-bordered (Esau 1977, Raven et al. 1999).

Bordered pits are thus called because the secondary wall overarches the pit chamber and the aperture is generally smaller and/or of a different shape than the pit chamber. The portion of the cell wall that is overarching the pit chamber is called the border (Figure 2.8, Figure 2.9A, and Figure 2.9D). When seen in face view, bordered pits often are round in appearance and look somewhat like a doughnut, with a small round or almond-shaped hole, the pit aperture, in the middle of the pit (Figure 2.9). When seen in sectional view, the pit often looks like a pair of V's with the open ends of the V's facing each other (Figure 2.9A and Figure 2.9D). In this case, the long stems of the V represent the borders, the secondary walls that are overarching the pit chamber. Bordered pits always occur between two conducting cells, and sometimes between other cells,

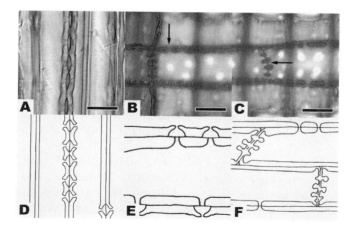

FIGURE 2.9 Light micrographs and sketches of the three types of pits. (A) Longitudinal section of bordered pits in *Xanthocyparis vietnamensis*. The pits look like a vertical stack of thick-walled letter V's. (B) Half-bordered pits in *Pseudotsuga mensiezii*. The arrow shows one half-bordered pit. (C) Simple pits on an end wall in *Pseudotsuga mensiezii*. The arrow indicates one of five simple pits on the end wall. Scale bars = 20 μm. (D–F) Sketches of the pits shown in A–C.

Structure and Function of Wood

typically those with thick cell walls. The structure and function of bordered pits, particularly those in softwoods (see next section), are much studied and known to be well suited to the safe and efficient conduction of sap. The status of the bordered pit (whether it is open or closed) has great importance in the fields of wood preservation, wood finishing, and bonding.

Simple pits are so called because they lack any sort of border (Figure 2.9C and Figure 2.9F). The pit chamber is straight-walled, and the pits are uniform in size and shape in each of the partner cells. Simple pits are typical between parenchyma cells, and in face view merely look like clear areas in the walls.

Half-bordered pits occur between a conducting cell and a parenchyma cell. In this case, each cell forms the kind of pit that would be typical of its type (bordered in the case of a conducting cell and simple in the case of a parenchyma cell) and thus one half of the pit pair is simple and one half is bordered (Figure 2.9B and Figure 2.9E). In the living tree, these pits are of great importance, as they represent the communication link between the conducting cells and the biochemically active cells.

2.11 THE MICROSCOPIC STRUCTURE OF SOFTWOODS AND HARDWOODS

As is no doubt clear by now, the fundamental differences between different kinds of woods are founded on the types, sizes, proportions, pits, and arrangements of different cells that compose the wood. Softwoods have a simpler basic structure than do hardwoods due to the presence of only two cell types, and relatively little variation in structure within these cell types. Hardwoods have greater structural complexity because they have both a greater number of basic cell types and a far greater degree of variability within the cell types. In each case, however, there are fine details of structure that could affect the use of a wood, and elucidating these details is the subject of the next portion of this chapter.

2.11.1 SOFTWOODS

The structure of a typical softwood is relatively simple. The axial or vertical system is composed mostly of axial tracheids and the radial or horizontal system, the rays, are composed mostly of ray parenchyma cells.

2.11.1.1 Tracheids

Tracheids are very long cells, often more than 100 times longer than wide, and they are the major component of softwoods, making up over 90% of the volume of the wood. They provide both conductive and mechanical functions to softwoods. In the transverse view or section (Figure 2.10A), tracheids appear as square or slightly rectangular cells in radial rows. Within one growth ring they can be thin-walled in the earlywood and thicker-walled in the latewood. In order for water to flow between tracheids it must pass through circular bordered pits that are concentrated in the long, tapered ends of the cells. Tracheids overlap with adjacent cells across both the top and bottom 20–30% of their length. Water flow thus must take a slightly zigzag path as it goes from one cell to the next through the pits. Because the pits have a pit membrane, there is a significant resistance to flow. The resistance of the pit membrane coupled with the narrow diameter of the lumina makes tracheids relatively inefficient conduits, compared to the conducting cells of hardwoods. Detailed treatments of the structure of wood in relation to its conductive functions can be found in the literature (Zimmermann 1983, Kozlowski and Pallardy 1997).

2.11.1.2 Axial Parenchyma and Resin Canal Complexes

Another cell type that is sometimes present is axial parenchyma. Axial parenchyma cells are similar in size and shape to ray parenchyma cells, but they are vertically oriented and stacked one on top

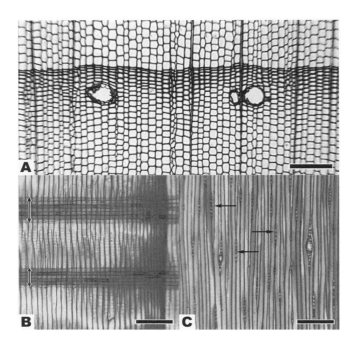

FIGURE 2.10 The microscopic structure of *Picea glauca*, a typical softwood. (A) Transverse section, scale bar = 150 µm. The bulk of the wood is made of tracheids, the small rectangles of various thicknesses. The three large, round structures are resin canals and their associated cells. The dark lines running from the top to the bottom of the photo are the ray cells of the rays. (B) Radial section showing two rays (arrows) running from left to right. Each cell in the ray is a ray cell, and they are low, rectangular cells. The rays begin on the left in the earlywood (thin-walled tracheids) and continue into and through the latewood (thick-walled tracheids), and into the next growth ring, on the right side of the photo. Scale bar = 200 µm. (C) Tangential section. Rays seen in end-view; they are mostly only one cell wide. Two rays are fusiform rays; there are radial resin canals embedded in the rays, causing them to bulge. Scale bar = 200 µm.

of the other to form a parenchyma strand. In transverse section (Figure 2.11A) they often look like an axial tracheid, but can be differentiated when they contain dark-colored organic substances in the lumen of the cell. In the radial or tangential section (Figure 2.11B) they appear as long strands of cells generally containing dark-colored substances. Axial parenchyma is most common in redwood, juniper, cypress, bald cypress, and some species of *Podocarpus*, but never makes up even 1% of the cells. Axial parenchyma is generally absent in pine, spruce, larch, hemlock, and species of *Araucaria* and *Agathis*.

In species of pine, spruce, Douglas fir, and larch structures commonly called resin ducts or resin canals are present vertically (Figure 2.12) and horizontally (Figure 2.12C). These structures are voids or spaces in the wood and are not cells. However, specialized parenchyma cells that function in resin production surround resin canals. When referring to the resin canal and all the associated parenchyma cells, the correct term is axial or radial resin canal (Wiedenhoeft and Miller 2002). In pine, resin canal complexes are often visible on the transverse section to the naked eye, but they are much smaller in spruce, larch, and Douglas fir, and a hand lens is needed to see them. Radial resin canal complexes are embedded in specialized rays called fusiform rays (Figure 2.10C and Figure 2.12C). These rays are much higher and wider than normal rays. Resin canal complexes are absent in the normal wood of other softwoods, but some species can form large tangential clusters of traumatic resin canals in response to significant injury.

Structure and Function of Wood

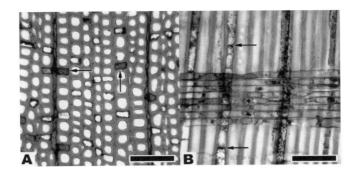

FIGURE 2.11 Axial parenchyma in *Podocarpus madagascarensis*. (A) Transverse section showing individual axial parenchyma cells. They are the dark-staining rectangular cells. Two are denoted by arrows, but many more can be seen. (B) Radial section showing axial parenchyma in longitundinal view. The parenchyma cells can be differentiated from the tracheids by the presence of end walls (arrows) in addition to the dark-staining contents. Scale bars = 100 µm.

2.11.1.3 Rays

The other cells in Figure 2.10A are ray parenchyma cells that are barely visible and appear as dark lines running in a top-to-bottom direction. Ray parenchyma cells are rectangular prisms or brick-shaped cells. Typically they are approximately 15 µm high by 10 µm wide by 150–250 µm long in the radial or horizontal direction (Figure 2.10B). These brick-like cells form the rays, which function primarily in the synthesis, storage, and lateral transport of biochemicals and, to a lesser degree, water. In radial view or section (Figure 2.10B), the rays look like a brick wall and the ray parenchyma cells are sometimes filled with dark-colored substances. In tangential section (Figure 2.10C), the rays are stacks of ray parenchyma cells one on top of the other forming a ray that is only one cell in width, called a uniseriate ray.

When ray parenchyma cells intersect with axial tracheids, specialized pits are formed to connect the vertical and radial systems. The area of contact between the tracheid wall and the wall of the ray parenchyma cells is called a cross-field. The type, shape, size, and number of pits in the cross-field is generally consistent within a species and very diagnostic. Figure 2.13 illustrates several types of cross-field pitting.

Species that have resin canal complexes also have ray tracheids, which are specialized horizontal tracheids that normally are situated at the margins of the rays. These ray tracheids have bordered pits like vertical tracheids, but are much shorter and narrower. Ray tracheids also occur in a few

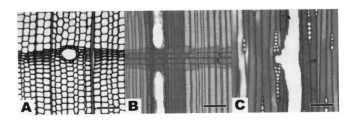

FIGURE 2.12 Resin canal complexes in *Pseudotsuga mensiezii*. (A) Transverse section showing a single axial resin canal complex. In this view the tangential and radial diameters of the canal can be measured accurately. (B) Radial section showing an axial resin canal complex embedded in the latewood. It is crossed by a ray that also extends into the earlywood on either side of the latewood. (C) Tangential section showing the anastomosis between an axial and a radial resin canal complex. The fusiform ray bearing the radial resin canal complex is in contact with the axial resin canal complex. Scale bars = 100 µm.

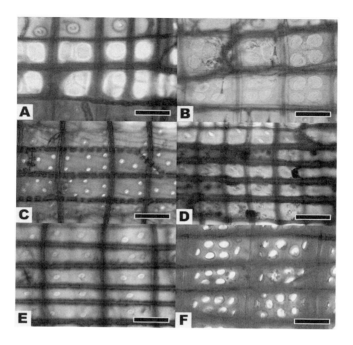

FIGURE 2.13 Radial sections showing a variety of types of cross-field pitting. All the pits are half-bordered pits, but in some cases the borders are difficult to see. (A) Fenestriform pitting in *Pinus strobus*. (B) Pinoid pitting in *Pinus elliottii*. (C) Piceoid pitting in *Pseudotsuga mensiezii*. (D) Cuppressoid pitting in *Juniperus virginiana*. (E) Taxodioid pitting in *Abies concolor*. (F) Araucarioid pitting in *Araucaria angustifolia*. Scale bars = 30 μm.

species that do not have resin canals. Alaska yellow cedar, (*Chamaecyparis nootkatensis*), hemlock (*Tsuga*), and rarely some species of true fir (*Abies*) have ray tracheids.

Additional detail regarding the microscopic structure of softwoods can be found in the literature (Phillips 1948, Kukachka 1960, Panshin and deZeeuw 1980, IAWA Committee 2004).

2.11.2 Hardwoods

The structure of a typical hardwood is much more complicated than that of a softwood. The axial or vertical system is composed of fibrous elements of various kinds, vessel elements in various sizes and arrangements, and axial parenchyma cells in various patterns and abundance. Like softwoods, the radial or horizontal system are the rays, which are composed of ray parenchyma cells, but unlike softwoods, hardwood rays are much more diverse in size and shape.

2.11.2.1 Vessels

The unique feature that separates hardwoods from softwoods is the presence of specialized conducting cells in hardwoods called vessels elements (Figure 2.14A). These cells are stacked one on top of the other to form vessels. Where the ends of the vessel elements come in contact with one another, a hole is formed, called a perforation plate. Thus hardwoods have perforated tracheary elements (vessel elements) for water conduction, whereas softwoods have imperforate tracheary elements (tracheids). On the transverse section, vessels appear as large openings and are often referred to as pores (Figure 2.2D).

Vessel diameters may be quite small (<30 μm) or quite large (>300 μm), but typically range from 50–200 μm. Their length is much shorter than tracheids and range from 100–1200 μm or

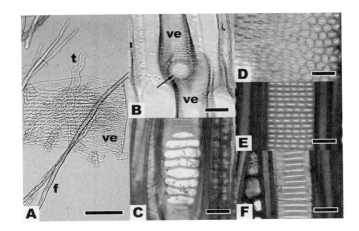

FIGURE 2.14 Vessel elements and vessel features. (A) Macerated cells of *Quercus rubra*. There are three types of cells labeled. There is a single vessel element (ve); note that it is wider than it is tall, and it is open on both ends. The fiber (f) is long, narrow, and thick-walled. The hardwood tracheid (t) is shorter than a fiber but longer than a vessel element, and it is contorted in shape. Scale bar = 200 μm. (B) A simple perforation plate in *Malouetia virescens*. There are two vessel elements (ve), and where they overlap there is an open hole between the cells, the perforation plate (arrow). As the perforation is completely open, it is called a simple perforation plate. (C) A scalariform perforation plate in *Magnolia grandiflora*. This perforation plate has eight bars crossing it (the eighth is very small), and it is the presence of bars that distinguishes this type of perforation plate from a simple plate. (D) Alternate intervessel pitting in *Hevea microphylla*. (E) Opposite intervessel pitting in *Liriodendron tulipifera*. (F) Linear (scalariform) intervessel pitting in *Magnolia grandiflora*. Note that these intervessel pits are not the same structures as the scalariform perforation plate seen in C. Scale bars in B–F = 30 μm.

0.1–1.2 mm. Vessels are arranged in various patterns. If all the vessels are the same size and more or less scattered throughout the growth ring, the wood is diffuse porous (Figure 2.6D). If the earlywood vessels are much larger than the latewood vessels, the wood is ring porous (Figure 2.6F). Vessels can also be arranged in a tangential or oblique arrangement, in a radial arrangement, in clusters, or in many combinations of these types (IAWA Committee 1989). In addition, the individual vessels may occur alone (solitary arrangement) or in pairs or radial multiples of up to five or more vessels in a row. At the end of the vessel element is a hole or perforation plate. If there are no obstructions across the perforation plate, it is called a simple perforation plate (Figure 2.14B). If bars are present, the perforation plate is called a scalariform perforation plate (Figure 2.14C).

Where the vessels elements come in contact with each other tangentially, intervessel or intervascular bordered pits are formed (Figure 2.14D, Figure 2.14E, and Figure 2.14F). These pits range in size from 2–16 μm in height and are arranged on the vessels walls in threes basic ways. The most common arrangement is alternate, in which the pits are more or less staggered (Figure 2.14D). In the opposite arrangement the pits are opposite each other (Figure 2.14E), and in the scalariform arrangement the pits are much wider than high (Figure 2.14F). Combinations of these can also be observed in some species. Where vessel elements come in contact with ray cells, often simple or bordered pits called vessel-ray pits are formed. These pits can be the same size and shape and the intervessel pits or much larger.

2.11.2.2 Fibers

Fibers in hardwoods function solely as support. They are shorter than softwood tracheids (200–1200 μm) and average about half the width of softwood tracheids, but are usually 2–10 times longer than vessel elements (Figure 2.15). The thickness of the fiber cell wall is the major factor governing

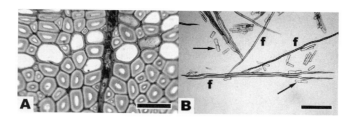

FIGURE 2.15 Fibers in *Quercus rubra*. (A) Transverse section showing thick-walled, narrow-lumined fibers. A ray is passing vertically through the photo, and there are nine axial parenchyma cells, the thin-walled, wide-lumined cells, in the photo. Scale bar = 30 μm. (B) Macerated wood. There are several fibers (f), two of which are marked. Also easily observed are parenchyma cells (arrows) both individually and in small groups. Note the thin walls and small rectangular shape compared to the fibers. Scale bar = 300 μm.

density and strength. Species with thin-walled fibers such as cottonwood (*Populus deltoides*), basswood (*Tilia americana*), ceiba (*Ceiba pentandra*), and balsa (*Ochroma pyramidale*) have a low density and strength, whereas species with thick-walled fibers such as hard maple (*Acer saccharum* and *Acer nigrum*), black locust (*Robinia pseudoacacia*), ipe (*Tabebuia serratifolia*), and bulletwood (*Manilkara bidentata*) have a high density and strength. The air-dry (12% moisture content) density of hardwoods varies from 100–1400 kg/m^3. The air-dry density of typical softwoods varies from 300–800 kg/m^3. Fiber pits are generally inconspicuous and may be simple or bordered. In some woods like oak (*Quercus*) and meranti/lauan (*Shorea*), vascular or vasicentric tracheids are present especially near or surrounding the vessels (Figure 2.14A). These specialized fibrous elements in hardwoods typically have bordered pits, are thin-walled, and are shorter than the fibers of the species. The tracheids in hardwoods function in both support and transport.

2.11.2.3 Axial Parenchyma

In softwoods, axial parenchyma is absent or only occasionally present as scattered cells, but in hardwoods there is a wide variety of axial parenchyma patterns (Figure 2.16). The axial parenchyma cells in hardwoods and softwoods is essentially the same size and shape, and they also function in the same manner. The difference comes in the abundance and specific patterns in hardwoods. There are two major types of axial parenchyma in hardwoods. Paratracheal parenchyma is associated with the vessels and apotracheal is not associated with the vessels. Paratracheal parenchyma is further divided into vasicentric (surrounding the vessels, Figure 2.16A), aliform (surrounding the vessel and with wing-like extensions, Figure 2.16C), and confluent (several connecting patches of paratracheal parenchyma sometimes forming a band, Figure 2.16E). Apotracheal parenchyma is also divided into diffuse (scattered), diffuse in aggregate (short bands, Figure 2.16B), and banded whether at the beginning or end of the growth ring (marginal, Figure 2.16F) or within a growth ring (Figure 2.16D). Each species has a particular pattern of axial parenchyma, which is more or less consistent from specimen to specimen, and these cell patterns are very important in wood identification.

2.11.2.4 Rays

The rays in hardwoods are much more diverse than those found in softwood. In some species, including willow (*Salix*), cottonwood, and koa (*Acacia koa*), the rays are exclusively uniseriate and are much like the softwood rays. In hardwoods most species have rays that are more than one cell wide. In oak and hard maple the rays are two-sized, uniseriate and over eight cells wide, and in oak several centimeters high (Figure 2.17A). In most species the rays are 1–5 cells wide and less than 1 mm high (Figure 2.17B) Rays in hardwoods are composed of ray parenchyma cells that are either procumbent or upright. As the name implies, procumbent ray cells are horizontal and are

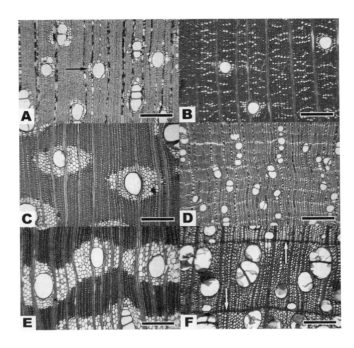

FIGURE 2.16 Transverse sections of various woods showing a range of hardwood axial parenchyma patterns. A, C, and E show woods with paratracheal types of parenchyma. (A) Vasicentric parenchyma (arrow) in *Licaria excelsa*. (C) Aliform parenchyma in *Afzelia africana*. The parenchyma cells are the light-colored, thin-walled cells, and are easily visible. (E) Confluent parenchyma in *Afzelia cuazensis*. B, D, and F show woods with apotracheal types of parenchyma. (B) Diffuse-in-aggregate parenchyma in *Dalbergia stevensonii*. (D) Banded parenchyma in *Micropholis guyanensis*. (F) Marginal parenchyma in *Juglans nigra*. In this case, the parenchyma cells are darker in color, and they delimit the growth rings (arrows). Scale bars = 300 μm.

similar in shape and size to the softwood ray parenchyma cells (Figure 2.17C). The upright ray cells are ray parenchyma cells turned on end so that their long axis is vertical (Figure 2.17D). Upright ray cells are generally shorter and sometimes nearly square. Rays that have only one type of ray cell, typically only procumbent cells, are called homocellular rays. Those that have procumbent and upright cells are called heterocellular rays. The number of rows of upright ray cells varies from one to more than five.

The great diversity of hardwood anatomy is treated in many sources throughout the literature (Metcalfe and Chalk 1950, Metcalfe and Chalk 1979, Panshin and deZeeuw 1980, Metcalfe and Chalk 1987, IAWA Committee 1989, Gregory 1994, Cutler and Gregory 1998, Dickison 2000, Carlquist 2001).

2.12 WOOD TECHNOLOGY

Though it is necessary to speak briefly of each kind of cell in isolation, the beauty and complexity of wood are found in the interrelationship between many cells at a much larger scale. The macroscopic properties of wood such as density, hardness, and bending strength, among others, are properties derived from the cells that compose wood. Such larger-scale properties are really the product of a synergy in which the whole is indeed greater than the sum of its parts, but are nonetheless based on chemical and anatomical details of wood structure (Panshin and deZeeuw 1980).

The cell wall is largely made up of cellulose and hemicellulose, and the hydroxyl groups on these chemicals make the cell wall very hygroscopic. Lignin, the agent cementing cells together,

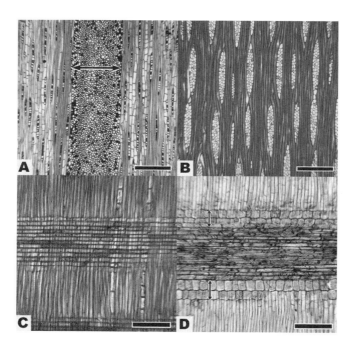

FIGURE 2.17 Rays in longitudinal sections. A and B show tangential sections, scale bars = 300 μm. (A) *Quercus rubra* showing very wide multiseriate ray (arrow) and many uniseriate rays. (B) *Swietenia macrophylla* showing numerous rays ranging from one to four cells wide. Note that in this wood the rays are arranged roughly in rows from side to side. C and D show radial sections, scale bars = 200 μm. (C) Homocellular ray in *Fraxinus americana*. All the cells in the ray are procumbent cells; they are longer radially than they are tall. (D) A heterocellular ray in *Khaya ivorensis*. The central portion of the ray is composed of procumbent cells, but the margins of the ray, both top and bottom, have two rows of upright cells (arrows), which are as tall as or taller than they are wide.

is a generally hydrophobic molecule. This means that the cell walls in wood have a great affinity for water, but the ability of the walls to take up water is limited, in part by the presence of lignin. Water in wood has a great effect on wood properties, and wood-water interactions greatly affect the industrial use of wood in wood products.

Often it is useful to know how much water is contained in a tree or a piece of wood. This relationship is called moisture content and is the weight of water in the cell walls and lumina expressed as a percentage of the weight of wood with no water (oven-dry weight). Water exists in wood in two forms: free water and bound water. Free water is the liquid water that exists within the lumina of the cells. Bound water is the water that is adsorbed to the cellulose and hemicellulose in the cell wall. Free water is only found when all sites for the adsorption of water in the cell wall are filled; this point is called the fiber saturation point (FSP). All water added to wood after the FSP has been reached exists as free water.

Wood of a freshly cut tree is said to be green; the moisture content of green wood can be over 100%, meaning that the weight of water in the wood is more than the weight of the dried cells. In softwoods the moisture content of the sapwood is much higher than that of the heartwood, but in hardwoods, the difference may not be as great and in a few cases the heartwood has a higher moisture content than the sapwood.

When drying from the green condition to the FSP (approximately 25–30% moisture content), only free water is lost, and thus no change in the cell wall volumes occurs. However, when the wood is dried further, bound water is removed from the cell walls and shrinkage of the wood begins.

Structure and Function of Wood

Some of the shrinkage that occurs from green to dry is irreversible; no amount of rewetting can swell the wood back to its original dimensions. After this process of irreversible shrinkage has occurred, however, shrinkage and swelling is reversible and essentially linear from 0% moisture content up to the FSP. Controlling the rate at which bound water is removed from green wood is the subject of entire fields of research. By properly controlling the rate at which wood dries, drying defects such as cracking, checking, honeycombing, and collapse can be minimized (Hillis 1996).

Density or specific gravity is one of the most important physical properties of wood (Desch and Dinwoodie 1996, Forest Products Laboratory 1999, Bowyer et al. 2003). Density is the weight of wood divided by the volume at a given moisture content. Thus the units for density are typically expressed as pounds per cubic foot (lbs/ft^3) or kilograms per cubic meter (kg/m^3). When density values are reported in the literature it is critical that the moisture content of the wood is also given. Often density values are listed as air-dry, which means 12% moisture content in North America and Europe, but air-dry sometimes means 15% moisture content in tropical countries.

Specific gravity is similar to density and is defined as the ratio of the density of wood to the density of water. Since 1 cm^3 of water weighs 1 g, density in g/cm^3 is numerically the same as specific gravity. Density in kg/m^3 must be divided by 1000 to get the same numerical number as specific gravity. Since specific gravity is a ratio, it does not have units. The term basic specific gravity (sometimes referred to as basic density) is defined as the oven-dry weight of wood divided by the volume of the wood when green (no shrinkage).

$$\text{Basic specific gravity} = \frac{\text{Density of wood (oven-dry weight/volume when green)}}{\text{Density of water}}$$

Specific gravity can be determined at any moisture content, but typically it is based on weight when oven-dry and when the volume is green or at 12% moisture content (Forest Products Laboratory 1999). However, basic specific gravity is generally the standard used throughout the world. The most important reason for measuring basic specific gravity is repeatability. The weight of wood can be determined at any moisture content, but conditioning the wood to a given moisture content consistently is difficult. The oven-dry weight (at 0% moisture content) is relatively easy to obtain on a consistent basis. Green volume is also relatively easy to determine using the water displacement method (ref). The sample can be large or small and nearly any shape. Thus basic specific gravity can be determined as follows:

$$\text{Basic specific gravity} = \frac{\text{Oven-dry weight}}{\text{Weight of displaced water}}$$

Specific gravity and density are strongly dependent on the weight of the cell wall material in the bulk volume of the wood specimen. In softwoods where the latewood is abundant (Figure 2.5A) in proportion to the earlywood, the specific gravity is high (e.g., 0.54 in longleaf pine, *Pinus palustris*). The reverse is true when there is much more earlywood than latewood (Figure 2.6B) (e.g., 0.34 in eastern white pine, *Pinus strobus*). To say it another way, specific gravity increases as the proportion of cells with thick cell walls increases. In hardwoods specific gravity is not only dependent on fiber wall thickness, but also on the amount of void space occupied by the vessels and parenchyma. In balsa the vessels are large (typically >250 μm in tangential diameter), and there is an abundance of axial and ray parenchyma. The fibers that are present are very-thin-walled and the basic specific gravity may be less than 0.20. In very dense woods the fibers are very-thick-walled, the lumina are virtually absent, and the fibers are abundant in relation to the vessels and parenchyma. Some tropical hardwoods have a basic specific gravity of greater than 1.0. In a general sense in all woods, the specific gravity is the relation between the volume of cell wall material to the volume of the lumina of those cells in a given bulk volume.

2.13 JUVENILE WOOD AND REACTION WOOD

Two key examples of the biology of the tree affecting the quality of the wood can be seen in the formation of juvenile wood and reaction wood. They are grouped together because they share several common cellular, chemical, and tree physiological characteristics, and each may or may not be present in a certain piece of wood.

Juvenile wood is the first-formed wood of the young tree, the rings closest to the pith. If one looks at the growth form of a tree, based on the derivation of wood from the vascular cambium, it quickly becomes evident that the layers of wood in a tree are concentric cones. In a tree of large diameter, the deflection of the long edge of the cone from vertical may be very close to zero, but in narrower-diameter trees, or narrower-diameter portions of a large tree, the angle of deflection is considerably greater. These areas of narrower diameter are typically chronologically younger portions of the tree, for example, the first 15–20 years of growth in softwoods are the areas where juvenile wood may form. Juvenile wood in softwoods is in part characterized by the production of axial tracheids that have a higher microfibril angle in the S_2 wall layer (Larson et al. 2001). A higher microfibril angle in the S_2 is correlated with drastic longitudinal shrinkage of the cells when the wood is dried for human use, resulting in a piece of wood that has a tendency to warp, cup, and check. The morphology of the cells themselves is often altered so that the cells, instead of being long and straight, are shorter and angled, twisted, or bent. The precise functions of juvenile wood in the living tree are not fully understood, but it must confer certain little-understood advantages.

Reaction wood is similar to juvenile wood in several respects, but is formed by the tree for different reasons. Almost any tree of any age will form reaction wood when the woody organ

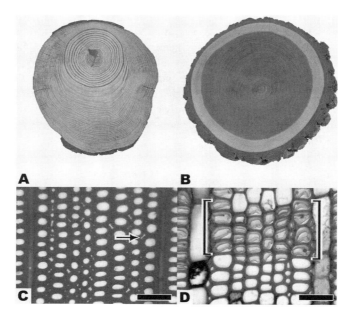

FIGURE 2.18 Macroscopic and microscopic views of reaction wood in a softwood and a hardwood. (A) Compression wood in *Pinus* sp. Note that the pith is not in the center of the trunk, and the growth rings are much wider in the compression wood zone. (B) Tension wood in *Juglans nigra*. The is nearly centered in the trunk, but the growth rings are wider in the tension wood zone. (C) Transverse section of compression wood in *Picea engelmannii*. The tracheids are thick-walled and round in outline, giving rise to prominent intercellular spaces in the cell corners (arrow). (D) Tension wood fibers (between the brackets) showing prominent gelatinous layers in *Hevea microphylla*. Rays run from top to bottom on either side of the tension wood fibers, and below them is a band of normal fibers with thinner walls. Scale bars (in C and D) = 50 μm.

Structure and Function of Wood

(whether a twig, a branch, or the trunk) is deflected from the vertical by more than one or two degrees. This means that all nonvertical branches form considerable quantities of reaction wood. The type of reaction wood formed by a tree differs in softwoods and hardwoods. In softwoods, the reaction wood is formed on the underside of the leaning organ, and is called compression wood (Figure 2.18A) (Timmel 1986). In hardwoods, the reaction wood forms on the top side of the leaning organ, and is called tension wood (Figure 2.18B) (Desch and Dinwoodie 1996, Bowyer et al. 2003). As just mentioned, the various features of juvenile wood and reaction wood are similar. In compression wood, the tracheids are shorter, misshapen cells with a large S_2 microfibril angle, a high degree of longitudinal shrinkage, and high lignin content (Timmel 1986). They also take on a distinctly rounded outline (Figure 2.18C). In tension wood, the fibers fail to form a proper secondary wall and instead form a highly cellulosic wall layer called the G layer, or gelatinous layer (Figure 2.18D).

2.14 WOOD IDENTIFICATION

The identification of wood can be of critical importance to primary and secondary industrial users of wood, government agencies, and museums, as well as to scientists in the fields of botany, ecology, anthropology, forestry, and wood technology. Wood identification is the recognition of characteristic cell patterns and wood features, and is generally accurate only to the generic level. Since woods of different species from the same genus often have different properties and perform differently under various conditions, serious problems can develop if species or genera are mixed during the manufacturing process and in use. Since foreign woods are imported into the U.S. market, it is imperative that both buyers and sellers have access to correct identifications and information about their properties and uses.

Lumber graders, furniture workers, and those working in the industry, as well as hobbyists, often identify wood with their naked eye. Features often used are color, odor, grain patterns, density, and hardness. With experience these features can be used to identify many different woods, but the accuracy of the identification is dependent on the experience of the person and the quality of the unknown wood. If the unknown wood is atypical, decayed, or small, often the identification is incorrect. Examining woods, especially hardwoods, with a 10–20X hand lens greatly improves the accuracy of the identification (Panshin and deZeeuw 1980, Hoadley 1990, Brunner et al. 1994). Foresters and wood technologists armed with a hand lens and sharp knife can accurately identify lumber in the field. They make a cut on the transverse surface and examine the vessel and parenchyma patterns to make an identification.

Scientifically rigorous accurate identifications require that the wood be sectioned and examined with a light microscope. With the light microscope even with only a 10X objective, many more features are available for use in making the determination. Equally as important as the light microscope in wood identification is the reference collection of correctly identified specimens to which unknown samples can be compared (Wheeler and Baas 1998). If a reference collection is not available, books of photomicrographs or books or journal articles with anatomical descriptions and dichotomous keys can be used (Miles 1978, Schweingruber 1978, Core et al. 1979, Gregory 1980, Ilic 1991, Miller and Détienne 2001). In addition to these resources, several computer-assisted wood identification packages are available and are suitable for people with a robust wood anatomical background.

Wood identification by means of molecular biological techniques is a field that is still in its infancy. Though technically feasible, there are significant population-biological limits to the statistical likelihood of a robust and certain identification for routine work (Canadian Forest Service 1999). In highly limited cases of great financial or criminal import and a narrowly defined context, the cost and labor associated with rigorous evaluation of DNA from wood can be warranted (Hipkins 2001). For example, if the question were, "Did this piece of wood come from this individual tree?" or, "Of the 15 species present in this limited geographical area, which one produced this root?" it is feasible to analyze the specimens with molecular techniques (Brunner et al. 2001). If, however, the question were, "What kind of wood is this, and from which forest did it come?" it would not be feasible at this point in time to analyze the specimen. Workers have shown that specific identification among six species of Japanese white oak can be accomplished using DNA (Ohyama et al. 2001), but the broad application of their methods is not likely for some time. As technological advances improve the quality, quantity, and speed with which molecular data can be collected, the difficulty and cost of molecular wood identification will decrease. At some point in the indefinite future it is reasonable to expect that molecular tools will be employed in the routine identification of wood, and that such techniques will greatly increase the specificity and accuracy of identification, but until that day comes, scientific wood identifications will rely on microscopic evaluation of wood anatomical features.

REFERENCES

Bowyer, J., Shmulsky, R., and Haygreen, J.G. (2003). *Forest Products and Wood Science: An Introduction* (4th ed.). Iowa State University Press, Des Moines.
Brunner, I., Brodbeck, S., Buchler, U., and Sperisen, C. (2001). Molecular identification of fine roots from trees from the Alps: Reliable and fast DNA extraction and PCR-RFLP analyses of plastid DNA. *Mol. Ecol.* 10:2079–2087.
Brunner, M., Kucera, L.J., and Zürcher, E. (1994). *Major Timber Trees of Guyana: A Lens Key*. Tropenbos Series 10. The Tropenbos Foundation, Wageningen, Netherlands.
Callado, C.H., Neto, A.J.d.S., Scarano, F.R., and Costa, C.G. (2001). Periodicity of growth rings in some flood-prone trees of the Atlantic rain forest in Rio de Janeiro, Brazil. *Trees* 15:492–497.
Canadian Forest Service, Pacific Forestry Centre. (1999). *Combating Tree Theft Using DNA Technology*. [Breakout session consensus.] Author, Victoria, BC, Canada.
Carlquist, S. (2001). *Comparative Wood Anatomy* (2nd ed.). Springer.
Chudnoff, M. (1984). *Tropical Timbers of the World*. USDA Agriculture Handbook # 607. U.S. Government Printing Office, Washington, DC.
Core, H.A., Côte, W.A., and Day, A.C. (1979). *Wood Structure and Identification* (2nd ed.). Syracuse University Press, Syracuse, NY.
Cutler, D.F. and Gregory, M. (1998). *Anatomy of the Dicotyledons* (2nd ed.). Vol. IV. Oxford University Press, New York.
DeMeijer, M., Thurich, K., and Militz, H. (1998). Comparative study on penetration characteristics of modern wood coatings. *Wood Sci. and Tech.* 32:347–365.
Desch, H.E. and Dinwoodie, J.M. (1996). *Timber Structure, Properties, Conversion and Use* (7th ed.). Macmillan Press, London.
Dickison, W. (2000). *Integrative Plant Anatomy*. Academic Press, New York.
Esau, K. (1977). *Anatomy of the Seed Plants* (2nd ed.). John Wiley & Sons, New York.
Forest Products Laboratory. (1999). *Wood Handbook: Wood as an Engineering Material*. USDA General Technical Report FPL-GTR-113. U.S. Department of Agriculture Forest Service, Madison, WI.
Gregory, M. (1980). Wood identification: An annotated bibliography. *IAWA Bull. n.s.* 1(1):3–41.
Gregory, M. (1994). Bibliography of systematic wood anatomy of dicotyledons. *IAWA J.* Suppl. 1.
Hillis, W.E. (1996). Formation of robinetin crystals in vessels of Intsia species. *IAWA J.* 17(4):405–419.
Hipkins, V. (2001). DNA profiling and identity analysis of Ponderosa pine evidence samples, in *NFGEL Annual Report*.
Hoadley, R.B. (1990). *Identifying Wood: Accurate Results with Simple Tools*. Taunton Press, Newtown, CT.
Hoadley, R.B. (2000). *Understanding Wood: A Craftsman's Guide to Wood Technology* (2nd ed.). Taunton Press, Newtown, CT.
IAWA Committee. (1989). IAWA list of microscopic features for hardwood identification, Wheeler, E.A., Baas, P., and Gasson, P. (Eds.). *IAWA Bull. n.s.* 10(3):219–332.
IAWA Committee. (2004). IAWA list of microscopic features of softwood identification. Richter, H.G., Grosser, D., Heinz, I., and Gasson, P. (Eds.). *IAWA J.* 25(1):1–70.
Ilic, J. (1991). *CSIRO Atlas of Hardwoods*. Crawford House Press, Bathurst, Australia.

Kozlowksi, T.T. and Pallardy, S.G. (1997). *Physiology of Woody Plants* (2nd ed.). Academic Press, San Diego, CA.

Kretschmann, D.E., Alden, H.A., and Verrill, S. (1998). Variations of microfibril angle in loblolly pine: Comparison of iodine crystallization and x-ray diffraction techniques, in *Microfibril Angle in Wood*, Butterfield, B.G. (Ed.). University of Canterbury, pp. 157–176.

Kukachka, B.F. (1960). Identification of coniferous woods. *Tappi* 43(11):887–896.

Lagenheim, J.H. (2003). *Plant Resins: Chemistry, Evolution, Ecology, and Ethnobotany.* Timber Press, Portland, OR.

Larson, P.R. (1994). *The Vascular Cambium, Development and Structure.* Springer-Verlag, Berlin.

Larson, P.R., Kretschmann, D.E., Clark, A., III, and Isenbrands, J.G. (2001). *Formation and Properties of Juvenile Wood in Southern Pines: A Synopsis.* USDA General Technical Report FPL-GTR–129. U.S. Government Printing Office, Washington, DC.

Metcalfe, C.R. and Chalk, L. (1950). *Anatomy of the Dicotyledons*, 2 vols. Clarendon Press, Oxford, UK.

Metcalfe, C.R. and Chalk, L. (1979). *Anatomy of the Dicotyledons* (2nd ed.). Vol. I. Oxford University Press, New York.

Metcalfe, C.R. (1987). *Anatomy of the Dicotyledons* (2nd ed.). Vol. III. Oxford University Press, New York.

Miles, A. (1978). *Photomicrographs of World Woods*, Building Research Establishment, Her Majesty's Stationery Office, London.

Miller, R.B. and Détienne, P. (2001). *Major Timber Trees of Guyana: Wood Anatomy.* Tropenbos Series 20. Tropenbos International, Wageningen, Netherlands.

Ohyama, M., Baba, K., and Itoh, T. (2001). Wood identification of Japanese *Cyclobalanopsis* species (Fagaceae) based on DNA polymorphism of the intergenic spacer between trnT and trnL 5' exon. *J. Wood Sci.* 47:81–86.

Panshin, A.J. and deZeeuw, C. (1980). *Textbook of Wood Technology* (4th ed.). McGraw-Hill, New York.

Phillips, E.W.J. (1948). Identification of softwoods by microscopic structure. *For. Prod. Res. Bull.* 22.

Raven, P., Evert, R., and Eichhorn, S. (1999). *Biology of Plants* (6th ed.). W.H. Freeman, New York.

Rijkaert, V., Stevens, M., de Meijer, M., and Militz, H. (2001). Quantitative assessment of the penetration of water-borne and solvent-borne wood coatings in Scots pine sapwood. *Holz als Roh- und Werkstoff* 59:278–287.

Schweingruber, F. (1978). *Microscopic Wood Anatomy.* Swiss Federal Institute for Foreign Research, Birmensdorf.

Simpson, W.T. (Ed.). (1991). *Dry Kiln Operator's Manual.* USDA Agriculture Handbook AH-188.

Taiz, L. and Zeiger, E. (1991). *Plant Physiology.* Benjamin/Cummings, Redwood City, CA.

Timmel, T.E. (1986). *Compression Wood in Gymnosperms.* Springer, Heidelberg, Germany.

Wheeler, E.A., and Baas, P. (1998). Wood Identification—A Review. *IAWA J.* 19(3):241–264.

Wiedenhoeft, A.C., and Miller, R.B. (2002). Brief comments on the nomenclature of softwood axial resin canals and their associated cells. *IAWA J.* 23(3):299–303.

Worbes, M. (1995). How to measure growth dynamics in tropical trees: a review. *IAWA J.* 16(4):337–351.

Worbes, M. (1999). Annual growth rings, rainfall-dependent growth and long-term growth patterns of tropical trees in the Capar Forest Reserve in Venezuela. *J. Ecol.* 87:391–403.

Zimmermann, M.H. (1983). *Xylem Structure and the Ascent of Sap.* Springer-Verlag, New York.

3 Cell Wall Chemistry

Roger M. Rowell[1,3], *Roger Pettersen*[1], *James S. Han*[1],
Jeffrey S. Rowell[2], *and Mandla A. Tshabalala*
[1]USDA, Forest Service, Forest Products Laboratory, Madison, WI
[2]Department of Forest Ecology and Management, University of Wisconsin, Madison, WI
[3]Department of Biological Systems Engineering, University of Wisconsin, Madison, WI

CONTENTS

3.1 Carbohydrate Polymers 37
    3.1.1 Holocellulose 37
    3.1.2 Cellulose 37
    3.1.3 Hemicelluloses 39
        3.1.3.1 Hardwood Hemicelluloses 41
        3.1.3.2 Softwood Hemicelluloses 42
    3.1.4 Other Minor Polysaccharides 43
3.2 Lignin 43
3.3 Extractives 45
3.4 Bark 46
    3.4.1 Extractives 46
        3.4.1.1 Chemical Composition of Extractives 47
    3.4.2 Hemicelluloses 48
    3.4.3 Cellulose 49
    3.4.4 Lignin 49
    3.4.5 Inorganics and pH 50
3.5 Inorganics 50
3.6 Distribution of Chemical Components in the Cell Wall 50
3.7 Juvenile Wood and Reaction Wood 52
3.8 Analytical Procedures 53
    3.8.1 Sampling Procedure 53
    3.8.2 Extraction 54
        3.8.2.1 Scope and Summary 54
        3.8.2.2 Sample Preparation 55
        3.8.2.3 Apparatus 61
        3.8.2.4 Reagents and Materials 61
        3.8.2.5 Procedures 61
    3.8.3 Ash Content (ASTM D-1102-84) 61
        3.8.3.1 Scope 62
        3.8.3.2 Sample Preparation 62
        3.8.3.3 Apparatus 62
        3.8.3.4 Procedure 62

	3.8.3.5	Report 62
	3.8.3.6	Precision 62
3.8.4	Preparation of Holocellulose (Chlorite Holocellulose) 62	
	3.8.4.1	Scope 62
	3.8.4.2	Sample Preparation 63
	3.8.4.3	Apparatus 63
	3.8.4.4	Reagents 63
	3.8.4.5	Procedure 63
3.8.5	Preparation of Alpha-Cellulose (Determination of Hemicelluloses) 63	
	3.8.5.1	Scope 63
	3.8.5.2	Principle of Method 63
	3.8.5.3	Apparatus 64
	3.8.5.4	Reagents 64
	3.8.5.5	Procedure 64
	3.8.5.6	Calculation and Report 64
3.8.6	Preparation of Klason Lignin 65	
	3.8.6.1	Scope 65
	3.8.6.2	Apparatus 65
	3.8.6.3	Reagent 65
	3.8.6.4	Procedure 65
	3.8.6.5	Additional Information 65
3.8.7	Determination of Methoxyl Groups 65	
	3.8.7.1	Scope 65
	3.8.7.2	Principle of Method 66
	3.8.7.3	Sample Preparation 66
	3.8.7.4	Apparatus 66
	3.8.7.5	Reagents 66
	3.8.7.6	Procedure 67
	3.8.7.7	Calculation and Report 67
3.8.8	Determination of Acetyl by Gas Liquid Chromatography 67	
	3.8.8.1	Scope 67
	3.8.8.2	Reagents 68
	3.8.8.3	Sample Preparation 68
	3.8.8.4	Gas Chromatography 68
	3.8.8.5	Reporting 68
References 72		

In chemical terms, wood is best defined as a three-dimensional biopolymer composite composed of an interconnected network of cellulose, hemicelluloses, and lignin with minor amounts of extractives and inorganics. The major chemical component of a living tree is water, but on a dry-weight basis, all wood cell walls consist mainly of sugar-based polymers (carbohydrates, 65–75%) that are combined with lignin (18–35%). Overall, dry wood has an elemental composition of about 50% carbon, 6% hydrogen, 44% oxygen, and trace amounts of inorganics. Simple chemical analysis can distinguish between hardwoods (angiosperms) and softwoods (gymnosperms) but such techniques cannot be used to identify individual tree species because of the variation within each species and the similarities among species. In general, the coniferous species (softwoods) have a higher cellulose content (40–45%), higher lignin (26–34%), and lower pentosan (7–14%) content as compared to deciduous species (hardwoods) (cellulose 38–49%, lignin 23–30%, and pentosans 19–26%). Table 3.1 shows a summary of the carbohydrates, lignin, and ash content of hardwoods and softwoods in the United States (Pettersen 1984).

Cell Wall Chemistry

TABLE 3.1
Summary of Carbohydrate, Lignin, and Ash Compositions for U.S. Hardwoods and Softwoods

Species	Holocellulose	Alpha Cellulose	Pentosans	Klason Lignin	Ash
Hardwoods	71.7 ± 5.7	45.4 ± 3.5	19.3 ± 2.2	23.0 ± 3.0	0.5 ± 0.3
Softwoods	64.5 ± 4.6	43.7 ± 2.6	−9.8 ± 2.2	28.8 ± 2.6	0.3 ± 0.1

Source: Pettersen, 1984.

A complete chemical analysis accounts for all of the components of wood. Vast amounts of data are available on the chemical composition of wood. The tables at the end of this chapter (Table 3.13 and Table 3.14) summarize data for wood species in North America (Pettersen 1984).

3.1 CARBOHYDRATE POLYMERS

3.1.1 HOLOCELLULOSE

The major carbohydrate portion of wood is composed of cellulose and hemicellulose polymers with minor amounts of other sugar polymers such as starch and pectin (Stamm 1964). The combination of cellulose (40–45%) and the hemicelluloses (15–25%) is called holocellulose and usually accounts for 65–70 percent of the wood dry weight. These polymers are made up of simple sugars, mainly, D-glucose, D-mannose, D-galactose, D-xylose, L-arabinose, D-glucuronic acid, and lesser amounts of other sugars such as L-rhamnose and D-fucose. These polymers are rich in hydroxyl groups that are responsible for moisture sorption through hydrogen bonding (see Chapter 4).

3.1.2 CELLULOSE

Cellulose is the most abundant organic chemical on the face of the earth. It is a glucan polymer of D-glucopyranose units, which are linked together by β-(1 → 4)-glucosidic bonds. The building block for cellulose is actually cellobiose, since the repeating unit in cellulose is a two-sugar unit (Figure 3.1).

The number of glucose units in a cellulose molecule is referred to as the degree of polymerization (DP). Goring and Timell (1962) determined the average DP for native celluloses from several sources using a nitration isolation procedure that minimizes depolymerization and maximizes yield. These molecular weight determinations, done by light-scattering experiments, indicate that wood cellulose has an average DP of at least 9,000–10,000 and possibly as high as 15,000. An average DP of 10,000 would correspond to a linear chain length of approximately 5 μm in wood. This would mean an approximate molecular weight for cellulose ranging from about 10,000 to 150,000. Figure 3.2 shows a partial structure of cellulose.

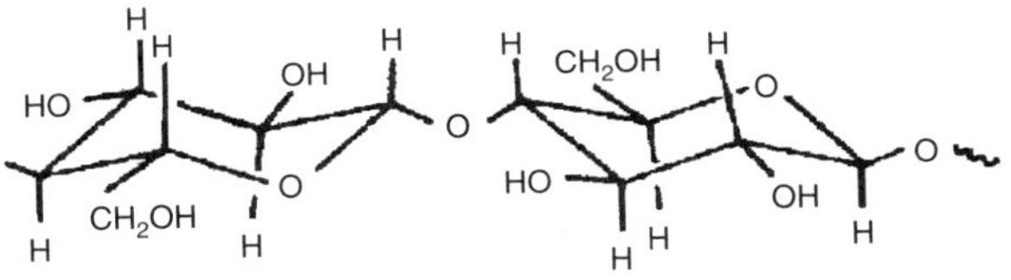

FIGURE 3.1 Chemical structure of cellobiose.

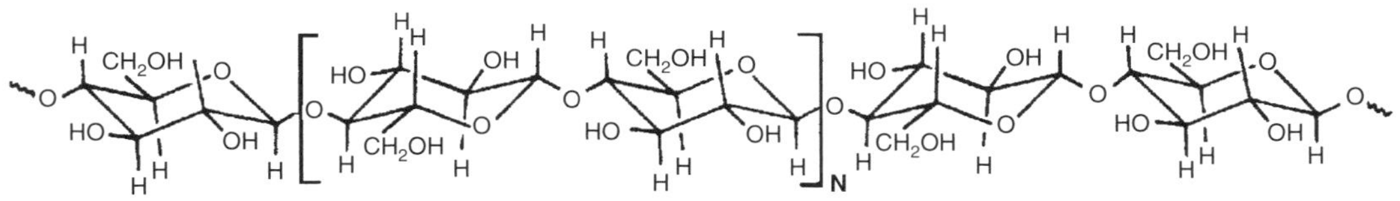

FIGURE 3.2 Partial structure of cellulose.

Cellulose molecules are randomly oriented and have a tendency to form intra- and intermolecular hydrogen bonds. As the packing density of cellulose increases, crystalline regions are formed. Most wood-derived cellulose is highly crystalline and may contain as much as 65% crystalline regions. The remaining portion has a lower packing density and is referred to as amorphous cellulose. X-ray diffraction experiments indicate that crystalline cellulose (*Valonia ventricosa*) has a space group symmetry where a = 16.34 Å and b = 15.72 Å (Figure 3.3, Gardner and Blackwell 1974). The distance of one repeating unit (i.e., one cellobiose unit) is c = 10.38 Å (Figure 3.4). The unit cell contains eight cellobiose moieties. The molecular chains pack in layers that are held together by weak van der Waals forces. The layers consist of parallel chains of anhydroglucopyranose units and the chains are held together by intermolecular hydrogen bonds. There are also intramolecular hydrogen bonds between the atoms of adjacent glucose residues (Figure 3.4). This structure is referred to as cellulose I or native cellulose.

There are several types of cellulose in wood: crystalline and noncrystalline (as described in the preceding paragraph) and accessible and nonaccessible. Accessible and nonaccessible refer to the availability of the cellulose to water, microorganisms, etc. The surfaces of crystalline cellulose are accessible but the rest of the crystalline cellulose is nonaccessible. Most of the noncrystalline cellulose is accessible but part of the noncrystalline cellulose is so covered with both hemicelluloses and lignin that it becomes nonaccessible. Concepts of accessible and nonaccessible cellulose are very important in moisture sorption, pulping, chemical modification, extractions, and interactions with microorganisms.

Cellulose II is another important type of cellulose used for making cellulose derivatives. It is not found in nature. Cellulose II is obtained by mercerization and regeneration of native cellulose. Mercerization is treatment of cellulose I with strong alkali. Regeneration is treatment with carbon disulfide to form a soluble xanthate derivative. The derivative is converted back to cellulose and reprecipitated as cellulose II. Cellulose II has space group a = 8.01 Å, b = 9.04 Å (Figure 3.3), and c = 10.36 Å (Figure 3.4).

There is also a cellulose III structure, which is formed by treatment of cellulose I with liquid ammonia at about –80°C followed by evaporation of the ammonia. Alkali treatment of cellulose III gives cellulose II. Cellulose IV is formed by heating cellulose III in glycerol at 260°C.

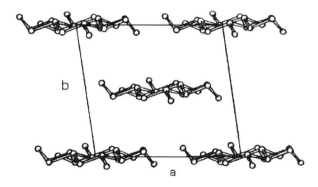

FIGURE 3.3 Axial projection of the crystal structure of cellulose I.

Cell Wall Chemistry

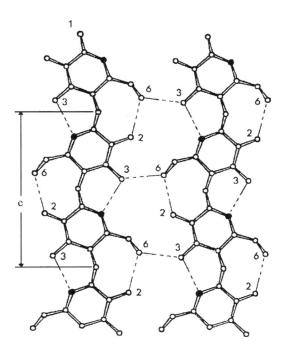

FIGURE 3.4 Planar projection of two cellulose chains showing some of the hydrogen bond between cellulose chains and within a single cellulose chain.

Another type of cellulose (based on the method of extraction from wood) often referred to in the literature is Cross and Bevan cellulose. It consists largely of cellulose I but also contains some hemicellulose. It is obtained by chlorination of wood meal, followed by washing with aqueous solutions of 3% sulfur dioxide (SO_2) and 2% sodium sulfite ($NaSO_3$).

Finally, there is another structure of cellulose referred to as Kürschner cellulose (also based on the method of isolation). Kürschner cellulose is obtained by refluxing wood meal three times for 1 hour with a 1:4 (v/v) mixture of nitric acid and ethyl alcohol. The water-washed and dried cellulose is referred to as Kürschner cellulose, which also contains some hemicelluloses. This method of cellulose isolation is not often used because it destroys some of the cellulose and the nitric acid–ethanol mixture is potentially explosive.

Cellulose I is insoluble in most solvents including strong alkali. Alkali will swell cellulose but not dissolve it. Cellulose dissolves in strong acids such as 72% sulfuric acid, 41% hydrochloric acid, and 85% phosphoric acid, but degradation occurs rapidly. It is difficult to isolate cellulose from wood in a pure form because it is intimately associated with lignin and hemicellulose. The analytical method for isolating cellulose is given in the analytical procedures section of this chapter (Section 3.8).

3.1.3 HEMICELLULOSES

In general, the hemicellulose fraction of woods consists of a collection of polysaccharide polymers with a lower DP than cellulose (average DP of 100–200) and containing mainly the sugars D-xylopyranose, D-glucopyranose, D-galactopyranose, L-arabinofuranose, D-mannopyranose, D-glucopyranosyluronic acid, and D-galactopyranosyluronic acid with minor amounts of other sugars. The structure of hemicelluloses can be understood by first considering the conformation of the monomer units. There are three entries under each monomer in Figure 3.5. In each entry, the letter designations D and L refer to the standard configurations for the two optical isomers of glyceraldehyde,

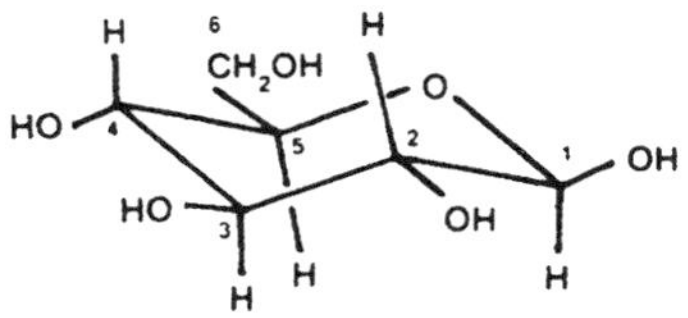

β-D-Glucose
β-D-Glucopyranose
β-D-Glup

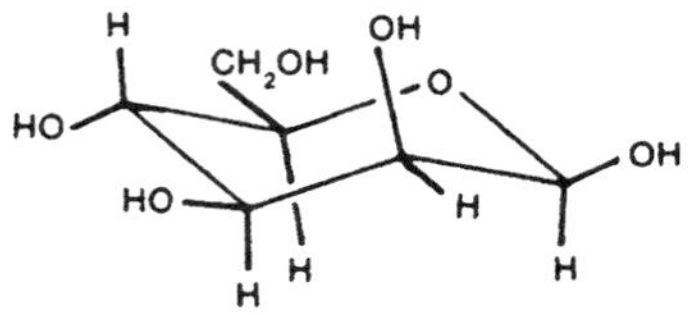

β-D-Mannose
β-D-Mannopyranose
β-D-Manp

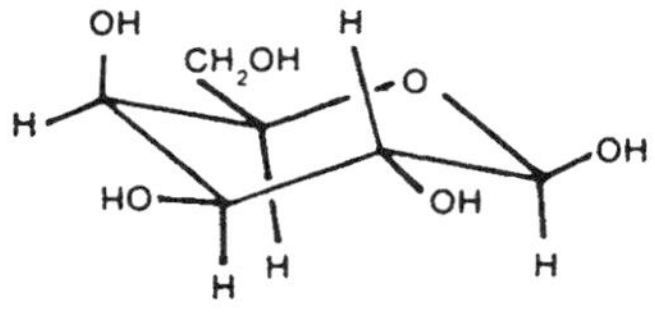

β-D-Galactose
β-D-Galactopyranose
β-D-Galp

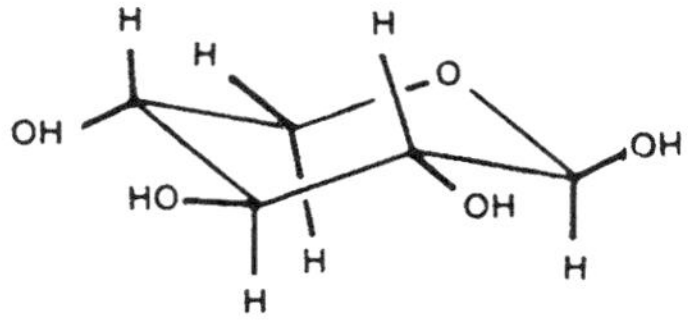

β-D-Xylose
β-D-Xylopyranose
β-D-Xylp

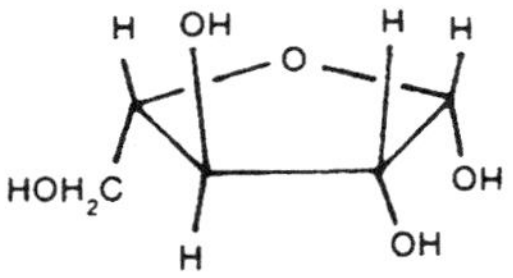

α-L-Arabinose
α-L-Arabinofuranose
α-L-Araf

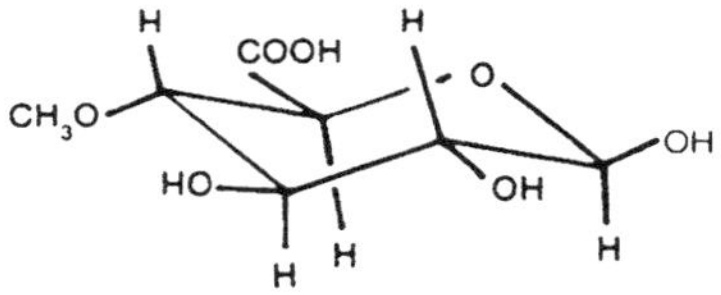

4-O-Methylgucuronic acid
4-O-Methylglucopyranosyluronic acid
4-O-Me-α-D-GlupA

FIGURE 3.5 Sugar monomer components of wood hemicellulose.

the simplest carbohydrate, and designate the conformation of the hydroxyl group at carbon 4 (C-4) for pentoses (xylose and arabinose) and C-5 for hexoses (glucose, galactose, and mannose). The Greek letters α and β refer to the configuration of the hydroxyl group on C-1. The two configurations are called anomers. The first name given for each structure is a shortened form of the sugar name. The second name given for each structure explicitly indicates the ring structure: Furanose refers to a five-membered ring and pyranose refers to a six-membered ring. The six-membered ring is usually in a chair conformation. The third name given for each structure is an abbreviation commonly used for a sugar residue in a polysaccharide (Whistler et al. 1962, Timell 1964, Timell 1965, Whistler and Richards 1970, Jones et al. 1979).

Hemicelluloses are intimately associated with cellulose and contribute to the structural components of the tree. Some hemicelluloses are present in very large amounts when the tree is under stress, e.g., compression wood has a higher D-galactose content as well as a higher lignin content (Timell 1982). They usually contain a backbone consisting of one repeating sugar unit linked β-(1→4) with branch points (1→2), (1→3), and/or (1→6).

Hemicelluloses usually consists of more than one type of sugar unit and are sometimes referred to by the sugars they contain, for example, galactoglucomanan, arabinoglucuronoxylan, arabinogalactan, glucuronoxylan, glucomannan, etc. The hemicelluloses also contain acetyl- and methyl-substituted groups. Hemicelluloses are soluble in alkali and are easily hydrolyzed by acids. A gradient elution at varying alkali concentrations can be used for a crude fractionation of the hemicelluloses from

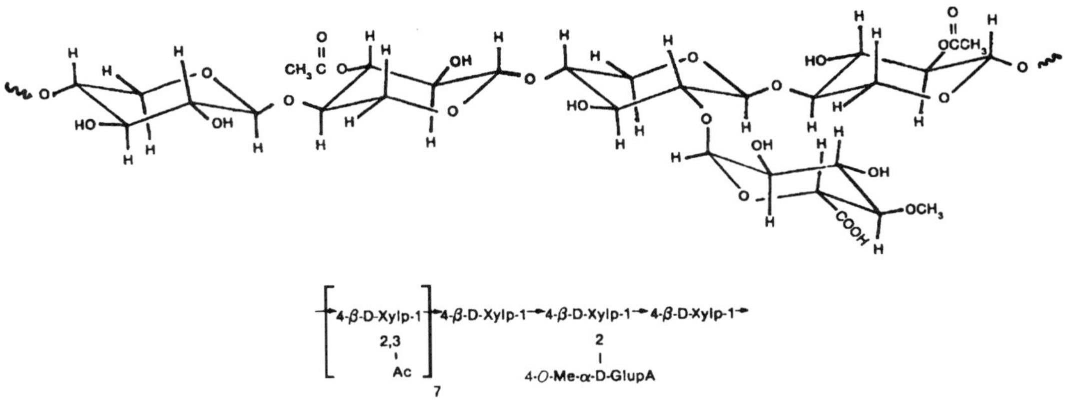

Cell Wall Chemistry

FIGURE 3.6 Partial molecular structure (top) and structure representation (bottom) of *O*-acetyl-4-*O*-methyl-glucuronoxylan.

wood. The hemicelluloses can then be precipitated from the alkaline solution by acidification using acetic acid. Further treatment of the neutralized solution with a neutral organic solvent such as ethyl alcohol results in a more complete precipitation (Sjöström 1981). The detailed structures of most wood hemicelluloses have not been determined. Only the ratios of sugars that these polysaccharides contain have been studied.

3.1.3.1 Hardwood Hemicelluloses

Figure 3.6 shows a partial structure of an *O*-acetyl-4-*O*-methyl-glucuronoxylan from a hardwood. This class of hemicelluloses is usually referred to as glucuronoxylans. This polysaccharide contains a xylan backbone of D-xylopyranose units linked β-(1→4) with acetyl groups at C-2 or C-3 of the xylose units on an average of seven acetyls per ten xylose units (Sjöström 1981). The xylan is substituted with sidechains of 4-*O*-methylglucuronic acid units linked to the xylan backbone α-(1→2) with an average frequency of approximately one uronic acid group per ten xylose units. The sidechains are quite short.

Hardwoods also contain 2–5% of a glucomannan composed of β-D-glucopyranose and β-D-mannopyranose units linked (1→4). The glucose:mannose ratio varies between 1:2 and 1:1 depending on the wood species. Table 3.2 shows the major hemicelluloses found in hardwoods.

TABLE 3.2
Major Hemicelluloses in Hardwoods

Hemicellulose Type DP	Percent in Wood	Units	Molar Ratio	Linkage
Glucuronoxylan 200	15–30	β-D-Xyl*p*	10	1→4
		4-*O*-Me-α-D-Glu*p*A	1	1→2
		Acetyl	7	
Glucomannan 200	2–5	β-D-Man*p*	1-2	1→4
		β-D-Glu*p*	1	1→4

TABLE 3.3
Major Hemicelluloses in Softwoods

Hemicellulose Type DP	Percent in Wood	Units	Molar Ratio	Linkage Avg
Galactoglucomannan 100	5–8	β-D-Man*p*	3	1→4
		β-D-Glu*p*	1	1→4
		α-D-Gal*p*	1	1→6
Galactoglucomannan 100	10–15	β-D-Man*p*	4	1→4
		β-D-Glu*p*	1	1→4
		α-D-Gal*p*	0.1	1→6
		Acetyl	1	
Arabinoglucuronoxylan 100	7–10	β-D-Xyl*p*	10	1→4
		4-*O*-Me-α-D-Glu*p*A	2	1→2
		α-L-Ara*f*	1.3	1→2
Arabinogalactan 200 (Larch wood)	5–35	β-D-Gal*p*	6	1→4
				1→6
		α-L-Ara*f*	2–3	1→6
		β-D-Ara*p*	1–3	1→3
		β-D-Glu*p*A	trace	1→6

3.1.3.2 Softwood Hemicelluloses

Table 3.3 shows the major hemicelluloses from softwoods. One of the main hemicelluloses from softwoods contains a backbone polymer of D-galactose, D-glucose, and D-mannose (Sjöström 1981). The galactoglucomannan is the principal hemicellulose (approximately 20%), with a linear or possibly slightly branched chain with β-(1→4) linkages (Figure 3.7). Glucose and mannose make up the backbone polymer with branches containing galactose. There are two fractions of these polymers, which differ in galactose content. The low-galactose fraction has a ratio of galactose:glucose:mannose of about 0.1:1:4 and the high-galactose fraction has a ratio of 1:1:3. The D-galactopyranose units are linked as a single-unit sidechain by α-(1→6) bonds. The C-2 and C-3 positions of the backbone polymer have acetyl groups substituted on them an average of every three to four hexose units.

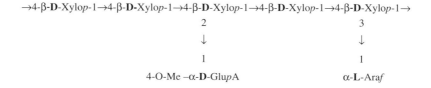

FIGURE 3.7 Partial structure of a softwood arabino 4-*O*-methylglucuronoxylan.

```
→4-β-D-Manop-1→4-β-D-Glup-1→4-β-D-Manop-1→4-β-D-Manop-1→4-β-D-Glup-1→
                 6                      2 or 3
                 ↓                        ↓
                 1                        1
              β-D-Galp                  Acetyl
```

FIGURE 3.8 Partial structure of a softwood *O*-acetyl-galacto-glucomannan.

Another major hemicellulose polymer in softwoods (5–10%) is an arabinoglucuronoxylan consisting of a backbone of β-(1→4) xylopyranose units with α-(1→2) branches of D-glucopyranosyluronic acid on an average of every two to ten xylose units and α-(1→3) branches of L-arabinofuranose, on average, every 1.3 xylose units (Figure 3.8).

Another hemicellulose that is found mainly in the heartwood of larches is an arabinogalactan. Its backbone is a β-(1→3)-linked D-galactopyranose polymer with almost every unit having a branch attached to C-6 of β-D-galactopyranose residues. In some cases this sidechain is β-L-arabinofuranose linked (1→3) or β-D-arabinopyranose linked (16).

There are other minor hemicelluloses in softwoods that mainly contain L-arabinofuranose, D-galactopyranose, D-glucopyranouronic acid, and D-galactopyroanuronic acid (Sjöström 1981).

3.1.4 OTHER MINOR POLYSACCHARIDES

Both softwoods and hardwoods contain small amounts of pectins, starch, and proteins. Pectin is a polysaccharide polymer made up of repeating units of D-galacturonic acid linked α-(1→4). Pectin is found in the membranes in the bordered pits between wood cells and in the middle lamella. Degradation of this membrane by microorganisms increases permeability of wood to water-based treatment chemicals such as fire retardants and wood preservatives. Pectins are found in high concentration in the parenchyma cell walls in the inner bark where they may act as a binder. L-Arabinofuranose and D-galactopyranose are often found as a minor part of the pectic substance. Pectin is also found as the methyl ester.

Starch is the principal reserve polysaccharide in plants. Small amount of starch can also be found in the wood cell wall. Starch normally occurs as granules and is composed of D-glucopyranose units linked α-(1→4) (amylose) or α-(1→4) with branches about every 25 glucopyraosyl units at α-(1→6) (amylopectin). Amylose occurs as a helix structure in the solid state due to the α-configuration in the polymer. Amylopectin is highly branched.

3.2 LIGNIN

Lignins are amorphous, highly complex, mainly aromatic polymers of phenylpropane units (Figure 3.9) that are considered to be an encrusting substance. The three-dimensional polymer is made up of C–O–C and C–C linkages. The precursors of lignin biosynthesis are *p*-coumaryl alcohol (Figure 3.9, structure 1), coniferyl alcohol (Figure 3.9, structure 2), and sinapyl alcohol (Figure 3.9, structure 3). Structure 1 is a minor precursor of both softwood and hardwood lignins, structure 2 is the predominate precursor of softwood lignin, and structures 2 and 3 are both precursors of hardwood lignin (Alder 1977).

Softwood lignin has a methoxyl content of 15–16%; hardwood lignin has a methoxyl content of 21%. Lignin does not have a single repeating unit of the hemicelluloses like cellulose does, but instead consists of a complex arrangement of substituted phenolic units.

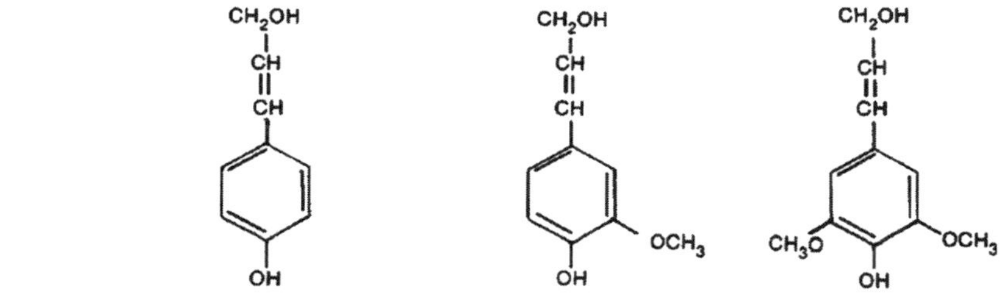

FIGURE 3.9 Chemical structures of lignin precursors: (1) *p*-coumaryl alcohol, (2) coniferyl alcohol, and (3) sinapyl alcohol.

Lignins can be classified in several ways, but they are usually divided according to their structural elements (Sjöström 1981). All wood lignins consist mainly of three basic building blocks of guaiacyl, syringyl, and *p*-hydroxyphenyl moieties, although other aromatic units also exist in many different types of woods. There is a wide variation of structures within different wood species. The lignin content of hardwoods is usually in the range of 18–25%, whereas the lignin content of softwoods varies between 25 and 35%. The phenylpropane can be substituted at the α, β, or γ positions into various combinations linked together both by ether and carbon to carbon linkages (Sakakibara 1991).

Lignins from softwoods are mainly a polymerization product of coniferyl alcohol and are called guaiacyl lignin. Hardwood lignins are mainly syringyl-guauacyl lignin, because they are a copolymer

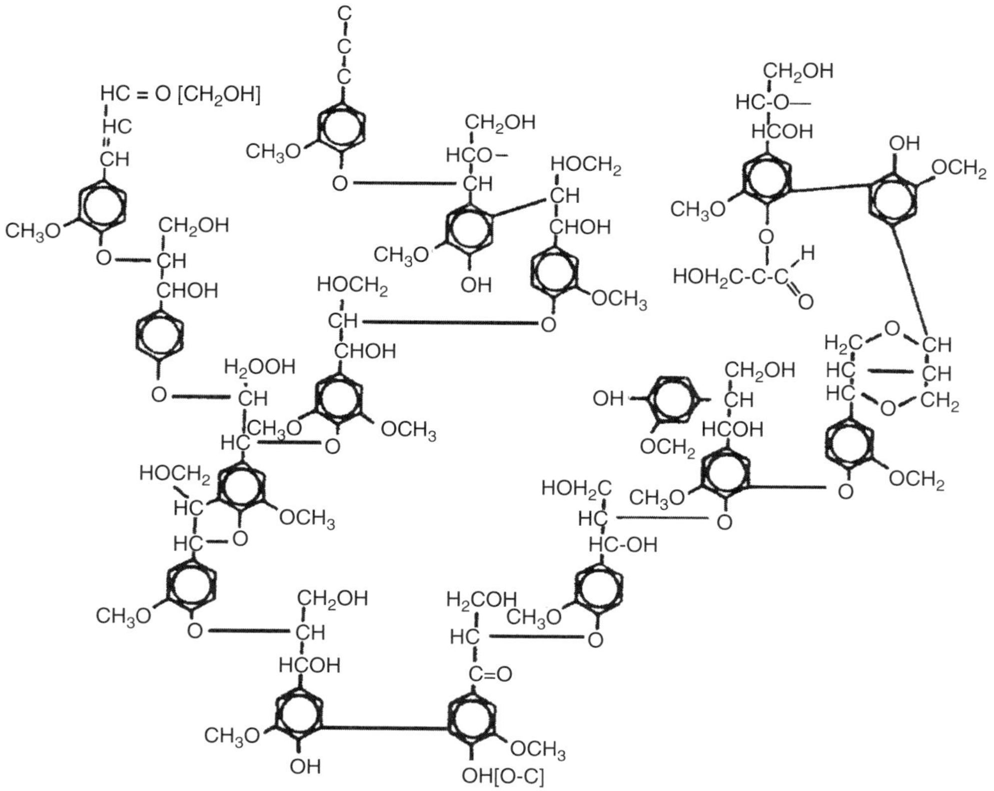

FIGURE 3.10 Partial structure of a softwood lignin.

Cell Wall Chemistry

of coniferyl and sinapyl alcohols. The ratio of these two types varies in different lignins from about 4:1 to 1:2 (Sarkanen and Ludwig 1971). A proposed structure for a hardwood lignin (*Fagus silvatica* L.) is shown in Figure 3.10 (Adler 1977).

Lignins found in woods contain significant amounts of constituents other than guaiacyl- and syringylpropane units (Sarkanen and Ludwig 1971). Lignin is distributed throughout the secondary cell wall, with the highest concentration in the middle lamella. Because of the difference in the volume of middle lamella to secondary cell wall, about 70% of the lignin is located in the cell wall.

Lignin can be isolated from wood in several ways. So-called Klason lignin is obtained after hydrolyzing the polysaccharides with 72% sulfuric acid. It is highly condensed and does not truly represent the lignin in its native state in the wood. The polysaccharides can be removed using enzymes to give an "enzyme lignin" that is much closer to a native lignin than Klason lignin. "Milled wood lignin" or Björkman lignin can be isolated by using a vibratory ball mill on fine wood flour and then extracting with suitable organic solvents (Björkman 1956, 1957). Approximately 30–50% of the native lignin is isolated using this procedure. This procedure is tedious but does isolate a lignin closer to a native lignin.

The molecular weight of lignin depends on the method of extraction. Klason lignin, since it is highly condensed, has molecular weights as low as 260 and as high as 50 million (Goring 1962). Björkman lignin has a molecular weight of approximately 11,000.

Lignins are associated with the hemicelluloses forming, in some cases, lignin–carbohydrate complexes that are resistant to hydrolysis even under pulping conditions (Obst 1982). There is no evidence that lignin is associated with cellulose.

3.3 EXTRACTIVES

As the name implies, extractives (also referred to as natural products) are chemicals in the wood that can be extracted using solvents. In some cases, the extractives are classified by the solvent used to extract them. For example, water-soluble or toluene-ethanol–soluble or ether-soluble extractives. Hundreds of extractives have been identified and in some cases their role in the tree is well understood. In other cases, it is not clear why they are present (Rowe 1989). Extractives, such as pine pitch and resins, have been used for centuries to waterproof wooden boats, in torches, and as a binder. They have also found application in medicine, cosmetics, and as a preservative (Hillis 1989). Some of the extractives in wood are precursors to other chemicals, some are formed in response to wounds, and some act as part of a defense mechanism.

The extractives are a group of cell wall chemicals mainly consisting of fats, fatty acids, fatty alcohols, phenols, terpenes, steroids, resin acids, rosin, waxes, and many other minor organic compounds. These chemicals exist as monomers, dimers, and polymers. In general, softwoods have a higher extractives content than hardwoods. Most of the extractives in both softwoods and hardwoods are located in the heartwood, and some are responsible for the color, smell, and durability of the wood. The qualitative difference in extractive content from species to species is the basis of chemotaxonomy (taxonomy based on chemical constituents).

Resins and fats are made up of resin acids and fatty acids, respectively. Fatty acids are esters with alcohols, such as glycerol, and mainly occur in sapwood. Resin acids have a free carboxylic acid functional group and are mainly found in heartwood (Kai 1991). Abietic acid (Figure 3.11, structure 1) is a common type of resin acid.

The most common terpenes in softwoods are pinene (Figure 3.11, structure 2) and other similar chemical structures. One of the most important polyphenols is pinosylvin (Figure 3.11, structure 3), which is very toxic and found in pine heartwood. Lignans are a combination of two phenylpropane units and are common in softwoods (Gottlieb and Yoshida 1989). Conidendrin (Figure 3.11, structure 4) is found in spruce and hemlock. Tannins in wood can be classified into three classes: gallotannins, ellagtannins, and condensed tannins (Hemingway 1989, Porter 1989). Gallotannins are polymeric esters of gallic acid (Figure 3.11, structure 5) and are usually associated with sugars

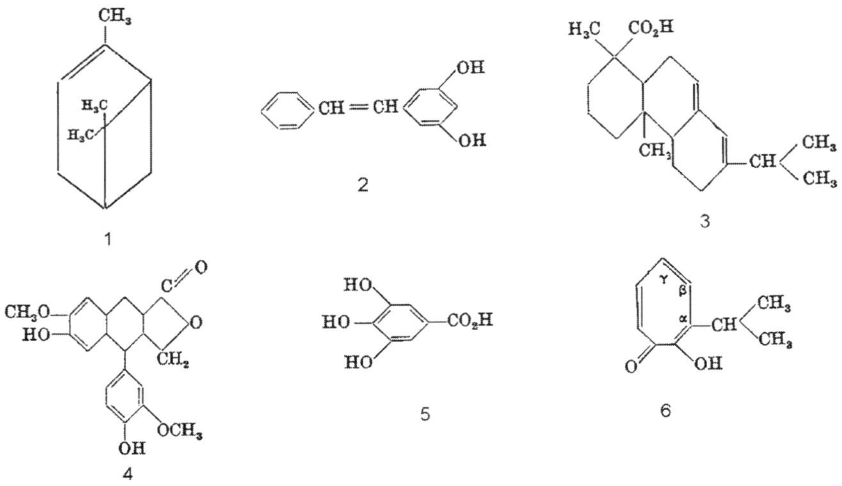

FIGURE 3.11 Chemical structures of some of the extractives in wood: (1) abietic acid, (2) α-pinene, (3) pinosylvin, (4) pineresinol, (5) gallic acid, (6) α-, β-, and γ-thujaplicin.

(Haslam 1989). Tropolones are responsible for the durability of cedar wood. Examples of this class of extractives include α-, β-, and γ-thujaplicin (Figure 3.11, structure 6) (Kollmann and Côté 1968).

3.4 BARK

Bark is a very complex tissue that is composed of two principal zones: the inner bark and the outer bark. The outer bark, which is sometimes referred to as rhytidome and is also known as the periderm, is made up of three layers: the phellem (cork cells), phellogen (cork cambium), and the phelloderm (cork skin). The thickness of the periderm varies greatly between and within species and with the age of the bark. The inner bark, which is referred to as the phloem or bast, is complex in structure and is composed of several types of cells including sieve tubes, fiber cells, albuminose cells, companion cells, parenchyma cells, ideoblasts, and lactifers. Not all cell types occur in every bark. The bark is divided from the wood or xylem by the vascular cambium layer (Sandved et al. 1992).

The chemical composition of bark is complex and varies between and within species, and also between the inner and outer bark. Proximate chemical analysis of bark from different species indicates that the chemical constituents of bark can be classified into four major groups: polysaccharides (cellulose, hemicellulose, and pectic materials); lignin and polyphenols; hydroxy acid complexes (suberin); and extractives (fats, oils, phytosterols, resin acids, waxes, tannins, terpenes, phlobaphenes, and flavonoids). Table 3.4 illustrates the variability of the chemical composition of bark between softwood and hardwood species, *Pinus pinaster* and *Quercus suber*, respectively.

3.4.1 EXTRACTIVES

The extractives content of bark is quite high compared to wood, but values reported in the literature can be very different even for the same species. These apparent differences depend on the method of extraction. For example, McGinnis and Parikh (1975) reported 19.9% extractives for loblolly pine bark using petroleum ether, benzene, ethanol, and cold and hot water. Labosky (1979) extracted loblolly pine bark with hexane, benzene, ethyl ether, ethanol, water, and 1% sodium hydroxide and reported 27.5% extractives.

The analysis methods developed for wood cannot be used for bark directly. There are many compounds in bark that are not found in wood that interfere with these analysis methods. For example,

TABLE 3.4
Average Chemical Composition of Softwood and Hardwood Bark

Component	Percent Oven-Dry Weight	
	Pinus pinaster[a]	*Quercus suber*[b]
Polysaccharides	41.7 ± 0.9	19.9 ± 2.6
Lignin and polyphenols	43.7 ± 2.4	23.0 ± 0.5
Suberin	1.5 ± 0.2	39.4 ± 1.7
Extractives	11.4 ± 2.2	14.2 ± 1.1
Ash	1.2 ± 0.6	1.2 ± 0.2

[a] Data obtained from Nunes et al. 1996.
[b] Data obtained from Pereira, 1988.

the presence of suberin in bark tends to limit access of delignification reagents to the lignin in the bark, and therefore may lead to a holocellulose that is not pure enough to permit fractionation of individual bark polysaccharides. Suberin, polyflavonoids, and other high-molecular-weight condensed tannins can also complicate analysis of bark lignin, resulting in false high values of lignin content in bark.

Because of the interference of the extractives in polysaccharide and lignin analysis, procedures for elucidation of the chemical composition of bark begin with an extraction protocol that consists of sequential extraction solvents of increasing polarity. A common protocol begins with a diethyl ether extraction step that yields waxes, fatty acids, fats, resin acids, phytosterols, and terpenes. This is followed by an ethyl alcohol extraction step that yields condensed tannins, flavonoids, and phenolics. The third step uses hot water, and yields condensed tannins and water-soluble carbohydrates. To release phenolic acids, hemicelluloses, and suberin monomers from the residue from the third step, 1% aqueous sodium hydroxide is used (Holloway and Deas 1973, Kolattukudy 1984).

The extract fractions from the above-mentioned steps are then subjected to further workup to separate each into easy-to-analyze mixtures of compounds. For example, partitioning the diethyl ether fraction against aqueous sodium bicarbonate separates the fatty acids and resin acids from the neutral components, tannins, terpenes, and flavonoids. The neutral fraction is then saponified to give the alcohols and salts of fatty acids, dicarboxylic, hydroxy-fatty, and ferulic acids. Ethanol extraction followed by hot water extraction of the insoluble ether fraction yields soluble simple sugars and condensed tannins. Sodium hydroxide extraction of the insoluble residue gives soluble suberin monomers, phenolic acids, and hemicelluloses. Sulfuric acid treatment of the insoluble fraction yields lignin (Chang and Mitchell 1955, Hemingway 1981, Laks 1991).

3.4.1.1 Chemical Composition of Extractives

The waxes in bark are esters of high-molecular-weight long-chain monohydroxy-alcohol fatty acids. A lot of research has been done on softwood waxes, but very little on hardwood waxes. At one time, hardwood waxes were produced commercially for use in polishes, lubricants, additives to concrete, carbon paper, fertilizers, and fruit coatings (Hemingway 1981).

Terpenes are a condensation of two or more five-carbon isoprene (2-methy-1,3-butadiene) units in a linear or cyclic structure. They can also contain various functional groups. The most common of the monoterpenes are α- and β-pinenes found in firs and pines. Birch bark can contain up to 25% terpenes, by total dry weight (Seshadri and Vedantham 1971).

Flavonoids are a group of compounds based on a 15-carbon hydroxylated tricyclic unit (Laks 1991). They are often found as glycosides. Many tree barks are rich in mono- and polyflavonoids (Hergert 1960, 1962). Their function seems to be as an antioxidant, pigment, and growth regulator (Laks 1991).

Hydrolyzable and condensed tannins are also major extractives from bark. The hydrolyzable tannins are esters of carboxylic acids and sugars that are easily hydrolyzed to give benzoic acid derivatives and sugars. Over 20 different hydrolyzable tannins have been isolated from oaks (Nonaka et al. 1985).

The condensed tannins are a group of polymers based on a hydroxylated C-15 flavonoid monomer unit. Low degree of polymerization tannins are soluble in polar solvents, whereas the high degree of polymerization tannins are soluble in dilute alkali solutions (Hemingway et al. 1983). It is difficult to isolate pure fractions of tannins and the structure can be altered by the extraction procedure.

Free sugars are also extracted from bark. Hot water extraction yields about 5% free sugar fraction, which is mainly composed of glucose and fructose; this amount varies depending on the growing season. For example, the free sugar content is low in early spring and increases during the growing season, reaching a maximum in the fall (Laks 1991). Other minor free sugars found in bark include galactose, xylose, mannose, and sucrose. Hydrolysis of the hot water extract of bark yields more free sugars, the most abundant one being arabinose. These sugars are tied up as glycosides or in the hemicelluloses. Other sugars released during hydrolysis are glucose, fructose, galactose, xylose, mannose, and rhamnose.

3.4.2 Hemicelluloses

The hemicellulose content of different barks varies from 9.3% for *Quercus robur* to 23.1% for *Fagus sylvatica* (Dietrichs et al. 1978). The main hemicellulose in conifer barks is a galactoglucomannan. Arabino-4-*O*-methyl-glucuronoxylan is the main hemicellulose in deciduous barks. In general, bark xylans and glucomannans are similar to ones found in wood. Other hemicelluloses that have been isolated from barks include 4-*O*-methy-glucuronoxylans, glucomannans, *O*-acetyl-galactoglucomannan, and *O*-acetyl-4-*O*-methyl-glucuronoxylan (Painter and Purves 1960, Jiang and Timell 1972, Dietrichs 1975). In the xylans, the xylose units are connected β-(1→4) and the glucuronic acid groups are attached to the xylan backbone α-(1→2). The ratio of xylose to GluU is 10:1 with a degree of polymerization of between 171 and 234 (Mian and Timell 1960). Glucomannans from deciduous barks contain mannose and glucose units in a ratio of from 1:1 to 1.4:1 (Timell 1982). In the mannans from the barks of aspen and willow, galactose units were found as sidechains. The ratio of mannose:glucose:galactose was 1.3:1:0.5 with an average degree of polymerization of 30 to 50 (Timell 1982).

Arabinans have been reported in the barks of aspen, spruce, and pine (Painter and Purves 1960). The backbone is α-(1→5)-arabinofuranose units and, in the case of pine, the average degree of polymerization is 95 (Timell 1982). A group of galacturonic acid polymers has been isolated from birch. One is a galacturonic acid backbone linked α-(1→4) with arabinose sidechains in a ratio of galacturonic acid to arabinose of 9:1, and another consists of galacturonic acid, arabinose, and galactose in a ratio of 7:3:1. Small amounts of glucose, xylose, and rhamnose were also found in these polymers (Mian and Timell 1960, Timell 1982).

A pectic substance that contains either galactose alone or galactose and arabinose units has also been isolated from barks (Toman et al. 1976). The pure galactan is water-soluble and consists of 33 β-(1→4)-linked galactose units with a sidechain at C-6 of the backbone. A highly branched arabinogalactan was found in the bark of spruce with a ratio of galactose to arabinose of 10:1 (Painter and Purves 1960).

In almost all cases, the hemicelluloses found in bark are similar to those found in wood, with some variations in composition. Table 3.5 shows the sugars present after hydrolysis of the polysaccharides in bark.

TABLE 3.5
Sugars Present in Hydrolyzates of Some Tree Barks

Species	Glu	Man	Gal	Xyl	Ara	Rha	UrA	Ac
Abies amabills	37.4	8.0	1.6	3.2	3.2	—	5.6	0.8
Picea abies	36.6	6.5	1.3	4.8	1.8	0.3	—	—
Picea engelmannii	35.7	2.9	2.4	3.8	3.3	—	8.0	0.5
Pinus contoria								
Inner bark	40.9	2.5	4.3	3.7	10.6	—	9.9	0.2
Outer bark	26.8	2.5	4.2	3.4	5.5	—	7.7	0.8
Pinus sylvestris	30.2	5.4	2.4	5.8	2.1	0.3	—	—
Pinus taeda								
Inner bark	21.3	2.5	3.1	2.1	5.6	0.3	4.6	—
Outer bark	15.8	2.6	2.5	3.8	1.8	0.1	2.1	—
Betula papyrifera								
Inner bark	28.0	0.2	1.0	21.0	2.7	—	2.2	—
Fagus sylvatica	29.7	0.2	3.1	20.1	3.1	1.2	—	—
Quercus robur	32.3	0.5	1.3	16.4	2.0	0.5	—	—

Source: Fengel and Wegener, 1984.

3.4.3 CELLULOSE

The cellulose content of barks ranges from 16–41% depending on the method of extraction. In unextracted bark, the cellulose content was between 20.2% for pine and 32.6% for oak (Dietrichs et al. 1978). The high extractives content, especially of suberin, requires harsh conditions to isolate the cellulose, so the cellulose content is usually low and the cellulose is degraded during the isolation process. The outer bark usually contains less cellulose than the inner bark (Harun and Labosky 1985).

Timell (1961a,b) and Mian and Timell (1960), found a number average degree of polymerization for bark cellulose of 125 (*Betula papyrifera*) to 700 (*Pinus contoria*), and a weight average of 4000 (*Abies amabilis, Populus grandidentata*) to 6900 (*Pinus contoria*). Bark cellulose has the same type of crystalline lattice (cellulose I) as normal wood, but the degree of crystallinity is less.

3.4.4 LIGNIN

As with other analyses involving bark components, literature values for lignin content can vary depending on the method of extraction (Kurth and Smith 1954, Higuchi et al. 1967). Bark contains high contents of condensed and hydrolyzable tannins and sulfuric acid-insoluble suberin that can give false high values of lignin content. For example, the Klason lignin from *Pinus taeda* bark is 46.0% when including both lignin and condensed tannins but only 20.4% when the bark is first extracted with alkali (McGinnis and Parikh 1975). Other researchers have found lignin contents from 38–58% (Labosky 1979). The elemental composition and functional group content of bark lignins are similar to those of the lignin from the wood of the same species (Sarkanen and Hergert 1971, Hemingway 1981). There is less lignin in the inner bark as compared to the outer bark.

There is a lower ratio of OCH_3 groups in aspen bark than in aspen wood and a higher ratio of phenolic OH groups to OCH_3 (Clermont 1970). There are more guaiacyl units in deciduous bark and more *p*-hydroxyphenyl units in coniferous bark as compared to the wood of the same species (Andersson et al. 1973). While there are some differences in the ratios of components, no structural difference have been found between most bark lignins and the corresponding wood.

3.4.5 INORGANICS AND pH

Bark is generally higher in inorganics than normal wood. The inorganic (ash) content can be as high as 13% and, in general, the inner bark contains more inorganics as compared to the outer bark (Young 1971, Choong et al. 1976, Hattula and Johanson 1978, Harper and Einspahr 1980). For example, the outer bark of willow contains 11.5% ash, the inner bark 13.1%, and the sapwood 0.9%; sweetgum outer bark is 10.4%, inner bark 12.8%, and sapwood 0.5% ash; red oak outer bark is 8.9%, inner bark 11.1%, and sapwood 0.9% ash; and ash outer bark is 12.3%, inner bark 12.1%, and sapwood 0.9% ash. The major inorganic elements in bark are Na, K, Ca, Mg, Mn, Zn, and P (Choong et al. 1976). There is more Na, K, Mg, Mn, Zn, and P in sapwood than in bark and more Ca in bark than in sapwood.

In general, the pH of bark is lower than that of the corresponding wood due to the higher inorganic content of bark compared to wood. For example, Martin and Gray (1971) reported pH values of southern pines ranging from about 3.1–3.8 with an average of 3.4–3.5 compared to a pH of 4.4–4.6 for sapwood. The outer bark has a lower pH than the inner bark, presumably due to a higher content of Ca in the outer bark (Volz 1971). The pH of bark decreases slightly with the age of the tree.

3.5 INORGANICS

The inorganic content of a wood is usually referred to as its ash content, which is an approximate measure of the mineral salts and other inorganic matter in the fiber after combustion at a temperature of $575 \pm 25°C$. The inorganic content can be quite high in woods containing large amounts of silica; however, in most cases, the inorganic content is less than 0.5% (Browning 1967). This small amount of inorganic material contains a wide variety of elements (Ellis 1965, Young and Guinn 1966). Ca, Mg, and K make up 80% of the ash in wood. These elements probably exist in the wood as oxalates, carbonates, and sulfates, or bound to carboxyl groups in pectic materials (Hon and Shiraishi 1991). Other elements present are Na, Si, B, Mn, Fe, Mo, Cu, Zn, Ag, Al, Ba, Co, Cr, Ni, Pb, Rb, Sr, Ti, Au, Ga, In, La, Li, Sn, V, and Zr (Ellis 1965). Some of these are essential for wood growth. Inorganic ions are absorbed into the tree through the roots and transported throughout the tree. Young and Guinn (1966) determined the distribution of 12 inorganic elements in various part of a tree (roots, bark, wood, and leaves) and concluded that both the total inorganic content and concentration of each element varied widely both within and between species. The inorganic content varies depending on the environmental conditions in which the tree lives. See Table 3.12 for a partial list of the inorganic content of some woods.

Saka and Goring (1983) studied the distribution of inorganics from the pith to the outer ring of black spruce (*Picea mariana* Mill) using EDXA. They found 15 different elements including Na, Mg, Al, S, K, Ca, Fe, Ni, Cu, Zn, and Pb. They also found that the inorganic content was higher in earlywood as compared to latewood.

The pH of wood varies from 4.2 (*Pinus sylvestris*) to 5.3 (*Fagus sylvatica*) with an average of approximately 4.7.

3.6 DISTRIBUTION OF CHEMICAL COMPONENTS IN THE CELL WALL

The content of cell wall components depends on the tree species and where in the tree the sample is taken. Softwoods are different from hardwoods, heartwood from sapwood, and latewood from earlywood. Table 3.6 shows the cell wall polysaccharides in earlywood compared to latewood (Saka 1991). Latewood contains more glucomannans as compared to earlywood, but earlywood contains more glucuronoarabinoxylans. Heartwood contains more extractives than sapwood, and as sapwood is transformed into heartwood, aspiration of the bordered pits takes place in softwoods

Cell Wall Chemistry

TABLE 3.6
Cell Wall Polysaccharides in Earlywood and Latewood in Pine

Cell Wall Component	Earlywood%	Latewood%
Cellulose	56.7	56.2
Galactan	3.4	3.1
Glucomannan	20.3	24.8
Arabinan	1.0	1.8
Glucuronoarabinoxylan	18.6	14.1

Source: Saka, 1991.

and encrustation of pit membranes with the formation of tyloses occurs in hardwoods. Earlywood contains more lignin than latewood.

Figure 3.12 shows the distribution of components across the cell wall of scotch pine. The middle lamella and primary wall is mainly composed of lignin (84%) with lesser amounts of hemicelluloses (13.3%) and even less cellulose (07%). The S_1 layer is composed of 51.7% lignin, 30.0% cellulose, and 18.3% hemicelluloses. The S_2 layer is composed of 15.1% lignin, 54.3% cellulose, and 30.6% hemicelluloses. The S_3 layer has little or no lignin, 13% cellulose, and 87% hemicelluloses. The content of xylan is lowest in the S_2 layer and higher in the S_1 and S_3 layers. The concentration of galactoglucomannan is higher in the S_2 than in the S_1 or S_3 layers. On a percentage basis, the middle lamella and primary wall contain the highest concentration of lignin but there is more lignin in the S_2 because it is a much thicker layer as compared to the middle lamella and primary wall. The lignin in the S_2 layer is evenly distributed throughout the layer.

The angle of the cellulose microfibrils in the various cell wall layers, in relation to the fiber axis, is known as the fibril angle. It is one of the most important structural parameters determining mechanical properties of wood. For normal wood, the microfibril angle of the cellulose in the S_2

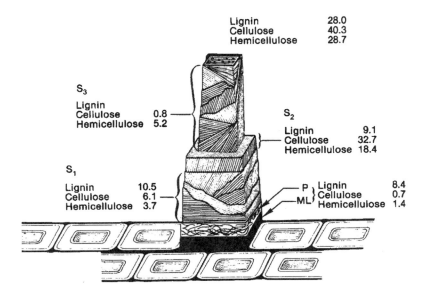

FIGURE 3.12 Chemical composition of the cell wall of scots pine.

layer is 14–19 degrees. It is because this angle is so low in the thick S_2 layer that wood does not swell or shrink to as large an extent in the longitudinal direction (0.1–0.3%).

A further discussion of the distribution of the hemicelluloses in the cell wall can be found in Chapter 15. Strength properties of wood are related to the distribution of hemicelluloses in the cell wall.

3.7 JUVENILE WOOD AND REACTION WOOD

Juvenile wood is the wood that develops in the early stages of tree growth. It physical properties are described in Chapter 2 part 13. Juvenile wood cells are shorter, have smaller cell diameter, larger microfibril angle (up to 55 degrees) and have a high content of compression wood as compared to mature wood. Juvenile wood has a lower density and strength than mature wood. Juvenile wood has less cellulose, more hemicelluloses and lignin compared to mature wood. There is a gradual increase in cellulose content as the cells mature and a gradual decrease in hemicellulose content. The lignin content decreases more rapidly as the cell mature.

Normal wood growth is erect and vertical. When a tree is forced out of this pattern either by wind or gravitational forces, abnormal woody tissue is formed in different parts of the tree to compensate for the abnormal growing conditions. The wood cells that are formed when softwoods and hardwoods are out of vertical are called reaction wood since these cells are reacting to the stressful conditions. In softwoods, irregular cells develop on the underside of a stem or branch and are referred to as compression wood. In hardwoods, irregular cells develop on the upper side of a stem or branch and are referred to as tension wood.

Table 3.7 shows the chemical composition of softwood compression wood (Panshin and de Zeeuw 1980, Timell 1982). Compression wood has a higher lignin content and a lower cellulose content as compared to normal wood. The cellulose in the S_2 layer has a lower degree of crystallization than normal wood and the lignin is largely concentrated in the S_2 layer as compared to normal wood. Forty percent of the lignin is is in the outer zone of the S_2 layer and an additional 40% is uniformly distributed over the remaining part of the S_2 layer (Panshin and de Zeeuw 1980). There are more galactoglucomannans in normal wood and more $1 \rightarrow 3$ linked glucans and galactans in compression wood. The midrofibril angle in the modified S_2 layer in compression wood is quite high (44–47°) and have more rounded tracheids that are 10 to 40% shorter than normal tracheids. Compression wood is weaker than normal wood and lower elastic properties. The reduced cellulose content and high microfibril angle is probably responsible for the reduction in mechanical properties (Panshin and de Zeeuw 1980).

TABLE 3.7
Chemical Composition of Compression Wood in Softwoods

Cell Wall Component	Normal Wood		Compression Wood	
	Range%	Average%	Range%	Average%
Lignin	24.2–33.3	28.8	30.9–40.9	37.7
Cellulose	37.7–60.6	44.6	27.3–53.7	34.9
Galactoglucomannan	—	18	—	9
1,3-Glucan	—	Trace	—	2
Galactan	1.0–3.8	2.2	7.1–12.9	10.0
Glucuronoarabinoxylan	—	8	—	8
Other polysaccharides	—	2	—	2

Data from Panshin and Zeeuw, 1980, and Timell, 1982.

TABLE 3.8
Chemical Composition of Tension Wood in Hardwoods

Cell Wall Component	Normal Wood	Compression Wood
Lignin	29%	14%
Cellulose	44%	57%
Pentosans	15%	11%
Acetyl	3%	2%
Galactosans	2%	7%

Source: Schwerin, 1958.

Table 3.8 shows the chemical composition of hardwood tension wood (Schwerin 1958). Tension wood has a lower lignin content and a higher cellulose content as compared to normal wood. There is a lower content of pentosans (xylans) and acetyls than in normal wood and more galactosans in tension wood. There is no S_3 layer in tension wood but rather what is known as a G layer or gelatinous layer. This layer is approximately 98% cellulose. The cellulose in the G layer is highly crystalline with a microfibril angle of only 5% and contains very little hemicelluloses or lignin. The G layer is as thick or thicker than the S_2 layer in normal wood and contains about the same quantity of inorganics. Tension wood has lower mechanical properties as compared to normal wood (Panshin and de Zeeuw 1980). For example, compression parallel and perpendicular to the grain, modulus of elasticity in bending, modulus of rupture in static bending and longitudinal shear are all reduced in tension wood as compared to normal wood.

3.8 ANALYTICAL PROCEDURES

Chemical composition varies from species to species and within different parts of the same wood species. Chemical composition also varies within woods from different geographic locations, ages, climates and soil conditions.

There are hundreds of reports on the chemical composition of wood material. In reviewing this vast amount of data, it becomes apparent that the analytical procedures used, in many cases, are different from lab to lab and a complete description of what procedure was used in the analysis is not clear. For example, many descriptions do not describe if the samples were pre-extracted with some solvent before analysis. Others do not follow a published procedure so comparison of data is not possible. The following section is composed of standard procedures used in many laboratories to determine the chemical components of the wood cell wall. Tables 3.9 through 3.12 give summaries of various types of chemical compositions of hardwoods and softwoods in the United States. This data has been collected from the analytical laboratories of the USDA, Forest Service, Forest Products Laboratory from 1927 to 1968.

3.8.1 SAMPLING PROCEDURE

In reporting the chemical content of a wood, it is very important to report as much information about the samples as possible. Since the chemical content of a given species may vary depending upon the growing conditions, harvesting times of the year, etc., it is critical to report these conditions along with the chemical analysis. It is also important to report the exact analytical conditions and procedures used. This way, it may be possible to reproduce the results by other workers in different laboratories. Without this information, it is not possible to compare data from different laboratories.

TABLE 3.9
Methoxyl Content of Some Common Hardwoods and Softwoods

Type of Wood	Methoxy Content (%)
Hardwoods	
Balsa	5.68
Basswood	6.00
Yellow birch	6.07
Shellbark hickory	5.63
Sugar maple	7.25
Mesquite	5.55
Tanoak	5.74
Softwoods	
Incense cedar	6.24
Alaska cedar	5.25
Douglas fir	4.95
Western larch	5.03
Longleaf pine	5.05
Western white pine	4.56
Redwood	5.21
White spruce	5.30

Source: Moore and Johnson, 1967.

The following information should accompany each chemical analysis:

1. Source of the wood
    a. Geographic location
    b. Part of the tree sampled
    c. Date sample was taken
2. Sampling
    a. Different anatomical parts
    b. Degree of biological deterioration, if any
    c. Sample size
    d. Drying method applied
3. Analytical procedure used
4. Calculations and reporting technique

All of the above-mentioned criteria could contribute in one way or another toward variations in chemical analyses.

3.8.2 EXTRACTION

3.8.2.1 Scope and Summary

This method describes a procedure for extraction of wood for further analysis, such as holocellulose, hemicellulose, cellulose, and lignin analysis.

$$\text{Wood materials} = \text{Extractives} + \text{holocellulose} + \text{lignin} + \text{inorganics (ash)}$$

TABLE 3.10
Acetyl Content of Some Common Hardwoods and Softwoods

Type of Wood	Acetyl Content (%)
Hardwoods	
Aspen	3.4
Balsa	4.2
Basswood	4.2
Beech	3.9
Yellow birch	3.3
White birch	3.1
Paper birch	4.4
American elm	3.9
Shellbark hickory	1.8
Red maple	3.8
Sugar maple	3.2
Mesquite	1.5
Overcup oak	2.8
Southern red oak	3.3
Tanoak	3.8
Softwoods	
Eastern white-cedar	1.1
Incense-cedar	0.7
Western red-cedar	0.5
Alaska-cedar	1.1
Douglas-fir	0.7
Balsam fir	1.5
Eastern hemlock	1.7
Western hemlock	1.2
Western larch	0.5
Jack pine	1.2
Loblolly pine	1.1
Longleaf pine	0.6
Western white pine	0.7
Redwood	0.8
White spruce	1.3
Tamarack	1.5

Source: Moore and Johnson, 1967.

Neutral solvents, water, toluene or ethanol, or combinations of solvents are employed to remove extractives in wood. However, other solvents ranging from diethyl ether to 1% NaOH, etc. could be applied according to the nature of extractives and sample type, i.e., bark, leaves, etc.

3.8.2.2 Sample Preparation

It is highly recommended to have a fresh sample. If this is not possible, keep the sample frozen or in a refrigerator to avoid fungal attack. Peel off the bark from the stem and separate the sample into component parts. Dry samples are oven dried for 24 hours (usually at 105°C) prior to milling. Wet samples can be milled while frozen in order to prevent oxidation or other undesirable chemical reactions. Samples are ground to pass 40 mesh (0.40 mm) using a Wiley Mill.

TABLE 3.11
Chemical Composition of North American Hardwoods and Softwoods

								Solubility		
Botanical Name	Common Name	Holo Cellulose	Alpha Cellulose	Pentosans	Klason Lignin	1% NaOH	Hot Water	EtOH/ Benzene	Ether	Ash
Hardwoods										
Acer macrophyllum	Bigleaf maple	—	46.0	22.0	25.0	18.0	2.0	3.0	0.7	0.5
Acer negundo	Boxelder	—	45.0	20.0	30.0	10.0	—	—	0.4	—
Acer rubrum	Red maple	77.0	47.0	18.0	21.0	16.0	3.0	2.0	0.7	0.4
Acer saccharinum	Silver maple	—	42.0	19.0	21.0	21.0	4.0	3.0	0.6	—
Acer saccharum	Sugar maple	—	45.0	17.0	22.0	15.0	3.0	3.0	0.5	0.2
Alnus rubra	Red alder	74.0	44.0	20.0	24.0	16.0	3.0	2.0	0.5	0.3
Arbutus menziesii	Pacific madrone	—	44.0	23.0	21.0	23.0	5.0	7.0	0.4	0.7
Betula alleghaniensis	Yellow birch	73.0	47.0	23.0	21.0	16.0	2.0	2.0	1.2	0.7
Betula nigra	River birch	—	41.0	23.0	21.0	21.0	4.0	2.0	0.5	—
Betula papyrifera	Paper birch	78.0	45.0	23.0	18.0	17.0	2.0	3.0	1.4	0.3
Carya cordiformus	Bitternut hickory	—	44.0	19.0	25.0	16.0	5.0	4.0	0.5	—
Carya glaubra	Pignut hickory	71.0	49.0	17.0	24.0	17.0	5.0	4.0	0.4	0.8
Carya ovata	Shagbark hickory	71.0	48.0	18.0	21.0	18.0	5.0	3.0	0.4	0.6
Carya pallida	Sand hickory	69.0	50.0	17.0	23.0	18.0	7.0	4.0	0.4	1.0
Carya tomentosa	Mockernut hickory	71.0	48.0	18.0	21.0	17.0	5.0	4.0	0.4	0.6
Celtis laevigata	Sugarberry	—	40.0	22.0	21.0	23.0	6.0	3.0	0.3	—
Eucalyptus gigantea	Alpine ash	72.0	49.0	14.0	22.0	16.0	7.0	4.0	0.3	0.2
Fagus grandifolia	American beech	77.0	49.0	20.0	22.0	14.0	2.0	2.0	0.8	0.4
Fraxinus americana	White ash	—	41.0	15.0	26.0	16.0	7.0	5.0	0.5	—
Fraxinus pennsylvanica	Green ash	—	40.0	18.0	26.0	19.0	7.0	5.0	0.4	—
Gleditsia triacanthos	Honey locust	—	52.0	22.0	21.0	19.0	—	—	0.4	—
Laguncularia racemosa	White mangrove	—	40.0	19.0	23.0	29.0	15.0	6.0	2.1	—
Liquidambar styraciflua	Sweetgum	—	46.0	20.0	21.0	15.0	3.0	2.0	0.7	0.3
Liriodendron tulipifera	Yellow poplar	—	45.0	19.0	20.0	17.0	2.0	1.0	0.2	1.0

Species	Common name								
Lithocarpus densiflorus	Tanoak	71.0	46.0	20.0	19.0	20.0	5.0	3.0	0.4	0.7
Milalenca quinquenervia	Cajeput	—	43.0	19.0	27.0	21.0	4.0	2.0	0.5	—
Nyssa aquatica	Water tupelo	—	45.0	16.0	24.0	16.0	4.0	3.0	0.6	0.6
Nyssa sylvatica	Black tupelo	72.0	45.0	17.0	27.0	15.0	3.0	2.0	0.4	0.5
Populus alba	White poplar	—	52.0	23.0	16.0	20.0	4.0	5.0	0.9	—
Populus deletoides	Eastern cottonwood	—	47.0	18.0	23.0	15.0	2.0	2.0	0.8	0.4
Populus tremoides	Quaking aspen	78.0	49.0	19.0	19.0	18.0	3.0	3.0	1.2	0.4
Populus trichocarpa	Black cottonwood	—	49.0	19.0	21.0	18.0	3.0	3.0	0.7	0.5
Prunus serotina	Black cherry	85.0	45.0	20.0	21.0	18.0	4.0	5.0	0.9	0.1
Quercus alba	White oak	67.0	47.0	20.0	27.0	19.0	6.0	3.0	0.5	0.4
Quercus coccinea	Scarlet oak	63.0	46.0	18.0	28.0	20.0	6.0	3.0	0.4	—
Quercus douglasii	Blue oak	59.0	40.0	22.0	27.0	23.0	11.0	5.0	1.4	1.4
Quercus falcata	Southern red oak	69.0	42.0	20.0	25.0	17.0	6.0	4.0	0.3	0.4
Quercus kelloggii	California black oak	60.0	37.0	23.0	26.0	26.0	10.0	5.0	1.5	0.4
Quercus lobata	Valley oak	70.0	43.0	19.0	19.0	23.0	5.0	7.0	1.0	0.9
Quercus lyrata	Overcup oak	—	40.0	18.0	28.0	24.0	9.0	5.0	1.2	0.3
Quercus marylandica	Blackjack oak	—	44.0	20.0	26.0	15.0	5.0	4.0	0.6	—
Quercus prinus	Chestnut oak	76.0	47.0	19.0	24.0	21.0	7.0	5.0	0.6	0.4
Quercus rubra	Northern red oak	69.0	46.0	22.0	24.0	22.0	6.0	5.0	1.2	0.4
Quercus stellata	Post oak	—	41.0	18.0	24.0	21.0	8.0	4.0	0.5	1.2
Quercus velutina	Black oak	71.0	48.0	20.0	24.0	18.0	6.0	5.0	0.2	0.2
Salix nigra	Black willow	—	46.0	19.0	21.0	19.0	4.0	2.0	0.6	—
Tilia heterophylla	Basswood	77.0	48.0	17.0	20.0	20.0	2.0	4.0	2.1	0.7
Ulmus americana	American elm	73.0	50.0	17.0	22.0	16.0	3.0	2.0	0.5	0.4
Ulmus crassifolia	Cedar elm	—	50.0	19.0	27.0	14.0	—	—	0.3	—

(Continued)

TABLE 3.11
Chemical Composition of North American Hardwoods and Softwoods (*Continued*)

Botanical Name	Common Name	Holo Cellulose	Alpha Cellulose	Pentosans	Klason Lignin	1% NaOH	Hot Water	Solubility EtOH/Benzene	Ether	Ash
Softwoods										
Abies amabilis	Pacific silver fir	—	44.0	10.0	29.0	11.0	3.0	3.0	0.7	0.4
Abies balsamea	Balsam fir	—	42.0	11.0	29.0	11.0	4.0	3.0	1.0	0.4
Abies concolor	White fir	66.0	49.0	6.0	28.0	13.0	5.0	2.0	0.3	0.4
Abies lasiocarpa	Subalpine fir	67.0	46.0	9.0	29.0	12.0	3.0	3.0	0.6	0.5
Abies procera	Noble fir	61.0	43.0	9.0	29.0	10.0	2.0	3.0	0.6	0.4
Chamaecyparis thyoides	Atlantic white cedar	—	41.0	9.0	33.0	16.0	3.0	6.0	2.4	—
Juniperus deppeana	Alligator juniper	57.0	40.0	5.0	34.0	16.0	3.0	7.0	2.4	0.3
Larix larcina	Tamarack	64.0	44.0	8.0	26.0	14.0	7.0	3.0	0.9	0.3
Larix occidentalis	Western larch	65.0	48.0	9.0	27.0	16.0	6.0	2.0	0.8	0.4
Libocedrus decurrens	Incense cedar	56.0	37.0	12.0	34.0	9.0	3.0	3.0	0.8	0.3
Picea engelmanni	Engelman spruce	69.0	45.0	10.0	28.0	11.0	2.0	2.0	1.1	0.2
Picea glauca	White spruce	—	43.0	13.0	29.0	12.0	3.0	2.0	1.1	0.3
Picea mariana	Black spruce	—	43.0	12.0	27.0	11.0	3.0	2.0	1.0	0.3
Picea sitchensis	Sitka spruce	—	45.0	7.0	27.0	12.0	4.0	4.0	0.7	—
Pinus attenuata	Knobcone pine	—	47.0	14.0	27.0	11.0	3.0	1.0	—	0.2
Pinus banksiana	Jack pine	66.0	43.0	13.0	27.0	13.0	3.0	5.0	3.0	0.3
Pinus clausa	Sand pine	—	44.0	11.0	27.0	12.0	2.0	3.0	1.0	0.4
Pinus contorta	Lodgepole pine	68.0	45.0	10.0	26.0	13.0	4.0	3.0	1.6	0.3
Pinus echinata	Shortleaf pine	69.0	45.0	12.0	28.0	12.0	2.0	4.0	2.9	0.4
Pinus elliottii	Slash pine	64.0	46.0	11.0	27.0	13.0	3.0	4.0	3.3	0.2
Pinus monticola	Western white pine	69.0	43.0	9.0	25.0	13.0	4.0	4.0	2.3	0.2

Cell Wall Chemistry

Pinus palustris	Longleaf pine	—	44.0	12.0	30.0	12.0	3.0	4.0	1.4	—
Pinus ponderosa	Ponderosa pine	68.0	41.0	9.0	26.0	16.0	4.0	5.0	5.5	0.5
Pinus resinosa	Red pine	71.0	47.0	10.0	26.0	13.0	4.0	4.0	2.5	—
Pinus sabiniana	Digger pine	—	46.0	11.0	27.0	12.0	3.0	1.0	—	0.2
Pinus strobes	Eastern white pine	68.0	45.0	8.0	27.0	15.0	4.0	6.0	3.2	0.2
Pinus sylvestris	Scotch or Scots pine	—	47.0	11.0	28.0	—	1.0	—	1.6	0.2
Pinus taeda	Loblolly pine	68.0	45.0	12.0	27.0	11.0	2.0	3.0	2.0	—
Pseudotsuga menziesii	Douglas fir	66.0	45.0	8.0	27.0	13.0	4.0	4.0	1.3	0.2
Sequoia sempervirens	Redwood old growth	55.0	43.0	7.0	33.0	19.0	9.0	10.0	0.8	0.1
	Redwood second growth	61.0	46.0	7.0	33.0	14.0	5.0	<1.0	0.1	0.1
Taxodium distichum	Bald cypress	—	41.0	12.0	33.0	13.0	4.0	5.0	1.5	—
Thuja occidentalis	Northern white cedar	59.0	44.0	14.0	30.0	13.0	5.0	6.0	1.4	0.5
Thuja plicata	Western red cedar	—	38.0	9.0	32.0	21.0	11.0	14.0	2.5	0.3
Tsuga Canadensis	Eastern hemlock	—	41.0	9.0	33.0	13.0	4.0	3.0	0.5	0.5
Tsuga heterophylla	Western hemlock	67.0	42.0	9.0	29.0	14.0	4.0	4.0	0.5	0.4
Tsuga mertensiana	Mountain hemlock	60.0	43.0	7.0	27.0	12.0	5.0	5.0	0.9	0.5

Source: Pettersen, 1984.

TABLE 3.12
Polysaccharide Content of Some North American Woods

Scientific Name	Common Name	Glu	Xyl	Gal	Arab	Mann	Uronic	Acetyl	Lignin	Ash
Hardwoods										
Acer rubrum	Red maple	46	19	0.6	0.5	2.4	3.5	3.8	24	0.2
Acer saccharum	Sugar maple	52	15	<0.1	0.8	2.3	4.4	2.9	23	0.3
Betula alleghaniensis	Yellow birch	47	20	0.9	0.6	3.6	4.2	3.3	21	0.3
Betlula papyrifera	White birch	43	26	0.6	0.5	1.8	4.6	4.4	19	0.2
Fagus grandifolia	Beech	46	19	1.2	0.5	2.1	4.8	3.9	22	0.4
Liquidambar styraciflua	Sweetgum	39	18	0.8	0.3	3.1	—	—	24	0.2
Platanus occidentalis	Sycamore	43	15	2.2	0.6	2.0	5.1	5.5	23	0.7
Populus deltoides	Eastern cottonwood	47	15	1.4	0.6	2.9	4.8	3.1	24	0.8
Populus tremuloides	Quaking aspen	49	17	2.0	0.5	2.1	4.3	3.7	21	0.4
Quercus falcata	Southern red oak	41	19	1.2	0.4	2.0	4.5	3.3	24	0.8
Ulmus americana	White elm	52	12	0.9	0.6	2.4	3.6	3.9	24	0.3
Softwoods										
Abies balsamea	Balsam fir	46	6.4	1.0	0.5	12	3.4	1.5	29	0.2
Ginkgo biloba	Ginkgo	40	4.9	3.5	1.6	10	4.6	1.3	33	1.1
Juniperus communis	Juniper	41	6.9	3.0	1.0	9.1	5.4	2.2	31	0.3
Larix decidua	Larch	46	6.3	2.0	2.5	11	4.8	1.4	26	0.2
Larix laricina	Tamarack	46	4.3	2.3	1.0	13	2.9	1.5	29	0.2
Picea abies	Norway spruce	43	7.4	2.3	1.4	9.5	5.3	1.2	29	0.5
Picea glauca	White spruce	45	9.1	1.2	1.5	11	3.6	1.3	27	0.3
Picea mariana	Black spruce	44	6.0	2.0	1.5	9.4	5.1	1.3	30	0.3
Picea rubens	Red spruce	44	6.2	2.2	1.4	12	4.7	1.4	28	0.3
Pinus banksiana	Jack pine	46	7.1	1.4	1.4	10	3.9	1.2	29	0.2
Pinus radiata	Radiata pine	42	6.5	2.8	2.7	12	2.5	1.9	27	0.2
Pinus resinosa	Red pine	42	9.3	1.8	2.4	7.4	6.0	1.2	29	0.4
Pinus rigida	Pitch pine	47	6.6	1.4	1.3	9.8	4.0	1.2	28	0.4
Pinus strobus	Eastern white pine	45	6.0	1.4	2.0	11	4.0	1.2	29	0.2
Pinus sylvestris	Scots pine	44	7.6	3.1	1.6	10	5.6	1.3	27	0.4
Pinus taeda	Loblolly pine	45	6.8	2.3	1.7	11	3.8	1.1	28	0.3
Pseudotsuga menziesti	Douglas fir	44	2.8	4.7	2.7	11	2.8	0.8	32	0.4

(Continued)

TABLE 3.12
Polysaccharide Content of Some North American Woods (*Continued*)

Thuja occidentalis	Northern white cedar	43	10.0	1.4	1.2	8.0	4.2	1.1	31	0.2
Tsuga canadensis	Eastern hemlock	44	5.3	1.2	0.6	11	3.3	1.7	33	0.2

Source: Pettersen, 1984.

3.8.2.3 Apparatus

Buchner funnel
Extraction thimbles
Extraction apparatus, extraction flask (500 ml), Soxhlet extraction tube
Heating device, heating mantle or equivalent
Boiling chips, glass beads, boilers, or any inert granules for taming boiling action
Chemical fume hood
Vacuum oven

3.8.2.4 Reagents and Materials

Ethanol (ethyl alcohol), 200 proof
Toluene, reagent grade
Toluene-ethanol mixture, 1:1 (v/v)

3.8.2.5 Procedures

Weigh 2 to 3 grams of each sample into covered preweighed extraction thimbles. Place the thimbles in a vacuum oven not exceeding 45°C for 24 hours, or to constant weight. Cool the thimbles in a desiccator for one hour and weigh. Then place the thimbles in Soxhlet extraction units. Place 200 ml of the toluene:ethanol mixture in a 500-ml round-bottom flask with several boiling chips to prevent bumping. Carry out the extraction in a well-ventilated chemical fume hood for 2 hours, keeping the liquid boiling so that siphoning from the extractor is no less than four times per hour. After extraction with the toluene:ethanol mixture, take the thimbles from the extractors, drain the excess solvent, and wash the samples with ethanol. Place them in the vacuum oven over night at temperatures not exceeding 45°C for 24 hours. When dry, remove them to a desiccator for an hour and weigh. Generally, the extraction is complete at this stage; however, the extractability depends upon the matrix of the sample and the nature of extractives. Second and the third extractions with different polarity of solvents may be necessary. Browning (1967) suggests 4 hours of successive extraction with 95% alcohol, however, two successive extractions, four hours with ethanol followed with distilled water for 1 hour can also be done. Pettersen (1984) extracted pine sample with acetone/water, followed by the toluene/ethanol mixture.

3.8.3 ASH CONTENT (ASTM D-1102-84)

3.8.3.1 Scope

The ash content of fiber is defined as the residue remaining after ignition at 575 ± 25°C for 3 hr, or longer if necessary to burn off all the carbon. It is a measure of mineral salts in the fiber, but it is not necessarily quantitatively equal to them.

3.8.3.2 Sample Preparation

Obtain a representative sample of the fiber, preferably ground to pass a 40-mesh screen. Weigh, to 5 mg or less, a specimen of about 5 g of moisture-free wood for ashing, preferably in duplicate. If the moisture in the sample is not known, determine it by drying a corresponding specimen to constant weight in a vacuum oven at $105 \pm 3°C$.

3.8.3.3 Apparatus

Crucible. A platinum crucible or dish with lid or cover is recommended. If platinum is not available, silica may be used.
Analytical balance having a sensitivity of 0.1 mg.
Electric muffle furnace adjusted to maintain a temperature of $575 \pm 25°C$.

3.8.3.4 Procedure

Carefully clean the empty crucible and cover, and ignite them to constant weight in a muffle furnace at $575 \pm 25°C$. After ignition, cool slightly and place in a desiccator. When cooled to room temperature, weigh the crucible and cover on the analytical balance.

Place all, or as much as practicable, of the weighed specimen in the crucible. Burn the sample directly over a low flame of a Bunsen burner (or preferably on the hearth of the furnace) until it is well carbonized, taking care not to blow portions of the ash from the crucible. If a sample tends to flare up or lose ash during charring, the crucible should be covered, or at least partially covered during this step. If the crucible is too small to hold the entire specimen, gently burn the portion added and add more as the flame subsides. Continue heating with the burner only as long as the residue burns with a flame. Place the crucible in the furnace at $575 \pm 25°C$ for a period of at least 3 hr, or longer if needed, to burn off all the carbon. When ignition is complete, as indicated by the absence of black particles, remove the crucible from the furnace, replace the cover and allow the crucible to cool somewhat. Then place in a desiccator and cool to room temperature. Reweigh the ash and calculate the percentage based on the moisture-free weight of the fiber.

3.8.3.5 Report

Report the ash as a percentage of the moisture-free wood to two significant figures, or to only one significant figure if the ash is less than 0.1%.

3.8.3.6 Precision

The results of duplicate determinations should be suspect if they differ by more than 0.5 mg. Since the ignition temperature affects the weight of the ash, only values obtained at $575 \pm 25°C$ should be reported as being in accordance with this method. Porcelain crucibles can also be used in most cases for the determination of ash. Special precautions are required in the use of platinum crucibles. There can be significant losses in sodium, calcium, irons and copper at temperatures over $600°C$.

3.8.4 PREPARATION OF HOLOCELLULOSE (CHLORITE HOLOCELLULOSE)

3.8.4.1 Scope

Holocellulose is defined as a water-insoluble carbohydrate fraction of wood materials. According to Browning (1967) there are three ways of preparing holocellulose and their modified methods (1) Chlorination method, (2) Modified chlorination method, (3) Chlorine dioxide and chlorite method. The standard purity of holocellulose is checked following lignin analysis.

3.8.4.2 Sample Preparation

The sample should be extractive and moisture free and prepared after Procedure 9.2. If Procedure 9.2 is skipped for some reason, the weight of the extractives should be accounted for in the calculation of holocellulose.

3.8.4.3 Apparatus

Buchner funnel
250 ml Erlenmeyer flasks
25 ml Erlenmeyer flasks
Water bath
Filter paper
Chemical fume hood

3.8.4.4 Reagents

Acetic acid, reagent grade
Sodium chlorite, $NaClO_2$, technical grade, 80%

3.8.4.5 Procedure

To 2.5 g of sample, add 80 ml of hot distilled water, 0.5 ml acetic acid, and 1 g of sodium chlorite in a 250-ml Erlenmeyer flask. An optional 25-ml Erlenmeyer flask is inverted in the neck of the reaction flask. The mixture is heated in a water bath at 70°C. After 60 minutes, 0.5 ml of acetic acid and 1 g of sodium chlorite are added. After each succeeding hour, fresh portions of 0.5 ml acetic acid and 1 g sodium chlorite are added with shaking. The delignification process degrades some of the polysaccharides and the application of excess chloriting should be avoided. Continued reaction will remove more lignin but hemicellulose will also be lost (Rowell 1980).

Addition of 0.5 ml acetic acid, and 1 g of sodium chlorite is repeated until the wood sample is completely separated from lignin. It usually takes 6 to 8 hours of chloriting and the sample can be left without further addition of acetic acid and sodium chlorite in the water bath for over night. At the end of 24 hours of reaction, cool the sample and filter the holocellulose on filter paper using a Buchner funnel until the yellow color (the color of holocellulose is white) and the odor of chlorine dioxide is removed. If the weight of the holocellulose is desired, filter the holocellulose on a tarred fritted dics glass thimble, wash with acetone, vacuum oven dry at 105°C for 24 hours, place in a desiccator for an hour and weigh. The holocellulose should not contain any lignin and the lignin content of holocellulose should be determined and subtracted from the weight of the prepared holocellulose.

3.8.5 PREPARATION OF ALPHA-CELLULOSE (DETERMINATION OF HEMICELLULOSES)

3.8.5.1 Scope

The preparation of α-cellulose is a continuous procedure from procedure 9.4. The term hemicellulose is defined as the cell wall components that are readily hydrolyzed by hot dilute mineral acids, hot dilute alkalies, or cold 5% sodium hydroxide.

3.8.5.2 Principle of Method

Extractive-free, lignin-free holocellulose is treated with sodium hydroxide and then with acetic acid, with the residue defined as α-cellulose. The soluble fraction represents the hemicellulose content.

3.8.5.3 Apparatus

A thermostat or other constant-temperature device will be required that will maintain a temperature of 20 ± 0.1°C in a container large enough to hold a row of at least three 250-ml beakers kept in an upright position at all times.

Filtering crucibles of Alundum or fritted glass thimbles of medium porosity.

3.8.5.4 Reagents

Sodium hydroxide (NaOH) solution, 17.5% and 8.3%
Acetic acid, 10% solution.

3.8.5.5 Procedure

Weigh out about 2 g of vacuum-oven dried holocellulose and place into a 250-ml glass beaker provided with a glass cover. Add 10 ml of 17.5% NaOH solution to the holocellulose in a 250-ml beaker, cover with a watch glass, and maintain at 20°C in a water bath. Manipulate the holocellulose lightly with a glass rod with a flat end so that all of the specimen becomes soaked with the NaOH solution. After the addition of the first portion of 17.5% NaOH solution to the specimen, at five minute intervals, add 5 ml more of the NaOH solution and thoroughly stir the mixture with the glass rod. Continue this procedure until the NaOH is consumed. Allow the mixture to stand at 20°C for 30 min. making the total time for NaOH treatment 45 min.

Add 33 ml of distilled water at 20°C to the mixture. Thoroughly mix the contents of the beaker and allow to stand at 20°C for 1 hour before filtering.

Filter the cellulose with the aid of suction into the tarred, alkali-resistant Alundum or fritted-glass crucible of medium porosity. Transfer all of the holocellulose residue to the crucible, and wash with 100 ml of 8.3% NaOH solution at 20°C. After the NaOH wash solution has passed through the residue in the crucible, continue the washing at 20°C with distilled water, making certain that all particles have been transferred from the 250-ml beaker to the crucible. Washing the sample in the crucible is facilitated by releasing the suction, filling the crucible to within 6 mm of the top with water, carefully breaking up the cellulose mat with a glass rod so as to separate any lumps present, and again applying suction. Repeat this step twice.

Pour 15 ml of 10% acetic acid at room temperature into the crucible, drawing the acid into the cellulose by suction but, while the cellulose is still covered with acid, release the suction. Subject the cellulose to the acid treatment for 3 min. from the time the suction is released; then apply suction to draw off the acetic acid. Without releasing the suction, fill the crucible almost to the top with distilled water at 20°C and allow to drain completely. Repeat the washing until the cellulose residue is free of acid as indicated by litmus paper. Give the cellulose a final washing by drawing, by suction, an additional 250 ml of distilled water through the cellulose in the crucible. Dry the crucible on the bottom and sides with a cloth and place it overnight in an oven to dry at 105°C. Cool the crucible and weighing bottle in a desiccator for 1 hr before weighing.

3.8.5.6 Calculation and Report

Calculate the percentage of α-cellulose on the basis of the oven-dried holocellulose sample, as follows:

$$\alpha\text{-cellulose, percent} = (W_2/W_1) \times 100$$

where
  W_2 = weight of the oven-dried α-cellulose residue
  W_1 = weight of the original oven-dried holocellulose sample

Cell Wall Chemistry

3.8.6 Preparation of Klason Lignin

3.8.6.1 Scope

Klason lignin gives a quantitative measure of the acid insoluble lignin and is not suitable for the study of lignin structures and some other lignins such as cellulolytic enzyme lignin, or Björkman (milled wood lignin) should be prepared (Sjöström 1981) for the study of lignin structure. This procedure is a modified version of ASTM D-1166-84. The lignin isolated using this procedure is also called sulfuric acid lignin.

3.8.6.2 Apparatus

Autoclave
Buchner funnel
100 ml centrifuge tube, Pyrex 8240
Desiccator
Glass rods
Water bath
Glass fiber
Filter paper, Whatman Cat No. 1827-021, 934-AH
Glass microfilter, 2.1 cm

3.8.6.3 Reagent

Sulfuric acid (H_2SO_4), 72% and 4% by volume
Fucose, 24.125% in 4% H_2SO_4 (w/w)

3.8.6.4 Procedure

Prepare samples by Procedure 9.2 and dry the sample at 45°C in a vacuum oven overnight. Accurately weigh out approximately 200 mg of ground vacuum-dried sample into a 100 ml centrifuge tube. To the sample in the 100 ml centrifuge tube, add 1 ml of 72% (w/w) H_2SO_4 for each 100 mg of sample. Stir and disperse the mixture thoroughly with a glass rod twice, then incubate the tubes in a water bath at 30°C for 60 min. Add 56 ml of distilled water. This results in a 4% solution for the secondary hydrolysis. Add 1 ml fucose internal standard (this procedure is required only if five sugars are to be analyzed by HPLC as a part of the analysis). Autoclave at 121°C and 15 psi, for 60 min. Remove the samples from the autoclave and filter off the lignin, with glass fiber filters (filters were rinsed into crucibles, dried and tarred) in crucibles using suction, keeping the solution hot. Wash the residue thoroughly with hot water and dry at 105°C overnight. Move to a desiccator, and let it sit an hour and weigh. Calculate Klason lignin content from weights.

3.8.6.5 Additional Information

Condensation reactions involving protein can cause artificially high Klason lignin measurements when tissues containing significant protein contents are analyzed. A nitrogen determination can be done to indicate possible protein content.

3.8.7 Determination of Methoxyl Groups

3.8.7.1 Scope

Methoxyl groups (–OCH3) are present in the lignin and lignin derivatives as side chains of aromatic phenylpropanes and in the polysaccharides mainly as methoxy uronic acids. Methoxyl content is determined using ASTM, D-1166-84.

3.8.7.2 Principle of Method

In the original method, methyl iodide was absorbed in an alcoholic solution of silver nitrate. The solution was diluted with water, acidified with nitric acid, and boiled. The silver iodide was removed by filtration, washed, and weighed in the manner usual for halide determinations. A volumetric modification is based on absorption of the methyl iodide in a known volume of standard silver nitrate solution and titration of the unused silver nitrate with standard potassium thiocyanate solution (ferric alum indicator solution). In this procedure, the methyl iodide is collected in an acetic acid solution of potassium acetate containing bromine.

$$CH_3I + Br_2 \rightarrow CH_3Br + IBr$$

$$IBr + 2Br_2 + 3H_2O \rightarrow HIO_3 + 5HBr$$

The excess bromine is destroyed by addition of acid, and the iodate equivalent of the original methoxyl content is determined by titration with sodium thiosulfate of the iodine liberated in the reaction:

$$HIO_3 + 5HI \rightarrow 3I_2 + 3H_2O$$

One methoxyl group is equivalent to six atoms of iodine and, consequently, a favorable analytical factor is obtained.

3.8.7.3 Sample Preparation

The sample is dried, ground, and extracted accordingly prior to analysis.

3.8.7.4 Apparatus

Reaction flask
Heat source
Vertical air-cooled condenser
Scrubber
Absorption vessels

3.8.7.5 Reagents

Bromine, liquid
Cadmium sulfate solution: Dissolve 67.2 g of $CdSO_4 \cdot 4H_2O$ in 1 liter of water.
Carbon dioxide gas
Formic acid, 90%
Hydroiodic acid
Phenol
Potassium acetate solution in acetic acid. Anhydrous potassium acetate (100 g) is dissolved in 1 l of glacial acetic acid.
Potassium iodide solution-Dissolve 100 g of KI in water and dilute to 1 l.
Sodium acetate solution-Dissolve 415 g of sodium acetate trihydrate in water and dilute to 1 l.
Sodium thiosulfate solution (0.1 N)-Dissolve 25 g of $Na_2S_2O_3 \cdot H_2O$ in 200 ml of water and dilute to 1 l.
Starch indicator solution (10 g/l)
Sulfuric acid: Mix one volume of H_2SO_4 (sp gr 1.84) with nine volumes of water.

3.8.7.6 Procedure

Weigh the sample, about 100 mg of wood or 50 mg of lignin and place in the reaction flask. Place in the reaction flask 15 ml of HI, 7 g of phenol, and a boiling tube. Place in the scrubber a mixture of equal volumes of $CdSO_4$ solution and $Na_2S_2O_3$. The volume of solution should be adjusted so that the inlet tube of the scrubber is covered to a depth of about 4 mm. Adjust the flow of CO_2 to about 60 bubbles per minute through the scrubber. Heat the flask and adjust the rate of heating so that the vapors of the boiling HI rise about 100 mm into the condenser. Heat the flask under these conditions for 30 to 45 minutes, or longer if necessary, to remove methoxyl-containing or other interfering substances which are usually present in the reagents.

Let the distilling flask cool below 100°C. In the meantime, add to 20 ml of the potassium acetate solution, about 0.6 ml of bromine, and mix. Add approximately 15 ml of the mixture to the first receiver and 5 ml to the second, and attach the receiver to the apparatus. Seal the ground-glass joint with a small drop of water from a glass rod.

Remove the distilling flask and introduce the test specimen. Immediately reconnect the flask and seal the ground-glass joint with a drop of molten phenol from a glass rod. Bring the contents of the flask to reaction temperature while passing a uniform stream of CO_2 through the apparatus.

Adjust the rate of heating so that the vapors of the boiling HI rise about 100 ml into the condenser. Continue the heating for a time sufficient to complete the reaction and sweep out the apparatus. Usually, not more than 50 minutes are required.

Wash the contents of both receivers into a 250-ml Erlenmeyer flask that contains 15 ml of sodium acetate solution. Dilute with water to approximately 125 ml and add 6 drops of formic acid. Rotate the flask until the color of the bromine is discharged, then add 12 more drops of formic acid and allow the solution to stand for 1 to 2 min. Add 10 ml of KI solution and 10 ml of H_2SO_4, and titrate the liberated iodine with $Na_2S_2O_3$ solution, adding 1 ml of starch indicator solution just before the end point is reached, continuing the titration to the disappearance of the blue color.

3.8.7.7 Calculation and Report

$$\text{Methoxyl, \%} = (VN \times 31.030 \times 100)/(G \times 1000 \times 6) = (VN/G) \times 0.517$$

where:
V = milliliters of $Na_2S_2O_3$ solution required for the titration,
N = normality of $Na_2S_2O_3$ solution, and
G = grams of moisture free sample.

Table 3.9 shows the methoxyl content of some common hardwoods and softwoods.

3.8.8 DETERMINATION OF ACETYL BY GAS LIQUID CHROMATOGRAPHY

3.8.8.1 Scope

The acetyl and formyl groups that are in the polysaccharide portion can be determined in one of three ways: (1) acid hydrolysis; sample is hydrolyzed to form acetic acid, (2) saponification; acetyl groups are split from polysaccharides with hot alkaline solution and acidified to form acetic acid, or (3) trans-esterificaion; sample is treated with methanol in acid or alkaline solution to form methyl acetate. Acetic acid and methyl acetate are analyzed by gas chromatography.

The procedure presented here is saponification and acetyl determined by gas chromatography.

$$CH_3COOR + NaOH \rightarrow CH_3COONa + ROH$$

$$CH_3COONa + H+ \rightarrow CH_3COOH$$

3.8.8.2 Reagents

Formic acid, 2%: Dilute 2 ml of 90% formic acid to 900 ml with deionized H_2O.
Internal standard stock solution: Weigh 25.18 grams of 99+% proprionic acid in 500 ml volumetric flask, make to volume with 2% formic acid.
Internal standard solution: Pipette 10 ml stock solution into a 200-ml volumetric flask, make to volume with deionized water.
Acetic acid standard solution: Weigh 100 mg 99.7% glacial acetic acid into a 100-ml volumetric flask, make to volume with deionized water.
NaOH solution 1 N: Weigh 4 grams sodium hydroxide, dissolve in 100 ml deionized water.

3.8.8.3 Sample Preparation

The amount of sample is based on the approximate acetyl content: Acetyl content (AC) 0–10%, 50 mg; AC 10%, 25 mg; AC 15%, 20 mg; AC 20%, 15 mg; AC 25%, 10 mg. Weigh an oven-dried sample in a long-handled weighing tube and transfer it to an acetyl digestion flask and add boiling chips. Pipette 2 ml 1 N NaOH solution to wash down the neck of the flask. Connect the reaction flask to a water cooled reflux condenser and reflux for 1 hour. Cool the reaction flask to room temperature and pipette 1 ml of propionic acid (internal standard) into a 10 ml volumetric flask. Quantitatively transfer the liquid from the reaction flask to the volumetric flask. Wash the reaction flask and the solid residue with several portions of distilled water. Add 0.2 ml of 85% phosphoric acid and make to volume with distilled water. This solution may be filtered through a small plug of glass wool to remove solid particles. Analyze the sample by GLC and determine the average ratio. Milligrams of acetic acid are determined from the calibration curve.

3.8.8.4 Gas Chromatography

Column: Supelco 60/80 Carbopack C/0.3% carbowax 20 M/0.1% H_3PO_4 - 3 ft 1/4 inch O. D. and 4 mm I. D.; Oven temperature 120°C; Injection port 150°C; F.I.D. 175°C; Nitrogen 20 ml/min.

The ratio of the area is determined by dividing the area of the acetic acid by the area of the propionic acid (internal standard). The average of the ratios is used to determine mg/ml of acetic acid from the calibration curve.

Preparation of a calibration curve: Pipette 1, 2, 4, 6, and 8 ml of standard acetic acid solution into 10 ml volumetric flasks. Pipette 1 ml of propionic acid internal standard into each sample, then add 0.2 ml 85% phosphoric acid. Make to volume with distilled water. Analyze each solution three times by GLC. Calculate the ratios by dividing the area of the acetic acid by the area of the propionic acid (internal standard). Plot the average ratios against milligrams per milliliter of acetic acid. Standard and sample solutions can be stored in the refrigerator for at least 1 week.

3.8.8.5 Reporting

Report the average, standard deviation and precision of each sample. The results may be reported as percent acetic acid or as percent acetyl:

$$\% \text{ acetic acid} = \frac{\text{mg/ml acetic acid found} \times 10 \text{ ml } \% \text{ } 100}{\text{sample weight in mg}}$$

$$\% \text{ acetyl} = \% \text{ acetic acid} \times 0.7172$$

TABLE 3.13
Chemical Composition of Selected Hardwoods from the Southeastern United States (Percent of Oven-Dry Wood)

Scientific Name	Common Name	Carbohydrates		Components of Hemicelluloses				Lignin	Total Ext	Ash
		Cell	Total Hemi	Gluco Mann	AcGlu UrXyl	Arab Gal	Pectin			
Acer rubrum	Red maple	40.7	30.4	3.5	23.5	1.6	1.9	23.3	5.3	0.3
Aesculus octandra	Yellow buckeye	40.6	25.8	3.6	18.6	1.0	2.6	30.0	3.1	0.5
Carya glabra	Pignut hickory	46.2	26.7	1.1	22.1	1.2	2.3	23.2	3.4	0.6
Carya illinoensis	Pecan	38.7	30.2	1.6	24.7	1.6	2.3	23.2	3.4	0.6
Carya tomentosa	Mockernut	43.5	27.7	1.5	21.5	1.3	3.5	23.6	5.0	0.4
Cornus florida	Flowering dogwood	36.8	35.4	3.4	27.2	1.0	5.0	21.8	4.6	0.3
Fagus grandifolia	American beech	36.0	29.4	2.7	23.5	1.3	1.8	30.9	3.4	0.4
Fraxinus americana	White ash	39.5	29.1	3.8	22.1	1.4	1.9	24.8	6.3	0.3
Gordonia lasianthus	Loblolly-bay	43.8	29.1	4.1	22.1	1.1	1.8	21.5	5.2	—
Liquidambar styraciflua	Sweetgum	40.8	30.7	3.2	21.4	1.3	4.9	22.4	5.9	0.2
Liriodendron tulipifera	Yellow poplar	39.1	28.0	4.9	20.1	0.7	2.4	30.3	2.4	0.3
Magnolia virginiana	Sweetbay	44.2	37.7	4.3	20.2	1.6	1.6	24.1	3.9	0.2
Nyssa aquatica	Water tupelo	45.9	24.0	3.5	18.6	0.8	1.1	25.1	4.7	0.4
Nyssa sylvatica	Black tupelo	42.6	27.3	3.6	18.0	1.0	4.8	26.6	2.9	0.6
Oxydendron arboreum	Sourwood	40.7	34.6	1.3	31.9	1.0	0.4	20.8	3.6	0.3
Persea borbonia	Redbay	45.6	25.6	1.0	23.2	0.9	0.5	23.6	5.0	0.2
Platanus occidentalis	Sycamore	43.0	27.2	2.3	22.3	1.4	1.2	25.3	4.4	0.1
Populus deltoids	Eastern cottonwood	46.5	26.6	4.4	16.8	1.6	1.8	25.9	2.4	0.6
Quercus alba	White oak	41.7	28.4	3.1	21.0	1.6	2.7	24.6	5.3	0.2
Quercus coccinea	Scarlet oak	43.2	29.2	2.3	23.3	1.4	2.2	20.9	6.6	0.1
Quercus falcata	Southern red oak	40.5	24.2	1.7	18.6	1.7	2.2	23.6	9.6	0.5

TABLE 3.13
Chemical Composition of Selected Hardwoods from the Southeastern United States (Percent of Oven-Dry Wood) *(Continued)*

		Carbohydrates		Components of Hemicelluloses							
Scientific Name	Common Name	Cell	Total Hemi	Gluco Mann	AcGlu UrXyl	Arab Gal	Pectin	Lignin	Total Ext	Ash
Quercus ilicifolia	Scrub oak	37.6	27.5	1.0	22.3	1.8	2.4	26.4	8.0	0.5
Quercus marylandica	Blackjack oak	33.8	28.2	2.0	21.0	2.3	2.9	30.1	6.6	1.3
Quercus nigra	Water oak	41.6	34.8	3.0	28.9	2.2	0.7	19.1	4.3	0.3
Quercus prinus	Chestnut oak	40.8	29.9	2.9	23.8	1.8	1.4	22.3	6.6	0.4
Quercus rubra	Northern red oak	42.2	33.1	3.3	26.6	1.6	1.6	20.2	4.4	0.2
Quercus stellata	Post oak	37.7	29.9	2.6	23.0	2.0	2.3	26.1	5.8	0.5
Quercus velutina	Black oak	39.6	28.4	1.9	23.2	1.1	1.9	25.3	6.3	0.5
Quercus virginiana	Live oak	38.1	22.9	1.0	18.3	1.7	1.9	25.3	13.2	0.6
Sassafras albidum	Sassafras	45.0	35.1	4.0	30.4	0.9	<0.1	17.4	2.4	0.2
Ulmus americana	American elm	42.6	26.9	4.6	19.9	0.8	1.6	27.8	1.9	0.8

Cell = Cellulose.

Source: Pettersen, 1984.

(Continued)

TABLE 3.14
Elemental Composition of Some Woods

Scientific Name	Common Name	Ca ppt	K ppt	Mg ppt	P ppt	Mn ppt	Fe ppm	Cu ppm	Zn ppm	Na ppm	Cl ppm
Abies balsamea	Balsam fir	0.8	0.8	0.27	—	0.13	13	17	11	18	—
Acer rubrum	Red maple	0.8	0.7	0.12	0.03	0.07	11	5	29	5	18
Betula papyrifera	White birch	0.7	0.3	0.18	0.15	0.03	10	4	28	9	10
Fraxinus americana	White ash	0.3	2.6	1.8	0.01	—	—	—	—	31	—
Liquidambar styraciflua	Sweetgum	0.55	0.3	0.34	0.15	0.08	—	—	19	81	—
Picea rubens	Red spruce	0.8	0.2	0.07	0.05	0.14	14	4	8	8	0.3
Pinus strobes	Eastern white pine	0.2	0.3	0.07	—	0.03	10	5	11	9	19
Populus deltoids	Eastern cottonwood	0.9	2.3	0.29	—	0.2	100	—	30	940	—
Populus tremuloides	Quaking aspen	1.1	1.2	0.27	0.10	0.03	12	7	17	5	—
Quercus alba	White oak	0.5	1.2	0.31	—	<0.01	—	—	—	21	15
Quercus falcate	Southern red oak	0.3	0.6	0.03	0.02	0.01	30	73	38	44	—
Tilia americana	Basswood	0.1	2.8	0.35	—	—	—	—	—	63	38
Tsuga canadensis	Eastern hemlock	1.0	0.4	0.11	0.12	0.15	6	5	2	6	—

Source: Pettersen, 1984.

REFERENCES

Alder, E. (1977). Lignin chemistry: Past, present and future. *Wood Sci. Technology,* 11:169–218.
Andersson, A., Erickson, M., Fridh, H., and Miksche, G.E. (1973). Gas chromatographic analysis of lignin oxidation products. XI. Structure of the bark lignins of hardwoods and softwoods. *Holzforschung* 27:189–193.
ASTM D1166-84. (1984). Standard test method for methoxyl groups in wood and related materials. ASTM International. Available at www.astm.org.
ASTM D1106-84. (1984). Standard test method for lignin in wood. ASTM International. Available at www.astm.org.
ASTM D1102-84. (1984). Standard test method for ash content in wood and wood-based materials. ASTM International. Available at www.astm.org.
Björkman, A. (1956). Studies on finely divided wood. Part 1. Extraction of lignin with neutral solvents. *Svensk Papperstid.* 59:477–485.
Björkman, A. (1957). Studies on finely divided wood. Part 2. The properties of lignins extracted with neutral solvents from softwoods and hardwoods. *Svensk Papperstid.* 60:158–169.
Browning, G.L. (1955). *Methods in Wood Chemistry,* Vol. 2. Wiley Interscience New York.
Chang, Y.P. and Mitchell, R.L. (1955). Chemical composition of common North American pulpwood parks. *Tappi* 38(5):315.
Choong, E.T., Abdullah, G., and Kowalczuk, J. (1976). Wood Utilization Notes, No. 29. Louisiana State University.
Clermont, L.P. (1970). Study of lignin from stone cells of aspen poplar inner bark. *Tappi* 53(1):52–57.
Dietrichs, H.H. (1975). Polysaccharides in bark. *Holz als Roh- und Werkstoff* 33:13–20.
Dietrichs, H.H., Graves, K., Behrenwsdorf, D., and Sinner, M. (1978). Studies on the carbohydrates in tree barks. *Holzforschung* 32(2):60–67.
Ellis, E.L. (1965). Inorganic elements in wood. *Cellular Ultrastructure of Woody Plants.* W.A. Côté Jr. (Ed.). Syracuse University Press, New York, pp. 181–189.
Fengel, D. and Wegener, G. (1984). *Wood: Chemistry, Ultrastructure and Reactions.* W. de. Gruyter, Berlin.
Gardner, K.H. and Blackwell, J. (1974). The hydrogen bonding in naïve cellulose. *Biochim. Biophys. Acta* 343:232–237.
Goring, D.A.I. (1962). The physical chemistry of lignin. *Pure Appl. Chem.* 5:233–254.
Goring, D.A.I. and Timell, T.E. (1962). Molecular weight of native celluloses. *Tappi* 45(6):454–460.
Gottlieb, O.R. and Yoshida, M. (1989). Lignins. In: Rowe, J.W. (Ed.), *Natural Products of Woody Plants.* Vol. I. Springer-Verlag, New York, pp. 439–511.
Harder, M.L. and Einspahr, D.W. (1980). Levels of some essential metals in bark. *Tappi* 63(2):110.
Harun, J. and Labosky, P. (1985). Chemical constituents of five northeastern barks. *Wood and Fiber* 17(2):274.
Haslam, E. (1989). Gallic acid derivatives and hydrolysable tannins. In: Rowe, J.W. (Ed.), *Natural Products of Woody Plants.* Vol. I. Springer-Verlag, New York, pp. 399–438.
Hattula, T. and Johanson, M. (1978). Determination of some of the trace elements in bark by neutron activation analysis and high resolution spectroscopy. *Radiochem. Radioanal. Letters* 32(1–2):35.
Hemingway, R.W. (1981). Bark: Its chemistry and prospects for chemical utilization. In: Goldstein, I.S. (Ed.), *Organic Chemicals and Biomass.* CRC Press, Boca Raton, FL, pp. 189–248.
Hemingway, R.W. (1989). Biflavonoids and proanthocyanidins. In: Rowe, J.W. (Ed.), *Natural Products of Woody Plants.* Vol. I. Springer-Verlag, New York, pp. 571–650.
Hemingway, R.W., Karchesy, J.J., McGraw, G.W., and Wielesek, R.A. (1983). Heterogeneity of interflavaniod bond located in loblolly pine bark procyanidins. *Phytochemistry* 22:275.
Hergert, H.T. (1960). Chemical composition of tannins and polyphenolics from conifer wood and bark. *Forest Products J.* 10:610–617.
Hergert, H.T. (1962). Economic importance of flavonoid compounds: Wood and bark. In: Geissmann, T.A. (Ed.), *The Chemistry of Flavonoid Compounds.* MacMillan, New York, Chapter 17, pp. 553–592.
Higuchi, T., Ito, Y., Shimada, M., and Kawamura, I. (1967). Chemical properties of bark lignins. *Cellulose Chem. Technol.* 1:585–595.
Hillis, W.E. (1973). Historical uses of extractives and exudates. In: Rowe, J.W. (Ed.), *Natural Products of Woody Plants.* Vol. I. Springer-Verlag, New York, pp. 1–12.
Holloway, P.J. and Deas, A.H.B. (1973). Epoxyoctadeconic acids in plant cutins and suberins. *Phytochemistry* 12:1721.

Hon, D.N.-S. and Shiraishi, N. (1991). *Wood and Cellulosic Chemistry.* Marcel Dekker, New York.
Jiang, K.S. and Timell, T.E. (1972). Polysaccharides in the bark of aspen (*Populus tremuloides*). III. The constitution of a galactoglucomann. *Cellulose Chem. Technol.* 6:503–505.
Jones, R.W., Krull, J.H., Blessin, C.W., and Inglett, G.E. (1979). Neutral sugars of hemicellulose fractions of pith from stalks of selected woods. *Cereal Chem.* 56(5):441.
Kai, Y. (1991). Chemistry of extractives. In: Hon, D.N.-S. and Shiraishi, N. (Eds.), *Wood and Cellulosic Chemistry.* Marcel Dekker, New York, pp. 215–255.
Kolattukudy, P.E. (1984). Biochemistry and function of cutin and suberin. *Can. J. Bot.* 62(12):2918.
Kollmann, F.P. and Côté, W.A. Jr. (1968). *Principles of Wood Science and Technology.* Springer-Verlag, New York.
Kurth, E.F. and Smith, J.E. (1954). The chemical nature of the lignin of Douglas-fir bark. *Pulp Paper. Mag. Canada.* 55:125.
Labosky, P. (1979). Chemical constituents of four southern pine parks. *Wood Science* 12(2):80–85.
Laks, P.E. (1991). Chemistry of bark. In: Hon, D.N.-S. and Shiraishi, N. (Eds.) *Wood and Cellulosic Chemistry.* Marcel Dekker, New York, pp. 257–330.
Martin, R.E. and Gray, G.R. (1971). pH of southern pine barks. *Forest Products J.* 21(3):49–52.
McGinnis, G.D. and Parikh, S. The chemical constituents of loblolly pine. *Wood Science* 7(4):295–297.
Mian, A.J. and Timell, T.E. (1960). Isolation and characterization of a cellulose from the inner bark of white birch (*Betula papyrifera*). *Can. J. Chem.* 38:1191–1198.
Moore, W. and Johnson, D. (1967). Procedures for the chemical analysis of wood and wood products, USDA, Forest Service, Forest Products Laboratory.
Nonaka, G., Nishimura, H., and Nishioka, I. (1985). Tannins and related compounds: Part 26. Isolation and structures of stenophyllanins A, B, and C, novel tannins from *Quercus stenophylla, J. Chem. Soc. Perkin Trans.* I. 163.
Nunes, E., Quilhó, T., and Pereira, H. (1996). Anatomy and chemical composition of *Pinus Pinaster* bark. *IAWA J.* 17(2):141–149.
Obst, J.R. (1982). Guaiacyl and syringyl lignin composition in hardwood cell components. *Holzforschung* 36(3):143–153.
Painter, T.J. and Purves, C.B. (1960). Polysaccharides in the inner bark of white spruce, *Tappi* 43:729–736.
Panshin, A.J. and de Zeeuw, D. (1980). *Textbook of wood technology,* McGraw-Hill, New York.
Pereira, H. (1988). Chemical composition and variability of cork from *Quercus suber L. Wood Sci. Technol.* 22: 211–218.
Pettersen, R.C. (1984). The chemical composition of wood. In: *The Chemistry of Solid Wood,* Rowell, R. M., Editor, Advances in Chemistry Series 20, American Chemical Society.
Porter, L.J. (1989). Condensed tannins. In: Rowe, J.W. (Ed.), Natural products of woody plants. I Chapter 7.7, 651–688, Springer-Verlag, New York.
Rowe, J.W. (Ed.) (1989). *Natural products of woody plants. I and II.* Springer-Verlag, New York.
Rowell, R.M. (1980). Distribution of reacted chemicals in southern pine modified with methyl isocyanate. *Wood Sci.* 13(2):102–110.
Saka, S. (1991). Chemical composition and distribution. In: Hon, D.N.-S. and Shiraishi, N. *Wood and Cellulosic Chemistry,* Marcel Dekker, Inc, New York, Chapter 2, 59–88.
Saka, S. and Goring. D.A.I. (1983). The distribution of inorganic constituents in black spruce wood as determined by TEM-EDXA. *Mokuzai Gakkaishi* 29:648.
Sakakibara, A. (1991). Chemistry of lignin. In: Hon, D.N.-S. and Shiraishi, N. Wood and Cellulosic Chemistry, Marcel Dekker, Inc, New York, Chapter 4, 113–175.
Sarkanen, K.V. and Hergert, H.L. (1971). Classification and distribution, In: Sarkanen, K.V and Ludwig, C.H. (Eds.), *Lignins, Occurrence, formation, structure and reactions,* Wiley-Interscience, New York.
Sarkanen, K.V and Ludwig, C.H. (Eds.) (1971). Lignins: Occurrence, formation, structure and reactions, Wiley-Interscience, New York.
Schwerin, G. (1958). The chemistry of reaction wood. II. The polysaccharides of *Eucalyptus goniocalyx* and *Pinus radiate. Holzforschung* 12:43–48.
Sandved, K.B., Prance, G.T., and Prance, A.E. (1992). Bark: the formation, characteristics, and uses of bark around the world. Timber Press, Portland, OR.
Seshadri, T.R. and Vedantham, T.N.C. (1971). Chemical examination of the barks and hearwood of *Betula* species of American origin. *Phytochem.* 10:897.
Sjöström, E. (1981). Wood polysaccharides. *Wood Chemistry, Fundamentals and Applications.* Academic Press, New York, pp. 51–67.

Stamm, A.J. (1964). *Wood and Cellulose Science*. The Ronald Press Co., New York.
Timell, T.E. (1961a). Characterization of celluloses from the bark of gymnosperms. *Svensk Papperestid.* 64:685–688.
Timell, T.E. (1961b). Isolation of polysaccharides from the bark of gymnosperms. *Svensk Papperstid.* 64: 651–661.
Timell, T.E. (1964). Wood hemicelluloses. Part 1. Advances in Carbohydrate Chemistry. W.W. Pigman and M.L. Wolfrom (Eds.), Academic Press, New York, NY. 19:247–302.
Timell, T.E. (1965). Wood hemicelluloses. Part 2. Advances in Carbohydrate Chemistry. W.W. Pigman and M.L. Wolfrom (Eds.), Academic Press, New York, NY. 20:409–483.
Timell, T.E. (1982). Recent progress in the chemistry and topochemistry of compression wood. *Wood Sci. Technology* 16:83–122.
Toman, R., Karacsonyi, S., and Kubackova, M. (1976). Studies on pectin present in the bark of white willow (*Salic alba*), structure of the acidic and neutral oligosaccharides obtained by partial acid hydrolysis. *Cellulose Chem. Technol.* 10:561.
Volz, K.R. (1971). Influence of inorganic content on the pH of bark. *Holz-Zentralbl.* 97:1783.
Whistler, R.L. and Richards, E.L. (1970). Hemicelluloses. In: The Carbohydrates (W. Pigman and D. Horton, eds.) 2nd ed. Vol 2A, 447–469.
Whistler, R.L., Wolfrom, M.L., and BeMiller, J.N. (Eds.). (1962). *Methods in Carbohydrate Chemistry*, Vol. 1. Academic Press, New York, NY.
Young, H.E. (1971). Preliminary estimates of bark percentages and chemical elements of bark percentages and chemical elements in complete trees of eight species in Main. *Forest Products Journal* 21(5):56–59.
Young, H.E. and Guinn, V.P. (1966). Chemical elements in complete mature trees of seven species in Maine. *Tappi* 49(5):190–197.

Part II

Properties

4 Moisture Properties

Roger M. Rowell
USDA, Forest Service, Forest Products Laboratory,
and Biological Systems Engineering Department, University
of Wisconsin, Madison, WI

CONTENTS

4.1   Moisture Content of Green Wood 77
4.2   Fiber Saturation Point 79
4.3   Equilibrium Moisture Content 80
4.4   Sorption Isotherms 81
        4.4.1   Effect of Temperature on Sorption and Desorption of Water 82
4.5   Swelling of Dry Wood in Water 82
4.6   Distribution of Moisture 83
4.7   Measuring Swelling 84
4.8   Rate of Water Sorption and Activation Energy 84
4.9   Cell Wall Elastic Limit 87
4.10 Swelling Pressure 87
4.11 Effects of Moisture Cycles 89
4.12 Effects on Vibrational Properties 90
4.13 Effects on Biological Properties 90
4.14 Effects on Insulation and Electrical Properties 90
4.15 Effects on Strength Properties 90
4.16 Water Repellency and Dimensional Stability 91
4.17 Swelling in Wood Composites 94
4.18 Swelling in Liquids Other than Water 94
References 97

Wood was designed by Nature over millions of years to perform in a wet environment. The wood structure is formed in a water-saturated environment in the living tree, and the water in the living tree keeps the wood elastic and able to withstand environmental strain such as high wind loads. We cut down a tree, dry the wood, and mainly use it in its dry state. But wood in use remains a hygroscopic resource. Wood's dimensions and mechanical, elastic, and thermal properties depend on the moisture content. Wood is also anisotropic, which means that its properties vary according to its growing direction (longitudinal [vertical or length direction], tangential [parallel to annual growth rings], and radial [perpendicular to the annual growth rings]). The mechanical properties depend very much on both moisture content and growing direction.

4.1 MOISTURE CONTENT OF GREEN WOOD

Moisture exists in wood as both liquid moisture in the cell voids or lumens (free water) and as moisture in the cell wall (bound water). The moisture content of green wood is defined as the total amount of free and bound water in the living tree. This is the maximum moisture content that can exist in a living tree. The moisture content of green wood varies from species to species and depends on the specific gravity. Lumen volume decreases as the specific gravity increases so the green moisture content decreases with increasing specific gravity. The maximum moisture content M_{max} can be calculated by the following:

$$M_{max} = \frac{100(1.54 - G_{bsg})}{1.54\, G_{bsg}}$$

where G_{bsg} is the basic specific gravity based on oven-dry weight and green volume and 1.54 is the specific gravity of the wood cell wall.

Using this equation, the maximum possible moisture content of green wood would be 267% with a basic specific gravity of 0.3 and the minimum possible moisture content would be 44% with a basic specific gravity of 0.9. The density of most woods falls between 320 and 720 kg/m³, although balsa is 160 kg/m³ and some imported hardwoods are 1040 kg/m³.

Table 4.1 shows some average moisture contents of green heartwood and sapwood of some common United States wood species. In some cases, the green moisture content is highest in heartwood and in others the sapwood is highest in moisture content.

TABLE 4.1
Average Moisture Content of Green Wood

Species	Moisture Content	
	Heartwood	Sapwood
Aspen	95	113
Basswood	81	133
Beech	55	72
Birch, Paper	89	72
Cedar, Incense	40	213
Cottonwood, Eastern	162	146
Douglas-fir, Coastal	37	115
Elm, American	95	92
Fir, Balsam	88	173
Hemlock, Western	85	170
Maple, Sugar	65	72
Oak, Red	83	75
Pine, Longleaf	31	106
Pine, Ponderosa	40	148
Pine, Sugar	98	219
Poplar, Yellow	83	106
Redwood, Old growth	86	210
Spruce, Sitka	41	142
Sweetgum	79	137
Sycamore, American	114	130
Walnut, Black	90	73

Source: USDA, 1999.

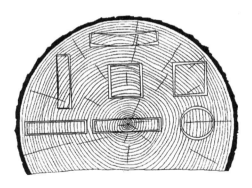

FIGURE 4.1 Shrinkage and distortion of wood upon drying.

4.2 FIBER SATURATION POINT

As water is lost in green wood, there is no change in the volume of the wood until it reaches the fiber saturation point (FSP). The FSP is defined as the moisture content of the cell wall when there is no free water in the voids and the cell walls are saturated with water. This point ranges from 20 to 50 percent weight gain depending on the wood species (Feist and Tarkow 1967). As moisture is removed below the FSP, the wood volume starts to shrink. As stated before, wood is anisotropic, so the shrinkage in wood is different in all three growing directions. Figure 4.1 shows the change in wood shape as a cross section of a log is dried below the FSP. It can be seen that, depending on where the piece of wood is located in the log, the wood will not only get smaller due to the loss of water but also will become distorted due to the anisotropic properties of wood. As will be discussed later, tangential shrinkage is about twice that of radial shrinkage, and longitudinal shrinkage, in most woods, is almost zero.

The shrinkage of wood upon drying depends on several variables, including specific gravity, rate of drying, and the size of the piece. As can be seen in Figure 4.1, the piece of wood cut from the center, left, and right middle is distorted the least (quarter sawing). Although quarter sawing is somewhat wasteful, it does result in minimum distortion in the cut lumber.

Table 4.2 shows the average shrinkage values for some common United States woods. It can be seen that most radial shrinkage values are less than about 6 percent, most tangential shrinkage values less than 10%, and most volumetric shrinkage values less than 15 percent.

To determine the approximate volumetric shrinkage that would occur at a moisture content greater than oven-dry but less than the FSP, the approximate volumetric value at a given moisture can be calculated using the following formula and using the data in Table 4.2:

$$S_m = S_o \times \frac{30 - M}{30}$$

where
S_m is the volumetric shrinkage at a given moisture content
S_o is the total volumetric shrinkage
M is the moisture content

Table 4.2 does not include any longitudinal shrinkage (shrinkage parallel to the grain) because it is usually less than 0.2% for almost all United States species. If a piece of wood is cut near the center of a tree that contains a large amount of juvenile wood or a piece containing reaction wood, the longitudinal shrinkage from green to oven-dry can be as high as 2%.

TABLE 4.2
Average Radial, Tangential, and Volumetric Shrinkage

Species	Shrinkage from Green to Over-Dry Moisture Content		
	Radial	Tangential	Volumetric
Aspen	3.5	6.7	11.5
Basswood	6.6	9.3	15.8
Beech	5.5	11.9	17.2
Birch, Paper	6.3	8.6	16.2
Cedar, Incense	3.3	5.2	7.7
Cottonwood, Eastern	3.9	9.2	13.9
Douglas-fir, Costal	4.8	7.6	12.4
Elm, American	4.2	9.5	14.6
Fir, Balsam	2.9	6.9	11.2
Hemlock, Western	4.2	7.8	12.4
Maple, Sugar	4.8	9.9	14.7
Oak, Red	4.7	11.3	16.1
Pine, Longleaf	5.1	7.5	12.2
Pine, Ponderosa	3.9	6.2	9.7
Pine, Sugar	2.9	5.6	7.9
Redwood, Old growth	2.6	4.4	6.8
Spruce, Sitka	4.3	7.5	11.5
Sweetgum	5.3	10.2	15.8
Sycamore, American	5.0	8.4	14.1
Walnut, Black	5.5	7.8	12.8

Source: USDA, 1999.

The size of the cell cavities remains almost the same size during the loss of water in the cell wall (Tiemann 1944). The thickness of the cell wall decreases in proportion to the moisture lost below the FSP, but the size of the cell lumen remains approximately constant. If this relationship is constant, the volumetric shrinkage V_s of a wood with water soak specific gravity SP_w can be calculated as follows (Stamm and Loughborough 1942):

$$V_s = (M)(SP_w) \quad \text{or} \quad M = V_s/SP_w$$

This ratio should be the approximate FSP for most woods. The value of M for 107 hardwood species was 27 and for 52 softwood species the value was 26 (Stamm and Loughborough 1942).

4.3 EQUILIBRIUM MOISTURE CONTENT

As the green wood loses moisture, it does not change dimensions until the FSP is reached and after that the dimensions change respective to the relative humidity (RH) of the wood surroundings. When the wood is in equilibrium with the surrounding RH, the wood is defined as being at its equilibrium moisture content (EMC). The moisture content of wood is a dynamic property in that the moisture content of wood is constantly changing as the surrounding moisture content changes. When the wood stays at one RH for long periods of time, the wood will reach an equilibrium moisture content. Test results show that, for small pieces of wood at a constant RH, the EMC is reached in about 14 days. A larger wood member may take several weeks to reach its EMC. Wax and extractive content of wood can have a large effect on the length of time it takes for wood to reach its EMC.

TABLE 4.3
Equilibrium Moisture Content of Pine and Aspen

Species	Equilibrium Moisture Content at			
	30%RH	65%RH	80%RH	90%RH
Southern pine	5.8	12.0	16.3	21.7
Aspen	4.9	11.1	15.6	21.5

Table 4.3 shows EMC experimental values for southern pine and aspen at 30, 65, 80, and 90% RH.

4.4 SORPTION ISOTHERMS

A sorption isotherm for wood is defined as the sorption of water in wood at a defined temperature. Figure 4.2 shows the sorption isotherm for Douglas fir at 32°C (Spalt 1958). It is a plot of moisture content (**M%**) vs. relative vapor pressure (**h** = relative humidity/100); note that as moisture is lost from green wood (**IN DES**—initial descending), it follows a different curve than both the rewetting curve (**ADS**—adsorbing) and the second redrying (**SEC DES**—secondary descending) curve. The second, third, fourth, and all subsequent redrying and the second, third, and subsequent rewetting will approximately follow the **SEC SEC** and **ADS** curves. The difference between these curves is referred to as sorption hysteresis for wood (Skaar 1984). Understanding of the difference in moisture content from wet to dry and from dry to wet is very important when it comes to mechanical properties of wood that will be discussed later. The adsorbing (A) curve is always lower than the desorbing (D) curve and the A/D ratio generally ranges from 0.8 and 0.9 depending on the relative humidity and wood species (Okoh and Skaar 1980). Most mechanical properties of wood are very dependent on moisture content and Figure 4.2 shows that at a given relative humidity of, for example, 60% (**h** = 0.6), the moisture content going from wet to dry is approximately 13%, whereas the same wood going from dry to wet would have a moisture content of approximately 10%. This difference of 3% can make a major difference in mechanical properties of the wood. Wax and extractive content of wood can have a large effect on the sorption isotherm.

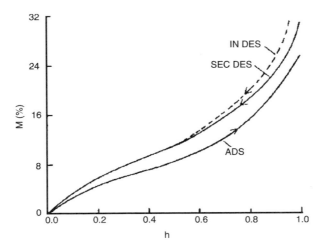

FIGURE 4.2 Sorption isotherm for wood.

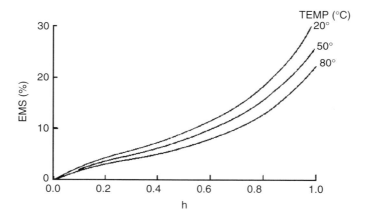

FIGURE 4.3 Effect of temperature on sorption isotherm.

4.4.1 EFFECT OF TEMPERATURE ON SORPTION AND DESORPTION OF WATER

Sorption isotherms for wood decrease with increasing temperature at temperatures about 0°C. Figure 4.3 shows the effect of temperature on the sorption isotherm of wood (Skaar 1984). The desorption isotherm follows the same trend. Because temperature effects sorption and desorption of moisture in wood, EMC values must be determined at a single constant temperature.

4.5 SWELLING OF DRY WOOD IN WATER

Theoretically, it is possible to have absolutely dry wood, wood with a zero moisture content. In actual fact, wood with zero moisture content has never been achieved. When we talk about "oven-dry" (dried above 100°C) wood, the actual moisture content is less than one percent but not zero. There is a small amount of water that is so tightly bound to wood that it is impossible to remove. Because of this, there is no such thing as the volume of zero moisture content wood. But, the amount of water in oven-dry wood is so small, that the water volume is considered to be negligible.

For wood to swell from the dry state, water or some other swelling agent must enter the cell wall. Entry may result from mass flow or diffusion of water vapor into the cell lumens and diffusion from there into the cell wall, or from a diffusion of bound water entirely within the cell wall. In most cases, both processes probably occur. Penetration by mass flow followed by diffusion into the cell wall is a much more rapid process than either vapor-phase or bound-water diffusion (Banks 1973).

Caulfield (1978) proposed a "zipper" model for the movement of water into the wood structure. That is, water moves through wood by forcing cell wall polymers apart as it moves deeper into the wood structure.

Wood is much more permeable in the longitudinal direction than in the radial or tangential directions. Because of this anisotropy, longitudinal flow paths are of major importance in the wetting of wood exposed to the weather (Miller and Boxall 1984). Wood is a hydroscopic resource, which means that the hydroxyl groups in the cell wall polymers are attracted to and form hydrogen bonds with environmental moisture. As water is added to the cell wall, wood volume increases nearly proportionally to the volume of water added (Stamm 1964). Swelling of the wood continues until the cell reaches the FSP, and water beyond the FSP is free water in the void structure and does not contribute to further swelling. This process is reversible, and wood shrinks as it loses moisture below the FSP. This dimensional instability restricts wood from many applications where movement of a material due to changes in moisture content cannot be tolerated.

Moisture Properties

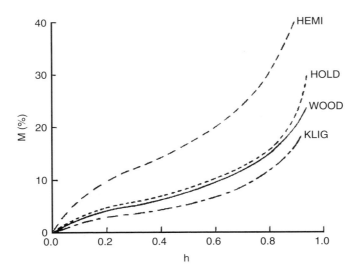

FIGURE 4.4 Sorption isotherms for wood cell wall components.

The amount of swelling that occurs in wood because of hygroscopic expansion is dependent on the density of the wood (Stamm 1964). The percent volumetric swelling V is a function of the dry density d (in g/cc) and the FSP (K_{fsp}, in cc/g) as shown in the following:

$$V = K_{fsp}d$$

This equation determines the approximate volumetric swelling of wood going from an oven-dry state to the FSP and the approximate volumetric shrinkage of wood going from the FSP to oven-dry. Deviations from this relationship usually occur in species high in natural extractives.

All of the cell wall polymers (cellulose, the hemicelluloses, and lignin) are hydroscopic. The order of hydroscopicity is hemicellulose (**HEMI**) > cellulose > lignin (**K LIG**), as shown in Figure 4.4 (Skaar 1984). The holocellulose isotherm is a combination of the hemicelluloses and cellulose polymers. The sorption of moisture by each cell wall polymer depends on not only its hydrophilic nature but also accessibility of water to the polymer's hydroxyl groups. Most, if not all, of the hydroxyl sites in the hemicelluloses are accessible to moisture and the same is probably true of the lignin. The noncrystalline portion of cellulose (approximately 40%) and the surfaces of the crystallites are accessible to moisture, but the crystalline part (approximately 60%) is not (Stamm 1964, Sumi et al. 1964).

4.6 DISTRIBUTION OF MOISTURE

According to the Dent sorption theory, water is added to the cell wall polymers in monolayers (Dent 1977). This theory is based on a modification of the BET model by Brunauer et al. (1938) that was based on a Langmuir model of 1918. The model also permits more than one layer of water on any particular sorption site. This means that when liquid water or water vapor comes into contact with wood, it does not concentrate in one spot but spreads out over the entire cell wall structure at equilibrium. Uneven swelling does occur when one part of the wood is wetter than another until equilibrium is reached.

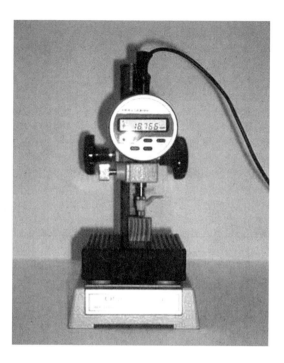

FIGURE 4.5 Measuring wood swelling using a flat-bed micrometer.

4.7 MEASURING SWELLING

There are several ways to measure swelling resulting from interaction with water or other solvents in wood. Some report volumetric swelling, which is a combination of radial, tangential, and longitudinal swelling, and some report just tangential swelling. Since tangential swelling is usually about twice that of radial, tangential swelling alone is representative of the swelling characteristics of each species.

In measuring the rate and extent of swelling in wood using a flat bed micrometer, it is best to cut specimens thin in the longitudinal direction for fast penetration of moisture into the wood with maximum tangential length (see Figure 4.5). It is critical that the tangential grain run parallel to the cut top and bottom edge. If the tangential grain is not parallel to the top and bottom edge, the specimen will go out of square when it swells and will not be measured accurately in the flat bed micrometer.

4.8 RATE OF WATER SORPTION AND ACTIVATION ENERGY

The rate of swelling of Sitka spruce wood at 23°C is shown in Table 4.4. The specimen size was small (2.5 cm square), but it takes some time for water to penetrate into the wood structure. Table 4.4 shows that more than half of the final tangential swelling occured in the first 15 minutes of liquid water contact with the wood. Since the swelling rate depends on specimen size, the rate of swelling would be faster with a thinner specimen (in the longitudinal direction) but the extent of swelling at equilibrium would be the same. Tangential swelling was determined as follows:

$$\text{Tangential swelling} = \frac{(\text{Swollen dimension} - \text{Oven-dry dimension})}{\text{Oven-dry dimension}}$$

TABLE 4.4
Rate of Tangential Swelling of Sitka Spruce in Water at 23°C

Time (minutes)	Tangential Swelling (%)
15	5.3
30	7.0
45	7.8
60	8.1
75	8.4
90	8.5
105	8.6
120	8.7
180	8.8
240	8.8
480	9.0
960	9.0

The rate of swelling of wood in water or other solvents is dependent on several factors, for example, hydrogen bonding ability, molecular size of the reagent, extractives content, temperature, and specimen size (Banks and West 1989). In the case of water, there is an initial induction period due to the diffusion of the water into the cell wall structure and then the water penetrates cell wall capillaries and moves from lumen to lumen in the fiber direction.

Using the data in Table 4.4, the swelling rate constant k can be derived from the slope of a plot of ln k vs. 1/T. From this data, the activation energy Ea can be calculated from the standard Arrhenius equation:

$$k = Ae^{-E_a/RT}$$

where
  A = constant
  R = gas constant
  T = temperature (in Kelvin)

Table 4.5 shows the maximum tangential swelling of wood in water, the swelling rate (k), and the activation energy of swelling (Mantanis et al. 1994a) for Sitka spruce, Douglas fir, and sugar maple. Of these three species, Sitka spruce swells the fastest and has the lowest activation energy. The rate of swelling increased with an increase in temperature (see Table 4.6). The higher density hardwood had greater activation energies as compared to softwoods, which Stamm had found earlier (1964).

The presence of extractives also has a great effect on the rate of swelling of wood in water and other liquids. Stamm and Loughborough (1942) discuss two types of extractives in wood: extractives deposited in the coarse capillary structure and extractives deposited in the cell wall structure. The extractives deposited in the cell wall structure have a great influence on the rate of swelling. The effect of extractives on swelling rate as a function of temperature is shown in Table 4.6. Extractives were removed using 80% ethanol in water for 2 hours at room temperature. In the case of Sitka spruce, the swelling rate greatly increases with the removal of extractives but the removal of extractives from Douglas fir and sugar maple has less effect.

TABLE 4.5
Rate of Swelling, Maximum Tangential Swelling, and Activation Energy of Wood in Water

Temp (C)	Maximum Tangential Swelling (%)		
	Sitka Spruce	Douglas Fir	Sugar Maple
23	6.5	7.6	10.0
40	6.9	7.7	10.8
60	6.9	7.7	11.5
80	7.2	8.0	11.5
100	7.3	8.1	12.3
Swelling Rate, k			
23	1.3	0.3	0.3
40	3.3	0.9	1.1
60	6.5	1.8	3.6
80	10.8	4.7	6.6
100	14.8	6.9	7.6
Activation energy, A_e, kJ/mole	32.2	38.9	47.6

TABLE 4.6
Rate of Swelling and Activation Energy for Unextracted and Extracted Woods

	Sitka Spruce		Douglas Fir		Sugar Maple	
	Unextracted	Extracted	Unextracted	Extracted	Unextracted	Extracted
Max Tangential Swelling (%) at T (C)						
23	6.5	7.4	7.6	9.0	10.0	10.7
40	6.9	7.5	7.7	9.1	10.8	11.6
60	6.9	7.7	7.7	9.1	11.5	11.4
80	7.2	7.7	8.0	9.1	11.5	11.2
100	7.3	7.7	8.1	9.2	12.3	11.3
Swelling Rate, k, at T (C)						
23	1.3	7.2	0.3	0.2	0.3	0.9
40	3.3	14.9	0.9	1.2	1.1	4.0
60	6.5	25.6	1.8	1.4	3.6	7.1
80	10.8	32.4	4.7	3.2	6.6	16.7
100	14.8	36.2	6.9	3.4	7.6	20.7
Activation Energy, E_a (KJ/mole)	32.2	23.3	38.9	41.5	47.6	42.3

TABLE 4.7
Water-Repellent Effectiveness, Contact Angle, and Time for Contact Angle to Fall to 90°

Treating Solution	Concentration (%)	WRE (%)	Contact Angle (θ)	Time for θ to Fall to 90 (Min)
Alkyd resin	10	78	130	425
Paraffin wax	0.5			
Hydrocarbon				
Resin	10	64	131	416
Paraffin wax	0.5			
Rosin ester	10	68	136	460
Paraffin wax	0.5			

Source: Rowell and Banks, 1985.

4.9 CELL WALL ELASTIC LIMIT

The orientation of the cell wall polymers in the S_1 layer determines the extent of swelling of wood in water or other solvent. Swelling from the dry state to the water saturation state continues until the cross banding of the cell wall polymers in the S_1 layer restricts further swelling. This point is defined as the elastic limit of the cell wall. In many cases, data on swelling in solvents other than water is given relative to the swelling in water so that the value of water is set at a value of 1, 10, or 100 and all other solvents are reported above or below this value. Some solvents do swell wood greater than water due to several factors, including extractive removal, plasticization, softening, or solubilization of one or more of the cell wall polymers (usually lignin or hemicelluloses) (see Tables 4.7, 4.8, and 4.9).

4.10 SWELLING PRESSURE

The maximum swelling pressure of wood has been measured (Tarkow and Turner 1958). It was measured by inserting wooden dowels into steel restraining rings equipped with a strain gage. The wooden dowels of different densities were then wetted with liquid water and the swelling pressure was measured as a function of density. Extrapolation of the curve of swelling pressure vs. density to the density of the cell wall (1.5) gave a value of 91 MPa (13,200 psi). A theoretical value based

TABLE 4.8
Antiswell and Antishrink Efficiency of Wood Treatments

Treatment[a]	Antiswell Efficiency 1st wetting	Antishrink Efficiency 1st drying	Antiswell Efficiency 2nd wetting	Antishrink Efficiency 2nd drying
Methyl Methacrylate 65 WPG	15	20	13	20
Polyethylene Glycol	85	15	10	10
Phenol-formaldehyde	83	80	82	81
Acetic Anhydride 21 WPG	82	85	83	85

[a] WPG = weight percent gain.

TABLE 4.9
Volumetric Swelling Coefficients for Southern Pine Sapwood in Various Solvents

Solvent	V 120°C, 1 hour	V 25°C, 24 hours
Methyl isocyanate	52.6[a]	5.1
n-Butylamine	15.5	15.2
Piperidine	13.3	0
Dimethyl solfoxide	13.3	11.7
Dimethyl formamide	12.8	12.5
Pyridine	11.3	13.1
Formaldehyde	12.3	12.3
Acetic anhydride	12.3*	1.5
Acetic acid	11.1	8.8
Aniline	11.0	0.5
Cellosolve	10.6	10.2
Methyl cellosolve	10.3	10.0
WATER	10.0	10.0
Methyl alcohol	9.0	9.3
Epichlorohydrin	6.9	5.9
Acrolein	6.7	7.0
1,4-Dioxane	6.5	0.6
Tetrahydrofuran	5.4	7.2
Propylene oxide	5.2	5.0
Acetone	5.1	5.6
Diethylamine	5.0	11.0
Acrylonitrile	4.6	4.5
Butylene oxide	4.1	0.7
Dichloromethane	3.8	3.3
Methyl ethyl ketone	3.6	5.0
N-Methyl aniline	2.6	0.8
Ethyl acetate	2.4	4.2
Cyclohexanone	2.3	0.5
N-Methylpiperidine	2.2	1.6
4-Methyl-2-pentanone	0.4	1.5
N,N-Dimethylaniline	0.3	0.5
Xylenes	0.1	0.2
Cyclohexane	0.1	0.1
Triethylamine	−0.1	2.1
Hexanes	−0.2	0.2

[a]Reaction occurred.

on an osmotic pressure theory gave a theoretical value of 158 MPa at room temperature. The difference between the actual and the theoretical values was thought to be due to hydroelastic factors. These factors are associated with the fact that restrained wood has a lower moisture content than unrestrained wood.

The best example of using the swelling pressure of wood to practical use goes back to the Egyptians. On a trip to Egypt in 2001, the author saw the evidence first hand. On the edge of a shear granite cliff, holes about 15 cm long, 5 cm wide and 10 cm deep had been chiseled into the rock. Dry wooden wedges the size of the cavities were driven into these holes. Water was then

Moisture Properties

FIGURE 4.6 Evidence of holes cut in wood to split the rock (Egypt 2001).

poured onto the dry wood and allowed to swell. The swelling pressure caused the granite to split down the chain of holes resulting in a giant obelisk or other large building blocks. Figure 4.6 shows the series of holes that were chiseled into a large rock and Figure 4.7 shows the result of the splitting process.

4.11 EFFECTS OF MOISTURE CYCLES

Drying and rewetting causes increases in both the rate of swelling and the extent of swelling. The degradation and extraction of hemicelluloses and extractives as well as some degradation of the cell wall structure during wetting, drying, and rewetting cycles results in more accessibility of water to the cell wall. This is especially evident in hot or boiling water extraction of wood, where significant amounts of cell wall polymers can be lost. A similar effect occurs with wood that is exposed to high relative humidity in repeated cycles. Even though no cell wall polymers are extracted, repeated humidity cycles result in a slight increase in moisture content with each cycle.

FIGURE 4.7 Rock split by using the swelling pressure of wood (Egypt 2001).

4.12 EFFECTS ON VIBRATIONAL PROPERTIES

Moisture has a great effect on vibrational properties of wood. Because wood is a viscoelastic material, vibrational properties are highly dependent on the elasticity of, as well as the internal friction within, the cell wall polymers and matrix. One way of studying the viscoelastic properties of wood is through vibrational analysis. A simple harmonic stress results in a phase difference between stress and strain. The ratio of dynamic Young's modulus (E') to specific gravity (γ) (i.e., E'/γ = specific modulus) and internal friction (tangent of the phase angle, tan δ) measurements can be used to study the viscoelastic nature of wood. The E'/γ ratio is related to sound velocity and tan δ to sound absorption or damping within the wood.

The sorption of water molecules between the wood cell wall polymers acts as a plasticizer to loosen the cell wall microstructure. This affects the tone quality of wooden musical instruments because as moisture content increases, the acoustic properties of wood, such as specific dynamic Young's modulus and internal friction, are reduced or dulled (James 1964, Sasaki et al. 1988, Yano et al. 1993). The decrease in cohesive forces in the cell wall also enhances the deformation of wooden parts under stress.

In practical terms, as the moisture content of a wooden musical instrument, such as an oboe, clarinet, or recorder, increases, the quality of the sound, the brightness of the tone, and the separation of sound between notes decreases.

4.13 EFFECTS ON BIOLOGICAL PROPERTIES

The ability of microorganisms to attack wood depends on the moisture content of the cell wall. The old saying, "dry wood does not rot," is basically true. Ten thousand year-old wood from tombs in China that has remained dry through the years is essentially the same wood today as it was when it was first used (Rowell and Barbour 1990). Termites may seem to attack dry wood, but, in fact, they bring their own moisture to the wood. White-rot fungi need the least water to attack, brown-rot fungi require more, and soft-rot fungi require the highest water content. But all of them require moisture at or near the FSP to be able to degrade wood (see Chapter 5).

4.14 EFFECTS ON INSULATION AND ELECTRICAL PROPERTIES

Thermal and electrical conductivity is very low in dry wood. Early log cabins were warmer in the winter and cooler in the summer due to the insulation properties of wood. Thermal conductivity increases with increasing moisture content. Heat transmission through dry wood is slow but heat transfer is much faster in moist wood using the water as the heat conduit.

The electrical resistance of wood is extremely sensitive to the wood's moisture content. Moisture meters that determine the moisture content of wood are based on this sensitivity. There is also a strong increase in resistance with a decrease in wood temperature. Moisture meters measure the resistance between pairs of pin electrodes driven into the wood to various depths. The meter is calibrated by using data obtained on a given species at room temperature. Meter readings are less reliable at moisture contents above about 25% and read high when used on hot wood and low on cold wood.

4.15 EFFECTS ON STRENGTH PROPERTIES

Changes in moisture content of the wood cell wall below the FSP have a major effect on the mechanical properties of wood (see Chapter 11). Mechanical properties change very little at moisture contents above the FSP. Mechanical properties increase with decreasing moisture content with compression parallel to the grain being the most affected.

4.16 WATER REPELLENCY AND DIMENSIONAL STABILITY

The terms water repellency and dimensional stability are often used interchangeably as if they were the same. They are very different concepts. Water repellency is a rate phenomenon and dimensional stability is an equilibrium phenomenon (Rowell and Banks 1985). Confusion over these two concepts has led to some product failures in service, costing contractors or owners considerable money.

A water repellent treatment is one that prevents or slows down the rate that moisture or liquid water is taken up by the wood. Examples of water repellents include coating, surface applied oils, or lumen filling. A dimensional stability treatment is one that reduces or prevents swelling in wood no matter how long it is in contact with moisture or liquid water. Examples of dimensional stability treatments include bulking the cell wall with polyethylene glycol, penetrating polymers, or bonded cell wall chemicals, or cross-linking cell wall polymers (Rowell and Youngs 1981).

An attractive force exists between a solid and a liquid in contact with it. The net value of this force is governed by the relative magnitudes of the cohesive forces within the liquid and the adhesive forces generated between the liquid and solid. Where the adhesion of liquid to solid is equal to or greater than the cohesion of the liquid, a drop of liquid in contact with the solid spreads spontaneously. That is, the angle between solid and liquid at the solid/liquid/air interface, the contact angle, is zero. If the liquid/solid adhesion is less than the liquid cohesion, an applied liquid droplet does not spread but stands on the surface making a finite contact angle with it (see Figure 4.8). The magnitude of the contact angle increases as the magnitude of the adhesive forces relative to the cohesion of the liquid decreases (Adam 1963). These relationships are expressed algebraically as the Young equation:

$$\gamma_S = \gamma_{SL} + \gamma_L \cos\theta$$

where
- γ_S = surface tension of the solid
- γ_{SL} = liquid/solid interfacial tension
- γ_L = surface tension of the liquid
- θ = contact angle between liquid and solid

Wood is a capillary porous medium. The pore structure is defined by the lumina of the cells and the cell wall openings (pits) interconnecting them. The primary routes for liquid penetration into wood are by these two routes. Except in the case where the contact angle is equal to 90° (cos θ = 0), any liquid contained in a cylindrical capillary of uniform bore has a curved surface. The pressure difference (Pc), often called the capillary pressure, across this curved surface is given by the following relationship, derived from the Kelvin equation:

$$Pc = \frac{-2\gamma_L \cos\theta}{r}$$

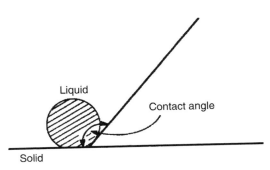

FIGURE 4.8 Contact angle of a water drop on wood.

where

γ_L = surface tension of the liquid
θ = contact angle between liquid and solid
r = capillary radius

The pressure gradient set up by the pressure difference acts in the sense that liquid is forced into the capillary spontaneously for values of θ less than 90°. Conversely, where θ is greater than 90°, external pressure larger than Pc must be applied to force liquid into the capillary. Although wood structure departs significantly from a simple cylindrical capillary model, the general principles of capillary penetration into its structure hold, and the magnitude of Pc remains functionally related to the cosine of the contact angle.

In systems involving water as the liquid phase, surfaces forming contact angles less than 90° are said to be hydrophilic, whereas those with a contact angle greater than 90° are said to be hydrophobic or water repellent.

The chemistry of surfaces giving rise to these properties is fairly well understood. Those surfaces presenting polar functional groups, especially those capable of forming hydrogen bonds with water, tend to be hydrophilic. In contrast, surfaces consisting of nonpolar moieties tend to be strongly hydrophobic.

Water repellent effectiveness (WRE) is also measured as a time dependent function of swelling as follows:

$$WRE = \frac{D_c - D_t \times 100}{D_c}$$

where

D_c = Swelling (or weight of water uptake) of control during exposure in water for t minutes
D_t = Swelling (or weight of water uptake) of treated specimen for the same t time.

For a given exposure time, considerable variation in WRE may result from variations in specimen geometry and in permeability to water of the wood species. To ensure good reproducibility of swelling or water uptake tests, conditions must be closely specified and carefully adhered to.

As was stated before, water repellents are applied to wood principally to prevent or reduce the rate of liquid water flow into the cellular structure and do not significantly alter the dimensions or water sorption observed at equilibrium. Usually the treatments involve the deposition of a thin layer of a hydrophobic substance onto external and, to some extent, internal cell lumen surfaces of wood. The measured WRE varies between 0 and 100 percent depending on the time the test specimens are exposed to water. In some cases, the time to reach equilibrium may be weeks, months, or even years but eventually, maximum swelling will be reached at equilibrium. Generally, the equilibrium moisture content is not altered by the water repellent treatments.

Water repellent treatments, such as impregnation with a wax, may fail due to a failure of the bond formed between the wax and the wood cell wall. This is usually due to degradation of the wood (Banks and Voulgaridis 1980). Wood flooring that has been treated with methyl methacrylate or a similar monomer and polymerized in situ also give a very high WRE. Moisture is physically blocked from entering lumens and penetration of the water must proceed by wicking through the cell wall. Moisture pickup can take a very long time so this type of treatment can be confused as a treatment for dimensional stability.

Table 4.7 shows the results of three resin/wax treatments on WRE, contact angle and the time it takes for the contact angle to fall to 90°. The data indicates that the hydrophobic effect (contact angle) only partly controls the water resistance of specimens treated with the resin/wax systems.

In contrast to water repellency, which is a measure of the rate of moisture pickup, dimensional stability is the measure of the extent of swelling resulting from ultimate moisture pickup. A variety of terms have been used to describe the degree of dimensional stability given to wood by various treatments. The most common term, antishrink efficiency or ASE, is misleading because it seems to apply to shrinking and not to swelling. The abbreviation is acceptable because ASE might also stand for antiswelling efficiency; this usage requires a statement whether the values were determined during swelling or shrinking, and, in addition because both are possible, whether they were obtained in liquid water or water vapor.

Changes in wood dimensions are a result of moisture uptake and can be measured as a single dimensional component, i.e., usually tangential alone, but can also be measured volumetrically taking into account all three dimensional changes. Calculations for dimensional stability are as follows:

$$S = \frac{V_2 - V_1 \times 100}{V_1}$$

where
 S = Volumetric swelling coefficient
 V_2 = Wood volume after humidity conditioning or wetting in water
 V_1 = Wood volume of oven-dried wood before conditioning or wetting

Then

$$ASE = \frac{S_2 - S_1 \times 100}{S_1}$$

where
 ASE = Antishrink efficiency or reduction in swelling resulting from a treatment
 S_2 = Volumetric swelling coefficient of oven-dried after treatment
 S_1 = Volumetric swelling coefficient of oven-dried before treatment

The test conditions for determining the efficiency of a treatment to reduce dimensional changes depend on the treatment as well as the application of the treated product. For a water-leachable treatment, such as polyethylene glycol, a test method based on water vapor is usually used. Humidity tests are also applied to products intended for indoor use. Nonleachable treatments and products intended for exterior use are usually tested in liquid water. For changes in wood dimensions resulting from changes in humidity, the test must be continued long enough to ensure that final equilibrium swelling has been reached.

For a series of humidity cycles, the specimens are placed back and forth in the different humidity conditions and the swelling coefficients are determined. The cycles can be repeated many times to show the efficiency of the treatment when exposed repeatedly to extremes of humidity.

If a liquid water test is used on a leachable chemical treatment, then a single water-swelling test may be run. Because the treating chemicals are leached out during the swelling test, cyclic water tests cannot be done.

Nonleachable treatments are usually tested for dimensional stability by water soaking. The swelling coefficients are determined on the first swelling, again on the first drying, and again on the second wetting. A series of water soaking and drying cycles give the best indication of the durability of the treatment (Rowell and Ellis 1978). This repeated water soaking and oven-drying test is very severe and may result in specimen checking and splitting. It should only be used to determine the effectiveness of treatments for outdoor applications under the harshest of conditions.

Since determination of dimensional stability is based on a comparison between an untreated and a treated specimen, it is critical that the treated specimen come from the same source as the control. Usually, specimens are cut from the same board, taking every other specimen for treatment and the other as controls. To illustrate this point, the S value for southern pine earlywood is 6 to 9, whereas southern pine latewood is 17 to 20. The average swelling coefficient for a sample, therefore, depends on the proportion of latewood and earlywood. If a control is used to compare with a treated sample that differs in percent of latewood, then the values obtained for ASE are nearly useless.

Treatments that have been used for improving dimensional stability include lumen filling (methyl methacrylate), non–cell wall bonded-leachable (polyethylene glycol), non–cell wall bonded-nonleachable (phenol-formaldehyde), and cell wall bonded (reaction with acetic anhydride). Table 4.8 shows the ASE values for these treatments. More information on the chemistry of these treatments is found in Chapter 14.

Complete dimensional stability of wood, i.e., ASE of 100, has only been accomplished through the process of petrification. Chemical treatments are unlikely ever to achieve this level of stabilization.

4.17 SWELLING IN WOOD COMPOSITES

The swelling that occurs in wood composites, such as flake-, particle-, or fiberboard, is much greater than in the wood itself. This is due to the release of compressive forces as well as normal wood swelling. The compressive forces are a result of the physical compression of the wood elements during pressing of the board. This type of swelling is known as irreversible swelling and is the release of the compressive forces during the first wetting of the composites. It is known as irreversible swelling because it is not reversible upon redrying. Reversible swelling also occurs during wetting, which is the swelling of wood elements as a result of moisture pickup. It is known as reversible swelling because the wood shrinks again as a result of redrying.

Most of the irreversible swelling occurs early in the wetting of a composite but the rate can be slow due to the presence of waxes or other chemicals added to retard the rate of moisture pickup.

4.18 SWELLING IN LIQUIDS OTHER THAN WATER

The maximum swelling of wood in various organic liquids is mainly influenced by three solvent properties: the solvent basicity, the molar size, and the hydrogen bonding capacity of the liquid. In addition to these are the extractives content, temperature, and specimen size. An increase in molecule size not only decreases the swelling rate but the swelling equilibrium is also decreased due to the larger molecule slow diffusion into the fine capillary structure of the wood. Swelling of wood in organic liquids is closely associated with the swelling of cellulose alone (Stamm 1935, 1964, Stamm and Tarkow 1950, West 1988, Banks and West 1989, Mantanis et al. 1994a,b).

Table 4.9 shows the decreasing volumetric swelling coefficient order of pine sapwood in various liquids at two different temperatures (Rowell 1984). Several organic liquids swell wood greater than water. It is believed that this is due to a partial softening and solublization of the lignin in the cell wall that allows the wood to swell to a larger volume than in water (Stamm 1964). The largest swelling occurs in methyl isocyanates at 120°C but this is due to a noncatalyzed reaction of the isocyanates with the wood cell wall hydroxyl groups forming a urethane bond. The reacted wood is much larger than the green volume resulting from cell wall rupture due to excess bonded chemical in the cell wall (see Chapter 14). Acetic anhydride also reacts with cell wall hydroxyl groups at 120°C but there is no rupture of the cell wall occurring with this reaction.

Table 4.9 also shows swelling coefficients at a high temperature (120°C) for one hour versus room temperature (25°C) for 24 hours in various liquids. Solvents such as *n*-butyl amine, dimethyl sulfoxide, dimethylformanide, pyridine, formaldehyde, cellosolve, and methyl cellosolve swell wood to the same extent at both the high and low temperature. Piperidine, aniline, 1,4-dioxane, and butylene oxide swell wood to a much greater extent at the high temperature as compared to

TABLE 4.10
Maximum Tangential Swelling of Wood at 23°C, 100 Days

Solvent	Sitka Spruce	Douglas Fir	Sugar Maple
Water	8.4	8.8	10.6
n-Butylamine	14.5	16.4	19.3
Dimethyl sulfoxide	13.9	13.7	14.5
Pyridine	12.2	11.4	13.3
Dimethyl formamide	11.8	11.5	13.2
Formamide	11.2	9.6	16.8
2-Methylpyridine	10.8	11.6	11.9
Diethylamine	10.1	10.4	11.0
Ethylene glycol	9.5	9.1	10.4
Acetic acid	8.7	7.6	10.2
Methyl alcohol	8.2	7.3	8.7
Pyrrole	7.6	6.9	12.0
Butyrolacetone	7.2	7.5	9.6
Ethyl alcohol	7.0	6.3	6.9
Propionic acid	6.4	6.3	10.0
Acetone	5.0	4.6	7.1
Dioxane	5.7	7.5	8.4
Furfural	5.5	5.4	7.6
Methyl acetate	5.0	4.7	5.6
Propyl alcohol	4.9	5.1	5.3
Nitromethane	4.5	4.0	5.4
2-Butanone	4.3	4.1	5.1
Benzyl alcohol	2.9	2.6	8.9
Ethyl acetate	2.6	2.7	3.7
Propyl acetate	2.2	1.4	3.1
Ethylene dichloride	2.1	2.1	4.6
Toluene	1.6	1.5	1.5
Isopropyl ether	1.5	1.3	1.0
Chloroform	1.4	1.5	3.9
Dibutylamine	1.3	1.0	0.8
2,6-Dimethyl pyridine	1.3	1.0	6.3
Carbon tetrachloride	1.2	1.3	1.1
Benzaldehyde	1.0	0.9	2.1
Benzyl benzoate	1.0	0.9	1.0
Piperidine	0.9	0.2	4.3
Octane	0.9	0.7	0.6
Butylaldehyde	0.7	0.7	1.2
Nitrobenzene	0.5	0.4	0.7
Quinoline	0.4	0.3	0.6

the low temperature. Pyridine, tetrahydrofuran, and diethylamine swell wood to a greater extent at the low temperature as compared to the high temperature. Liquids such as trithylamine and hexanes cause a slight shrinkage of the wood at the high temperature.

Table 4.10 shows the maximum tangential swelling of Sitka spruce, Douglas fir, and sugar maple at 23°C (Mantanis et al. 1994b). The three different woods swell to different extents in the same organic liquid. For example, 2-methyl pyridine swells sugar maple almost twice as much as it does Douglas fir. Sugar maple also swells greater than Douglas fir or Sitka spruce in acetic acid, pyrrole, propionic acid, benzyl alcohol, and 2,6-dimethyl pyridine.

TABLE 4.11
Maximum Tangential Swelling of Wood at 23°C, 100 Days

	Sitka Spruce		Sugar Maple	
Solvent	Unextracted	Extracted	Unextracted	Extracted
WATER	5.9	6.3	9.5	10.6
n-Butylamine	11.1	13.3	17.7	19.6
Dimethyl sulfoxide	8.6	9.1	15.7	16.0
Pyridine	8.4	8.6	12.3	14.3
Dimethyl formamide	8.0	8.2	12.5	14.4
Formamide	8.1	8.0	13.3	13.9
2-Methylpyridine	7.8	10.8	11.6	13.2
Diethylamine	7.9	8.6	11.1	11.6
Ethylene glycol	6.9	7.1	10.0	10.8
Acetic acid	5.6	5.7	9.4	11.4
Methyl alcohol	5.8	6.1	9.2	10.2
Butyrolacetone	5.0	6.5	8.3	10.0
Ethyl alcohol	5.0	5.6	8.2	9.7
Propionic acid	4.1	5.3	7.9	9.5
Acetone	4.6	4.7	6.4	8.0
Dioxane	5.3	6.7	8.3	9.9
Furfural	4.0	5.7	7.6	9.4
Methyl acetate	3.7	4.2	5.2	6.9
Propyl alcohol	4.3	4.4	5.6	6.7
Nitromethane	2.7	2.8	5.3	6.6
2-Butanone	4.2	4.2	5.2	7.0
Ethyl acetate	2.8	3.1	4.0	6.1
Propyl acetate	1.1	2.0	2.3	4.6
Ethylene dichloride	2.2	2.1	3.3	4.6
Toluene	1.2	1.3	1.5	2.3
Isopropyl ether	1.7	1.9	1.4	1.8
Chloroform	2.6	3.4	4.3	6.4
Dibutylamine	0.8	0.8	0.5	0.5
2,6-Dimethyl pyridine	1.1	1.9	1.7	10.0
Carbon tetrachloride	1.1	1.5	1.4	1.7
Benzaldehyde	1.6	2.2	1.7	8.5
Benzyl benzoate	1.4	1.5	1.0	1.6
Piperidine	1.3	2.3	1.2	10.6
Octane	1.1	1.2	0.7	0.8
Butylaldehyde	0.4	0.6	1.0	2.2
Nitrobenzene	1.8	1.9	1.7	3.7

Table 4.11 shows the effects of the extractives on the maximum tangential swelling of Sitka spruce and sugar maple at 23°C in various organic liquids (Mantanis et al. 1995a). Swelling is usually larger in the extracted wood. In the case of sugar maple in 2,6-dimethyl pyridine and piperidine, the swelling is much greater in the extracted wood as compared to the unextracted wood.

Table 4.12 shows the effect of the organic liquid basicity, hydrogen bonding potential, molar volume, and molecular weight on wood swelling (Mantanis et al. 1995b). Nayer (1948) and Stamm (1964) speculated that only hydrogen bonding ability alone could explain the relative swelling of wood in different organic liquids. There are, however, several significant exceptions to this correlation, including di-n-butyl amine, tri-n-butyl amine, and benzaldehyde (Mantains et al. 1994b).

TABLE 4.12
Properties of Swelling Solvents Affecting Wood Swelling

Solvent	Basicity[a] Kcal/mole	Hydrogen Bonding[b]	Molar Volume[c] (cc)	Molecular Weight
Butyl amine	57.0	16.8	98.80	73.10
Pyridine	33.1	18.1	80.40	79.10
Dimethyl sulfoxide	29.8	7.7	71.00	78.10
Dimethyl formamide	26.6	11.7	77.00	73.10
Acetic acid	25.0	9.7	57.10	60.00
Formaldehyde	24.0	21.5	39.90	45.04
Ethyl alcohol	20.0	18.7	58.50	46.07
Methyl alcohol	19.0	18.7	40.70	32.04
Ethylene glycol	18.0	20.6	55.80	62.10
WATER	18.0	39.0	18.05	18.02
Propyl alcohol	18.0	18.7	75.00	60.10
Propyl acetate	17.4	8.6	115.7	102.10
Ethyl acetate	17.1	8.4	98.50	88.10
Acetone	17.0	9.7	74.00	58.10
Methyl acetate	16.5	8.4	79.70	74.10
Dioxane	14.8	9.7	85.70	88.10
Chloroform	7.0	1.5	80.70	120.40
Nitrobenzene	4.4	2.8	102.30	123.10
Carbon tetrachloride	3.0	0.0	97.10	153.80
Toluene	3.0	4.5	106.40	92.10
Octane	0.0	0.0	162.50	114.20

[a]*Source:* Gutmann 1976.
[b]Wave number shift × 10, Crowley et al. 1966; Gory 1939, 1941.
[c]Robertson 1964, Handbook of Chemistry and Physics, 2003–4.

Swelling of wood in organic liquids can be predicted much more accurately using the four parameters of basicity, hydrogen bonding ability, molar volume, and molecular weight.

REFERENCES

Adam, N.K. (1963). Water proofing and water repellency. In: Moilliet, J.L. (Ed.), *Principles of Water Repellency*. Elsevier, London.

Banks, W.B. (1973). Water uptake by Scots pie and its restriction by the use of water repellents. *Wood Sci. and Tech.* 7:271–284.

Banks, W.B. and Voulgaridis, E. (1980). The performance of water repellents in the control of moisture absorption by wood exposed to the weather. *Records of the Annual Convention, British Wood Preservation Association;* June 24–27, Cambridge. British Wood Preservation Association, 43–53.

Banks, W.B. and West, H. (1989). A chemical kinetics approach to the process of wood swelling. In Schuerch, C. (Ed.), *Proc. Tenth Cellulose Conf.*, John Wiley & Sons, New York.

Brunauer, S., Emmett, P.H., and Teller, E.J. (1938). Adsorption of gases in multi molecular Layers, *J. Am. Chem. Soc.* 60:309–319.

Caulfield, D.F. (1978). The effect of cellulose on the structure of water. In: *Fibre-Water Interactions in Paper-Making*. Clowes and Sons, Ltd. London.

Crowley, J.D., Teague, G.S., and Lowe, J.W. (1966). A three-dimensional approach to solubility. *J. Paint Technol.* 38:269–280.

Dent, R.W. (1977). A multiplayer theory for gas sorption. Part 1: Sorption of a single gas. *Text. Res. J.* 47(2):145–152.

Feist, W.C. and Tarkow, H. (1967). A new procedure for measuring fiber saturation points. *Forest Prod. J.* 17(10):65–68.

Gordy, W.J. (1939). Spectroscopic comparison of the proton-attracting properties of liquids. *J. Phys. Chem.* 7:93–101.

Gutmann, V. (1976). Empirical parameters for donor and acceptor properties of solvents. *Electrochimica Acta* 21:661–670.

James, W.L. (1964). Vibration, static strength, and elastic properties of clear Douglas fir at various levels of moisture content. *Forest Prod. J.* 14(9):409–413.

Langmuir, I. (1918). The adsorption of gases on plane surfaces of glass, mica and platinum. *J. Am. Chem. Soc.* 40:1361.

Lide D.R. (Ed.) (2003–4). *Handbook of Chemistry and Physics. Molar Volum.* (84th ed) CRC Press, Boca Raton, FL.

Mantanis, G.I., Young, R. A., and Rowell, R.M. (1994a). Swelling of wood: Part 1. Swelling in water. *Wood Sci. Technol.* 28:119–134.

Mantanis, G.I., Young, R.A., and Rowell, R.M. (1994b). Swelling of wood: Part 2. Swelling in organic liquids. *Holzforschung* 48:480–490.

Mantanis, G.I., Young, R.A., and Rowell, R.M. (1995a). Swelling of wood: Part 3. Effect of temperature and extractives on rate and maximum swelling. *Holzforschung* 49:239–248.

Mantanis, G.I., Young, R.A., and Rowell, R.M. (1995b). Swelling of wood: Part 4. A statistical model for prediction of maximum swelling of wood in organic liquids. *Wood and Fiber Sci.* 27(1):22–24.

Miller, E.R. and Boxall, J. (1984). The effectiveness of end-grain sealers in improving paint performance on softwood joinery. *Holz als Roh und Werkstoff* 42(1):27–34.

Nayer, A.N. (1948). *Swelling of wood in various organic liquids.* Ph.D. thesis, University of Minnesota, Mineapolis.

Okoh, K.A.I and Skaa, C. (1980). Moisture sorption isotherms of wood and inner bark of ten southern U.S. hardwoods. *Wood and Fiber* 12(2):98–111.

Robertson, A.A. (1964). Cellulose-liquid interactions. *Pulp Paper Mag. Canada* 65:T171–T178.

Rowell, R.M. (1984). Penetration and reactivity of wood cell wall components. In R.M. Rowell (Ed.), American Chemical Society Advances in Chemistry Series No. 207, Washington, DC, pp. 175–210.

Rowell, R.M. and Banks, W.B. (1985). Water repellency and dimensional stability of wood. USDA Forest Service General Technical Report FPL 50. Forest Products Laboratory, Madison, WI.

Rowell, R.M. and Barbour, J. (Eds.) (1990). *Archaeological Wood: Properties, Chemistry, and Preservation.* American Chemical Society Advances in Chemistry Series 225, Washington, DC.

Rowell, R.M. and Ellis, W.D. (1978). Determination of dimensional stabilization of wood using the water-soak method. *Wood and Fiber* 10(2):104–111.

Rowell, R.M. and Youngs, R.L. (1981). Dimensional stabilization of wood in use. USDA Forest Serv. Res. Note. FPL-0243, Forest Products Laboratory, Madison, WI.

Sasaki, T., Norimoto, M., Yamada, T., and Rowell, R.M. (1988). Effect of moisture on the acoustical properties of wood. *J. Japan Wood Res. Soc.* 34(10):794–803.

Skaar, C. (1984). Wood-water relationships. In: *The Chemistry of Solid Wood*, R.M. Rowell (Ed.), Advances in Chemistry Series, American Chemical Society, Washington, DC, 207, 127–174.

Spalt, H.A. (1958). Fundamentals of water vapor sorption by wood. *Forest Prod. J.* 8(10):288–295.

Stamm. A.J. (1935). Shrinking and swelling of wood. *Ind. Eng. Chem.* 27:401–406.

Stamm, A.J. (1964). *Wood and Cellulose Science*, The Ronald Press Company, New York.

Stamm. A.J. and Loughborough, W.K. (1942). Variation in shrinking and swelling of wood. *Trans. Amer. Soc. Mech. Eng.*, 64:379–386.

Stamm. A.J. and Tarkow, H. (1950). Penetration of cellulose fibers. *J. Phys. Colloid Chem.* 54:745–753.

Sumi, Y. Hale, R.D., Meyer, J.A., Leopold, B., and Ranby, B.G. (1964). Accessibility of wood and wood carbohydrates measured with tritiated water. *TAPPI* 47(10):621–624.

Tarkow, H. and Turner, H.D. (1958). The swelling pressure of wood. *Forest Prod. J.* 8(7):193–197.

Tiemann, H.D. (1944). *Wood Technology.* Second Edition, Pitman Publishing Company, New York.

USDA (1999). *Wood Handbook—Wood as an engineering material.* Gen. Tech. Rep. FPL-GTE-113, U.S. Department of Agriculture, Forest Service, Forest Products Laboratory, Madison, WI.

West, H. (1988). *Kinetics and mechanism of wood-isocyanate reactions.* Ph.D. Thesis. University of North Wales, Bangor, UK.

Winandy, J.E. and Rowell, R.M. (1984). In: Rowell, R.M. (Ed.) *Chemistry of Solid Wood,* American Chemical Society Advances in Chemistry Series No. 207, Washington, DC, 211–255.

Yano, H., Norimoto, M., and Rowell, R.M. (1993). Stabilization of acoustical properties of wooden musical instruments by acetylation. *Wood and Fiber Sci.* 25(4):395–403.

5 Biological Properties

Rebecca E. Ibach
USDA, Forest Service, Forest Products Laboratory, Madison, WI

CONTENTS

5.1 Biological Degradations	100
5.1.1 Bacteria	100
5.1.2 Mold and Stain	100
5.1.3 Decay Fungi	101
5.1.3.1 Brown-Rot Fungi	102
5.1.3.2 White-Rot Fungi	102
5.1.3.3 Soft-Rot Fungi	103
5.1.4 Insects	103
5.1.4.1 Termites	105
5.1.4.1.1 Subterranean Termites	105
5.1.4.1.2 Formosan Subterranean Termites	107
5.1.4.1.3 Nonsubterranean (Drywood) Termites	107
5.1.4.1.4 Dampwood Termites	107
5.1.4.2 Carpenter Ants	107
5.1.4.3 Carpenter Bees	108
5.1.4.4 Beetles	108
5.1.4.4.1 Lyctid Powder-Post Beetles	108
5.1.4.4.2 Anobiid Powder-Post Beetles	108
5.1.4.4.3 Flatheaded Borers	108
5.1.4.4.4 Cerambycids	109
5.1.4.4.4.1 Long-Horned Beetles	109
5.1.4.4.4.2 Old-House Borers	109
5.1.5 Marine Borers	110
5.1.5.1 Shipworms	110
5.1.5.2 Pholads	110
5.1.5.3 Crustaceans	110
5.1.5.3.1 Gribbles	110
5.1.5.3.2 Pillbugs	111
5.2 Prevention or Protection of Wood	111
5.2.1 Wood Preservatives	111
5.2.2 Timber Preparation and Conditioning	113
5.2.3 Treatment Processes	115
5.2.3.1 Pressure Processes	115
5.2.3.2 Nonpressure Processes	115
5.2.4 Purchasing and Handling of Treated Wood	115
References	120

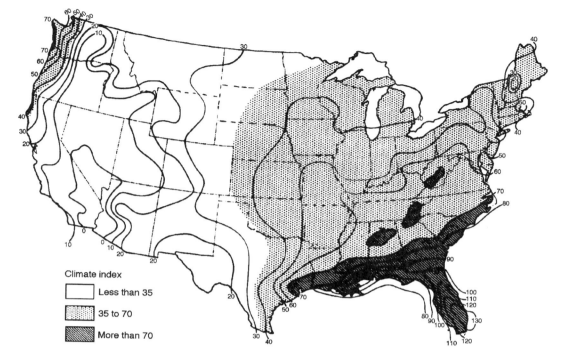

FIGURE 5.1 Climate index for decay potential for wood in service. Higher numbers (darker areas) have greater decay hazard.

There are numerous biological degradations that wood is exposed to in various environments. Biological damage occurs when a log, sawn product, or final product is not stored, handled, or designed properly. Biological organisms, such as bacteria, mold, stain, decay fungi, insects, and marine borers, depend heavily on temperature and moisture conditions to grow. Figure 5.1 gives the climate index for decay hazard for the United States of America. A higher number means a greater decay hazard. The southeastern and northwest coasts have the greatest potential and the southwest has the lowest potential for decay. This chapter will focus on the biological organisms and their mechanism of degradation and then prevention measures. If degradation cannot be controlled by design or exposure conditions, then protection with preservatives is warranted.

5.1 BIOLOGICAL DEGRADATIONS

5.1.1 Bacteria

Bacteria, the early colonizers of wood, are single-celled organisms that can slowly degrade wood that is saturated with water over a long period of time. They are found in wood submerged in seawater and freshwater, aboveground exposure, and in-ground soil contact. Logs held under water for months may have a sour smell attributed to bacteria. Bacteria usually have little effect on the properties of wood except over a long time period. Some bacteria can make the wood more absorptive, which can make it more susceptible to decay. When dried, the degraded area develops a cross checking on the tangential face. The sapwood is more susceptible than the heartwood and the earlywood more than the latewood.

5.1.2 Mold and Stain

Mold and stain fungi cause damage to the surface of wood, and only differ on their depth of penetration and discoloration. Both grow mainly on sapwood and are various colors. Molds are usually fuzzy or powdery growth on the surface of wood and range in color from different shades of green, to black or

FIGURE 5.2 Radial penetration of sapstain fungi in a cross section of pine.

light colors. On softwoods, the fungal hyphae penetrate into the wood, but it can usually be brushed or planed off. On the other hand, on large pored hardwoods, staining can penetrate too deeply to be removed.

The main types of fungus stains are called sapstain or bluestain. They penetrate deeply into the wood and cannot be removed by planing. They usually cause blue, black, or brown darkening of the wood, but some can also produce red, purple, or yellow colors. Figure 5.2 shows the discoloration on a cross section of wood that appears as pie-shaped wedges that are oriented radially.

The strength of wood is usually not altered by molds and stains (except for toughness or shock resistance), but the absorptivity can be increased, which makes the wood more susceptible to moisture and then decay fungi. Given moist and warm conditions, mold and stain fungi can establish on sapwood logs shortly after they are cut. To control mold and stain, the wood should be dried to less than 20 percent moisture content or treated with a fungicide. Wood logs can also be sprayed with water to increase the moisture content to protect wood against fungal stain, as well as decay.

5.1.3 Decay Fungi

Decay fungi are single-celled or multicellular filamentous organisms that use wood as food. Figure 5.3 shows the decay cycle of wood. The fungal spores spread by wind, insects, or animals. They germinate on moist, susceptible wood, and the hyphae spread throughout the wood. These hyphae secrete enzymes that attack the cells and cause wood to deteriorate. After serious decay, a new fruiting body may form. Brown-, white-, and soft-rot fungi all appear to have enzymatic systems that demethoxylate lignin, produce endocellulases, and with some fungi from each group, use single electron oxidation systems to modify lignin (Eaton and Hale 1993).

In the early or incipient stage of wood decay, serious strength losses can occur before it is even detected (see Chapter 10). Toughness, or impact bending, is most sensitive to decay. With incipient decay the wood may become discolored on unseasoned wood, but it is harder to detect on dry wood. The advanced stages of wood decay are easier to detect. Decayed wet wood will break across the grain, whereas sound wood will splinter.

Decay fungi need food (hemicellulose, cellulose, and lignin), oxygen (air), the right temperature (10 to 35°C; optimum 24 to 32°C), and moisture (above the fiber saturation point; about 30% moisture content) to grow. Free water must be present (from rain, condensation, or wet ground contact) for the fiber saturation point to be reached and decay to occur. Air-dried wood will usually have no more than 20% moisture content, so decay will not occur. But there are a few fungi, water-conducting

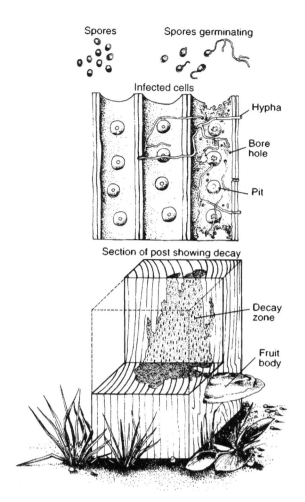

FIGURE 5.3 The wood decay cycle.

fungi, that transport water to dry wood and cause decay called dry-rot. When free water is added to wood to attain 25 to 30% moisture content or higher, decay will occur. Yet wood can be too wet or too dry for decay. If wood is soaked in water, there is not enough air for the fungi to develop.

5.1.3.1 Brown-Rot Fungi

Brown-rot fungi decompose the carbohydrates (i.e., the cellulose and hemicelluloses) from wood, which leaves the lignin remaining, making the wood browner in color, hence the name. Figure 5.4 shows the dark color and cross-grain checking of Southern pine wood caused by brown-rot decay. Brown-rot fungi mainly colonize softwoods, but they can be found on hardwoods as well. Because of the attack on the cellulose, the strength properties of brown-rot decayed wood decrease quickly, even in the early stages. When extreme decay is attained, the wood becomes a very dark, charred color. After the cross-grain cracking, the wood shrinks, collapses, and finally crumbles. Brown-rot fungi first use a low molecular weight system to depolymerize cellulose within the cell wall and then use endocellulases to further decompose the wood.

5.1.3.2 White-Rot Fungi

White-rot fungi decompose all the structural components (i.e., the cellulose, hemicellulose, and lignin) from wood. As the wood decays it becomes bleached (in part from the lignin removal) or

Biological Properties

FIGURE 5.4 Brown-rot decay of Southern pine wood.

white with black zone lines. White-rot fungi occur mainly on hardwoods but can be found on softwoods as well. The degraded wood does not crack across the grain until it is severely degraded. It keeps its outward dimensions but feels spongy. The strength properties decrease gradually as decay progresses, except toughness. White-rot fungi have a complete cellulase complex and also the ability to degrade lignin.

5.1.3.3 Soft-Rot Fungi

Soft-rot fungi are related to molds and occur usually in wood that is constantly wet, but they can also appear on surfaces that encounter wet-dry cycling. The decayed wood typically is shallow in growth and soft when wet, but the undecayed wood underneath is still firm. Upon drying, the decayed surface is fissured. Figure 5.5 shows surface checking of soft-rotted wood when dry. The wood becomes darker (dull-brown to blue-gray) when decayed by soft-rot fungi. Soft-rot fungi have a system to free the lignin in the wood to then allow the cellulases access to the substrate.

5.1.4 Insects

Insects are another biological cause of wood deterioration. Both the immature insect and the adult form may cause wood damage, and they are often not present when the wood is inspected. Therefore, identification is based on the description of wood damage as described in Table 5.1. Figure 5.6 shows pictures of four types of insect damage caused by termites, powder-post beetles, carpenter ants, and beetles.

FIGURE 5.5 Soft-rot decay of a treated pine pole.

TABLE 5.1
Description of Wood Damage Caused by Insects

Type of Damage	Description	Causal Agent	Damage Begins	Damage Ends
Pin holes	0.25 to 6.4 mm (1/100 to 1/4 in.) in diameter, usually circular			
	Tunnels open:			
	Holes 0.5 to 3 mm (1/50 to 1/8 in.) in diameter, usually centered in dark streak or ring in surrounding wood	Ambrosia beetles	In living trees and unseasoned logs and lumber	During seasoning
	Holes variable sizes; surrounding wood rarely dark stained; tunnels lined with wood-colored substance	Timber worms	In living trees and unseasoned logs and lumber	Before seasoning
	Tunnels packed with usually fine sawdust:			
	Exit holes 0.8 to 1.6 mm (1/32 to 1/16 in.) in diameter, in sapwood of large-pored hardwoods; loose floury sawdust in tunnels	Lyctid powder-post beetles	During or after seasoning	Reinfestation continues until sapwood destroyed
	Exit holes 1.6 to 3 mm (1/16 to 1/8 in.) in diameter, primarily in sapwood, rarely in heartwood; tunnels loosely packed with fine sawdust and elongate pellets	Anobiid powder-post beetles	Usually after wood in use (in buildings)	Reinfestation continues; progress of damage very slow
	Exit holes 2.5 to 7 mm (3/32 to 9/32 in.) in diameter; primarily sapwood of hard woods, minor in softwoods; sawdust in tunnels fine to coarse and tightly packed	Bostrichid powder-post beetles	Before seasoning or if wood is rewetted	During seasoning or redrying
	Exit holes 1.6 to 2 mm (1/16 to 1/12 in.) in diameter, in slightly damp or decayed wood; very fine sawdust or pellets tightly packed in tunnels	Wood-boring weevils	In slightly damp wood in use	Reinfestation continues while wood is damp
Grub holes	3 to 13 mm (1/8 to 1/2 in.) in diameter, circular or oval			
	Exit holes 3 to 13 mm (1/8 to 1/2 in.) in diameter, circular; mostly in sapwood; tunnels with coarse to fibrous sawdust or it may be absent	Roundheaded borers (beetles)	In living trees and unseasoned logs and lumber	When adults emerge from seasoned wood or when wood is dried
	Exit holes 3 to 13 mm (1/8 to 1/2 in.) in diameter; mostly oval; in sapwood and heartwood; sawdust tightly packed in tunnels	Flatheaded borers (beetles)	In living trees and unseasoned logs and lumber	When adults emerge from seasoned wood or when wood is dried
	Exit holes ~6 mm (~1/4 in.) in diameter; circular; in sapwood of softwoods, primarily pine; tunnels packed with very fine sawdust	Old house borers (a roundheaded borer)	During or after seasoning	Reinfestation continues in seasoned wood in use
	Exit holes perfectly circular, 4 to 6 mm (1/16 to 1/4 in.) in diameter; primarily in softwoods; tunnels tightly packed with coarse sawdust, often in decay softened wood	Woodwasps	In dying trees or fresh logs	When adults emerge from seasoned wood, usually in use, or when kiln dried

(Continued)

TABLE 5.1
Description of Wood Damage Caused by Insects (*Continued*)

Type of Damage	Description	Causal Agent	Damage Begins	Damage Ends
	Nest entry hole and tunnel perfectly circular ~13 mm (~1/2 in.) in diameter; in soft softwoods in structures	Carpenter bees	In structural timbers, siding	Nesting reoccurs annually in spring at same and nearby locations
Network of galleries	Systems of interconnected tunnels and chambers	Social insects with colonies		
	Walls look polished; spaces completely clean of debris	Carpenter ants	Usually in damp partly decayed, or soft-textured wood in use	Colony persists unless prolonged drying of wood occurs
	Walls usually speckled with mud spots; some chambers may be filled with "clay"	Subterranean termites	In wood structures	Colony persists
	Chambers contain pellets; areas may be walled-off by dark membrane	Dry-wood termites (occasionally damp wood termites)	In wood structures	Colony persists
Pitch pocket	Openings between growth rings containing pitch	Various insects	In living trees	In tree
Black check	Small packets in outer layer of wood	Grubs of various insects	In living trees	In tree
Pith fleck	Narrow, brownish streaks	Fly maggots or adult weevils	In living trees	In tree
Gum spot	Small patches or streaks of gum-like substances	Grubs of various insects	In living trees	In tree
Ring distortion	Double growth rings or incomplete annual layers of growth	Larvae of defoliating insects or flatheaded cambium borers	In living trees	In tree
	Stained area more than 25.4 mm (1 in.) long introduced by insects in trees or recently felled logs	Staining fungi	With insect wounds	With seasoning

5.1.4.1 Termites

Termites are the size of ants and live in colonies. Figure 5.7 is a map of the United States showing the northern limit of the subterranean termites, which live in the ground and the northern limit of the drywood termites or nonsubterranean, which live in wood.

5.1.4.1.1 Subterranean Termites

The native subterranean termites live in colonies in the ground, have 3 stages of metamorphosis (egg, nymph, and adult), and have 3 different castes (reproductives, workers, and soldiers). They can have winged and wingless adults living in one colony at the same time. Two reproductives (swarmers) are needed to start a colony. Figure 5.8 shows the difference between a winged termite (A) and a winged ant (B). The termite has longer wings and no waist indentation. They are light tan to black with 4 wings, 3 pairs of legs, 1 pair of antennae, and 1 pair of large eyes and they are about 8 to 13 mm long. Thousands of the swarmers are released from a colony during the daylight

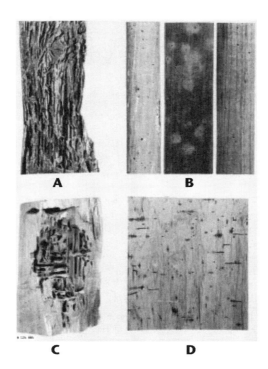

FIGURE 5.6 Types of insect damage caused by (A) termites, (B) powder-post beetles, (C) carpenter ants, and (D) beetles.

hours in the spring or early summer. They fly a short way and then lose their wings. Females attract the males, they find a nesting site, and eggs are laid within several weeks. The worker members are the ones that cause the destruction of the wood.

Moisture is critical for the termites, from either their nest in the soil or the wood they are feeding on. They form shelter tubes made of particles of soil, wood, and fecal material. These shelter tubes protect the termites and allow them to go from their nest in the soil to the wood above ground. Termites prefer the softer earlywood to the harder latewood.

FIGURE 5.7 Map of termite location in the United States of America: (A) subterranean northern limit and (B) drywood termites northern limit.

Biological Properties

To protect a house from termites, the soil should be treated with an insecticide and good design and construction practices should be followed, such as building the foundation with concrete. If termites get into a building, then a termite control specialist from a national pest control operator association should be contacted.

5.1.4.1.2 Formosan Subterranean Termites

The Formosan subterranean termite is originally from the Far East. It moved to Hawaii, other Pacific islands, California, Texas, and the southeastern United States. The Formosan termite multiplies and causes damage quicker than the native subterranean species. Infestation control measures are the same as for the native species, but treatment should be performed within a few months.

5.1.4.1.3 Nonsubterranean (Drywood) Termites

Nonsubterranean termites are found in the southern edge of the continental United States from California to Virginia, the West Indies, and Hawaii (see Figure 5.7). Drywood termites do not multiply as quickly as the subterranean termites, but they can live in dry wood without outside moisture or ground contact. Infestations can enter a house in wood products, such as furniture. Prevention includes examining all wood and cellulose-based materials before bringing inside, removing woody debris from the outside, and using preservative-treated lumber. Infestations should be treated or fumigated by a professional licensed fumigator. Call the state pest control association.

5.1.4.1.4 Dampwood Termites

Dampwood termites colonize in damp and decaying wood and do not need soil to live if the wood is wet enough. They are most prevalent on the Pacific Coast. Keeping wood dry is the best protection for preventing colonization and damage by dampwood termites.

5.1.4.2 Carpenter Ants

Carpenter ants use wood for shelter instead of food. They prefer wood that is soft or decayed. They can be black or brown and live in colonies. There are several casts: winged and unwinged queens, winged males, and different sizes of unwinged workers. Figure 5.8 shows that carpenter ants have a narrow waist and wings of two different sizes. The front wings are larger than the hind ones. They create galleries along the grain of the wood and around annual rings. They attack the earlywood first, and only attack the latewood to access between the galleries. Once a nest is established, it can extend into sound wood. The inside of the gallery is smooth and clean because the ants keep removing any debris, unlike termites.

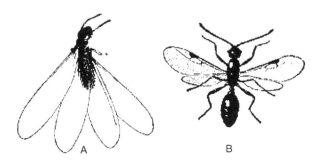

FIGURE 5.8 A winged termite with long wings (A), and a winged ant with a waist indentation (B).

One way to keep carpenter ants from colonizing wood is to keep moisture out and decay from happening. If they do get into the house, then the damaged wood should be removed and the new wood should be kept dry. If it is not possible to keep the wood dry, then a preservative-treated lumber should be used. To treat indoors with insecticides, the state pest control association should be called.

5.1.4.3 Carpenter Bees

Carpenter bees are like large bumblebees, but they differ in that their abdomens shine because the top is hairless. The females make 13-mm tunnels into unfinished softwood to nest. The holes are partitioned into cells, and each cell holds one egg, pollen, and nectar. Carpenter bees reuse nests year after year, therefore some tunnels can be quite long with many branches. They can nest in stained, thinly painted, light preservative salt treatments, and bare wood. To control carpenter bees, an insecticide can be injected in the tunnel, plugged with caulk, and then the entry hole surface treated so that the bees do not use it again the next year. A thicker paint film, pressure preservative treatments, screens, and tight-fitting doors can help prevent nesting damage.

5.1.4.4 Beetles

Table 5.1 describes the types of wood damage that result from various beetles.

5.1.4.4.1 Lyctid Powder-Post Beetles

Lyctid beetles cause significant damage to dry hardwood lumber, especially lumber with large pores, such as oak, hickory, and ash. They are commonly called powder-post beetles because they make a fine, powdery sawdust during infestation. Activity and damage is greatest when the moisture content of wood is between 10 to 20 percent, but activity can occur from 8 to 32% moisture content. Infestation can be detected after the first generation of winged adult beetles emerges from the wood, which produces small holes (0.8 to 1.6 mm diameter), and a fine wood powder falls out.

5.1.4.4.2 Anobiid Powder-Post Beetles

Anobiid beetles are found on older and recently seasoned hardwoods or softwoods throughout the United States of America. They prefer the sapwood that is closest to the bark and their exit holes are 1.6 to 3 mm in diameter. Their life cycle is 2 to 3 years and they need about 15% moisture content. If the infestation is old, then there may be very small round (0.8 mm) emergence holes from parasitic wasp larvae that feed on the beetle larvae.

There are several approaches to try to control powder-post beetles. One is to control the environmental conditions by lowering the moisture content of the wood through ventilation, insulation, and vapor barriers, as well as good building design. Another is to use chemical treatment by brushing or spraying the wood with insecticides, using boron diffusion treatments, or fumigating if infestation is extensive. Using pressure-treated wood can prevent beetle attack. Another approach is to just eliminate or reduce the beetle population.

5.1.4.4.3 Flatheaded Borers

Flatheaded borers are metallic-colored beetles that vary in size, but a hammer-headed shape produced by an enlarged, flattened body region behind the head characterizes the larvae. The adult flatheaded borer emerges causing 3- to 13-mm oval or elliptical exit holes in sapwood and heartwood of living trees and unseasoned logs and lumber. Powdery, pale-colored sawdust is found tightly packed in oval to flattened tunnels or galleries in softwoods and hardwood. The adult females lay

Biological Properties

eggs singly or in groups on the bark or in crevices in the bark or wood. The larvae or young borers mine the inner bark or wood. Since most infestations occur in trunks of weakened trees or logs, the best method of control is to spot treat the local infestations, which can be done by applying insecticides to the surface of the wood. This may prevent reinfestation or kill the larvae that feed close to the surface.

5.1.4.4.4 *Cerambycids*

The long-horned beetles and old-house borers are collectively called the Cerambycids, or the roundheaded beetles.

5.1.4.4.4.1 *Long-Horned Beetles*

The long-horned beetle or roundheaded beetle is the common name of the Asian Cerambycid beetle, *Anoplophora glabripennis*, which is indigenous to southern China, Korea, Japan, and the Isle of Hainan. It is extremely destructive to hardwood tree species and there is no known natural predator in the United States. It attacks not just stressed or aging trees but healthy trees of any age, and it produces new adults each year, instead of every 2 to 4 years like other longhorn beetles. The beetle bores into the heartwood of a host tree, eventually killing the tree.

The beetle is believed to have hitchhiked into the United States in wooden crating of a cargo ship in the early 1990s. It was discovered in the United States in August 1996 in Brooklyn, New York. Within weeks another infestation was found on Long Island in Amityville, New York. Two years later in July of 1998 the beetle was found in Chicago, Illinois. It attacks many healthy hardwood trees, including maple, horsechestnut, birch, poplar, willow, and elm.

The adult beetles have large bodies, are black with white spots, and have very long black and white antenna. They make large circular holes (3 to 13 mm diameter) upon emergence and can occur anywhere on the tree, including branches, trunks, and exposed roots. Oval to round, darkened wounds in the bark may also be observed, and these are oviposition sites where adult females chew out a place to lay their eggs. The larvae chew banana-shaped tunnels or galleries into the wood, causing heavy sap flow from wounds and sawdust accumulation at tree bases. These galleries interrupt the flow of water from the roots to the leaves. They feed on, and over-winter in, the interiors of trees. Quarantine is usually imposed on firewood and nursery stock in a known infected area and all infested trees are immediately destroyed.

5.1.4.4.4.2 *Old-House Borers*

Another roundheaded borer is called the old-house borer. The adult is a large (18 to 25 mm), black to dark brown, elongated beetle that burrows in structural wood, old and new, seasoned and unseasoned, and softwood lumber but not hardwoods. It is capable of reinfesting wood in use and is found along the Atlantic seaboard.

The adults lay their eggs in the cracks and crevices of wood and they hatch in about 2 weeks. The larvae can live in seasoned softwood for several years. They feed little during the winter months of December through February, but when the larvae are full grown, which usually takes about five years, they emerge through oval holes (6 to 9 mm) in the surface of the wood. Moisture content of 15 to 25% encourages growth. Emergence happens during June and July. During the first few years of feeding, the larvae cannot be heard, but when they are about four years old chewing sounds can be heard in wood during the spring and summer months. Damage depends on the number of larvae feeding, the extent of the infestation, how many years, and whether there has been a reinfestation. To control old-house borers insecticides can be applied to the surface of wood. When there is an extensive and active infestation of the old-house borer, then fumigation may be the best control method. To prevent reinfestation, small infestations can be controlled by applying insecticides to the surface of the wood, which kills the larvae that may feed close to the surface and contact the chemical just below the surface.

5.1.5 MARINE BORERS

Marine-boring organisms in salt or brackish waters can cause extensive damage to wood. Attack in the United States is significant along the Pacific, Gulf, and South Atlantic Coasts and slower along the New England Coast because of cold-water temperatures. The marine borers that cause the most damage in the United States are shipworms, Pholads, crustaceans, and pillbugs.

5.1.5.1 Shipworms

Shipworms are wormlike mollusks that cause great damage to wooden boats, piers, and structures. They belong in the family Teredinidae and the genera *Teredo* or *Bankia*. They are found in salt water along the United States coastal waters, but some can adapt to less saline conditions and live in many of the estuaries. The young larvae swim to wood and bury themselves using a pair of boring shells on their head. They have a tail that has 2 siphons: one to draw in water containing microscopic organisms for food and oxygen for respiration; and the second siphon to expel waste and for reproduction. The larvae eat the wood and organisms from the ocean, and grow wormlike bodies, but they never leave the wood. The shipworms grow in length and diameter, but their entrance holes are only the size of the young larvae (1.6 mm). The inside of the wood becomes honeycombed and severely degraded. Adults from the genus *Teredo* grow to be 0.3 to 0.6 m in length and 13 mm in diameter, whereas those of the genus *Bankia* grown to be 1.5 to 1.8 m in length and 22 mm in diameter. To protect wood from shipworms, a marine grade preservative treatment is used, such as creosote (400 kg/m^3), chromated copper arsenate (CCA, 40 kg/m^3), or ammoniacal copper citrate (CC, 40 kg/m^3).

5.1.5.2 Pholads

Pholads are also wood-boring mollusks but are different in that they resemble clams, i.e., they are encased in a double shell. They belong in the family Pholadidae with two familiar species, the *Martesia* and the *Xylophaga*. They enter the wood when they are very young and grow inside the wood, similar to the shipworms. Their entrance holes are about 6 mm in diameter, and most of the damage to the wood is close to the surface. Pholads grow no bigger than 64 mm long and 25 mm in diameter, but they can cause extensive damage to wood. They are found in Hawaii, San Diego, California, the Gulf Coast, and from South Carolina southward. To protect wood from Pholads, either marine-grade creosote (400 kg/m^3) or a dual treatment of CCA and then creosote are effective.

5.1.5.3 Crustaceans

The crustaceans include gribbles, in the family Limnoridae, genus *Limnoria*, and pillbugs, from the family Spaeromatidae, genus *Spaeroma*. They are related to lobsters and shrimp and differ from the other marine borers in that they are not imprisoned, so they can move from place to place. The boreholes made are shallow, therefore the borers, combined with water erosion, degrade the surface of the wood.

5.1.5.3.1 Gribbles

Gribbles or *Limnoria* are quite small (3 to 4 mm) and their boreholes are usually only about 13 mm deep, but with water erosion, the borers continually bore in deeper. They prefer earlywood, and the attack is usually located between half tide and low tide levels that result in an hourglass shape. Protection with preservative treatment against gribbles depends on where and what species is present in the water. Two recommended treatments are either a dual treatment of first CCA (16 to 24 kg/m^3) and then marine grade creosote (320-kg/m^3), or just using a higher concentration of just CCA (40 kg/m^3) or just marine-grade creosote (320 to 400 kg/m^3). To get more information, check with current American Wood Preservers Association (AWPA) Standards (AWPA 2003).

Biological Properties

5.1.5.3.2 Pillbugs

Pillbugs or *Spaeroma* are longer (13 mm long) and wider (6 mm wide) than *Limnoria* and look like a pill bug that lives in damp places. They use the wood for shelter and prefer softer woods. *Spaeroma* are found along the south Atlantic and Gulf Coasts and from San Francisco southward on the West Coast. It is common to find them in Florida estuaries. Dual treatment with CCA and then creosote is the best protection because they are tolerant to CCA and with time tolerant to creosote.

5.2 PREVENTION OR PROTECTION OF WOOD

To protect wood from biological degradation, chemical preservatives are applied to the wood either by nonpressure or pressure treatment (Eaton and Hale 1993). Penetration and retention of a chemical will depend on wood species and the amount of heartwood (more difficult to treat) or sapwood (easier to treat). The objective of adding wood preservatives is to obtain long-term effectiveness for the wood product, thus sequestering carbon.

Starting January 2004, the U.S. Environmental Protection Agency (EPA) no longer allows the most widely used wood preservative, chromated copper arsenate (CCA), for products for any residential use (i.e., play structures, decks, picnic tables, landscaping timbers, residential fencing, patios, walkways, and boardwalks). However, it has not concluded that arsenic-containing CCA-treated wood poses unreasonable risks to the public from the wood being used around or near their homes (EPA 2002). Alternative preservatives such as ammoniacal copper quat (ACQ) and copper azole (CBA) have replaced CCA for residential use (EPA 2002; PMRA 2002). Looking beyond these replacements for CCA may be wood protection systems not based on toxicity, but rather nontoxic chemical modifications to prevent biological degradation. Chemical modification alters the chemical structure of the wood components thereby reducing the biodegradability of wood, as well as increasing its dimensional stability when in contact with moisture (Rowell 1991) (see Chapter 14).

5.2.1 WOOD PRESERVATIVES

Wood preservatives work by being toxic to the biological organisms that attack wood. The active ingredients in wood preservative formulations are many and varied and each has its own mode of action, some of which are still unknown or unreported. In general, mechanisms of toxicity involve denaturation of proteins, inactivation of enzymes, cell membrane disruption causing an increase in cell permeability, and inhibition of protein synthesis.

The degree of protection of a particular preservative and treatment process depends on 4 basic requirements: toxicity, permanence, retention, and depth of penetration into the wood. Toxicity refers to how effective the chemical is against biological organisms, such as decay fungi, insects, and marine borers. Permanence refers to the resistance of the preservative to leaching, volatilization, and breakdown. Retention specifies the amount of preservative that must be impregnated into a specific volume of wood to meet standards and ensure that the product will be effective against numerous biological agents.

Wood preservatives can be divided into two general classes: Oil-type, such as creosote and petroleum solutions of pentachlorophenol, and waterborne salts that are applied as water solutions, such as CCA, ACQ, and CBA. The effectiveness of each preservative can vary greatly depending on its chemical composition, retention, depth of penetration, and ultimately the exposure conditions of the final product. Three exposure categories are *ground contact* (i.e., high decay hazard; usually pressure treated), *aboveground contact* (i.e., low decay hazard; not usually used for pressure treatment), and *marine exposure* (i.e., high decay hazard; often needs a dual treatment). The degree of protection needed will depend on geographic location and potential exposures of the wood, expected service life, structural and nonstructural applications, and replacement costs. Wood preservatives should always be used when exposed to ground (soil) contact and marine (salt-water) exposure.

Oilborne preservatives such as creosote and solutions with heavy, less volatile petroleum oils often help to protect wood from weathering but may adversely influence its cleanliness, odor, color, paintability, and fire performance. Waterborne preservatives are often used when cleanliness and paintability of the treated wood are required. In seawater exposure, a dual treatment (waterborne copper-containing salt preservatives followed by creosote) is most effective against all types of marine borers.

Exposure conditions and length of product lifetime need to be considered when choosing a particular preservative treatment, process, and wood species (Cassens, Johnson et al. 1995). The consensus technical committees consider all these factors in setting reference levels required in the American Wood Preservers' Association (AWPA), the American Society for Testing and Materials International (ASTM), and the Federal Specification Standards. For various wood products, preservatives, and their required retention levels see Federal Specification TT-W-571 and 572, the AWPA Book of Standards, or the UDSA Forest Service Forest Products Laboratory (FPL) Wood Handbook, Chapter 14 (USFSS 1968, 1969; FPL 1999; ASTM 2000; AWPA 2003). Table 5.2 gives the retention levels of creosote and some waterborne preservatives for lumber, timbers, and plywood exposed to various conditions. The retention specifies the amount of preservative that must be impregnated into a specific volume of wood to meet standards and to ensure that the product will be effective against numerous biological agents.

Evaluation for efficacy of preservative-treated wood is first performed on small specimens in the laboratory and then larger specimens with field exposure (ASTM 2000). The USDA Forest Service FPL has had in-ground stake test studies on southern pine sapwood ongoing since 1938 in Saucier, Mississippi, and Madison, Wisconsin (Gutzmer and Crawford 1995). Table 5.3 shows

TABLE 5.2
Retention Levels of Creosote and Some Waterborne Preservatives for Lumber, Timbers, and Plywood Exposed to Various Conditions[a]

	Preservative Retention (kg/m^3 (lb/ft^3))		
	Salt Water[b]	Ground Contact and Fresh Water	Above Ground
Creosote	400 (25)	160 (10)	128 (8)
CCA (Types I, II, or III)	40 (2.50)	6.4 (0.40)	4.0 (0.25)
ACQ (Types B or D)	NR	6.4 (0.40)	4.0 (0.25)
CDDC as Cu	NR	3.2 (0.20)	1.6 (0.10)
CC	40 (2.50)	6.4 (0.40)	4.0 (0.25)
CBA (Type A)	NR	NR	3.27 (0.20)

CCA, chromated copper arsenate; ACQ, ammoniacal copper quat; CDDC, copper bis(dimethyldithiocarbamate); CC, ammoniacal copper citrate, CBA, copper azole.

[a] Retention levels are those included in Federal Specification TT-W-571 and Commodity Standards of the American Wood Preservers' Association. Refer to the current issues of these specifications for up-to-date recommendations and other details. In many cases, the retention is different depending on species and assay zone. Retentions for lumber, timbers, and plywood are determined by assay of borings of a number and location as specified in Federal Specification TT-W-571 or in the Standards of the American Wood Preservers' Association. Unless noted, all waterborne preservative retention levels are specified on an oxide basis. NR is not recommended.

[b] Dual treatments are recommended when marine borer activity is known to be high.

TABLE 5.3
Results of the Forest Products Laboratory Studies on 5- by 10- by 46-cm (2- by 4- by 18-in) Southern Pine Sapwood Stakes, Pressure-Treated with Commonly Used Wood Preservatives, Installed at Harrison Experimental Forest, Mississippi[a]

Preservative	Average Retention kg/m^3 (lb/ft^3)	Average Life or Condition at Last Inspection
CCA-Type III	6.41 (0.40)	No failures after 20 years
Coal-tar Creosote	160.2 (10.0)	90% failed after 51 years
Copper naphthenate (0.86% copper in No. 2 fuel oil)	1.31 (0.082)	29.6 years
Oxine copper (copper-8-quinolinolate) (in heavy petroleum)	1.99 (0.124)	No failures after 28 years
No preservative treatment		1.8 to 3.6 years

[a]*Source:* Gutzmer and Crawford, 1995.

results of the Forest Products Laboratory studies on 5- by 10- by 46-cm (2- by 4- by 18-in) Southern Pine sapwood stakes, pressure-treated with commonly used wood preservatives, installed at Harrison Experimental Forest, Mississippi. A comparison of preservative treated small wood panels exposed to a marine environment in Key West, Florida has been evaluated (Johnson and Gutzmer 1990). Outdoor evaluations such as these compare various preservatives and retention levels under each exposure condition at each individual site. These preservatives and treatments include creosotes, waterborne preservatives, dual treatments, chemical modification of wood, and various chemically modified polymers.

5.2.2 Timber Preparation and Conditioning

Preparing the timber for treatment involves carefully peeling the round or slabbed products to enable the wood to dry quickly enough to avoid decay and insect damage and to allow the preservative to penetrate satisfactorily. Drying the wood before treatment is necessary to prevent decay and stain and to obtain preservative penetration, but when treating with waterborne preservatives by certain diffusion methods, high moisture content levels may be permitted. Drying the wood before treatment opens up the checks before the preservative is applied, thus increasing penetration and reducing the risk of checks opening up after treatment and exposing unpenetrated wood.

Treating plants that use pressure processes can condition green material by means other than air and kiln drying, thus avoiding a long delay and possible deterioration. When green wood is to be treated under pressure, one of several methods for conditioning may be selected. The steaming and vacuum process is used mainly for southern pines, and the Boulton (or boiling-under-vacuum) process is used for Douglas fir and sometimes hardwoods.

Heartwood of some softwood and hardwood species can be difficult to treat (see Table 5.4) (Mac Lean 1952). Wood that is resistant to penetration by preservatives, such as Douglas fir, western hemlock, western larch, and heartwood, may be incised before treatment to permit deeper and more uniform penetration. Incision involves passing the lumber or timbers through rollers that are equipped with teeth that sink into the wood to a predetermined depth, usually 13 to 19 mm (1/2 to 3/4 in.). The incisions open cell lumens along the grain that improve penetration but can result in significant strength reduction. As much cutting and hole boring of the wood product as is possible should be done before the preservative treatment, otherwise untreated interiors will allow ready access of decay fungi or insects.

TABLE 5.4
Penetration of the Heartwood of Various Softwood and Hardwood Species[a]

Ease of Treatment	Softwoods	Hardwoods
Least difficult	Bristlecone pine (*Pinus aristata*)	American basswood (*Tilia americana*)
	Pinyon (*P. edulis*)	Beech (white heartwood) (*Fagus grandifolia*)
	Pondersosa pine (*P. pondersosa*)	Black tupelo (blackgum) (*Nyssa sylvatica*)
	Redwood (*Sequoia sempervirens*)	Green ash (*Fraxinus pennsylvanica* var. *lanceolata*)
		Pin cherry (*Prunus pensylvanica*)
		River birch (*Betula nigra*)
		Red oaks (*Quercus spp.*)
		Slippery elm (*Ulmus fulva*)
		Sweet birch (*Betula lenia*)
		Water tupelo (*Nyssa aquatica*)
		White ash (*Fraxinus americana*)
Moderately difficult	Bald cypress (*Taxodium distichum*)	Black willow (*Salix nigra*)
	California red fir (*Abies magnifica*)	Chestnut oak (*Quercus montana*)
	Douglas fir (coast) (*Pseudotsuga taxifolia*)	Cottonwood (*Populus sp.*)
	Eastern white pine (*Pinus strobus*)	Bigtooth aspen (*P. grandidentata*)
	Jack pine (*P. banksiana*)	Mockernut hickory (*Carya tomentosa*)
	Loblolly pine (*P. taeda*)	Silver maple (*Acer saccharinum*)
	Longleaf pine (*P. palustris*)	Sugar maple (*A. saccharum*)
	Red pine (*P. resinosa*)	Yellow birch (*Betula lutea*)
	Shortleaf pine (*P. echinata*)	
	Sugar pine (*P. lambertiana*)	
	Western hemlock (*Tsuga heterophylla*)	
Difficult	Eastern hemlock (*Tsuga canadensis*)	American sycamore (*Platanus occidentalis*)
	Engelmann spruce (*Picea engelmanni*)	Hackberry (*Celtis occidentalis*)
	Grand fir (*Abies grandis*)	Rock elm (*Ulmus thomoasi*)
	Lodgepole pine (*Pinus contorta* var. *latifolia*)	Yellow-poplar (*Liriodendron tulipifera*)
	Noble fir (*Abies procera*)	
	Sitka spruce (*Picea sitchensis*)	
	Western larch (*Larix occidentalis*)	
	White fir (*Abies concolor*)	
	White spruce (*Picea glauca*)	
Very difficult	Alpine fir (*Abies lasiocarpa*)	American beech (red heartwood) (*Fagus grandifolia*)
	Corkbark fir (*A. lasiocarpa* var. *arizonica*)	American chestnut (*Castanea dentata*)
	Douglas fir (Rocky Mountain) (*Pseudotsuga taxifolia*)	Black locust (*Robinia pseudoacacia*)
	Northern white-cedar (*Thuja occidentalis*)	Blackjack oak (*Quercus marilandica*)
	Tamarack (*Larix laricina*)	Sweetgum (redgum) (*Liquidambar styraciflua*)
	Western red cedar (*Thaja plicata*)	White oaks (*Quercus spp.*)

[a] As covered in MacLean (1952).

5.2.3 TREATMENT PROCESSES

There are two general types of wood-preserving methods: pressure processes and nonpressure processes. During pressure processes wood is impregnated in a closed vessel under pressure above atmospheric. In commercial practice wood is put on cars or trams and run into a long steel cylinder, which is then closed and filled with preservative. Pressure forces are then applied until the desired amount of preservative has been absorbed into the wood.

5.2.3.1 Pressure Processes

Three pressure processes are commonly used: full-cell, modified full-cell, and empty-cell. The full-cell process is used when the retention of a maximum quantity of preservative is desired. The steps include the following: (1) The wood is sealed in a treating cylinder and a vacuum is applied for a half-hour or more to remove air from the cylinder and wood, (2) the preservative (at ambient or elevated temperature) is admitted to the cylinder without breaking the vacuum, (3) pressure is applied until the required retention, (4) the preservative is withdrawn from the cylinder, and (5) a short final vacuum may be applied to free the wood from dripping preservative. The modified full-cell process is basically the same as the full-cell process except for the amount of initial vacuum and the occasional use of an extended final vacuum.

The goal of the empty-cell process is to obtain deep penetration with relatively low net retention of preservative. Two empty-cell processes (the Rueping and the Lowry) use the expansive force of compressed air to drive out part of the preservative absorbed during the pressure period. The Rueping empty-cell process is often called the empty-cell process with initial air. Air pressure is forced into the treating cylinder, which contains the wood, and then the preservative is forced into the cylinder. The air escapes into an equalizing or Rueping tank. The treating pressure is increased and maintained until desired retention is attained. The preservative is drained and a final vacuum is applied to remove surplus preservative. The Lowry process is the same as the Rueping except that there is no initial air pressure or vacuum applied. Hence, it is often called the empty-cell process without initial air pressure.

5.2.3.2 Nonpressure Processes

There are numerous nonpressure processes and they differ widely in their penetration and retention of a preservative. Nonpressure methods consist of (1) surface applications of preservative by brushing or brief dipping, (2) cold soaking in preservative oils or steeping in solutions of waterborne preservative, (3) diffusion processes with waterborne preservatives, (4) vacuum treatment, and (5) various other miscellaneous processes.

5.2.4 PURCHASING AND HANDLING OF TREATED WOOD

The EPA regulates pesticides, and wood preservatives are one type of pesticide. Preservatives that are not restricted by EPA are available to the general consumer for nonpressure treatments, whereas the sale of others is restricted only to certified pesticide applicators. These preservatives can be used only in certain applications and are referred to as restricted-use. Restricted-use refers to the chemical preservative and not to the treated wood product. The general consumer may buy and use wood products treated with restricted-use pesticides; EPA does not consider treated wood a toxic substance nor is it regulated as a pesticide.

Consumer Safety Information Sheets (EPA-approved) are available from retailers of treated wood products. The sheets provide users with information about the preservative and the use and disposal of treated-wood products. There are consumer information sheets for three major groups of wood preservatives (see Table 5.5): (1) creosote pressure-treated wood, (2) pentachlorophenol pressure-treated wood, and (3) inorganic arsenical pressure-treated wood.

TABLE 5.5
EPA-Approved Consumer Information Sheets for Three Major Groups of Preservative Pressure-Treated Wood

Preservative Treatment	Inorganic Arsenicals	Pentachlorophenol	Creosote
CONSUMER INFORMATION	*This wood has been preserved by pressure-treatment with an EPA-registered pesticide containing inorganic arsenic to protect it from insect attack and decay. Wood treated with inorganic arsenic should be used only where such protection is important. *Inorganic arsenic penetrates deeply into and remains in the pressure-treated wood for a long time. However, some chemical may migrate from treated wood into surrounding soil over time and may also be dislodged from the wood surface upon contact with skin. Exposure to inorganic arsenic may present certain hazards. Therefore, the following precautions should be taken both when handling the treated wood and in determining where to use or dispose of the treated wood.	*This wood has been preserved by pressure-treatment with an EPA-registered pesticide containing pentachlorophenol to protect it from insect attack and decay. Wood treated with pentachlorophenol should be used only where such protection is important. *Pentachlorophenol penetrates deeply into and remains in the pressure-treated wood for a long time. Exposure to pentachlorophenol may present certain hazards. Therefore, the following precautions should be taken both when handling the treated wood and in determining where to use and dispose of the treated wood.	*This wood has been preserved by pressure-treatment with an EPA-registered pesticide containing creosote to protect it from insect attack and decay. Wood treated with creosote should be used only where such protection is important. *Creosote penetrates deeply into and remains in the pressure-treated wood for a long time. Exposure to creosote may present certain hazards. Therefore, the following precautions should be taken both when handling the treated wood and in determining where to use the treated wood.
HANDLING PRECAUTIONS	*Dispose of treated wood by ordinary trash collection or burial. Treated wood should not be burned in open fires or in stoves, fireplaces, or residential boilers because toxic chemicals may be produced as part of the smoke and ashes. Treated wood from commercial or industrial use (e.g., construction sites) may be burned only in commercial or industrial incinerators or boilers in accordance with state and Federal regulations. *Avoid frequent or prolonged inhalation of sawdust from treated wood. When sawing and machining treated wood, wear a dust mask. Whenever possible, these operations should be performed outdoors to avoid indoor accumulations of airborne sawdust from treated wood.	*Dispose of treated wood by ordinary trash collection or burial. Treated wood should not be burned in open fires or in stoves, fireplaces, or residential boilers because toxic chemicals may be produced as part of the smoke and ashes. Treated wood from commercial or industrial use (e.g., construction sites) may be burned only in commercial or industrial incinerators or boilers rated at 20 million btu/hour or greater heat input or its equivalent in accordance with state and Federal regulations. *Avoid frequent or prolonged inhalation of sawdust from treated wood. When sawing and machining treated wood, wear a dust mask.	*Dispose of treated wood by ordinary trash collection or burial. Treated wood should not be burned in open fires or in stoves, fireplaces, or residential boilers, because toxic chemicals may be produced as part of the smoke and ashes. Treated wood from commercial or industrial use (e.g., construction sites) may be burned only in commercial or industrial incinerators or boilers in accordance with state and Federal regulations. *Avoid frequent or prolonged inhalation of sawdust from treated wood. When sawing and machining treated wood, wear a dust mask. Whenever possible these operations should be performed outdoors to avoid indoor accumulations of airborne sawdust from treated wood.

	*When power-sawing and machining, wear goggles to protect eyes from flying particles. *Wear gloves when working with the wood. After working with the wood, and before eating, drinking, toileting, and use of tobacco products, wash exposed areas thoroughly. *Because preservatives or sawdust may accumulate on clothes, they should be laundered before reuse. Wash work clothes separately from other household clothing.	*Whenever possible, these operations should be performed outdoors to avoid indoor accumulations of airborne sawdust from treated wood. *Avoid frequent or prolonged skin contact with pentachlorophenol-treated wood. When handling the treated wood, wear long-sleeved shirts and long pants and use gloves impervious to the chemicals (for example, gloves that are vinyl-coated). *When power-sawing and machining, wear goggles to protect eyes from flying particles. *After working with the wood, and before eating, drinking, and use of tobacco products, wash exposed areas thoroughly. *If oily preservatives or sawdust accumulate on clothes, launder before reuse. Wash work clothes separately from other household clothing.	*Avoid frequent or prolonged skin contact with creosote-treated wood; when handling the treated wood, wear long-sleeved shirts and long pants and use gloves impervious to the chemicals (for example, gloves that are vinyl-coated). *When power-sawing and machining, wear goggles to protect eyes from flying particles. *After working with the wood and before eating, drinking, and use of tobacco products, wash exposed areas thoroughly. *If oily preservatives or sawdust accumulate on clothes, launder before reuse. Wash work clothes separately from other household clothing.
USE SITE PRECAUTIONS	*All sawdust and construction debris should be cleaned up and disposed of after construction. *Do not use treated wood under circumstances where the preservative may become a component of food or animal feed. Examples of such sites would be use of mulch from recycled arsenic-treated wood, cutting boards, counter tops, animal bedding, and structures or containers for storing animal feed or human food. *Only treated wood that is visibly clean and free of surface residue should be used for patios, decks, and walkways. *Do not use treated wood for construction of those portions of beehives that may come into contact with honey. *Treated wood should not be used where it may come into direct or indirect contact with drinking water, except for uses involving incidental contact such as docks and bridges.	*Logs treated with pentachlorophenol should not be used for log homes. Wood treated with pentachlorophenol should not be used where it will be in frequent or prolonged contact with bare skin (for example, chairs and other outdoor furniture) unless an effective sealer has been applied. *Pentachlorophenol-treated wood should not be used in residential, industrial, or commercial interiors except for laminated beams or building components which are in ground contact and are subject to decay or insect infestation and where two coats of an appropriate sealer are applied. Sealers may be applied at the installation site. Urethane, shellac, latex epoxy enamel, and varnish are acceptable sealers for pentachlorophenol-treated wood.	*Wood treated with creosote should not be used where it will be in frequent or prolonged contact with bare skin (for example, chairs and other outdoor furniture) unless an effective sealer has been applied. *Creosote-treated wood should not be used in interiors of industrial buildings. Creosote-treated wood in interiors of industrial building components that are in ground contact and are subject to decay or insect infestation and wood-block flooring. For such uses, two coats of an appropriate sealer must be applied. Sealers may be applied at the installation site. *Wood treated with creosote should not be used in the interiors of farm buildings where there may be direct contact with domestic animals or livestock that may crib (bite) or lick the wood.

(*Continued*)

TABLE 5.5
EPA-Approved Consumer Information Sheets for Three Major Groups of Preservative Pressure-Treated Wood (Continued)

Preservative Treatment	Inorganic Arsenicals	Pentachlorophenol	Creosote
		*Wood treated with pentachlorophenol should not be used in the interiors of farm buildings where there may be direct contact with domestic animals or livestock that may crib (bite) or lick the wood. *In interiors of farm buildings where domestic animals or livestock are unlikely to crib (bite) or lick the wood, pentachlorophenol-treated wood may be used for building components that are in ground contact and are subject to decay or insect infestation and where two coats of an appropriate sealer are applied. Sealers may be applied at the installation site. *Do not use pentachlorophenol-treated wood for farrowing or brooding facilities. *Do not use treated wood under circumstances where the preservative may become a component of food or animal feed. Examples of such sites would be structures or containers for storing silage or food. *Do not use treated wood for cutting-boards or countertops.	*In interiors of farm buildings where domestic animals or livestock are unlikely to crib (bite) or lick the wood, creosote-treated wood may be used for building components that are in ground contact and are subject to decay or insect infestation if two coats of an effective sealer are applied. Sealers may be applied at the installation site. Coal tar pitch and coal tar pitch emulsion are effective sealers for creosote-treated wood-block flooring. Urethane, epoxy, and shellac are acceptable sealers for all creosote-treated wood. *Do not use creosote-treated wood for farrowing or brooding facilities. *Do not use treated wood under circumstances where the preservative may become a component of food or animal feed. Examples of such use would be structures or containers for storing silage or food. *Do not use creosote-treated wood for cutting-boards or countertops.

Biological Properties

* Only treated wood that is visibly clean and free of surface residue should be used for patios, decks, and walkways.
* Do not use treated wood for construction of those portions of beehives that may come into contact with the honey.
* Pentachlorophenol-treated wood should not be used where it may come into direct or indirect contact with public drinking water, except for uses involving incidental contact such as docks and bridges.
* Do not use pentachlorophenol-treated wood where it may come into direct or indirect contact with drinking water for domestic animals or livestock, except for uses involving incidental contact such as docks and bridges.

* Only treated wood that is visibly clean and free of surface residues should be used for patios, decks, and walkways.
* Do not use treated wood for construction of those portions of beehives that may come into contact with the honey.
* Creosote-treated wood should not be used where it may come into direct or indirect contact with public drinking water, except for uses involving incidental contact such as docks and bridges.
* Do not use creosote-treated wood where it may come into direct or indirect contact with drinking water for domestic animals or livestock, except for uses involving incidental contact such as docks and bridges.

There are two important factors to consider depending upon the intended end use of preservative-treated wood: the grade or appearance of the lumber, and the quality of the preservative treatment in the lumber. The U.S. Department of Commerce American Lumber Standard Committee (ALSC), an accrediting agency for treatment quality assurance, has an ink stamp or end tag for each grade stamp and quality mark. These marks indicate that the producer of the treated-wood product subscribes to an independent inspection agency. The stamp or end tag contains the type of preservative or active ingredient, the retention level, and the intended exposure conditions. Retention levels are usually provided in pounds of preservatives per cubic foot of wood and are specific to the type of preservative, wood species, and intended exposure conditions. Be aware that suppliers often sell the same type of treated wood by different trade names. Depending on your intended use and location, there will be different types of treated wood available for residential use. Also, be aware that some manufacturers add colorants (such as brown) or water repellents (clear) into some of their preservative treatments. When purchasing treated wood, ask the suppliers for more information to determine what preservative and additives were used, as well as any handling precautions.

Note that mention of a chemical in this article does not constitute a recommendation; only those chemicals registered by the EPA may be recommended. Registration of preservatives is under constant review by EPA and the U.S. Department of Agriculture. Use only preservatives that bear an EPA registration number and carry directions for home and farm use. Preservatives, such as creosote and pentachlorophenol, should not be applied to the interior of dwellings that are occupied by humans. Because all preservatives are under constant review by EPA, a responsible state or Federal agency should be consulted as to the current status of any preservative.

REFERENCES

ASTM. (2000). *Annual Book of ASTM Standards*. American Society for Testing and Materials, West Conshohocken, PA.
AWPA. (2003). *AWPA 2003 Book of Standards*. American Wood-Preservers' Association, Selma, AL.
Cassens, D.L., Johnson, B.R., Feist, W.C., and De Groot, R.C. (1995). *Selection and Use of Preservative-Treated Wood,* Forest Products Society, Madison, WI.
Eaton, R.A. and Hale, M.D. (1993). *Wood: Decay, Pests and Protection.* Chapman & Hall, New York.
EPA. (2002). Whitman announces transition from consumer use of treated wood containing arsenic, U.S. Environmental Protection Agency.
FPL. (1999). *Wood Handbook: Wood as an Engineering Material*. U.S. Department of Agriculture, Forest Service, Forest Products Laboratory, Madison, WI.
Gutzmer, D.I. and Crawford, D.M. (1995). *Comparison of Wood Preservatives in Stake Tests*. U.S. Department of Agriculture, Forest Service, Forest Products Laboratory, Madison, WI, 124.
Johnson, B.R. and Gutzmer, D.I. (1990). *Comparison of Preservative Treatments in Marine Exposure of Small Wood Panels*. U.S. Department of Agriculture, Forest Service, Forest Products Laboratory, Madison, WI, 28.
Mac Lean, J.D. (1952). *Preservation of Wood by Pressure Methods*. U.S. Department of Agriculture, Forest Service, Washington, DC, 160.
PMRA. (2002). *Chromated Copper Arsenate (CCA).* Canadian Pest Management Regulatory Agency.
Rowell, R.M. (1991). Chemical modification of wood. In: Hon, D.N.-S. and Shiraishi, N., *Handbook on Wood and Cellulosic Materials*. Marcel Dekker, New York, pp. 703–756.
USFSS. (1968). *Wood Preservation Treating Practices*. U.S. Federal Supply Service, Washington, DC.
USFSS. (1969). *Fungicide: Pentachlorophenol*. U.S. Federal Supply Service, Washington, DC.

6 Thermal Properties

Roger M. Rowell[1,2] and Susan L. LeVan-Green[1]
[1]USDA, Forest Service, Forest Products Laboratory, Madison, WI and
[2]Department of Biological Systems Engineering, University
of Wisconsin, Madison, WI

CONTENTS

6.1	Pyrolysis and Combustion	122
6.2	Fire Retardancy	128
6.3	Testing Fire Retardants	128
	6.3.1 Thermogravimetric Analysis (TGA)	128
	6.3.2 Differential Thermal Analysis (DTA) and Differential Scanning Calorimetry (DSC)	129
	6.3.3 Tunnel Flame-Spread Tests	129
	6.3.4 Critical Oxygen Index Test	129
	6.3.5 Other Tests	130
6.4	Fire Retardants	131
	6.4.1 Chemicals that Promote the Formation of Increased Char at a Lower Temperature than Untreated Wood Degrades	131
	6.4.2 Chemicals which Act as Free Radical Traps in the Flame	133
	6.4.3 Chemicals Used to Form a Coating on the Wood Surface	134
	6.4.4 Chemicals that Increase the Thermal Conductivity of Wood	134
	6.4.5 Chemicals that Dilute the Combustible Gases Coming from the Wood with Non-Combustible Gases	134
	6.4.6 Chemicals that Reduce the Heat Content of the Volatile Gases	134
	6.4.7 Phosphorus-Nitrogen Synergism Theories	135
	6.4.8 Fire-Retardant Formulations	135
	6.4.8.1 Phosphorus	136
	6.4.8.2 Boron	136
	6.4.9 Leach Resistant Fire-Retardants	136
References		137

The traditional question at the start of a class on thermal properties of wood is, "Does wood burn?" The students have all been warmed in front of a wood-burning fire before, so they are sure the answer is yes—but since the professor asked the question, there must be some hidden trick to the obvious answer. Going with their experience, their answer is "yes, wood burns." But, the actual answer is no, wood does not burn. In short, wood undergoes thermal degradation as it heats up, giving rise to volatile, flammable gases which burn when they contact a source of ignition. So it is the flammable gases that burn, not the wood itself. This is, of course, an oversimplified explanation of the pyrolysis and burning processes which are the subject of this chapter.

Lignocellulosic materials decompose on heating and when exposed to an ignition source by two different mechanisms. The first, dominant at temperatures below 300°C, degrades polymers by the breaking of internal chemical bonds; dehydration (elimination of water); formation of free radicals, carbonyl, carboxyl, and hydroperoxide groups; formation of carbon monoxide and carbon dioxide; and finally, the formation of reactive carbonaceous char. Oxidation of the reactive char results in smoldering or glowing combustion, and further oxidation of the combustible volatile gasses gives rise to flaming combustion (Antal 1985, Bridegwater 1999, Czernik et al. 1999, Shafizadeh 1984).

The second mechanism, which takes over at temperatures above 300°C, involves the cleavage of secondary bonds and formation of intermediate products such as anhydromonosaccharides, which are converted into low molecular weight products (oliosaccharides and polysaccharides), which lead to carbonized products (Kawamoto et al. 2003).

6.1 PYROLYSIS AND COMBUSTION

The simplest method of evaluating the thermal properties of wood is by thermogravametric analysis (TGA). Figure 6.1 shows a schematic of the equipment used for this analysis. A sample is placed in a metal pan in a furnace tube. Nitrogen gas is passed through the system and the furnace tube is slowly heated at a constant rate. The percent of weight loss is measured as function of temperature and is printed out on a strip chart. The temperature is usually raised to 500–600°C in nitrogen at a rate of a few degrees Celsius per minute. The temperature is then lowered to around 300°C, oxygen is introduced into the system, and the temperature is increased again. This second scan shows the combustion of the char in oxygen.

A simple thermogram run in nitrogen is shown in Figure 6.2 for pine (Shafizadeh 1984). As the wood is heated from room temperature to 100°C, very few chemical reactions take place. At approximately 100°C, any moisture in the wood is vaporized out. As the temperature of the wood increases, very little degradation occurs until about 200°C, when chemical bonds start to break via dehydration and, possibly, free radical mechanisms to eliminate water and produce volatile gases. In the absence of oxygen, or in the presence of limited amounts, this thermal degradation process

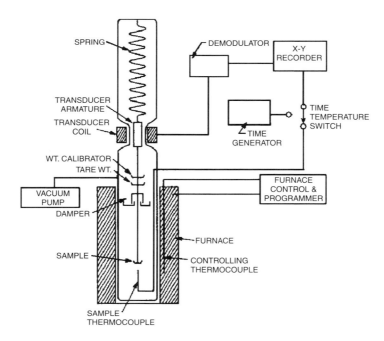

FIGURE 6.1 Schematic diagram of a simple thermogravimetric analysis system.

Thermal Properties

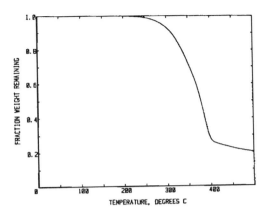

FIGURE 6.2 Thermogravimetric analysis of pine.

is called *pyrolysis*. The volatile gases produced diffuse out of the wood into the surrounding atmosphere. Figure 6.3 shows the first derivative of the TGA curve for pine.

Whole wood starts to thermally degrade at about 250°C (see Figure 6.4). Between about 300–375°C, the majority of the carbohydrate polymers have degraded and only lignin remains. The hemicellulose components start to decompose at about 225°C and are almost completely degraded by 325°C. The cellulose polymer is more stable to thermal degradation until about 370°C, and then decomposes almost completely over a very short temperature range. Both the acid lignin and milled wood lignin start to decompose at about 200°C, but are much more stable to thermal degradation as compared to the carbohydrate polymers. The curve for whole wood represents the results of each of the cell wall components. The pyrolysis products given off when wood is thermally degraded are shown in Table 6.1 (Shafizadeh 1984).

Since the major cell wall polymer is cellulose, the thermal degradation of cellulose dominates the chemistry of pyrolysis (Shafizadeh and Fu 1973). The decomposition of cellulose leads mainly to volatile gases, while lignin decomposition leads mainly to tars and char. Figure 6.5 shows the pyrolysis and combustion of cellulose (Shafizadeh 1984). In the early stages of cellulose degradation (below 300°C), the molecular weight is reduced by depolymerization caused by dehydration reactions. The main products are CO, CO_2 produced by decarboxylation and decarbonylation, water, and char residues. In the presence of oxygen, the char residues undergo glowing ignition. CO and CO_2 form much faster in oxygen than in nitrogen, and this rate accelerates as the temperature

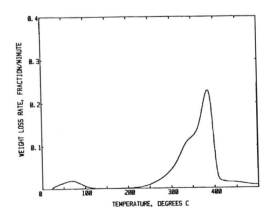

FIGURE 6.3 Derivative thermogravimetric analysis of pine.

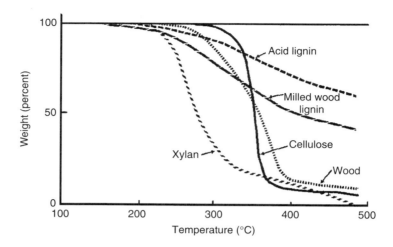

FIGURE 6.4 Thermogravimetric analysis of cottonwood and its cell wall components.

increases. Table 6.2 shows the rate constants for the depolymerization of cellulose in air and nitrogen (Shafizadeh 1984).

Cellulose also produces such combustible volatiles acetaldehyde, propenal, methanol, butanedione, and acetic acid. When the combustible volatiles mix with oxygen and are heated to the ignition temperature, exothermic combustion occurs. The heat from these reactions in the vapor phase transfers back into the wood, increasing the rate of pyrolysis in the solid phase. When the burning mixture accumulates sufficient heat, it emits radiation in the visible spectrum. This phenomenon is known as *flaming combustion* and occurs in the vapor phase. At 300°C the cellulose molecule is highly flexible and undergoes depolymerization by transglycosylation to create such products as anhydromonosaccharides, which include levoglucosan (1,6-anhydro-β-D-glycopyranose)

TABLE 6.1
Pyrolysis Products of Wood

Product	Percent in Mixture
Acetaldehyde	2.3
Furan	1.6
Acetone	1.5
Propenal	3.2
Methanol	2.1
2,3-Butanedione	2.0
1-Hydroxy-2-propanone	2.1
Glyoxal	2.2
Acetic acid	6.7
5-Methyl-2-furaldehyde	0.7
Formic acid	0.9
2-Furfuryl alcohol	0.5
Carbon dioxide	12.0
Water	18.0
Char	15.0
Tar (at 600°C)	28.0

Thermal Properties

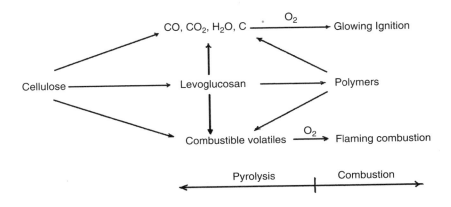

FIGURE 6.5 Pyrolysis and combustion of cellulose.

and 1,6-anhydro-β-D-glyucofuranose. These are converted into low molecular weight products, randomly linked oliosaccharides and polysaccharides, which lead to carbonized products.

Figure 6.6 shows the formation of levoglucosan from cellulose. When a single unit of cellulose (glucose) undergoes dehydration, it forms levoglucosan, levoglucosenone, and 1,4:3,6-dianhydro-α-D-glucopyranose. Products such as the 1,2-anhydride and the 1,4-anhydride, 3-deoxy-D-erythro-hexosulose, 5-hydroxymethyl-2-furaldehyde, 2-furaldehyde (furfural), other furan derivatives, 1,5-anhydro-4-deoxy-D-hex-1-ene-3-ulose, and other pyran derivatives also form (Shafizadeh 1984). These dehydration derivatives are important in the intermediate formation of char compounds.

The intermolecular and intramolecular transglycosylation reactions are accompanied by dehydration, followed by fission or fragmentation of the sugar units and disporportionation reactions in the gas phase (Shafizadeh 1982). The anhydromonosaccharides can recombine to form polymers, which can then degrade to CO, CO_2, water and char residues, or combustible volatiles.

Above 300°C, the rate of tar-forming reactions increases and the formation of char decreases. Table 6.3 shows the percent of char and tar products formed from cellulose as a function of temperature (Shafizadeh 1984). At 300°C, there are about 28% tar and 20% char products. As the

TABLE 6.2
Rate Constants for the Depolymerization of Cellulose in Air and Nitrogen

Temperature	Condition	$K_o \times 10^7$ (mol/162 g min)*
150	N_2	1.1
	Air	6.0
60	N_2	2.8
	Air	8.1
170	N_2	4.4
	Air	15.0
180	N_2	9.8
	Air	29.8
190	N_2	17.0
	Air	48.9

*One mol of glucose

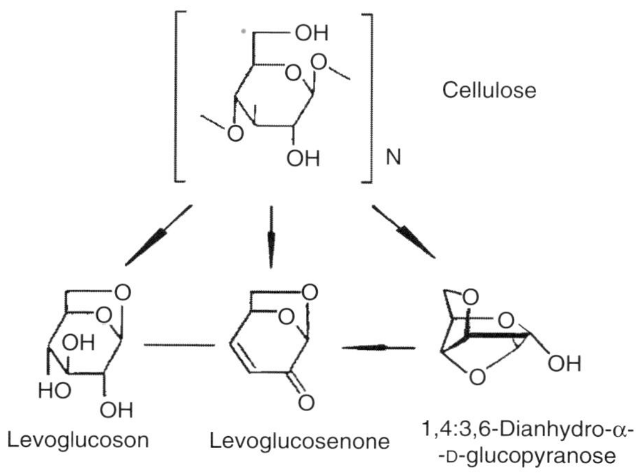

FIGURE 6.6 Formation of levogluosan and other monomers from cellulose.

temperature increases to 350°C, the tar product yield has increased to 38% and the char to 8%. At 500°C, the tar product yield has stayed about the same, but the char yield has decreased to 2%.

The tar products include anhydro sugar derivatives that can hydrolyze to reducing sugars. The evaporation of levoglucosan and other volatile pyrolysis products is highly endothermic. These reactions absorb heat from the system before the highly exothermic combustion reactions take place.

Table 6.4 shows the char yield of cellulose, wood, and lignin at various temperatures. The highest char yield for cellulose (63.3%) occurs at 325°C; by 400°C, the yield has decreased to 16.7%. The char yield for whole wood at 400°C is 24.9%, and is 73.3% for isolated lignin. The carbon, hydrogen, and oxygen analysis for the various char yields show that the highest carbon content char occurs at 500°C—but the char yield is only 8.7%. The table shows that the lignin component gives the highest char yield (Sharizadeh 1984). Lignin mainly contributes to char formation and cellulose; the hemicelluloses mainly create volatile pyrolysis products that are responsible for flaming combustion.

The intensity of combustion can be expressed as:

$$I_R = -\Delta H \frac{dw}{dt}$$

TABLE 6.3
Tar and Char Yields from Cellulose Pyrolysis at Different Temperatures

Temperature (°C)	Tar Yield (%)	Char Yield (%)
300	28	20
325	37	10
350	38	8
375	38	5
400	38	5
425	39	4
450	38	4
475	37	3
500	38	2

TABLE 6.4
Char Yield from Cellulose, Wood, and Lignin and Chemical Composition of the Char*

Material	Temperature (°C)	Char Yield (%)	Carbon Analysis (%)	Hydrogen Analysis (%)	Oxygen Analysis (%)
Cellulose	Control	—	42.8	6.5	50.7
	325	63.3	47.9	6.0	46.1
	350	33.1	61.3	4.8	33.9
	400	16.7	73.5	4.6	21.9
	450	10.5	78.8	4.3	16.9
	500	8.7	80.4	3.6	16.1
Wood	Control	—	46.4	6.4	47.2
	400	24.9	73.2	4.6	22.2
Lignin	Control	—	64.4	5.6	24.8
	400	73.3	72.7	5.0	22.3

* Isothermal pyrolysis, 5 minutes at temperature.

where:

I_R = reaction density,
$-\Delta H$ = heat of combustion
dw/dt = rate mass of fuel loss

Table 6.5 shows the heat of combustion for cellulose, whole wood, bark, and lignin. The highest level of volatiles are produced by cellulose, which has the lowest heat of combustion and the lowest percentage of char formation. The next level of volatiles are produced by whole wood, having slightly higher heat of combustion and char yield as compared to cellulose. Bark has a higher heat of combustion as compared to wood, with a char yield of 47% and 52% volatiles. Lignin has the highest heat of combustion and also the highest char yield and the lowest percent of volatiles.

The high rate of heat released upon flaming combustion provides the energy needed to gasify the remaining wood elements and propagate the fire. Oxidation of the residual char after flaming combustion results in glowing combustion. If the intensity of the heat and the concentration of combustible volatiles fall below the minimum level for flaming combustion, gradual oxidation of the reactive char initiates smoldering combustion. The smoldering combustion process releases noncombustible or unoxidized volatile products and usually occurs in low-density woods.

TABLE 6.5
Heat of Combustion of Wood, Cellulose, and Lignin, Char and Combustible Volatile Yields*

Fuel	Heat of Combustion ΔH 25 C (cal/g)	Char Yield (%)	Combustible Volatiles (%)
Cellulose	−4143	14.9	85.1
Wood (Poplar)	−4618	21.7	78.3
Bark (Douglas-fir)	−5708	47.1	52.9
Lignin (Douglas-fir)	−6371	59.0	41.0

*Heating rate 200°C/min to 400°C/min and held for 10 minutes.

FIGURE 6.7 Wood beam survives fire in casein plant in Frankfort, NY.

6.2 FIRE RETARDANCY

Wood has been used for many applications because it has poor thermal conductivity properties. In a fire, untreated wood forms a char layer—an insulation barrier protecting the wood below the burning layer (self-insulating). The comparative fire resistance of wood and metal was never more visibly demonstrated than in pictures taken after many hours of burning in the 1953 fire at a casein plant in Frankfort, New York (Figure 6.7). The steel girders softened, failed at high temperatures, and fell across the 12 × 16 inch laminated wood beams that were charred, but still strong enough to hold the steel girders.

To improve the fire resistance of wood, fire retardants have been developed. The use of fire retardants for wood can be documented back to the first century A.D., when the Romans treated their ships with alum and vinegar for protection against fire (LeVan 1984). Later, Gay-Lussac used ammonium phosphates and borax to treat cellulosic textiles. The U.S. Navy specified the use of fire retardants for their ships starting in 1895, and the City of New York required the use of fire retardants in building over twelve stories high starting in 1899 (Eickner 1966).

Building and fire codes specify fire-safety standards for structures. Building codes include area and height of the rooms, firestops, doors and other exits, automatic sprinklers, fire detectors, and type of construction. Fire codes include materials combustibility, flame spread, and fire endurance. In most residential construction, fire retardants may not be required but in public buildings, fire retardants are usually required.

6.3 TESTING FIRE RETARDANTS

6.3.1 THERMOGRAVIMETRIC ANALYSIS (TGA)

There are several ways to test the efficiency of a fire retardant. The most common is to run TGA analysis as described earlier. TGA involves weighing a finely ground sample and exposing it to a heated chamber in the presence of nitrogen. The sample is suspended on a sensitive balance that measures the weight loss of the sample as the system is heated (Figure 6.1). Nitrogen or another gas flows around the sample to remove the pyrolysis or combustion products. Weight loss is recorded as a function of time and temperature (see Figures 6.2 and 6.4). In isothermal TGA, the change in weight of the sample is recorded as a function of time at a constant temperature. With the use of a derivative computer, the rate of weight loss as a function of time and temperature can also be measured (Figure 6.3). This is referred to as *derivative thermogravimetry* (Slade and Jenkins 1966).

6.3.2 Differential Thermal Analysis (DTA) and Differential Scanning Calorimetry (DSC)

DTA measures the amount of heat liberated or absorbed by a wood sample as it moves from one physical transition state to another (such as melting or vaporization) or when it undergoes any chemical reaction. This heat is determined by measuring the temperature differences between the sample and an inert reference. DTA can be used to measure heat capacity, provide kinetic data, and give information on transition temperatures. The test device consists of sample and reference pans exposed to the same heat source. The temperature is measured using thermocouples embedded in the sample and the reference pan. The temperature difference between the sample and reference is recorded against time as the temperature is increased at a linear rate. For calorimetry, the equipment is calibrated against known standards at several temperatures (Slade and Jenkins 1966).

DSC is similar to DTA, except the actual differential heat flow is measured when the sample and reference temperature are equal. In DSC, both the sample and reference are heated by separate heaters. If a temperature difference develops between the sample and reference because of exothermic or endothermic reactions in the sample, the power input is adjusted to remove this difference. Thus, the temperature of the sample holder is always kept identical to that of the reference.

6.3.3 Tunnel Flame-Spread Tests

Building standards designed to control fire growth often require certain flame-spread ratings for various parts of a building. For code regulations, flame-spread ratings are determined by a 25-foot tunnel test, which is an approved standard test method (ASTM E 84). For research, 2- and 8-foot tunnel tests can also be done. All tunnel tests measure the surface flame spread of the wood, although each differs in the method of the exposure. A specimen is exposed to an ignition source, and the rate at which the flames travel to the end of the specimen is measured. In the past, red oak flooring was used as a standard and given a rating of 100.

The severity of the exposure and the time a specimen is exposed to the ignition source are the main differences between the various tunnel test methods. The 25-foot tunnel test, where the specimen is exposed for 10 minutes, is the most severe exposure. An extended test of 30 minutes is performed on fire-retardant treated products. Because the 25-foot tunnel test is the most severe exposure, it is used as the standard for building materials. The 2-foot tunnel test is the least severe, but because small specimens can be used, it is a valuable tool for development work on fire retardants. Table 6.6 shows average values for the flame spread index of several wood species (White and Dietenberger 1999).

6.3.4 Critical Oxygen Index Test

The oxygen index test measures the minimum concentration of oxygen in an oxygen-nitrogen mixture that will just support flaming combustion of a test specimen. Highly flammable materials have a low oxygen index, and less flammable materials have high values (White 1979). One advantage of this test is that very small specimens can be used. Another is that it can be used to study the retardant mechanism in the gas phase when TGA, DTA, or DSC cannot be used (because they only measure properties in the solid phase).

Table 6.7 shows the effect of inorganic additives on oxygen index and the yield of levoglucosan (Fung et al. 1972). Phosphoric acid is the most effective treatment of wood to increase the oxygen index and decrease the formation of levoglucosan. Ammonium dihydrogen orthophosphate, zinc chloride, and sodium borate are also very effective in reducing the yield of levoglucosan.

TABLE 6.6
Flame Spread Index for Different Woods Using the 25-Foot Tunnel Test

Species	Flame Spread Index
Douglas-fir	70–100
Western hemlock	60–75
Lodgepole pine	93
Western red cedar	70
Redwood	70
Sitka spruce	74–100
Yellow birch	105–110
Cottonwood	115
Maple	104
Red oak	100
Walnut	130–140
Yellow poplar	170–185

6.3.5 OTHER TESTS

Other tests that can be run on wood and fire-retardant–treated wood can determine smoke production, toxicity of smoke, and rate of heat release (ASTM 2002). The production of smoke can be a critical problem with some types of fire retardants. The 25-foot tunnel test uses a photoelectric cell to measure the density of smoke evolved. The effect of fire retardants on smoke production varies depending on the chemical used. Chemicals such as zinc chloride and ammonium phosphate generate much larger amounts of smoke as compared to borates. The toxicity of the smoke is also a critical consideration for fire-retardant–treated wood. A large percentage of fire victims are not touched by flames but are overcome from exposure to toxic smoke (Kaplan et al. 1982). The heat of combustion of wood varies depending on the species, resin content, moisture content, and other factors. The contribution to fire exposure depends on these factors along with the fire exposure and degree of combustion. Although the heat of combustion of wood does not change, fire retardants reduce the rate of heat release and extend the time at which the heat release begins to be measurable.

TABLE 6.7
Effect of Inorganic Additives on Oxygen Index and Levoglucosan Yield

Chemical	Oxygen Index (%)	Levoglucosan Yield (%)
Untreated	17.3	10.1
Potassium dihydrogen phosphate	18.5	0.9
Potassium hydrogen phosphate	18.6	0.2
Sodium borate	19.3	<0.1
Zinc chloride	19.6	0.3
Ammonium dihydrogen orthophosphate	19.6	0.8
Phosphoric acid	20.5	<0.1

Source: Fung 1972.

6.4 FIRE RETARDANTS

Fire-retardant treatments for wood can be classified into one of six classes: chemicals that promote the formation of increased char at a lower temperature than untreated wood degrades, chemicals which act as free-radical traps in the flame, chemicals used to form a coating on the wood surface, chemicals that increase the thermal conductivity of wood, chemicals that dilute the combustible gases coming from the wood with non-combustible gasses, and chemicals that reduce the heat content of the volatile gases.

In most cases, a given fire retardant operates by several of these mechanisms and much research has been done to determine the magnitude and role of each of these mechanisms in fire retardancy.

6.4.1 Chemicals that Promote the Formation of Increased Char at a Lower Temperature than Untreated Wood Degrades

Most of the evidence relating to the mechanism of fire retardancy in the burning of wood indicates that retardants alter fuel production by increasing the amount of char, reducing the amount of volatile, combustible vapors, and decreasing the temperature where pyrolysis begins. Figure 6.8 shows a TGA of untreated wood along with wood that has been treated with several inorganic fire retardants. Fire-retardant chemicals such as ammonium dihydrogen orthophosphate greatly increase the amount of residual char and lower the initial temperature of thermal decomposition. The amount of noncondensable gases increases at the expense of the flammable tar fraction. The chemical mechanism for the reduction of the combustible volatiles involves not only the ability of the fire retardant to inhibit the formation of levoglucosan (see the earlier discussion) and to catalyze dehydration of the cellulose to more char and fewer volatiles, but also its potential to enhance the condensation of the char to form cross-linked and thermally stable polycyclic aromatic structures (Shafizadeh 1984). Nanassy (1978) showed that Douglas-fir treated with either sodium chloride or ammonium dihydrogen orthophosphate increased both the char yield and the aromatic carbon content of the char (Table 6.8).

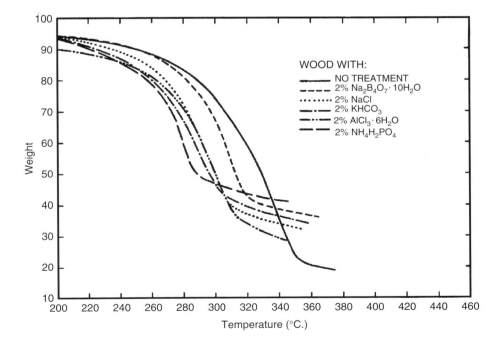

FIGURE 6.8 Thermogravimetric analysis of wood treated with various inorganic additives.

TABLE 6.8
Effect of Inorganic Additives on Char Yield and Aromatic Carbon Content of the Char

Chemical	Char Yield (wt%)	Aromatic Carbon in Char (wt%)
Untreated	15.3	13.7
Sodium chloride	17.5	16.8
Ammonium dihydrogen orthophosphate	28.9	18.6

Sources: Schafizadeh, 1984 and Nanassy, 1978.

Figure 6.9 shows the TGA of untreated cellulose along with cellulose that has been treated with several inorganic fire retardants. The thermal decomposition pattern for cellulose is similar to whole wood, except that pure cellulose is more thermally stable with the hemicelluloses removed.

All inorganic fire retardants reduce the amount of levoglucosan, regardless of the relative effectiveness of the fire retardant. This includes the effect of acidic, neutral, and basic additives on the levoglusosan yield. The acid treatment has the most pronounced effect on the reduction in the formation of levoglucosan.

During the heating of cellulose treated with either borax or ammonium dihydrogen orthophosphate, the degree of polymerization (DP) of the cellulose decreased (Fung et al. 1972). The DP decreased from 1110 to 650 after only two minutes of heating at 150°C with wood treated with ammonium dihydrogen orthophosphate. The DP dropped from 1300 to 700 after one hour of heating borax treated wood at 150°C. Both of these chemical treatments suppressed the formation of levoglucosan (see Table 6.7).

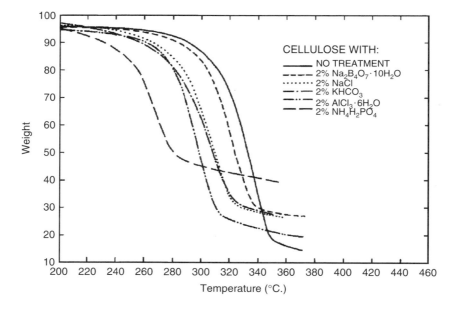

FIGURE 6.9 Thermogravimetric analysis of cellulose treated with various inorganic additives.

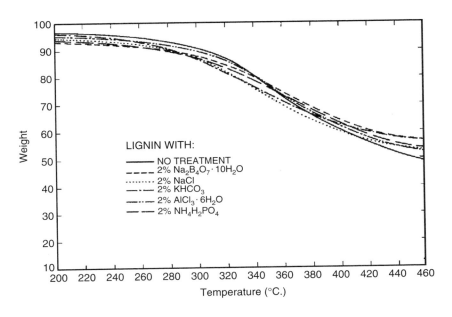

FIGURE 6.10 Thermogravimetric analysis of lignin treated with various inorganic additives.

Figure 6.10 shows the TGA of untreated lignin along with lignin that has been treated with several inorganic fire retardants. Treating lignin with the fire retardants has very little effect on the thermal decomposition of lignin.

6.4.2 Chemicals which Act as Free Radical Traps in the Flame

Certain fire retardants affect vapor-phase reactions by inhibiting the chain reactions as shown below. Halogens such as bromine and chlorine are good free radical inhibitors and have been studied extensively in the plastics industry. Generally, high concentrations of halogen are required (15–30% by weight) to attain a practical degree of fire retardancy. The efficiency of the halogen decreases in the order Br > Cl > F. A mechanism for the inhibition of the chain branching reactions using HBr as the halogen is:

$$H\cdot + HBr \rightarrow H_2 + Br\cdot$$

$$OH\cdot + HBr \rightarrow H_2O + Br\cdot$$

The hydrogen halide consumed in these reactions is regenerated to continue the inhibition.

An alternate mechanism has been suggested for halogen inhibition that involves recombination of oxygen atoms (Creitz 1970):

$$O\cdot + Br_2 \rightarrow BrO\cdot + Br\cdot$$

$$O\cdot + OBr\cdot \rightarrow Br\cdot + O_2$$

Thus, the inhibitive effect results from the removal of active oxygen atoms from the vapor phase. Additional inhibition can result from removal of OH radicals in the chain-branching reactions:

$$BrO\cdot + \cdot OH \rightarrow HBr + O_2$$

$$BrO\cdot + \cdot OH \rightarrow Br\cdot + HO_2$$

Some phosphorus compounds have also been found to inhibit flaming combustion by this mechanism.

6.4.3 CHEMICALS USED TO FORM A COATING ON THE WOOD SURFACE

A physical barrier can retard both smoldering and flaming combustion by preventing the flammable products from escaping and preventing oxygen from reaching the substrate. These barriers also insulate the combustible substrate from high temperatures. Common barriers include sodium silicates and coatings that intumesce (release a gas at a certain temperature that is trapped in the polymer coating the surface). *Intumescent systems* swell and char on exposure to fire to form a carbonaceous foam. They consist of several components, including a char-producing compound, a blowing agent, a Lewis-acid dehydrating agent, and other chemical components.

In the intumescent systems, the char-producing compound, such as a polyol, will normally burn to produce CO_2 and water vapor and leave flammable tars as residues. However, the compound can esterify when it reacts with certain inorganic acids, usually phosphoric acid. The acid acts as a dehydrating agent and leads to increased yield of char and reduced volatiles. Such char is produced at a lower temperature than the charring temperature of the wood. Blowing agents decompose at determined temperatures and release gases that expand the char. Common blowing agents include dicyandiamide, melamine, urea, and guanidine; they are selected on the basis of their decomposition temperatures. Many blowing agents also act as a dehydrating agent. Other chemicals can be added to the formulation to increase the toughness of the surface foam.

6.4.4 CHEMICALS THAT INCREASE THE THERMAL CONDUCTIVITY OF WOOD

A metal alloy with a melting point of 105°C can be used to treat wood. Upon heating, the temperature rise in the metal-alloy–treated wood is slower than non-treated wood until the melt temperature is reached (Browne 1958). Above the melt temperature of the metal alloy, the rise in temperature is the same for treated and non-treated wood.

Another thermal theory suggests that fire retardants cause chemical and physical changes so that heat is absorbed by the chemical to prevent the wood surface from igniting. This theory is based on chemicals that contain a lot of water of crystallization. Water will absorb latent heat of vaporization from the pyrolysis reactions until all of the water is vaporized. This serves to remove heat from the pyrolysis zone, thereby slowing down the pyrolysis reactions. This is why wet wood burns more slowly than dry wood. Once the water is removed, the wood undergoes pyrolysis independent of the past moisture content of the wood.

6.4.5 CHEMICALS THAT DILUTE THE COMBUSTIBLE GASES COMING FROM THE WOOD WITH NON-COMBUSTIBLE GASES

Chemicals such as dicyandiamide and urea release large amounts of non-combustible gases at temperatures below the temperature at which the major pyrolysis chemistries start. Chemicals such as borax release large amounts of water vapor. Any reduction in the percentage of flammable gases would be beneficial because it increases the volume of combustible volatiles needed for ignition. Also, the movement of gases away from the wood may dilute the amount of oxygen near the boundary layer between the wood and the vapor-phase reaction.

6.4.6 CHEMICALS THAT REDUCE THE HEAT CONTENT OF THE VOLATILE GASES

As previously seen in Section 4.1, the addition of inorganic additives lowers the temperature at which active pyrolysis begins, and this resulting decomposition leads to increased amounts of char and reduced amounts of volatiles. This is due to the increased dehydration reactions, mainly in the cellulose component of wood. However, other competing reactions also occur, such as decarbonylation, decomposition of simpler compounds, and condensation reactions. All of these reactions compete with each other. As a result, shifts favoring one reaction over another also change the overall heat of reaction. Differential thermal analysis is used to determine these changes in heats of reactions and can help gain an understanding about these competing reactions.

DTA of wood in helium shows two endothermic reactions followed by a smaller exothermic one. The first endothermic reaction, which peaks around 125°C, is a result of evaporation of water and desorption of gases; the second, peaking between 200–325°C, indicates depolymerization and volatilization. At around 375°C these endothermic reactions are replaced with a small exothermic peak. When the wood sample is run in oxygen, these endothermic peaks are replaced by strong exothermic reactions. The first exotherm, around 310°C for wood and 335°C for cellulose, is attributed to the flaming of volatile products; the second exotherm, at 440°C for wood and 445°C for both cellulose and lignin, is attributed to glowing combustion of the residual char.

DTA of inorganic fire-retardant–treated wood in oxygen shifts the peak position temperatures and/or the amount of heat released. Sodium tetraborate, for example, reduces the volatile products exotherm considerably, increases the glowing exotherm, and shows a second glowing peak around 510°C. Sodium chloride also reduces the first exotherm and increases the size of the second, but does not produce a second glowing exotherm as did the sodium tetraborate. Wood treated with ammonium phosphate is the most effective in both reducing the amount of volatile products and also reducing the temperature where these products are formed. Ammonium phosphate almost eliminates the glowing exotherm.

Fire retardant treatments of this type reduce the average heat of combustion for the volatile pyrolysis products released at the early stage of pyrolysis below the value associated with untreated wood at comparable stages of volatilization. At 40% volatilization, untreated wood has a 29% release of volatile products' heat of combustion; treated wood has only released 10–19% of this total heat. Of all the chemicals tested, only sodium chloride, which is known to be an ineffective fire retardant, does not reduce the heat content.

6.4.7 Phosphorus-Nitrogen Synergism Theories

One role phosphoric acid and phosphate compounds play in the fire retardancy of wood is to catalyze the dehydration reaction to produce more char. This reaction pathway is just one of several that are taking place all at once, including decarboxylation, condensation, and decomposition. The effectiveness of fire retardants containing both phosphorus and nitrogen is greater than the effectiveness of each of them by themselves.

The interaction of phosphorus and nitrogen compounds produces a more effective catalyst for the dehydration because the combination leads to further increases in the char formation and greater phosphorus retention in the char (Hendrix and Drake 1972). This may be the result of increased cross-linking of the cellulose during pyrolysis through ester formation with the dehydrating agents. The presence of amino groups results in retention of the phosphorus as a nonvolatile amino salt, in contrast to some phosphorus compounds that may decompose thermally and be released into the volatile phase. It is also possible that the nitrogen compounds promote polycondensation of phosphoric acid to polyphosphoric acid. Polyphosphoric acid may also serve as a thermal and oxygen barrier because it forms a viscous fluid coating.

6.4.8 Fire-Retardant Formulations

Many chemicals have been evaluated for their effectiveness as fire retardants. The major fire retardants used today include chemicals containing phosphorus, nitrogen, boron, and a few others. Most fire-retardant formulations are water leachable and corrosive, so research continues to find more leach-resistant and less corrosive formulations.

TABLE 6.9
Effects of Inorganic Additives on Thermogravimetric Analysis

Additive	Percent Weight Loss at 500°C
Phosphoric acid	61
Ammonium dihydrogen orthophosphate	66
Zinc chloride	74
Sodium hydroxyde	79
Boric acid	81
Sodium chloride	82
Tin chloride	84
Diammonium sulfate	86
Sodium tetraborate decahydrate	89
Sodium phosphate	91
Ammonium chloride	93
Untreated wood	93

6.4.8.1 Phosphorus

Chemicals containing phosphorus are one of the oldest classes of fire retardants. Monoammonium and diammonium phosphates are used with nitrogen compounds, since the synergistic effect allows for less chemical to be used (Hendrix and Drake 1972, Langley et al. 1980, Kaur et al. 1986). Organophosphorus and polyphosphate compounds are also used as fire retardants. Ammonium polyphosphate at loading levels of 96 kg/m^3 gives a flame-spread index of 15 according to ASTM E84 (Holmes 1977). This treatment generates a low smoke yield but it is corrosive to aluminum and mild steel. Other formulations containing phosphorus are mixture of guanyl urea phosphate and boric acid, and phosphoric acid, boric acid, and ammonia. Table 6.9 shows the effectiveness of some of the fire retardants in terms of the percent of weight loss at 500°C. The most effective chemical is phosphoric acid, with a weight loss of 61% as compared to 93% weight loss for untreated wood.

6.4.8.2 Boron

Borax (sodium tetraborate decahydrate) and boric acid are the most often used fire retardants. The borates have low melting points and form glassy films on exposure to high temperatures. Borax inhibits surface flame spread but also promotes smoldering and glowing. Boric acid reduces smoldering and glowing combustion but has little effect on flame spread. Because of this, borax and boric acid are usually used together. The alkaline borates also result in less strength loss in the treated wood and is less corrosive and hydroscopic (Middleton et al. 1965). Boron compounds are also combined with other chemicals such as phosphorus and amine compounds to increase their effectiveness. Table 6.9 shows that wood treated with boric acid shows a weight loss of 81% and borax an 89% weight loss at 500°C, which is not as effective as phosphorus compounds.

6.4.9 LEACH RESISTANT FIRE-RETARDANTS

A fire-retardant treatment that is resistant to water leaching is a requirement in some building codes today. Fires have spread from home to home due to wood shake roofs, and some states now require wood-based roofing materials to be treated with a leach-resistant fire retardant.

The most widely studied leach-resistant fire retardant system is based on amino-resins (Goldstein and Dreher 1964). Basically, the resin system consists of a combination of a nitrogen source (urea, melamine, guanidine, or dicyandiamide) with formaldehyde to produce a methylolated amine.

The product is then reacted with a phosphorus compound such as phosphoric acid. Other formulations include mixtures of dicyandiamide, melamine, formaldehyde, and phosphoric acid, or dicyandiamide, urea, formaldehyde, phosphoric acid, formic acid, and sodium hydroxide. Leach resistance is attributed to polymerization of the components within the wood (Goldstein and Dreher 1961). Another formulation uses a urea and melamine amino-resin (Juneja and Fung 1974). The stability of these resins is controlled by the rate of methylolation of the urea, melamine and dicyandiamide. The optimum mole ratio for stability of these solutions is 1:3:12:4 for urea ormelamine, dicyandiamide, formaldehyde, and orthophosphoric acid. Lee et al. (2004) bonded phosphoramides to wood by reacting phosphorus pentoxide with amines *in situ*. Leach resistance was greatly improved and the mechanism of effectiveness was said to be due to an increase in the dehydration mechanism.

Wood has been reacted with fire-retardant chemicals such as phosphorus pentoxideamine complexes (Lee et al. 2004) or glucose diammonium phosphate (Chen 2002) that results in treatments that are leach resistant (Rowell et al. 1984) (see Chapter 14, Section 5.10).

REFERENCES

American Society of Testing and Materials (2002). Surface burning characteristics of building materials. E 84-1979a. West Conshohocken, PA.

Antal, M.J. Jr. (1985). Biomass pyrolysis: A review of the literature. Part 2: Lignocellulose pyrolysis. *Adv. Solar Energy* 2:175–256.

Bridgewater, A.V. (1999). Principles and practice of biomass fast pyrolysis processes for liquids. *J. Anal. Appl. Pyrolysis* 51:3–22.

Browne, F.L. (1958). Theories of the combustion of wood and its control. Forest Service Report No. 2136. Madison, WI: Forest Products Laboratory.

Chen, G.C. (2002). Treatment of wood with glucose-diammonium phosphate for fire and decay protection. *Proceedings of the 6th Pacific Rim Bio-Based Composite Symposium.* Volume 2. 616–622.

Creitz, E.C. (1970). Literature survey of the chemistry of flame inhibition. *J. Res. Natl. Bur. Stand.* Section A, 74(4):521–530.

Czernik, S., Maggi, R., and Peacocke, G.V.C. (1999). A review of physical and chemical methods of upgrading biomass-derived fast pyrolysis liquids. In: Overend, R.P. and Cornet, E. (Eds.) *Biomass: Proceedings of the 4th Biomass Conference of the Americas,* Volume 2. New York: Pergamon. pp. 1235–1240.

Eickner, H.W. (1966). Fire-retardant treated wood. *J. Mater.* 1(3):625–644, 1966.

Fung, D.P.C., Tsuchiya, Y., and Sumi, K. (1972). Thermal degradation of cellulose and levoglucosan. Effect of inorganic salts. *Wood Sci.* 5(1):38–43.

Goldstein, I.S. and Dreher, W.A. (1961). A non-hygroscopic fire retardant treatment for wood. *Forest Prod. J.* 11(5):235–237.

Goldstein, I.S., and Dreher, W.A. (1964). Method of imparting fire retardance to wood and the resulting product. U.S. Patent 3,159,503.

Hendrix, J.S. and Drake, G.L. Jr. (1972). Pyrolysis and combustion of cellulose. III. mechanistic basis for synergism involving organic phosphates and nitrogenous bases. *J. Appl. Polymer Sci.* 16:257–274.

Holmes, C.A. (1977). *Wood Technology: Chemical Aspects.* Goldstein, I.S. (Ed.), Am. Chemical Soc. Symposium Series 43, Washington, DC, 82–106.

Kaplan, H.L., Grand, A.F., and Hartzell, G.E. (1982). A critical review of the state-of-the-art of combustion toxicology. San Antonio, TX: Southwest Research Institute.

Kaur, B, Gur, I.S., and Bhatnagar, H.L. (1986). Studies on thermal degradation of cellulose and cellulose phosphoramides. *J. Appl. Polymer Sci.* 31:667–683.

Kawamoto, H., Murayama, M., and Saka, S. (2003). Pyrolysis behavior of levoglucosan as an intermediate in cellulose pyrolysis: Polymerization into polysaccharide as a key reaction to carbonized product formation. *J. Japanese Wood Soc.* 49:469–473.

Juneja, S.C. and Fung, D.P.C. (1974). Stability of amino resin fire retardants. *Wood Sci.* 7(2):160–163.

Langley, J.T., Drews, M.J., and Barkeer, R.H. (1980). Pyrolysis and combustion of cellulose. VII. Thermal analysis of the phosphorylation of cellulose and model carbohydrates during pyrolysis in the presence of aromatic phosphates and phosphoramides. *J. Appl. Polymer Sci.* 25:243–262, 1980.

Lee, H.L., Chen, G.C., and Rowell, R.M. (2004). Thermal properties of wood reacted with a phosphorus pentoxide-amine system. *J. Appl. Polymer Sci.* 91(4):2465–2481.

LeVan, S.L. (1984). Chemistry of fire retardancy. In: Rowell, R.M. (Ed.), *The Chemistry of Solid Wood*. Advances in Chemistry Series, Number 207. Washington, DC: American Chemical Society. Chapter 14: pp. 531–574.

Middleton, J.C., Draganov, S.M., and Winters, F.T. Jr. (1965). Evaluation of borates and other inorganic salts as fire retardants for wood products. *For. Prod. J.* 15(12):463–467.

Nanassy, A.J. (1978). Treatment of Douglas-fir with fire retardant chemicals. *Wood Sci.* 11(2):111–117.

Rowell, R.M., Susott, R.A., DeGroot, W.F., and Shafizadeh, F. (1984). Bonding fire retardants to wood. Part 1. Thermal behavior of chemical bonding agents. *Wood and Fiber Sci.* 16(2):214–223.

Shafizadeh, F. (1982). Introduction to pyrolysis of biomass. *J. Anal. Appl. Pyrolysis* 3:283–305.

Shafizadeh, F. (1984). The chemistry of pyrolysis and combustion. In: Rowell, R.M., (Ed.), *The Chemistry of Solid Wood*. Advances in Chemistry Series, Number 207. Washington, DC: American Chemical Society. Chapter 13: 489–529.

Shafizadeh, F. and Fu, Y.L. (1973). Pyrolysis of cellulose. *Carbohydrate Res.* 29:113–122.

Slade, P.E. Jr. and Jenkins, L.T. (1966). *Techniques and Methods of Polymer Evaluation*. New York: Marcel Dekker, Inc.

White, R.H. (1979). Oxygen index evaluation of fire-retardant-treated wood. *Wood Sci.* 12(2):113–121.

White, R.H. and Dietenberger, M.A. (1999). In: *Wood Handbook—Wood as an Engineering Material*. Fire Safety. Chapter 17, Gen. Tech. Rep. FPL-GTE-113. Madison, WI: Department of Agriculture, Forest Service, Forest Products Laboratory. 17.1–16.

7 Weathering of Wood

R. Sam Williams
USDA, Forest Service, Forest Products Laboratory, Madison, WI

CONTENTS

7.1 Background 142
    7.1.1 Macroscopic Properties 142
        7.1.1.1 Specific Gravity 142
        7.1.1.2 Earlywood and Latewood 144
        7.1.1.3 Texture 145
        7.1.1.4 Juvenile Wood 145
        7.1.1.5 Compression Wood 146
        7.1.1.6 Heartwood and Sapwood 146
    7.1.2 Anatomical Structure of Wood 146
    7.1.3 Chemical Nature of Polysaccharides, Lignin, and Extractives 149
    7.1.4 UV Spectrum 149
    7.1.5 Wavelength Interactions with Various Chemical Moieties 151
7.2 Chemical Changes 152
    7.2.1 Free Radical Formation 154
    7.2.2 Hydroperoxides 157
7.3 Reaction Products and Chemical Analysis 157
    7.3.1 Depth of Degradation 162
    7.3.2 Acid Effects 163
7.4 Physical Aspects of Degradation 164
    7.4.1 Microscopic Effects 164
        7.4.1.1 Destruction of Middle Lamella 164
        7.4.1.2 Destruction of Bordered Pits and Cell Wall Checking 165
    7.4.2 Macroscopic Effects 165
        7.4.2.1 Loss of Fiber 165
        7.4.2.2 Grain Orientation 170
        7.4.2.3 Water Repellency 170
        7.4.2.4 Checks and Raised Grain 170
    7.4.3 Weathering of Wood/Wood Composites 171
    7.4.4 Weathering of Wood/Plastic Composites 172
    7.4.5 Effects of Biological Agents 173
    7.4.6 UV Degradation of Tropical Woods 173
    7.4.7 Paint Adhesion 173
7.5 Chemical Treatments to Retard Weathering 174
    7.5.1 Chromic Acid 174
    7.5.2 Chromated Copper Arsenate Preservatives 175
    7.5.3 Copper-Based Preservatives 175
    7.5.4 Chemically Bonded Stabilizers 175
    7.5.5 Commercial Stabilizers 175

7.5.6 Chemical Modification 176
7.5.7 Water-Repellent Preservatives 177
7.5.8 Paints and Stains..177
7.6 Summary and Future Considerations..178
References 178

Weathering is the general term used to define the slow degradation of materials exposed to the weather. The degradation mechanism depends on the type of material, but the cause is a combination of factors found in nature: moisture, sunlight, heat/cold, chemicals, abrasion by windblown materials, and biological agents. Tall mountains weather by the complex and relentless action of these factors. All natural and man-made materials weather; for polymeric materials, the weathering rate is considerably faster than the degradation of mountains. Many of the materials we depend on for clothing and shelter undergo degradation by the weathering process.

Wood is a material that has been used for countless centuries to provide people with shelter. Today we still depend on wood and wood-based products to provide this shelter. Our houses are usually made of wood, and the outermost barrier to the weather is often wood or a wood-based product (siding, windows, decks, roofs, etc.). If these wood products are to achieve a long service life, we must understand the weathering process and develop wood treatments to retard this degradation. Failure to recognize the effects of weathering can lead to catastrophic failure of wood products and other products used with wood. For example, if wood siding is left to weather for as little as one to two weeks before it is painted, the surface of the wood will degrade. During this short exposure period, the surface of the wood will not appear to have changed very much, but damage has occurred. Application of paint after one to two weeks of weathering will not give a durable coating. The surface of the wood has been degraded and it is not possible to form a good paint bond with the degraded surface. The paint will show signs of cracking and peeling within a few years. As the paint peels from the surface, the wood grain pattern can easily be seen on the back side of the paint. The peeling paint has lifted the damaged layer of wood from the sound wood underneath. The reasons for this will become apparent as we discuss the chemistry and degradation processes of wood weathering.

We see many examples of weathering. The rough, gray appearance of old barns, wood shake roofs, and drift wood are typical examples of weathered wood. In the absence of biological attacks, the weathering of wood can give a beautiful bright gray patina.

How does weathering differ from decay? Weathering is surface degradation of wood that is initiated primarily by solar radiation, but other factors are also important. The wetting and drying of wood through precipitation, diurnal and seasonal changes in relative humidity (RH), abrasion by windblown particulates, temperature changes, atmospheric pollution, oxygen, and human activities such as walking on decks, cleaning surfaces with cleaners and brighteners, sanding, and power-washing all contribute to the degradation of wood surfaces. However it is primarily the ultraviolet (UV) portion of the solar spectrum that initiates the process we refer to as weathering. It is a photo-oxidation or photochemical degradation of the surface. The weathering process affects only the surface of the wood. The degradation starts immediately after the wood is exposed to sunlight. First the color changes, then the surface fibers loosen and erode, but the process is rather slow. It can take more than 100 years of weathering to decrease the thickness of a board by 5–6 mm. In addition to the slow erosion process, other processes also occur. The wood may develop checks and a raised grain. Mildew will colonize the surface and discolor the wood. If boards contain compression or juvenile wood, cross-grain cracking may develop. The boards may warp and cup, particularly in decking applications. These other weathering factors such as mildew growth, checking, splitting, and warping, are often more important than the photo-oxidation, but these processes often act in concert to degrade the surface (see Figure 7.1). Note that the figure depicts 100 years

Weathering of Wood

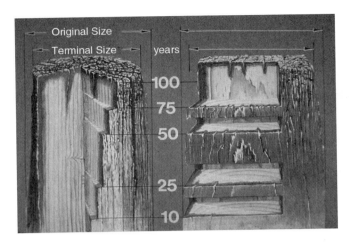

FIGURE 7.1 Simulation of 100 years of weathering of posts showing checking and erosion of the surface.

of weathering, and in addition to the slow loss of wood materials as depicted by the decreased size near the top, the post has also developed severe checking.

On the other hand, wood decay is a process that affects the whole thickness or bulk of the wood. It is caused by decay fungi that infect the wood. *Decay fungi* are plants that grow through the wood cells and release enzymes that break down the wood components that they then metabolize for food. Whereas weathering can take many decades to remove a few millimeters of wood from the surface, decay fungi can completely destroy wood in just a few years if the conditions are favorable for their growth. The critical factor for deterring their growth is to limit the water available to them. Wood cannot decay unless there is free water available in the wood cells. Free water is not necessary for weathering to occur, however the presence of water can help accelerate the process by causing splitting and checking of the wood (see Figure 7.2).

How does weathering differ from light-induced color change? Weathering is caused by the UV radiation portion of sunlight. The UV radiation has sufficient energy to chemically degrade wood structural components (lignin and carbohydrates). The visible portion of sunlight also causes surface changes in wood, but except for minimal damage at the short wavelengths of visible light, the

FIGURE 7.2 Cross-section of a board showing checking caused by weathering.

changes do not involve degradation of the wood structure. The chemicals that give wood its color are extractives: organic compounds of various types that may contain halogens, sulfur, and nitrogen. Chemical moieties containing these elements can undergo photo degradation reactions at lower energies than lignin and carbohydrates. So in addition to the UV radiation, visible light has sufficient energy to degrade extractives. They fade much the same as the dyes in textiles. Wood exposed outdoors will undergo a rather rapid color change in addition to the UV-induced degradation of lignin. In addition, rain will leach the water-soluble chemicals from the wood surface. Very little UV radiation can penetrate common window glass; therefore, wood does not undergo UV-catalyzed weathering indoors. The color change that occurs to wood when it is exposed indoors is caused by visible light. The visible light causes the organic dyes in the wood to fade. The color change indoors is not caused by UV light. The use of a UV stabilizer in interior finishes has little effect for achieving color stability. A few recent publications on color change considered pertinent to the mechanism of weathering are included in the chapter.

What is the risk to wood materials if weathering is not understood? Various wood-based products weather in different ways. Some wood products can be allowed to weather naturally to achieve a driftwood gray patina. For example Eastern White Cedar (*Thuja occidentalis*) and Western Red Cedar (*Thuja plicata*) shakes are often left unfinished to weather naturally. However, other wood products, such as plywood, can fail catastrophically within several years if they are not protected from weathering. By understanding the mechanism of weathering—the chemical changes, the effects of degradation on the physical properties, and methods for retarding or inhibiting degradation—it is possible to maximize the service life of all types of wood products in any type of climate.

The purpose of this chapter is to describe the chemical and physical changes that occur to wood during weathering and explore methods for preventing this degradation. Current literature back to 1980 has been reviewed and is included in this chapter. Literature prior to 1980 was reviewed in detail by Feist and Hon (1984) and by Feist (1990).

7.1 BACKGROUND

Understanding the chemistry of UV degradation of wood requires knowledge of its macroscopic properties, anatomical structure, chemical nature of polysaccharides, lignin, and extractives, the UV spectrum, and the interactions of UV radiation with various chemical moieties in wood.

7.1.1 MACROSCOPIC PROPERTIES

Wood is a natural biological material and as such, its properties vary not only from one species to another but also within the same species. Some differences can even be expected in boards cut from the same tree. Within a species, factors that affect the natural properties of wood are usually related to growth rate. Growth rate in turn is determined by climatic factors, geographic origin, genetics, tree vigor, and competition—factors over which we currently have little control. In addition to the natural properties, manufacturing influences the surface properties of wood: the grain angle, surface roughness, and amount of earlywood/latewood and heartwood/sapwood.

7.1.1.1 Specific Gravity

The properties of wood that vary greatly from species to species are specific gravity (density), grain characteristics (presence of earlywood and latewood), texture (hardwood or softwood), presence of compression wood, presence and amount of heartwood or sapwood, and the presence of extractives, resins, and oils. The specific gravity of wood is one of the most important factors that affect weathering characteristics. Specific gravity varies tremendously from species to species (see Table 7.1) and it is important because "heavy" woods shrink and swell more than do "light" woods. This dimensional change in lumber and, to a lesser extent, in reconstituted wood products and

TABLE 7.1
Characteristics of Selected Woods for Painting

Wood Species	Specific Gravity[a] Green/Dry	Shrinkage (%)[b] Flat Grain	Shrinkage (%)[b] Vertical Grain
Softwoods			
Bald cypress	0.42/0.46	6.2	3.8
Cedars			
Incense	0.35/0.37	5.2	3.3
Northern white	0.29/0.31	4.9	2.2
Port-Orford	0.39/0.43	6.9	4.6
Western red	0.31/0.32	5	2.4
Yellow	0.42/0.44	6	2.8
Douglas-fir[c]	0.45/0.48[d]	7.6	4.8
Larch, western	0.48/0.52	9.1	4.5
Pine			
Eastern white	0.34/0.35	6.1	2.1
Ponderosa	0.38/0.42	6.2	3.9
Southern	0.47/0.51[e]	8	5
Sugar	0.34/0.36	5.6	2.9
Western white	0.36/0.38	7.4	4.1
Redwood, old growth	0.38/0.40	4.4	2.6
Spruce, Engelmann	0.33/0.35	7.1	3.8
Tamarack	0.49/0.53	7.4	3.7
White fir	0.37/0.39	7.0	3.3
Western hemlock	0.42/0.45	7.8	4.2
Hardwoods			
Alder	0.37/0.41	7.3	4.4
Ash, white	0.55/0.60	8	5
Aspen, bigtooth	0.36/0.39	7	3.5
Basswood	0.32/0.37	9.3	6.6
Beech	0.56/0.64	11.9	5.5
Birch, yellow	0.55/0.62	9.5	7.3
Butternut	0.36/0.38	6.4	3.4
Cherry	0.47/0.50	7.1	3.7
Chestnut	0.40/0.43	6.7	3.4
Cottonwood, eastern	0.37/0.40	9.2	3.9
Elm, American	0.46/0.50	9.5	4.2
Hickory, shagbark	0.64/0.72	11	7
Magnolia, southern	0.46/0.50	6.6	5.4
Maple, sugar	0.56/0.63	9.9	4.8
Oak			
White	0.60/0.68	8.8	4.4
Northern red	0.56/0.63	8.6	4.0
Sweetgum	0.46/0.52	10.2	5.3
Sycamore	0.46/0.49	8.4	5
Walnut	0.51/0.55	7.8	5.5
Yellow-poplar	0.40/0.42	8.2	4.6

[a] Specific gravity based on weight ovendry and volume at green or 12% moisture content.
[b] Value obtained by drying from green to ovendry.
[c] Lumber and plywood.
[d] Coastal Douglas-fir.
[e] Loblolly, shortleaf, specific gravity of 0.54/0.59 for longleaf and slash.

plywood occurs as wood, particularly in exterior applications, gains or loses moisture with changes in the relative humidity and from periodic wetting caused by rain and dew. Excessive dimensional change in wood stresses the surface to cause checking.

The amount of warping and checking that occurs as wood changes dimensions and during the natural weathering process is also directly related to wood density. Cupping is probably the most common form of warp. *Cupping* is the distortion of a board that causes a deviation from flatness across the width of the piece. Wide boards cup more than do narrow boards. Boards may also twist or warp from one end to the other, deviating from a straight line along the length of the piece. Warping is generally caused by uneven shrinking or swelling within the board. Furthermore, checks, or small ruptures along the grain of the piece may develop from stress set up during the drying process or from stresses caused by the alternate shrinking and swelling that occurs during service. High-density (heavy) woods such as southern yellow pine tend to warp and check more than do the low-density (light) woods such as redwood.

7.1.1.2 Earlywood and Latewood

The presence and amount of latewood in softwood (conifer) lumber is closely related to wood density (see Figures 7.3a and 7.3b). Each year, tree species growing in temperate climates add one

(a)

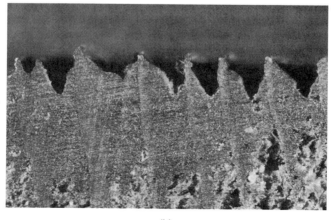

(b)

FIGURE 7.3 Micrographs of cross-sections of weathered boards showing earlywood and latewood bands (3X): a) southern pine; b) western redcedar.

(a) (b)

FIGURE 7.4 Surface texture of diffuse porous sugar maple (a) and ring porous red oak (b): top, end grain; middle, radial grain; bottom, tangential grain.

growth increment or ring to their diameter. For most species, this ring shows two distinct periods of growth and therefore two bands, called earlywood (springwood) and latewood (summerwood). Latewood is denser, harder, smoother, and darker than earlywood, and its cells have thicker walls and smaller cavities. The wider the latewood band, the denser the wood. Wide latewood bands are normally absent from edge-grained cedar and redwood, however, they are prominent in southern yellow pine and Douglas-fir, two of the most common species used for general construction and for the production of plywood.

7.1.1.3 Texture

Texture refers to the general coarseness of the individual wood cells and is often used in reference to hardwoods (see Figures 7.4a and 7.4b). Figure 7.4a shows fine-textured diffuse-porous sugar maple (*Acer saccharum*) and Figure 7.4b shows coarse-grained ring-porous red oak (*Quercus rubra*). Hardwoods are primarily composed of relatively short, small-diameter cells (fibers) and large-diameter pores (vessels); softwoods, in contrast, are composed of longer small-diameter cells (tracheids). The size and arrangement of the pores may outweigh the effect of density and grain pattern on weathering. Hardwoods with large pores, such as oak and ash, may erode more quickly at the pores than the surrounding fibers.

7.1.1.4 Juvenile Wood

Juvenile wood forms during the first eight to ten years of a tree's growth. It differs form normal wood in that the fibrils in the wood cell wall are oriented more across the width of the cell rather than along its length. This change in orientation gives juvenile wood a rather large longitudinal dimensional change as it changes moisture content. Whereas normal wood shrinks only about 0.1–0.2% as it changes moisture content from green to oven dry, juvenile wood and compression

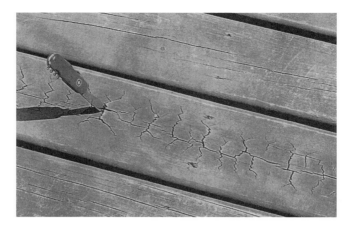

FIGURE 7.5 Cross-grain checking in juvenile wood.

wood can shrink as much as 2% (*Wood Handbook* 1999). This can cause serious warping and cross-grain cracking as wood weathers (see Figure 7.5).

7.1.1.5 Compression Wood

Compression wood is formed on the lower side of leaning softwood trees. The part of the growth ring with compression wood is usually wider than the rest of the ring and has a high proportion of latewood. As a result, the tree develops an eccentrically-shaped stem and the pith is not centered. Compression wood, especially the latewood, is usually duller and more lifeless in appearance than the rest of the wood. Compression wood presents serious problems in wood manufacturing because it is much lower in strength than is normal wood of the same density. Also, it tends to shrink excessively in the longitudinal direction, which causes cross-grain checking during weathering.

7.1.1.6 Heartwood and Sapwood

As trees mature, most species naturally develop a darker central column of wood called heartwood. The darker color is caused by the deposition of colored extractives. To the outside of the heartwood is a lighter cylinder of wood called sapwood. The sapwood is composed of live cells that serve to transport water and nutrients from the roots to the leaves and to provide mechanical support for the tree. The heartwood, on the other hand, serves only as support. Heartwood is formed as the individual cells die and are impregnated with extractives, pitch, oil, and other extraneous materials. Older trees have a higher percentage of heartwood as compared to younger trees. Some species such as southern yellow pine have a much wider sapwood zone than do species like the cedars and redwood. During the early stage of weathering, it is the heartwood that quickly loses its color because of leaching of the extractives. Other than the presence of extractives, the anatomy (and therefore the components) of cellulose, hemicellulose, and lignin are the same for heartwood and sapwood.

7.1.2 ANATOMICAL STRUCTURE OF WOOD

Typical anatomical structure for softwoods and hardwoods show distinct differences. Hardwoods contain specialized cells (pores or vessels) for liquid transport and these can occur more or less evenly spaced (diffuse porous; see Figure 7.6a) or bunched near the early wood in some temperate hardwoods (ring porous; see Figure 7.6b) as shown for Yellow Poplar (*Liriodendron tulipifera* L.) and White Oak (*Quercus alba* L.). Water transport in softwoods takes place via the tracheids and flows from one tracheid to another through bordered pits (holes in the side of the tracheid).

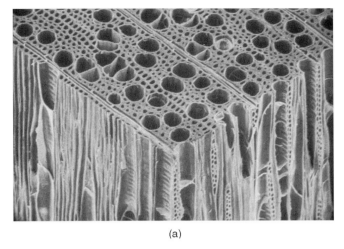

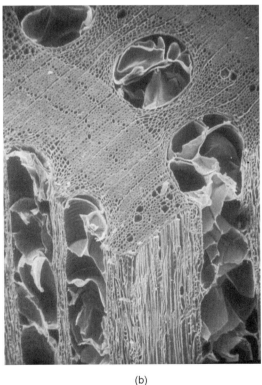

FIGURE 7.6 Micrographs of hardwoods: a) diffuse-porous; b) ring-porous.

As discussed earlier in Chapter 2, wood contains other types of specialized cells; however, these cells do not have any special effect on weathering.

An expanded view of a typical tracheid shows the primary wall and S_1, S_2, and S_3 layers (see Figure 3.12, Chapter 3). A micrograph of a typical softwood cross-section is also shown for comparison (see Figure 7.7). Note that the fibril angle with respect to the length of the tracheid is almost uniaxial. The area between wood cells is the *middle lamella* and the center hollow area is the *lumen*. When viewed in cross-section, the largest component of the cell wall is the S_2 layer.

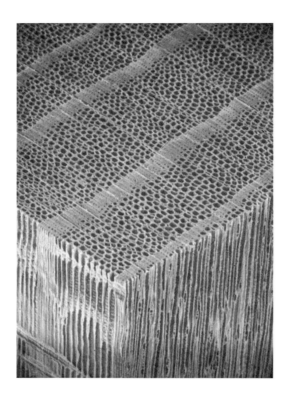

FIGURE 7.7 Micrograph of a softwood.

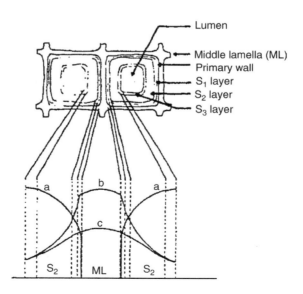

FIGURE 7.8 Diagram showing the relative amounts of cellulose, hemicellulose, and lignin across a cross-section of two wood cells: a) cellulose; b) lignin; c) hemicellulose.

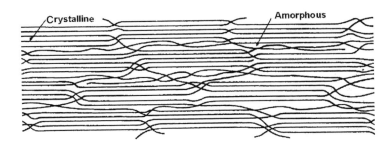

FIGURE 7.9 Diagram showing amorphous and crystalline regions of cellulose.

The components of the various layers differ in the chemical constituents and in the orientation of the fibrils within each layer. The fibril orientation can also differ for juvenile wood (see Section 1.1.4 of this chapter on juvenile wood). The fibril orientation and the chemical constituents greatly affect the way in which wood weathers. Figure 3.12 in Chapter 3 shows the relative ratios of the three main polymeric constituents of the cell wall. The concentration of cellulose is highest in the S_2 layer, whereas the lignin concentration is highest in the middle lamella (see Figure 7.8).

7.1.3 Chemical Nature of Polysaccharides, Lignin, and Extractives

Approximately 2/3 of the mass of wood is comprised of sugars. (Kollmann and Côté 1968) Cellulose is a linear polymer of (β-1 $\rightarrow$ 4)-D-glucopyranose and occurs primarily in the S_2 layer of the cell wall. It can be crystalline or amorphous, and a single cellulose chain may run through several alternating amorphous and crystalline regions (see Figure 7.9). Wood cellulose is about 60–70% crystalline. The glucopyranose forms high molecular-weight polymers through the 1 $\rightarrow$ 4 glycocidic bond to form straight chains having all three hydroxyls in the equatorial plane (see Figure 3.2, Chapter 3). It is the alternating β-linkage and the equatorial hydroxyls that permit cellulose to form crystalline regions. Hemicelluloses do not have this property. They are rather small macromolecules of approximately 150–200 sugar units (primarily L-arabinose, D-galactose D-glucose, D-mannose, D-xylose and 4-O-methyl-D-glucuronic acid). The polymers are essentially linear with numerous short side chains. The hemicelluloses in hardwoods and softwoods are quite different: hardwoods contain primarily glucurono-xylan and glucomannan, whereas softwoods contain arabinoxylzn and galactoglucomannan. The composition of the various polysaccharide in hardwoods and softwoods is shown in Table 7.2. The polysaccharides are comprised of simple sugars and they contain no conjugated systems or carbonyl groups.

Lignin is a three-dimensional polymer comprised of phenyl propane units, but it has no regular structure. Lignin cannot be isolated from wood without degrading it, so it has not been possible to determine its molecular weight. Estimates range as high as 50 million Daltons. One of the common representations for softwood lignin shows a structure with multiple conjugated systems and carbonyl groups (see Figure 3.10, Chapter 3). It should be emphasized that this is not a structure but merely a representation of typical groups in softwood lignin. The basic unit is a phenyl propane (see Figure 7.10a), and in softwoods, there is usually an oxygen and methoxy group giving a methoxy-phenyl propane (guaiacyl propane, see Figure 7.10b). Hardwoods have an additional methoxy group (see Figure 7.10c). In addition, there are approximately 20 carbonyls per each 100 guaiacyl propane. For additional information on lignin structure, refer to Kollmann and Côté (1968).

7.1.4 UV Spectrum

The UV and visible solar radiation that reaches the earth's surface is limited to the range between 295–800 nm. Wavelengths from 800 to about 3000 are infrared radiation. The radiation from 295–3000 nm comprises distinct ranges that affect weathering: UV radiation, visible

TABLE 7.2
Cellulose and Hemicellulose Composition of Hardwoods and Softwoods[a]

Polysaccharide	Occurrence	Percent of Extractive-Free Wood
Cellulose	All woods	42 ± 2
O-Acetyl-4-O-methyl-glucuronoxylan	Hardwoods	20–35
Glucomannan	Hardwoods	3–5
Arabino-4-O-methyl-glucuronoxylan	Softwoods	10–5
Galactoglucomannan (Water-soluble)	Softwoods	5–10
Galactoglucomannan (Alkali-soluble)	Softwoods	10–15
Arabinogalactan	Larch wood	10–20

[a] Abstracted from Table 2.6, Kollmann and Côté, 1968.

light, and infrared radiation (IR) (see Table 7.3). The energy from the sun that reaches the earth's surface as discrete bundles of energy called *photons* and their energy can be calculated from:

$$E = h\nu = hc/\lambda \tag{7.1}$$

where
 h = Planck's constant
 v = frequency
 c = velocity of radiation
 λ = wavelength of radiation.

From this equation, the photon energy is inversely proportional to the wavelength of the radiation (see Figure 7.11). The energy of the photon will become important when we consider the photochemical reactions that this radiation can initiate.

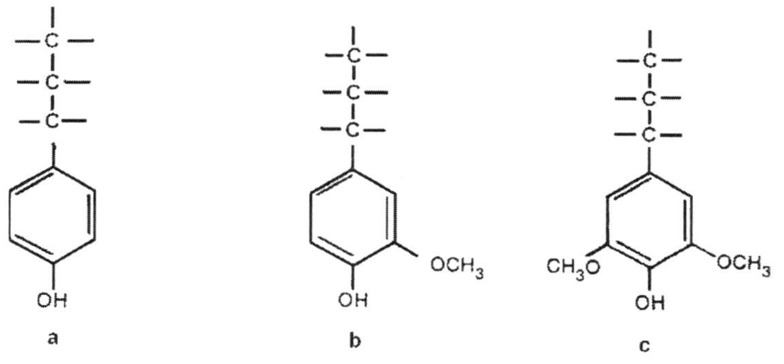

FIGURE 7.10 Units of lignin.

Weathering of Wood

TABLE 7.3
Percent of Total Spectral Irradiance for UV, Visible, and IR Radiation

Radiation	Wavelength Range	% of Total Irradiance
UV radiation	295-400 nm	6.8
Visible light	400-800 nm	55.4
Infrared radiation	800-3000 nm	37.8

Several terms need to be defined before continuing the discussion of solar energy:

Irradiance—the radiant flux per surface area (Watts/m² (W/m²))
Spectral irradiance—irradiance measured at a wavelength (W/m²/nm)
Radiant exposure—irradiance integrated over time (Joules/m² (J/m²))
Spectral radiant exposure—radiant exposure measured at a wavelength (J/m²/nm).

When using the irradiance term, it is necessary to define the spectral range (for example, the total solar irradiance (295–3000 nm) or the total UV irradiance (295–400 nm)). For exposure or measurement at a particular wavelength (such as 340 nm), the spectral irradiance would be expressed as W/m² at 340nm. The radiant exposure and spectral radiant exposure are irradiance and spectral irradiance integrated over time.

By plotting the spectral irradiance (W/m²/nm) as a function of wavelength, one gets the spectral power distribution (see Figure 7.12). The plot gives the spectral irradiance at each wavelength. By integrating the areas under the curve for UV radiation (295–400 nm), visible light (400–800 nm), and IR radiation (not shown in the figure), the percent of total irradiance for each component can be calculated.

As seen in Figure 7.12 and Table 7.3, the energy in the UV portion of the spectral power distribution is quite low compared to the total energy in the spectral power distribution. This means that although the energy for photons at 300 nm is quite high, there aren't very many of them, so the spectral irradiance is quite low. As the wavelength increases to 400 nm, the spectral irradiance increases (lower energy photons, but more of them).

7.1.5 Wavelength Interactions with Various Chemical Moieties

In order for a photochemical reaction to occur, sufficient energy to disrupt a chemical bond (bond dissociation energy) must be absorbed by some chemical moiety in the system. The bond

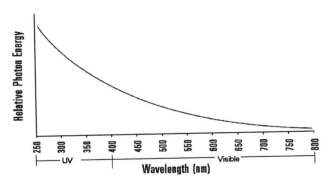

FIGURE 7.11 Relative photon energy versus wavelength for UV radiation and visible light.

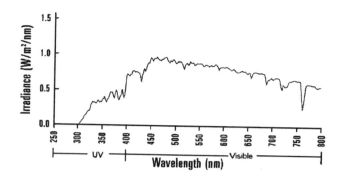

FIGURE 7.12 Spectral power distribution for UV radiation and visible light at the Earth's surface.

dissociation energy available for radiation in the UV and visible range is given in Figure 7.13. The absorbed energy may not result in a degrading chemical reaction, but the absorption is a necessary condition. The bond dissociation energies and corresponding wavelengths having the necessary energy for breaking these bonds for several chemical moieties are listed in Table 7.4. Using Equation 7.1, the energy for UV radiation at a wavelength of 295 nm is about 97 Kcal/mole and for 400 nm is about 72 Kcal/mole. Several of the chemical moieties have bond dissociation energies well above the energy of terrestrial UV radiation and therefore cannot be affected by natural UV radiation. The bond dissociation energy must be below 97 kcal/mole for the chemical moiety to absorb radiation. It can be seen from this table that the bond dissociation energies for many of carbon-oxygen moieties commonly found in lignin fall within the UV radiation range (295–400 nm).

7.2 CHEMICAL CHANGES

There have been many studies to investigate the mechanism of wood weathering, and it has been clearly shown that the absorption of a UV photon can result in the formation of a free radical, and that through the action of oxygen and water, a hydroperoxide is formed. Both the free radical and hydroperoxide can initiate a series of chain scission reactions to degrade the polymeric components of wood. Despite many studies spanning several decades, the mechanism is still not well defined and can only be represented in a general way (see Figure 7.14). The absorption of a photon can

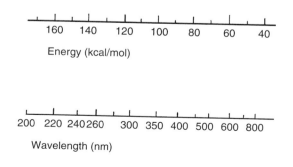

FIGURE 7.13 Bond-dissociation energy compared with wavelength of UV and visible solar radiation.

TABLE 7.4
Bond Dissociation Energies and Radiation Wavelength[a]

Bond	Bond Dissociation Energy (Kcal/mol)	Wavelength (nm)
C–C (Aromatic)	124	231
C–H (Aromatic)	103	278
C–H (Methane)	102	280
O–H (Methanol)	100	286
C–O (Ethanol)	92	311
C–O (Methanol)	89	321
CH_3COO–C (Methyl ester)	86	333
C–C (Ethane)	84	340
C–Cl (Methyl chloride)	82	349
C–$COCH_3$ (Acetone)	79	362
C–O (Methyl ether)	76	376
CH_3–SH (Thiol)	73	391
C–Br (Methyl bromide)	67	427
N–N (Hydrazine)	57	502
C–I (Methyl iodide)	53	540

[a] Bond energies abstracted from Table 2.1 (Rånby, B. and J.F. Rabek 1975).

cause the formation of a free radical that initiates a number of reaction pathways. Many unanswered questions remain:

Is there a wavelength dependence on the rate of the degradation?
If there is wavelength dependence, what functional moieties are affected?
How do studies using UV radiation <290 nm relate to the actual UV spectrum?
What is the rate-limiting step in this series of reactions?
Does the degradation follow the "law of reciprocity"?
Once the free radical is formed, is there a rate-limiting chemical?
Does the diffusion rate of oxygen affect the reaction?
Is water necessary for the reaction?
Is hydrolysis of hemicellulose an important component of weathering?
What is the temperature dependence?

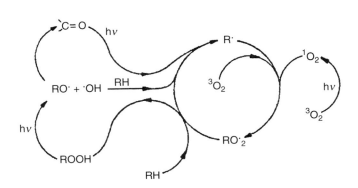

FIGURE 7.14 Mechanism of photodegradation of wood (Feist and Hon 1984).

As was previously discussed, UV radiation comprises only a small part of the total irradiance (spectral power distribution) that strikes the Earth's surface; however, the energy per photon is higher for UV radiation. The energy per photon increases as the wavelength decreases. The energy required to break chemical bonds depends on the type of chemical bond (see Table 7.4). The photon energy per wavelength is shown in Figure 7.13. By comparing the energy available from the photons in the UV range of the spectrum, it is apparent that there is sufficient energy to break bonds in the chemicals that comprise wood. However, in order for a bond to break, energy must be absorbed by some component of the wood. This is the first law of photochemistry (the Grotthus-Draper Principle). In addition, a particular molecule in the wood can absorb only one quantum of radiation (the Stark-Einstein Principle) (McKellar and Allen 1979). The absorbed energy puts the molecule in a higher energy (excited) state that can be dissipated through a number of paths. The most benign would be a return to the ground state through dissipation of heat. Other alternatives would involve chemical reactions.

7.2.1 Free Radical Formation

In early work by Kalnins (1966), he proposed a free-radical initiation and the necessity of oxygen. He isolated volatile degradation products, noted the decrease in lignin content, characterized the IR spectrum of the wood surface following irradiation, noted the post-irradiation reactions, evaluated the effect of extractives, and analyzed surface and interior cellulose and lignin contents of nine wood species. His work established a basis for subsequent studies by others. The results showed in a qualitative way many of the important aspects of weathering, but the light sources did not represent the UV light at the Earth's surface. About 85% of the energy of the lamp was at wavelengths below 295 nm.

Studies to elucidate free-radical formation in wood by the absorption of photons were done by Hon and his collaborators and are covered in detail in Chapter 8 of *Developments in Polymer Degradation—3* and references therein (Hon 1981a). Through a series of experiments, it was clearly shown that the absorption of a photon by wood results in formation of free radicals. In all of these early studies, the light source had UV wavelengths down to 254nm. The energy at this wavelength is approximately 135 Kcal/mole, about 30 kcal/mole higher than the most energetic photons found at the Earth's surface (see Figure 7.13). It is difficult to relate these higher energies to the exact chemical moiety important in the degradation; however, the work clearly showed the importance of free radicals in the degradation process.

One of the common chemical reaction paths following the absorption of a quantum of energy is chemical dissociation to form a free radical. Since wood does not normally have free radicals, their presence following UV irradiation signals the dissociation of a chemical bond (Hon et al. 1980, Hon 1981a, Zhao et al. 1995). These free radicals can easily be detected using electron spin resonance (ESR). A simple ESR spectrum of wood irradiated with UV radiation of different intensities is shown in Figure 7.15. The ESR signal intensity for various exposure times and storage at ambient conditions for different UV radiation sources showed a dependence on the radiation intensity (Figure 7.16). In simple radicals such as a methyl radical, the spin of the free electron interacts with the hydrogen to produce splitting. This splitting can be used to infer the chemical structure. More information on the technique can be found in many texts on photo degradation, such as *Photodegradation, Photo-oxidation and Photostabilization of Polymers* by Rånby and Rabek (1975).

In studies of wood surfaces and model compounds using UV radiation >254nm, Hon showed that the formation and decay rate of free radicals was temperature dependent (Hon 1981a). Interpretation of ESR spectra of lignin was not possible because the splitting patterns were extremely complex. The reactive moieties in lignin include various carbonyls, carboxyls, and ethers, and the ESR signal may be comprised of several types of free radicals. Several model compounds were studied (Figure 7.17) and it was found that compounds a, b, and c were cleaved at the carbon-carbon bond adjacent to the -carbonyl via a Norrish Type I reaction. The ESR spectrum for

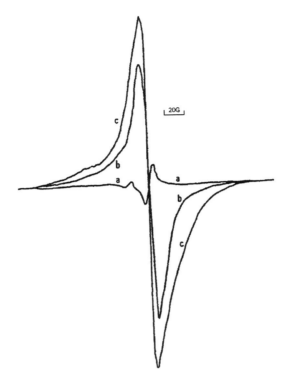

FIGURE 7.15 Electron spin resonance (ESR) signals from wood irradiated with different radiation sources (77°K for 60 min): a) fluorescent light; b) sunlight; c) UV radiation (Feist and Hon 1984).

compound-a showed a seven-line signal for an ethyl radical superimposed on a singlet for an acyl radical. Other compounds decomposed to form phenoxy radicals (see Figure 7.18). From the work on these model compounds, Hon concluded the following (Hon 1981a).

- "Phenoxy radicals are readily produced from phenolic hydroxy groups by the action of light.
- Carbon-carbon bonds adjacent to α-carbonyl groups are photo-disassociated via the Norrish Type I reaction.
- The Norrish Type I reaction does not occur efficiently in those compounds with ether bonds adjacent to the α-carbonyl Group. Photo-dissociation takes place at the ether bond.

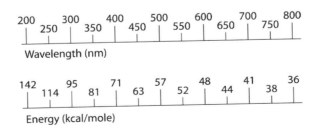

FIGURE 7.16 Electron spin resonance (ESR) signals from wood irradiated for a period of time then stored without radiation: 1) vacuum/control; 2) vacuum/fluorescent lamp; 3) air/control; 4) air/fluorescent lamp; 5) vacuum/sunlight; 6) air/sunlight (Feist and Hon 1984).

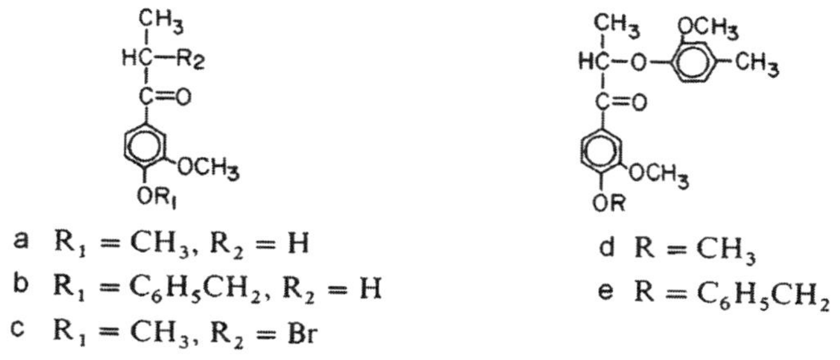

a $R_1 = CH_3$, $R_2 = H$
b $R_1 = C_6H_5CH_2$, $R_2 = H$
c $R_1 = CH_3$, $R_2 = Br$

d $R = CH_3$
e $R = C_6H_5CH_2$

FIGURE 7.17 Lignin model compounds (Hon 1981).

- Compounds bearing benzoyl alcohol groups are not susceptible to photo-dissociation except when photosensitizers are present.
- α-Carbonyl groups function as photosensitizers in the photo-degradation."

On the basis of the work with these model compounds, Hon concluded that the phenoxy radicals were the major intermediate formed in the photo degradation of lignin, and that these intermediates react with oxygen and demethylate to form an *o*-quinonoid structure (see Figure 7.19). In recent work by Kamoun et al. (1999) using ESR to evaluate photo degradation of lignin extracts from radiata pine, the formation of a phenoxy radical was confirmed. They proposed that the phenoxy radical is resonance stabilized by radical transfer reactions and confers stabilization to wood.

Two of the most interesting reports from early studies of lignin degradation involved measuring the yellowing of lignin model compounds (Lin and Kringstad 1970, Lin 1982). Twenty-seven

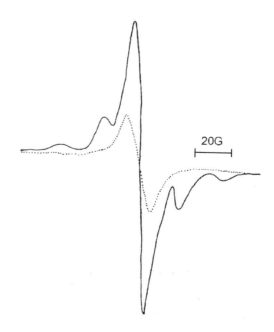

FIGURE 7.18 Electron spin resonance (ESR) spectrum of model compound d: solid line, signal following UV irradiation at 254 nm (77°K for 60 min); dotted line, signal after the irradiated specimen warmed to 298°K.

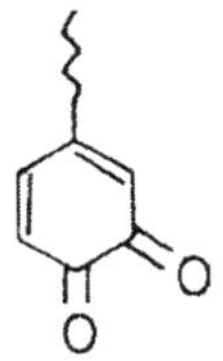

FIGURE 7.19 O-quinonoid moiety.

compounds were irradiated in a UV-transparent solvent. The UV radiation source had a wavelength of 305–420 nm with peak intensity at 350 nm. They reported that lignin structural units having a saturated propane side chain do not absorb UV radiation; therefore, moieties such as Guaiacylgycerl guaiacyl ether, phenylcoumaran, and pinoresinol probably are not involved in photochemical degradation. The α-carbonyl moiety was the most labile, followed by biphenyl and ring-conjugated double bond structures.

7.2.2 Hydroperoxides

In cooperation with Chang and Feist, Hon showed that the reaction with oxygen to form a hydroperoxide was an integral part of the photo degradation process (Hon et al. 1982). Using singlet oxygen generators and quenchers to investigate the interaction of oxygen under photochemical conditions, they concluded that singlet oxygen was involved in the degradation process and that singlet oxygen quenchers could preclude the formation of hydroperoxides, thereby stabilizing wood against photodegradation. The investigation of the hydroperoxides continued, and in 1992 Hon and Feist reported differences in the hydroperoxide formation at two different UV radiation distributions (>254 nm or >300 nm). Using DRIFT spectroscopy (a combination of diffuse reflectance spectroscopy and Fourier transform infrared spectroscopy), they analyzed the formation and reactions of the hydroperoxides that formed on wood surfaces. The formation of hydroperoxides and carbonyls as well as the destruction of cellulose ether linkages was tracked for up to 180 days of UV radiation exposure. They discussed the energy requirements for bond cleavage in terms of the dissociation energy and proposed mechanisms for the reactions of the hydroperoxides. They also noted that differences in hydroperoxide formation depended on wood species and whether the surface was tangential or radial. These differences were attributed to difference in the wood components and lignin concentration for the various species and grain angles.

In summary, terrestrial UV radiation (295–400 nm) has sufficient energy to cause bond dissociation of lignin moieties having α-carbonyl, biphenyl, or ring-conjugated double bond structures. A free radical is formed, which then reacts with oxygen to form a hydroperoxide. Additional reactions result in the formation of carbonyls. The degradation process depends on the surface composition of the wood. Lignin absorbs UV radiation throughout the UV radiation spectrum and into the visible light spectrum; however it is only the absorption above 295 nm that is important for weathering of wood (see Figure 7.20). UV radiation at wavelengths shorter than 295 nm are absorbed by the Earth's ozone layer and do not reach its surface.

7.3 REACTION PRODUCTS AND CHEMICAL ANALYSIS

UV radiation at wavelengths below 295 nm can cause degradation not found in normal UV exposure. Barta et al. (1998) used a UV-laser at a wavelength of 248 nm to degrade eight wood species and reported increased carbonyl absorption at 1710–1760 cm^{-1} and decrease aromatic absorptions at 1276 and 1510 cm^{-1}. They also found decreased absorption at 1396, 1465, and 1539 cm^{-1} and

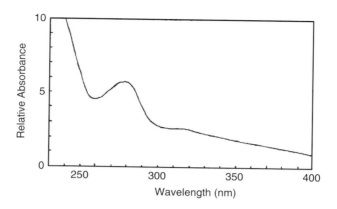

FIGURE 7.20 UV absorption spectra for lignin.

attributed them to changes in lignin, but these changes were not found using conventional xenon arc exposure (UV radiation having filters to approximate the UV radiation at the Earth's surface). Papp (1999) reported that a UV-laser at a wavelength of 248 nm gives completely different results than traditional light sources. Chang (1985) exposed wood to UV radiation and visible light at >220 nm, >254 nm, >300 nm, >350 nm, >400 nm, and >540 nm and reported no lignin degradation at wavelengths above 400 nm.

The infrared spectral analysis of Southern pine (*Pinus sp.*) surfaces at various times as specimens were exposed to UV radiation ($\lambda > 220$ nm) in the laboratory showed a progressive increase in the carbonyl absorption at 1720 and 1735 cm^{-1} and a decrease in absorption at 1265 and 1510 cm^{-1} (see Figure 7.21) (Hon 1983, Hon and Chang 1984). The carbonyl absorption was attributed to the oxidation of cellulose and lignin. The decrease in the absorption at 1265 and 1510 cm^{-1} was attributed to loss of lignin, and this loss was also confirmed by UV absorption spectra of the water-soluble extracts following exposure (see Figure 7.22). Infrared spectra of similar specimens exposed outdoors were distinctly different from the laboratory-exposed specimens (see Figure 7.23) (Hon 1983). The decreased carbonyl absorption and lack of absorption at 1265 and 1510 cm^{-1} was attributed to leaching of the surface by rain as the exposure progressed. No explanation was given for the change at about 1200 cm^{-1}. Gel permeation chromatography of the water soluble extract from the laboratory-exposed specimens showed a molecular weight (M_n) of about 900 Daltons that consisted of carbonyl-conjugated phenolic hydroxyls. They concluded that lignin was the major degradation product. It should be noted that the carbonyl absorptions are rather broad and poorly defined, and one would expect this from the degradation of a complex mixture of polymers. The oxidation undoubtedly occurs at a number of slightly different parts of the lignin.

Periodic FTIR surface analysis of radiata pine during 30 days of outdoor exposure showed perceptible lignin loss in as little as four hours and substantial lignin loss after six days (Evans et al. 1996). Evans (1988) attributed weight loss of specimens during weathering to lignin degradation, not to leaching of water-soluble extractives. Pandey and Pitman (2002) reported degradation in as little as one day of outdoor exposure. Substantial delignification of the surface of radiata pine (*Pinus radiata*) was found after as few as three days of outdoor exposure (Evans et al. 1996).

Hon and Feist (1986) studied the weathering of several hardwood species using UV radiation of ($\lambda > 220$ and > 254 nm) and as with softwoods, there was oxidation of the surface to form carboxyls, carbonyls, quinones, and loss of lignin. The oxidation was confirmed using electron spectroscopy for chemical analysis (ESCA) and showed chemical shifts of C_{1s} from 285.0 eV (carbon-carbon bond) to 287.0 eV and 289.5 eV (carbon-oxygen ether bond and carbon-oxygen carbonyl, respectively).

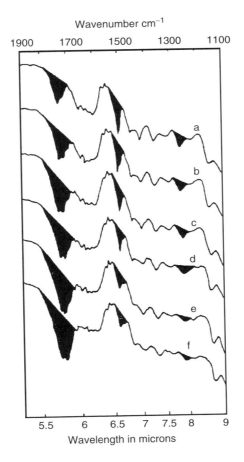

FIGURE 7.21 Infrared (IR) spectra of wood surface following UV irradiation at ≥254 nm; a) control; b) one day; c) four days; d) 10 days; e) 20 days; f) 40 days (Hon 1983).

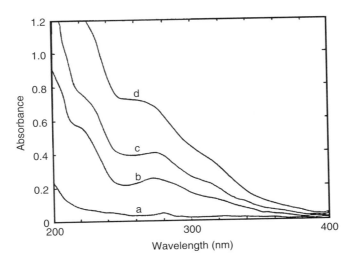

FIGURE 7.22 UV absorption spectra of water extracts of wood following UV irradiation; a) control: b) one day; c) four days; d) 20 days.

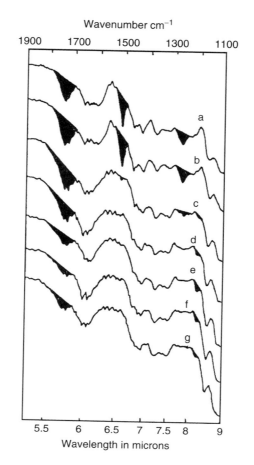

FIGURE 7.23 Infrared (IR) spectra of wood following outdoor exposure: a) control; b) no water leaching; c) 30 days; d) 60 days; e)180 days; f) 300 days; g) 480 days.

Li (1988) evaluated the weathering of basswood (*Tilia amurensis* Rupr.) using ESCA, FTIR, and SEM and found similar results.

Using a xenon light source having borosilicate filters, much sharper carbonyl peaks were observed for the photodegradation of southern pine and western redcedar (*Thuja plicata* Donn) (Horn et al. 1992). The light source used in this work closely matched the natural UV spectra. They also reported the decrease in lignin absorption after exposure and showed that leaching by water was an important component of the weathering. Anderson et al. (1991a, b) also used a xenon light source having borosilicate filters to approximate natural UV radiation. They measured the surface degradation over 2400 hours of UV light exposure with daily water spray of 4 hours, or light without the water spray. Matching specimens were subjected to just water spray for 400 hours. The surface degradation of three softwoods (western redcedar, southern pine, and Douglas-fir) and four hardwoods [yellow poplar, quaking aspen (*Populus tremuloides*), white oak (*Quercus alba*), and hard maple (*Acer saccharum*)] was evaluated using diffuse reflectance FTIR. For the softwoods, the spectra for the three species were quite different before weathering and during weathering with light and water (Anderson et al. 1991a). The absorption at 1730–1740 cm^{-1} increased in intensity during the early exposure then decreased, and the 1514 cm^{-1} decreased. They also reported a rapid

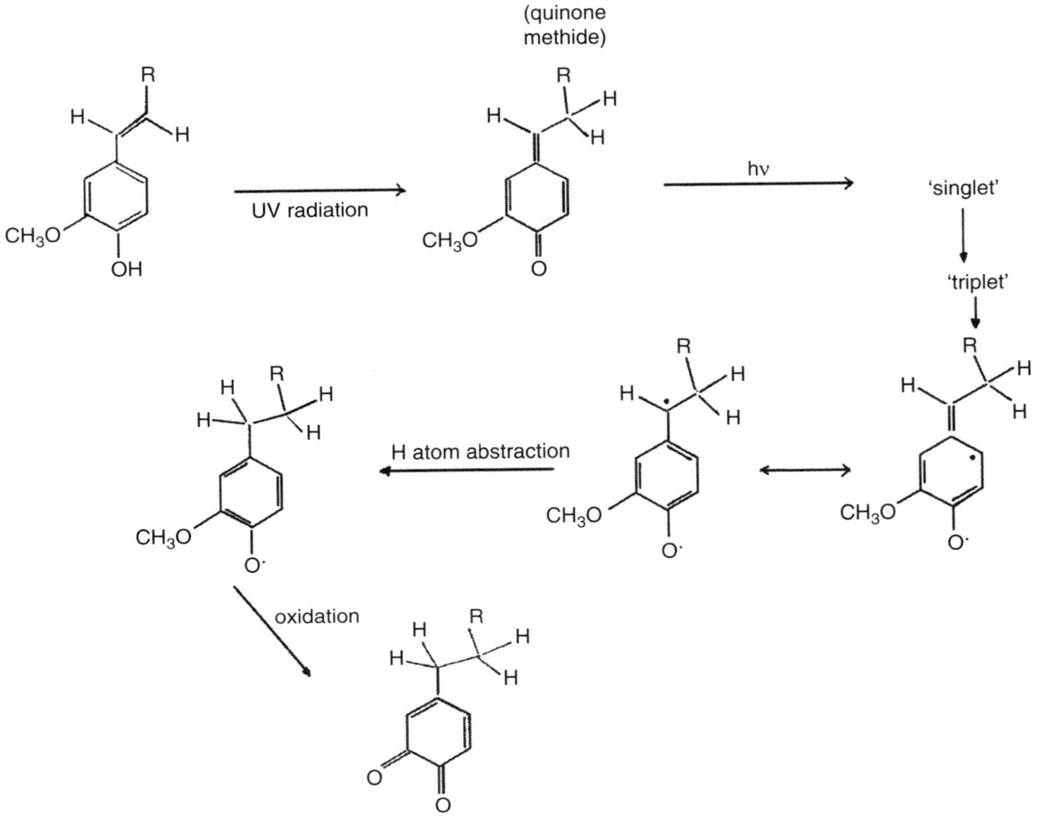

FIGURE 7.24 Lignin photo-oxidation mechanism (Anderson et al. 1991b).

increase in the intensity at 1650 cm^{-1} early in the exposure period, followed by a rapid decrease in intensity; they attributed this to the formation of quinones and quinone methides. All softwood had a distinctly cellulosic spectra following 2400 hours of accelerated weathering (light and water), indicating a loss of lignin. The four hardwoods were slightly different (Anderson et al. 1991b). Yellow poplar and quaking aspen weathered much the same as the softwoods; however, white oak and hard maple weathered slower, probably because of their higher density. As with the softwood, lignin was removed from the surface. A mechanism was proposed (see Figure 7.24). Németh and Faix (1994) used DRIFT FTIR (bands at 1510, 1600, and 1740 cm^{-1}) to quantify the degradation of locust (*Robinia pseudoacacia*) and quaking aspen. In a subsequent study of hardwoods and softwoods, Tolvaj and Faix (1995) reported that the carbonyl absorption was comprised of two sub-bands (1763 and 1710 cm^{-1}) for softwoods, but not for hardwoods. Detailed absorption bands for pine (*Pinus sylvestris*), spruce (*Picea abies*) larch (*Larix decidua*), locust, and poplar (*Populus euramericana*) were reported. Wang and Lin (1991) evaluated the weathered surface of nine Taiwan species following seven years of outdoor weathering in Taiwan using IR spectroscopy. Powders from various depths of the degraded surface were analyzed by transmission (KBr method). IR absorption bands were tabulated.

Košíková and Tolvaj (1998) irradiated *Populus grandis* for 50 hours (UV wavelength was not reported) and isolated the lignin using a series of dioxane, neutral, acid, and base extractions. Difference FTIR spectra (irradiated and unirradiated) of the neutral extracts showed an increase in

OH bands (3500 cm^{-1}), decrease in ring-conjugated carbonyl (1666 cm^{-1}), increase in non-conjugated carbonyl (1747 cm^{-1}), and a decrease in aromatic ring content (1514 and 1593 cm^{-1}). Acid and alkaline extracts showed an increase of carbonyl band (1740 cm^{-1}), and a decrease of aromatic and methoxy groups (1510 and 1270 cm^{-1} respectively).

Mitsui et al. (2001), in studies of color change of spruce and Japanese cypress (*Chamaecyparis obtusa* Sieb. St Zucc), showed that color change of UV irradiated wood was greater when heated than at low temperatures, but at low temperatures the color change was more rapid at high humidity. In a later study, Mitsui (2004) used various filters to study the effect of wavelength on heat-induced color change. He noted differences for color change with and without heat treatment and attributed this to different chemical reactions for UV irradiation at ambient temperature and UV at elevated temperature. The degree of color change increased with decreasing wavelength. Some color change was observed at 400–500 nm and it was attributed to color change of extractives.

It appears that lignin decomposition follows first-order kinetics and is dependent on the wavelength of radiation. High humidity seems to accelerate the degradation.

7.3.1 Depth of Degradation

On the basis of the depth of color change, Browne and Simonson (1957) reported degradation of wood as deep as 2500 μm following exposure to weathering. Later work by Hon and Ifju (1978) and Hon (1981b) showed that this depth was beyond the limit for generation of free radicals. They measured UV radiation and visible light transmission through radial and tangential sections of Douglas-fir (*Pseudotsuga menziesii* [Mirb.] Franco), redwood (*Sequoia sempervirens* [D. Don] Endl.), Southern pine, and western redcedar sections of various thicknesses in 25-μm steps from 25–300 μm, and the penetration of UV radiation was determined from the presence of free radicals using ESR. They reported that the UV radiation penetrated only 75 μm, whereas visible light penetrated 200 μm. More recent research has shown degradation products beyond the 75 μm limit reported by Hon and Ifju (1978). Horn et al. (1992) studied the penetration of UV radiation and visible light into western redcedar and southern pine. They measured the chemical change using FTIR in 10 μm steps and reported chemical change to about 120 μm. They also noted distinct differences with and without water spray; the water spray removed the degradation products. Kataoka and Kiguchi (2001) examined cross-sections of sugi (*Cryptomeria japonica*) following xenon arc exposure using a micro FTIR technique. Areas as small as 50 μm in cross-section were measured, and the change in the carbonyl (1730 cm^{-1}) and aromatic (1510 cm^{-1}) absorptions were reported at depths of 600–700 μm (see Figure 7.25). Park et al. (1996) reported degradation at depths of 750–850 μm after weathering of hinoki (*Chamecyparis obtusa* Endl.). Wang and Lin (1991) reported the limit to the depth of degradation of 900 μm. Yata and Tamura (1995) found that the depth of wood degradation remained constant after six months of outdoor weathering.

The differences in the reported depth of degradation may be reconciled by considering some of the factors that affect the penetration of UV radiation into wood. The penetration depends on the wood density and the wavelength distribution of the UV radiation and visible light. Denser wood is penetrated less by UV radiation, and shorter wavelengths also penetrate less. The 75 μm limit for UV radiation penetration reported by Hon and Ifju (1978) was determined using a light source having radiation well below 300 nm, whereas the work by Kataoka and Kiguchi (2001) used a UV lamp with radiation above 300 nm. The penetration for these two lights would be quite different, and this might explain the observed differences. The penetration is also dependent on the density, which would be different depending on species and the amount of earlywood and latewood. It was not possible in earlier research to determine UV radiation penetration on a fine enough scale to show the effects of wood anatomy (i.e. earlywood/latewood, radial/tangential gain). It is also possible that the reactions following the generation of the free radical might involve

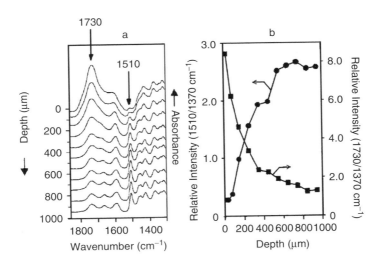

FIGURE 7.25 Fourier transform infrared (FTIR) spectroscopy and selected peak ratios: a) FTIR spectroscopy depth profile of weathered *Cryptomeria japanica*; b) relative peak intensities at 1510 cm^{-1} and 1730 cm^{-1} compared with the intensity at 1370 cm^{-1} (Kataoka and Kiguchi 2001).

chemical moieties deeper into the wood than the depth at which the free radical was generated. The degradation products could also be carried deeper into the wood by the action of water. It was clearly shown that the water-soluble reaction products could be washed from the surface; they might also be washed deeper into the wood. UV radiation sources and filters are now available to do more detailed analysis in future work.

Using FTIR to evaluate grand fir (*Abies grandis*) following exposure to UV radiation, Dirckx et al. (1992) reported different reaction products depending on the wavelength of the UV radiation. They also reported that the hydroxyl absorption decreased as the carbonyl absorption increased, and that the change was dependent on oxygen concentration.

7.3.2 Acid Effects

Williams (1987, 1988) studied the effect of various concentration of sulfuric, sulfurous, and nitric acids on the weathering of western redcedar and found that pH of 2–3 increased the rate of weathering. Park et. al (1996) did a similar study using pH 2 sulfuric acid on hinoki (*Chamaecyparis obtusa* Endl) and evaluated the degradation using SEM. Acid-treated specimens had 1.5 times more degradation of the middle lamella and cell walls than specimens treated with water. Hon (1993) exposed southern pine to UV radiation (≥223 nm) and sulfuric acid spray (pH = 4.4, 3.0, or 2.0) at 65°C and ambient temperatures. He reported increased carbonyl formation and decreased lignin with acid treatment and a slight increase in degradation with increased temperature. His results were similar to previous studies. For most areas of the country, the effect of acid appears to be minor except at pH of 2–3.

In summary, FTIR analysis of degraded wood surfaces show an increase in carbonyl and a decrease in hydroxyl and aromatic content. The oxidation results in cleavage of the lignin at specific locations such as α-carbonyls, and with sufficient degradation, the lignin products are washed from the surface. Oxygen appears to be essential for the reaction, but it was not determined whether it is rate-limiting for the degradation. The increase in the rate of erosion of wood with acid treatment is likely caused by increased hydrolysis of the carbohydrates.

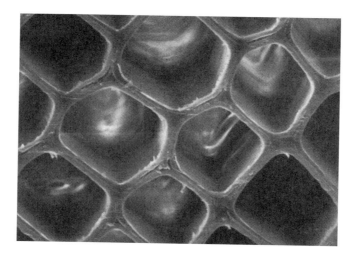

FIGURE 7.26 Southern pine cross-section (1000×).

7.4 PHYSICAL ASPECTS OF DEGRADATION

7.4.1 MICROSCOPIC EFFECTS

7.4.1.1 Destruction of Middle Lamella

As was shown in Chapter 3, Figure 3.8, because the lignin content is higher in the middle lamella than in the cell wall, therefore the photo degradation occurs preferentially in this area of the wood surface. This is particularly noticeable in micrographs of a southern pine cross-section before (Figure 7.26) and after UV exposure (≥220 nm) (Figure 7.27). Hardwood cross-sections (yellow poplar) showed similar degradation (Hon and Feist 1986). Both softwoods and hardwoods degraded at the middle lamella. In addition, the hardwoods degraded at the pores. Yellow poplar was also exposed outdoors and showed similar degradation. The radiant exposure for the laboratory- and outdoor-exposed specimens was not measured; therefore it is difficult to compare the two exposures.

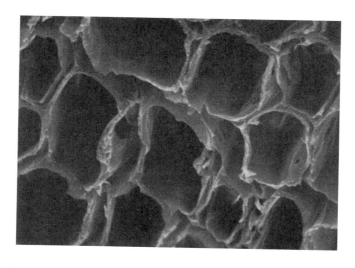

FIGURE 7.27 Southern pine cross-section following 1000 h of UV radiation ($\lambda > 220$ nm, 1000×).

Weathering of Wood

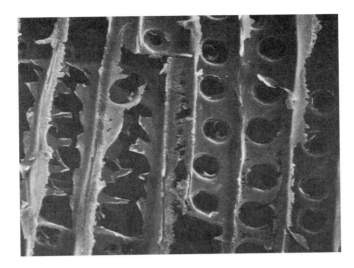

FIGURE 7.28 Deterioration of bordered pits of radial section of southern pine following 1000h of UV radiation (≥220 nm, 1000×).

It should be noted that the wavelength for the laboratory exposure for some of these studies extended well below 295 nm, and the contribution of these high-energy photons can not be determined from these micrographs. Qualitatively, both have degraded in a similar manner. The micrographs from both types of exposure support the premise that the lignin is the photosensitive component in wood.

A number of researchers have shown detailed micrographs of wood at various stages during the weathering of wood. (Leukens and Sell 1972). Kuera and Sell (1987) reported that the degradation around the ray cells in flat-sawn beech was caused by differential dimensional changes between the ray cells and surrounding wood. Kuo and Hu (1991) tracked UV radiation-peak = 254 nm induced degradation of red pine (*Pinus resinosa* Ait.) for 3 to 40 days and showed the progressive degradation from the corners of the middle lamella to the cell wall.

7.4.1.2 Destruction of Bordered Pits and Cell Wall Checking

Micrographs of tangential and radial surface of wood also show lignin degradation (see Figure 7.28). Checks form at the bordered pits and extend aligned with the fibril orientation at a diagonal to the axis of the tracheid. It appears that the lignin binding of the fibrils has been degraded. There is also a separation between the cell walls of two adjacent cells. With extended weathering, cell wall components develop severe checking and the fibrils and tracheids loosen and become detached from the surface. Cross-sectional view of several wood species following more than ten years of outdoor exposure show similar results: There is greater erosion from the earlywood bands and a tendency to check along the earlywood latewood interface for the softwoods (see Figures 7.29a–c). Diffuse porous hardwoods tend to show more even erosion and less checking (see Figure 7.29d).

7.4.2 MACROSCOPIC EFFECTS

7.4.2.1 Loss of Fiber

Much of the information on the rates of wood degradation has been obtained from artificial weathering studies. Futo (1976) exposed wood specimens to UV radiation or thermal treatment and evaluated degradation by weight loss. Using SEM, he noted differences in the microstructure of thermally- and UV-degraded wood (Futo 1976). Williams and Feist (1985) used artificial weathering to evaluate the effects of chromic acid and chromium nitrate treatment of wood surfaces to retard weathering.

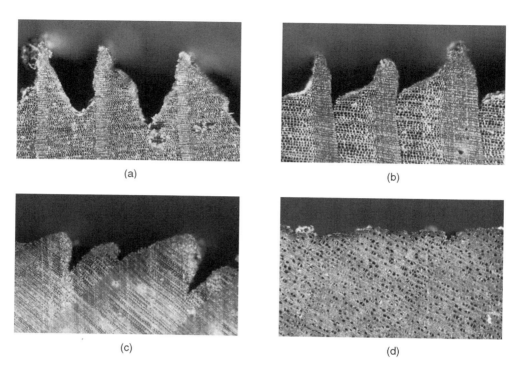

FIGURE 7.29 Micrographs of cross-sections of weathered wood following more than 10 years of outdoor exposure; a) western redcedar; b) Douglas-fir; c) Engelmann spruce; d) red alder.

Williams (1987) used artificial UV radiation to determine the effects of acid on the rate of erosion; degradation was determined by measuring the change in wood mass. Arnold et al. (1991) measured wood erosion of European yew (*Taxus baccata*), Norway spruce (*Picea abies*), southern pine, western redcedar, and white ash (*Fraxinus americana*) during 2400 hours of artificial weathering using xenon arc with borosilicate filters and fluorescent UV chambers. Both were equipped with water spray. They obtained similar results with both chambers and noted that water spray was essential for simulating natural weathering. Derbyshire et al. (1997) used artificial weathering to determine the activation energies for several wood species; wood degradation was determined by loss of tensile strength.

Derbyshire and Miller (1981) exposed thin strips of Scots pine and lime (*Tilia vulgaris*) for up to 24 weeks outdoors under various filters to give wavelengths >300 nm, >350 nm, >375 nm, and >400 nm and measured the tensile strength. They reported that visible light (≥400 nm) contributed to the degradation. From measurement of the cellulose disperse viscosity of weathered and unweathered specimens, they concluded that there was cellulose degradation on specimens exposed to the total solar spectrum. Scanning electron micrographs of the failed specimens clearly show the degraded wood components. Derbyshire et al. (1995a) also used accelerated weathering to degrade specimens and used short-span tensile tests to evaluate degradation. In a second study by Derbyshire et al. (1996), 0-span and 10mm-span tensile tests were conducted on six softwood species. The drop in the strength was plotted against UV radiation dose and the difference in strength between the 0-span and 10mm-span was attributed to degradation of different components of wood. The 0-span test measured the cellulose strength, whereas the 10-mm span measured the wood strength. An exponential expression for the degradation was proposed and compared with the experimental data. In a third study (Derbyshire et al. 1997), they showed that the degradation of ten wood species was temperature dependent; linear Arrhenius plots had activation energies of 5.9–24.8 kJ/mol. On the basis of 0-span and 10mm-span tensile tests and SEM data, Derbyshire et al. (1995b, 1996) attributed the degradation to three phases in which the first phase was a slow surface structural change and the other two phases were rapid degradation of lignin and cellulose. They also compared the degradation of six softwood species and developed a mathematical expression that fit strength loss of thin sections of wood following short period of natural weathering. (Derbyshire et al. 1996)

Researchers have compared the results of natural and accelerated weathering. Feist and Mraz (1978) found good correlation between erosion rates from natural and accelerated weathering. These researchers also found good correlation between erosion rate and wood density. They reported similar erosion rates for earlywood and latewood for some species after an initial two-year period. Deppe (1981) compared 12- to 60-week accelerated weathering with 3- to 8-year natural weathering of wood-based composites, but he was primarily interested in water absorption, thickness swelling, and strength properties. Derbyshire et al. (1995a, b) compared natural and artificial weathering by assessing the degradation of small strips of wood. Evans (1988) also evaluated the degradation of thin wood veneers exposed outdoors for up to 100 days, using "weight loss" as the unit of measurement. Sell and Leukins (1971) weathered 20 wood species outdoors for one year at 45° south, but their main interest was in the discoloration of wood. Evans (1989) used SEM to show the loss of wood, primarily degradation of the middle lamella, following two years of natural weathering. Yoshida and Taguchi (1977) noted loss of strength in plywood exposed to natural weathering for seven years, and Ostman (1983) measured the surface roughness of several wood and wood-based products after four years of natural weathering. Bentum and Addo-Ashong (1977) evaluated cracking and surface erosion of 48 timber species, primarily tropical hardwoods, after five years of outdoor exposure in Ghana. Weathering characteristics of tropical hardwoods from Taiwan during both natural and accelerated weathering have also been reported (Wang 1981, 1990; Wang et al. 1980). Williams et al. (2001d) used accelerated weathering to evaluate the erosion rates of several tropical hardwoods from Bolivia.

In a series of outdoor studies spanning 16 years, Williams and coauthors reported on the effects of surface roughness, grain angle, exposure angle, and earlywood/latewood on weathering rates of several softwood and hardwood species (Williams et al. 2001a, b, and c). Erosion rates for earlywood and latewood were reported for several plywood and lumber species (see Table 7.5). For all wood species, erosion rates for earlywood and latewood differed greatly during the first seven years of weathering. For Douglas-fir and southern pine, significant differences continued after seven years. However, for western redcedar and redwood, erosion rates were generally the same after seven years. Because of this change in the erosion rate at seven years, the rates from 8–16 years may be more representative of long-term erosion rates (see Table 7.6). The erosion rate of vertical-grained lumber was considerably higher than that of flat-grained plywood. Only slight differences were observed for saw-textured as opposed to smooth plywood. The erosion rates confirmed the effect of wood-specific gravity, showing that more dense species weather more slowly. The density difference also affects the erosion rates of earlywood and latewood (see Figures 7.29a–c). The figure clearly shows more earlywood erosion for western redcedar than for Douglas-fir and Engelmann spruce. The erosion rate for vertical-grained wood varied from 4.5 µm/yr for southern pine to 9.5 µm/yr for western redcedar and plywood was slightly less.

In the second study in the series by Williams et al. (2001b), the erosion rates for ponderosa pine (*Pinus ponderosa*, Dougl. ex Laws), lodgepole pine (*Pinus contorta*, Dougl. ex Loud.), Engelmann spruce (*Picea engelmannii*, Parry ex Engelm.), western hemlock (*Tsuga heterophylla*, (Raf.) Sarg.), and red alder (*Alnus rubra*, Bong.) measured over ten years are shown in Table 7.7. The average erosion rates for earlywood and latewood for ponderosa pine, lodgepole pine, Engelmann spruce, and western hemlock varied from 40 to 50 µm/yr. This is slightly less than the erosion rate usually given for softwoods. The erosion rate for earlywood alone was less than 60 µm/yr. The erosion rate for red alder earlywood, 58 µm/year, was the same as that for western hemlock earlywood, which is not surprising given the similar specific gravity of these species. The thin latewood bands of red alder broke off as the earlywood eroded and had little effect on the rate of

TABLE 7.5
Erosion of Earlywood and Latewood on Smooth-Planed Surfaces of Various Wood Species after Outdoor Exposure near Madison, Wisconsin[a]

Wood Species	Avg SG[b]	Erosion (μm) after Various Exposure Times[c]											
		4 years		8 years		10 years		12 years		14 years		16 years	
		LW[d]	EW[d]	LW	EW	LW	EW	LW	EW	LW	EW	LW	EW
Western redcedar plywood	—	170	580	290	920	455	1,09	615	1,16	805	1,355	910	1,475
Redwood plywood	—	125	440	295	670	475	800	575	965	695	1,070	845	1,250
Douglas-fir plywood	—	110	270	190	390	255	500	345	555	425	770	515	905
Douglas-fir	0.46	105	270	210	720	285	905	380	980	520	1,300	500	1,405
Southern pine	0.45	135	320	275	605	315	710	335	710	445	1,180	525	1,355
Western redcedar	0.31	200	500	595	1,09	765	1,32	970	1,56	1,16	1,801	1,380	1,945
Redwood	0.36	165	405	315	650	440	835	555	965	670	1,180	835	1,385

[a] Specimens were exposed vertically facing south. Radial surfaces were exposed with the grain vertical.
[b] SG is specific gravity.
[c] All erosion values are averages of nine observations (three measurements of three specimens).
[d] EW denotes earlywood; LW, latewood.

TABLE 7.6
Typical Erosion Rates for Various Species and Grain Angles for 8–16-Year Outdoor Exposure near Madison, Wisconsin[a]

Species, Wood, and Orientation	Earlywood Erosion[b] (μm/year)	Latewood Erosion[b] (μm/year)	Avg Erosion (μm) per Year[c]	Erosion (mm) per 100 Years
Western redcedar				
Lumber, vertical	101.5	91.5	95	9.5
Plywood, flat	66.5	75.5	70	7.0[d]
Redwood				
Lumber, vertical	84.5	67.5	75	7.5
Plywood, flat	65.0	56.0	60	6.5[d]
Douglas-fir				
Lumber, vertical	77.0	44.0	60	6.0
Plywood, flat	60.0	43.0	50	5.0[d]
Southern pine				
Lumber, vertical	59.5	34.0	45	4.5[e]

[a] For Southern Pine, erosion rates were determined from slope of regression line from 0- to 12-year data.
[b] Average of erosion rates for vertical and horizontal (flat) grain exposures.
[c] Average of earlywood and latewood erosion rates rounded off to nearest 5 units.
[d] The face veneer would be gone long before 100 years had passed.
[e] Specimens were decayed after 12 years of exposure.

TABLE 7.7
Typical Erosion Rates for Earlywood and Latewood for Various Species Measured during 10 Years of Outdoor Exposure near Madison, Wisconsin[a]

Species	Earlywood (μm/yr)	Latewood (μm/yr)	Average Erosion (μm) per Year[b]	Erosion (μm) per 100 Years[c] (mm)
Ponderosa pine	42	35	40	4.0
Lodgepole pine	49	33	40	4.0
Engelmann spruce	54	38	45	4.5
Western hemlock	58	39	50	5.0
Red alder	58	—	60	6.0

[a] Data from vertical and horizontal grain exposures were combined to compute earlywood and latewood erosion rates.
[b] Average erosion of earlywood and latewood rounded to nearest 5 units.
[c] Extrapolated from average earlywood and latewood erosion rates rounded to nearest 0.5 mm.

erosion (see Figure 7.29d). Differences in surface morphology were shown to affect weathering of the surface. Wood species with wide latewood bands weathered differently than did species with thin latewood bands (see Figure 7.29a–d).

Erosion rates were also determined for Douglas-fir, loblolly pine (*Pinus taeda* L.), southern pine, western redcedar, northern red oak (*Quercus rubra* L.), and yellow-poplar for different angles of exposure (see Table 7.8) (Williams et al. 2001c). The erosion rates of all species were considerably higher at the 45° angle of exposure than at 90°. For many species, erosion was about twice as fast for the 45° exposure. Most species showed little difference in erosion between the 0° and 45° exposures. Although UV radiation is higher for horizontal specimens (0° exposure), the authors hypothesize that the decrease in the washing action of water and the build-up of degradation

TABLE 7.8
Typical Erosion Rates for Earlywood and Latewood of Various Species during 6 Years of Outdoor Exposure near Madison, Wisconsin, at Different Exposure Angles[a]

Species	Earlywood (μm/year)			Latewood (μm/year)			Average Erosion[b] per 100 Years (mm)		
	90°	45°	0°	90°	45°	0°	90°	45°	0°
Douglas-fir	42	97	81	20	42	48	3	—	—
Loblolly pine	34	78	89	20	44	39	2.5	—	—
Southern pine	62	123	102	20	48	44	—	—	—
Western redcedar	131	177	135	26	96	77	—	—	—
Northern red oak	35	67	114	30	80	105	3	7.5	11
Yellow-poplar	48	117	209	—	—	—	—	—	—

[a] Data from vertical grain (radial surface) exposure of earlywood and latewood.
[b] Extrapolation of average of earlywood and latewood erosion rounded to nearest 5 units. Where the difference in erosion rate between earlywood and latewood was large, average erosion is not given.

products and dirt on the surface probably protected these specimens to some extent. More rapid degradation at 0–45° would be expected to have a great effect on the service life of wood products used for roofing and decks. The average erosion rate for Douglas-fir and loblolly pine (earlywood and latewood) exposed at 90° was about 30 and 25 μm/year respectively, considerably less than that for softwoods such as ponderosa pine, lodgepole pine, Englemann spruce, and western hemlock.

Evans measured weight loss of thin veneers of radiata pine following 50 days of outdoor exposure at various exposure angles and reported the following: 90° (vertical), 17.0%; 70°, 26.5%; 60°, 29.4%; 45°, 31.8%; and 0°, 34.1% weight loss. Seasonal effects cause slower erosion in the winter. (Raczkowski 1980; Evans 1996). Groves and Banana (1986) used SEM studies to determine the degradation of radiata pine exposed up to six months in Australia.

7.4.2.2 Grain Orientation

In a detailed study of the effects of grain angle on weathering of Scots pine and Norway spruce (*Picea abies* Karst.) following 33 months outdoors, Sandberg (1999) reported the following: Pine had 13 and spruce 6 times more total checking per unit area on the tangential versus radial surface; tangential surfaces had deeper checks; CCA and linseed oil treatment had only minimal effect to decrease checking; on radial surfaces, checking occurred primarily at the latewood/earlywood border; degradation of bordered pits was a distinct difference between radial and tangential surfaces; and the microfibrils in the S_2 layer are the most weather resistant component.

7.4.2.3 Water Repellency

As wood weathers, the extractives are leached from the surface and it becomes less water-repellent. The loss of lignin also makes the surface more hydrophilic. Contact angle measurements on weathered western redcedar dropped from 77° to 55° after four weeks of outdoor weathering (Kalnins and Feist 1993). Kang et al. (2002) also reported increased wettability for Sitka spruce exposed to xenon arc radiation and water spray.

7.4.2.4 Checks and Raised Grain

In addition to the slow erosion of the wood surface as wood weathers, the surface also developed checks and raised grain. This type of degradation is often more severe than erosion. For example, wood decks are often replaced long before their expected service life because of raised grain and splitting. This degradation is caused primarily by moisture. A cross-section view of weathered wood surfaces for several wood species after ten years of outdoor weathering clearly shows the formation of checks (Figures 7.29b and c). The check often forms at the earlywood/latewood interface as shown for southern pine and Douglas-fir. On flat-grained surfaces, checking occurs predominately on the bark side; however, the raised grain on the pith side can be a more severe problem. Both checking and raised grain result from the wetting and drying of wood that occurs from precipitation and daily and seasonal changes in RH. As was found from the study of Bolivian hardwoods, the erosion rate for many of the species was extremely slow, but after 20–30 wet/dry cycles, they developed severe checking (Williams et al. 2001d).

Water in the absence of UV radiation or light can cause strength loss of wood veneers (Banks and Evans 1984; Evans and Banks 1988; Voulgardis and Banks 1981). Scots pine (*Pinus sylvestris*) and lime (*Tilia vulgaris*) veneers were soaked in deionized water at 50°C and 65°C; SEM of specimens following tension tests showed different failure modes for the soaked specimens compared with controls that had not been soaked. Controls failed by fracture across the cell wall, whereas the veneers that had been soaked failed by inter-fibril sheer. Chemical analysis of wood flour that had been placed in 65°C water for 50 days showed a loss of hemicellulose and lignin, but not cellulose (Evans and Banks 1990). It seemed likely that hydrolysis of the hemicelluloses had occurred. In subsequent studies using radiata pine (Evans et al. 1992), mannose and xylose were isolated from water extracts from weathered wood. Analysis of the holocellulose following weathering showed depletion of mannose, xylose, arabinose, and galactose. Following outdoor exposure, methanol and hot water extracts of weathered wood yielded 6.4% and 18.0% soluble components respectively, whereas unweathered wood had only 1.6% and 2.2%. They attributed the increased solubility to degradation of the lignocellulose matrix.

During outdoor weathering, water mechanically abrades the surface and washes away degradation products. In addition, water probably hydrolyzes hemicelluloses, particularly at the surface. As the lignin degrades, the hemicelluloses become more vulnerable to hydrolysis. Sudiyani et al. (1999b) reported that water was instrumental in the decomposition of the lignin in addition to washing away the reaction products.

The effects of iron in contact with wood during laboratory weathering that included moisture, freezing, heat, UV radiation, and fatigue cycling caused 25% more creep in specimens exposed to iron than those without iron contact. (Helinska-Raczkowska and Raczkowski 1978). It seems the iron compounds can catalyze moisture-induced degradation. Raczkkowski (1982) reported that iron in contact with beech (*Fagus sylvatica*) increased the degradation during laboratory exposure to cyclic moisture, temperature, and UV radiation, but that beech modified by *in situ* polymerization of styrene retarded degradation. Hussey and Nicholas (1985) reported that hindered amine light stabilizers inhibited the iron/water degradation of wood. In tensile tests of specimens exposed to water, water/iron, water/iron/stabilizer and controls, the stabilized specimens had greater tensile strength than both the water- and water/iron-exposed specimens.

In summary, the erosion rate depends on the anatomy of the wood, grain angle, density, and angle of exposure. In general, most softwoods erode at a rate of 6 mm/century and hardwoods at 3 mm/century. In addition to UV radiation, water causes degradation. Water abrades the surface, washes away reaction products, hydrolyzes carbohydrates, and causes checking and raised grain.

7.4.3 Weathering of Wood/Wood Composites

Wood composites, such as plywood, fiberboard, flake-board, and particleboard, are vulnerable to degradation. As with solid wood, the surface of these products undergoes photochemical degradation. Composites comprised of wood bonded to wood, whether made with veneers, flakes, or particles, are much more vulnerable to moisture cycles than is solid wood.

Exterior grades of plywood are manufactured from veneers that are rotary-peeled from logs. This process yields a veneer that is flat-grained, and the peeling process forms lathe checks in the veneer. When the surface veneers are put in place to from the plywood, the surface having lathe checks is placed facing the inner veneer. The lathe checks are not visible on the surface, but the veneer has internal flaws. As plywood weathers, these lathe checks grow to the surface to form parallel-to-grain cracks in the surface veneer. As shown in Tables 7.5 and 7.6, plywood surfaces erode slightly slower than solid wood of the same spe cies, but it is the development of surface cracks that limit the service life of unprotected plywood. It is the moisture cycles that cause these cracks to form. If plywood is not protected with a finish, it can fail within ten years (see Figure 7.30). The figure shows a western redcedar plywood panel after eight years. The left part of the panel was protected by a batten and is not weathered. About half the thickness of the surface veneer has eroded and there are numerous cracks in the face veneer. Plywood must be protected with a finish. Surface checking and cracking of surface veneers also occurs on laminated veneer lumber (Hayashi et al. 2002). Yoshida and Taguchi (1977a, b) determined the decrease in thickness and mechanical properties of three- and five-ply red luan (*Shorea sp.*) and three-ply kapur (*Dryobalanops* sp.), Shina (basswood: *Tilia sp.*), and kaba (birch: *Betula sp.*) following seven years of outdoor exposure. Surface checks primarily affected shear strength of the panels. The erosion of the surface decreased static and impact bending strength. Five-ply panels had slightly higher erosion rates, compared with three-ply panels of the same species.

FIGURE 7.30 Western redcedar plywood following 8 years of outdoor exposure (left-half was protected by a batten).

As with plywood, flake-board, fiberboard, and other types of particle boards are extremely vulnerable to moisture cycles. As they weather, the surface flakes or particles debond and fall off. They must be finished with a multi-coat paint system if they are to be used outdoors.

7.4.4 Weathering of Wood/Plastic Composites

Wood/plastic composites are less susceptible to moisture cycles than wood composites, but are degraded by UV radiation and are more sensitive to heat (see Chapter 13). During weathering of wood/high density polyethylene (wood/HDPE) composites, such as decking, thermal degradation of the polyethylene component is an important factor and can be described by first-order kinetics with an activation energy of 23.2 kJ/mol (Li 2000). In addition to the thermal effects on the HDPE component, wetting and drying of the wood component causes the HDPE to crack. Li reported that these effects cause more damage than photo-oxidation. Bending tests of kanaf-filled HDPE following 2000 hours of xenon arc weathering showed significant decreases in bending strength and stiffness (24% and 42%, respectively) (Lundin et al. 2002). Wood flour-filled HDPE showed similar results; strength and stiffness decreased 20% and 33% respectively.

Stark and Matuana (2002) exposed various formulations of wood-flour/HDPE composites and HDPE to 2000 hours of xenon arc weathering. They reported that hindered amine light stabilizers (HALS) did improve color stability and mechanical properties. The wood/HDPE composites maintained flexural modulus of elasticity (MOE) until about 2000 hours of weathering, whereas the HDPE showed decreased MOE after 1000 hours. Matuana et al. (2001) reported that rutile titanium dioxide was an effective UV stabilizer for wood/PVC composites.

Matuana and Kamdem (2002) reported that the wood in wood/polyvinyl chloride (wood/PVC) composites was an effective chromosphore; wood/PVC degraded faster than PVC (QUV weathering with water spray). Degradation was evaluated using DRIFT-FTIR and ESCA. In contrast, Lundin et al. (2002) reported that weathered wood/PVC specimens showed no loss in strength or stiffness (tension tests), but the unfilled PVC lost almost 50% of its strength after only 400 hours of weathering. They attributed the improved mechanical properties to load-bearing properties and stress transfer of the wood flour. Stark et al. (2004) compared the weathering of injection-molded, extruded, and extruded then planed 50% wood/HDPE specimens. They showed that the manufacturing method affected the surface composition. Planed specimens had a higher wood component at the surface than the injection-molded or extruded specimens and had higher degradation during the first 1000 hours of xenon arc weathering with water spray.

Rowell et al. (2000) used xenon-arc/water-spray weathering cycle (2000 hours) to evaluate the effect of fiber content in an aspen/polyethylene composites (fiber loading of 0, 30–60%, by weight). As the fiber content increased, the weight loss increased, but at the conclusion of the 2000 hours, after the degraded surface was scraped from the surface, the weight loss was greater for the polyethylene with no wood fiber. It seems that the polyethylene degradation products are not washed from the surface during weathering as are the wood degradation products.

These studies exemplify the importance of the surface composition on weathering and the importance of water in the process. Composites having a high concentration of wood fibers at the surface can absorb water, whereas surfaces having fibers encapsulated by HDPE cannot absorb water until the HDPE has degraded.

7.4.5 Effects of Biological Agents

It is generally accepted that mold (mildew) such as *Aureobasidium pullulans*, the most common microorganism found on weathered wood, does not have the enzymes to degrade lignin or polysaccharides. However, the blue stain fungus (*Diplodia natalensis*) caused more strength loss to Lime (*Tilia vulgaris* Hayne) specimens weathered for 30 days compared with specimens that had not been inoculated with the fungi. (Evans and Banks 1986) In general, molds cause more appearance problems than actual degradation of wood; however, aggressive cleaning methods to remove them using strong chemicals and/or power-washing can greatly accelerate the loss of wood fiber from wood surfaces.

7.4.6 UV Degradation of Tropical Woods

Researchers have reported the degradation of several tropical wood species. Fuwape and Adeduntan (1999) reported the weight loss of Gmelina (*Gmelina arborea* Roxb.) weathered outdoors. Nakamura and Sato (1978) evaluated the degradation of red-lauan, kapur, shinanoki, and kaba (botanical names not given in paper) plywood after seven years of outdoor exposure. Onishi et al. (2000) reported on the degradation of 12 species following 2 years of outdoor exposure. Wang (1981) and Wang et al. (1980) reported strength loss of 18 Taiwan wood species exposed to four years outdoor exposure and accelerated weathering. The degradation was measured by static and impact bending tests. The strength decreased linearly with time of exposure.

7.4.7 Paint Adhesion

Williams et al. (1987) preweathered western redcedar for 0, 1, 2, 4, 8, and 16 weeks prior to painting. They reported that two to four weeks of weathering of western redcedar prior to painting caused up to 50% loss in paint adhesion. Matched specimens were placed outdoors. After 17 years exposure, the paint on the controls (0 weeks preweathering) was in almost perfect condition, whereas the paint on all preweathered panels was degraded. The longer the wood was preweathered, the poorer the paint (Williams and Feist 2001). Underhaug et al. (1983) reported that one month of wood weathering prior to painting lead to paint failure within 6 to 10 months. Kleive (1986) reported poor performance of paint when refinished after up to 4 years weathering. Williams and Feist (1993) reported premature failure of paints and solid-color stains that were weathered prior to finishing. Williams and Feist (1994) reported on the service life of semitransparent stains, acrylic latex paints, alkyd paints, and solid-color stains applied to Engelmann spruce (*Picea engelmannii*), yellow poplar (*Liriodendron tulipifera*), southern pine (*Pinus* sp.), and sweetgum (*Liquidambar styraciflua*). Wood substrates were weathered for 4, 8, and 12 weeks prior to applying the finishes. The type of finish had the greatest effect on service life, followed by surface roughness, type of substrate, and finally the amount of preweathering.

7.5 CHEMICAL TREATMENTS TO RETARD WEATHERING

7.5.1 CHROMIC ACID

Chromic acid is an aqueous solution of chromium trioxide (CrO_3); the chromium is in its +6 oxidation state. Treatment of wood with 5–10% chromic acid gives a surface that is highly resistant to photochemical degradation. The chromic acid treatment also gives a surface that is water repellent. Aqueous solutions of other chromium VI compounds such as sodium dichromate also retarded the weathering of wood. The benefits of the treatment were investigated in detail by Black and Mraz (1974) and later by Feist (1979) and the mechanism was studied by Pizzi (1981a,b) and Williams and Feist (1984). Numerous references to other work are contained in these publications. More recent work has shown similar effects (Li et al. 1989, Pandey and Khali 1998; Pandey and Pitman 2002). Chromium oxidizes wood components and in the process is reduced to chromium III (+3 oxidation state). Cr (VI) forms tetrahedral coordination compounds and when it is reduced to Cr (III), it becomes octahedral. With ample sites for coordination with either lignin or carbohydrates, chromium forms a highly insoluble complex with wood, cellulose, or lignin model compounds. In essence, it becomes a integral part of the wood surface. This complex greatly retards photochemical degradation. Chang et al. (1982) compared chromic acid treated and untreated southern pine following exposure to artificial UV radiation (see Figures 7.26 and 7.27). The untreated specimens clearly show degradation of the middle lamella and cell wall. Chromic acid-treated specimens show almost no degradation (see Figure 7.31). The mechanism by which chromic acid treatment protects wood has not been determined; however, the important characteristics are that a chromium complex is formed at the surface, the complex is insoluble, and the complex interferes with the photochemical pathway. Williams and Feist (1984) used ESCA to evaluate the oxidation of the surface of wood and cellulose and showed that cellulose was slightly degraded by the chromic acid treatment and gave off CO_2 following chromic acid treatment. ESCA evaluations of wood treated with chromic acid were also reported by de Lange et al. (1992). They also reported decarboxylation of cellulose and confirmed the stabilization of lignin by chromium complexes.

Guajacol was used as a lignin model compound to study the reactions of chromic acid with wood. It forms an insoluble polymeric complex and Pizzi and Mostert (1980) attributed the waterproofing of wood by chromic acid to the formation of chromium complexes with lignin. Kubel and Pizzi (1981) reported similar waterproofing with other metal oxides that form water-insoluble complexes with

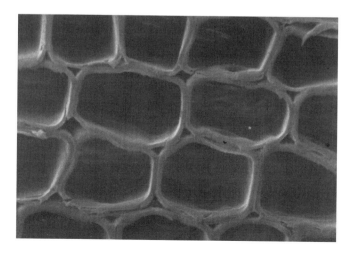

FIGURE 7.31 Southern pine cross-section treated with 5% chromic acid following 1000 hours UV radiation (≥220 nm, 1000×).

guajacol. The reaction kinetics have been determined and the rate constant and activation energies agree well with the rates of chromic acid reactions with wood. Pizzi and Mostert (1980) and Schmalzl et al. (1995) characterized the reaction products of guaiacol with aqueous solutions of ferric chloride and chromic trioxide. Both compounds gave complex mixtures of guaiacol oligomers, however chromium complexes were highly insoluble, thermally stable, and acted as a screen against UV radiation (Schmalzl et al. 2003).

Chromic acid also reacts with pure cellulose to form a waterproof complex. (Williams and Feist 1984) It seems that chromic acid can react with both lignin and polysaccharides to waterproof wood.

Evans and Schmalzl (1989) and Derbyshire et al. (1996) used 0-span tensile tests of thin veneers to determine strength loss. The strength loss following outdoor weathering was attributed to degradation of the cellulose components in wood. Evans and Schmalzl (1989) compared the strength loss with weight loss following weathering and found that chromic acid-treated wood had about the same 0-span strength loss as untreated controls, but the weight loss was significantly less. Image analysis scanning election micrographs of checking at bordered and half-bordered pits following weathering was used as a means to quantify the effect of chromic acid and aqueous solutions of ferric chloride and ferric nitrate treatment of wood. (Evans et al. 1994) It is known that chromic acid causes degradation of cellulose (Williams and Feist 1984) and the 0-span tensile test confirms this. The chromic acid protects the lignin; therefore, the weight loss is less for the chromic acid-treated specimens.

7.5.2 Chromated Copper Arsenate Preservatives

Southern pine pressure treated with chromated copper arsenate (CCA) eroded slower than similar wood treated with brush-applied chromic acid. (Feist and Williams 1991) The higher penetration and retention of chromium in the pressure-treated wood gave better protection. Others showed similar effects (Chang et al. 1982, Williams and Feist 1984, Hon and Chang 1984, Evans 1992).

7.5.3 Copper-Based Preservatives

Using FTIR to monitor the lignin degrade (1510 cm^{-1}) and formation of carbonyls (1720–1740 cm^{-1}), Liu et al. (1994) reported that 2% ammoniacal copper quat (ACQ) preservative treatment of southern pine sapwood retarded wood degradation more than 2% CCA treatment. The specimens were exposed for 35 days outdoors and evaluated weekly. In a similar study, treatment with didecyldimethyl ammonium chloride (DDAC) performed worse than the controls, while the ACQ retarded degradation (Jin et al.1991).

7.5.4 Chemically Bonded Stabilizers

On the basis of the studies using chromic acid, Williams (1983) investigated organic stabilizers that bonded to wood. Chemically reacting a UV absorber having an epoxy moiety [2-hydroxy-4-(2,3 epoxypropoxy)benzophenone] to the wood surface retarded the erosion rate and gave improved performance to clear coating applied over the modified surface. Kiguchi (1992) did additional work using the same treatment on several Japanese wood species and reported similar results.

7.5.5 Commercial Stabilizers

Hon et al. (1985) prepared benzophenone-containing monomers using glycidyl methacrylate to form 2-hydroxy-4 (3-methacryloxy-2-hydroxy propoxy) benzophenone. Polymeric coatings using this monomer improved color stability of southern pine. The UV stabilizer was not bound directly to the wood. They also prepared mixtures of several commercial stabilizers in the coatings. Treatments also included several penetrating mixtures of PEG-400 with 1-octadecanol, which improved color retention. It should be noted that the UV exposure included wavelengths below 295 nm and did not have water spray. Additional work was done using these chemicals and exposing the specimens at the same wavelengths by Panda and Panda (1996). They reached the same conclusions.

7.5.6 CHEMICAL MODIFICATION

Rowell et al. (1981) and Feist and Rowell (1982) reacted southern pine sapwood with either butylene oxide or butyl isocyanate as a cell wall modification (see Chapter 14) or with *in situ* polymerized methyl methacrylate as a lumen filler (see Chapter 15). Specimens that had only butylene oxide or butyl isocyanate cell wall modification degraded at the same rate as the unmodified specimens under xenon arc accelerated weathering, but the methyl methacrylate lumen fill decreased the rate. The best results were obtained using both lumen fill and cell wall modification—these specimens degraded at half the rate of the controls. The cell wall modification made the wood dimensionally stable, and the lumen fill helped hold degradation products in place, thus improving UV resistance. Similar results were obtained using acetic anhydride to modify the cell wall and methyl methacrylate to fill the lumen of aspen (Feist et al. 1991).

Using ESCA to evaluate the oxidation of the surface after exposure to UV fluorescent lamps, Kiguchi (1992) reported a slight improvement in weather resistance for butylene oxide-treated Sugi (*Cryptomeria japonica* D. Don) compared with methylation of phenolic hydroxyls and acetylation. Kiguchi (1997) used ESCA to evaluate the UV resistance of Sugi modified with acetic anhydride and butylene oxide, and reported no improvement in the weatherability of the uncoated wood, but an improvement in the performance of transparent coatings applied over the modified wood. He noted a slight improvement in color retention for the modified wood.

Hon (1995) found that acetylation of southern pine gave only temporary improvement in color stability and, on the basis of ESR studies of a model compound [(4-methyl-2-methoxphenoxy)–hydroxypropiovanillone], suggested that the cleavage of the β-O-4 linkage in the model compound and the acetyl group formed phenoxy radicals. Hon attributed the formation of chromophoric groups to these phenoxy radicals.

Denes and Young (1999) incorporated various combinations of zinc oxide, graphite, benzotriazole, 2-hydroxybenzophenone, and phthalocyanine into polydemethylsiloxane, coated wood with these mixtures, and modified the wood surface using oxygen plasma. Following two weeks of artificial weathering (filtered xenon arc, water spray), some of the modified woods had decreased weight loss compared with untreated controls.

Hill et al. (2001) investigated chemical modification and polymeric grafting of methacrylic anhydride/styrene to Scots pine, and exposed the specimens to UV radiation and water using QUV artificial weathering apparatus (Q-Panel Co). They reported no positive effect for the treatment.

Imamura (1993) exposed acetylated wood (weight gains up to 20%) to natural weathering. SEM observation of the treated and untreated controls showed that both the acetylated wood and the controls degraded at the middle lamella, had enlarged bordered pits, and displayed checking along the microfibril angle of the S_2 layer of the cell wall, but the checking was less severe in the treated wood. The treated wood tended to hold the degraded fibers in place longer. Kalnins (1984) also reported no improvement in weather resistance for acetylated, methylated, or phenylhydrazine-modified wood. Ohkoshi et al. (1996) reported that acetylation combined with polyethylene glycol methacrylate treatment of wood prior to clear-coating improved the coating performance (decreased color change, peeling, and checking). Onoda (1989) modified buna (*Fagus crenata* Blume), sawara (*Chamaecyparis pisifera* Endl), and hinoki (*Chamaecyparis obtusa* Endl) with alkyl ketene dimers; the treatment caused wood degradation, but the treated wood was more resistant to weathering than the untreated ones. Pandey and Pitman (2002) reported improved weathering resistance of acetylated rubberwood.

Chang and Chang (2001) modified China fir (*Cunninghamia lanceolata* var. *lanceolata*) with acetic anhydride, succinic anhydride, maleic anhydride, and phthalic anhydride and evaluated photodegradation (color change) following QUV irradiation (315–400 nm, with peak radiation at 351). Acetylation was most effective in decreasing color change and phthalic anhydride acted as a photosensitizer.

Kiguchi (1990) modified sugi surfaces by cyanoethylation, benzylation, or allylation. The modified surfaces did not improve weather resistance, but the benzylation and allylation as a modified surface gave improved performance of an acrylic-silicon coating. Murakami and Matsuda (1990) reported that phthalic anhydride- and epichlorohydrin-treated wood degraded less in accelerated weathering exposure. Benzoylation of wood to about 70% weight gain was found to decrease wood weathering and ESR spectroscopy showed that the treatment decreased the amount of free radicals during exposure to UV radiation (Evans et al. 2002).

Rapp and Peek (1999) treated Scots pine sapwood, Norway spruce (*Picea abies* L.), English oak heartwood (Quercus robur L.), and Douglas-fir heartwood with varnish or melamine resin and exposed them outdoors for two years. Compared with untreated controls, the melamine resin provided protection against weathering, but not against checking or absorption of moisture.

Sudiyani et al. (1996, 1999a, 2002) treated albizzia (*Paraserianthes falcata* Becker) and sugi (*Cryptomeria japonica* D.Don) with acetic anhydride (AA), propylene oxide, paraformaldehyde, dimethyrol dihydroxy ethylene urea (DMDHEU), and phenol-formaldehyde (PF) resins and evaluated their weathering following one year outdoors in the tropics or 1080 hours of carbon arc (with water spray). They reported the best improvement for PF-modified wood, followed by AA-modified wood. The outdoor exposure was much more severe. They attributed the more severe weathering to high rainfall and possible decay in the tropical exposure.

It seems that cell wall modification such as acetylation, benzylation, etc., and lumen filling with methyl methacrylate give only modest protection against UV degradation.

7.5.7 WATER-REPELLENT PRESERVATIVES

Water repellents give slight protection during early stage of weathering by decreasing the amount of degradation products washed from the surface (Feist 1987, Minemura et al. 1983). Wang and Tsai (1988) reported a decrease in weathering rate because of the water repellent. However, it was also reported that water repellent effectiveness is lost as wood weathers, and reapplications of the water repellent do not perform as well on weathered wood as on wood that has not previously been weathered (Voulgaridis and Banks 1981). The weathered substrate doesn't give a firm base for the water repellent to be effective.

7.5.8 PAINTS AND STAINS

Weathering of wood gives a surface that cannot hold paint very well, but the weathered surface can accept penetrating stains extremely well. Although the stains will not peel as does paint from weathered wood, the increase in the amount of stain does not improve the service life. (Arnold et al. 1992) A tape-peel test was used to determine adhesion of acrylic latex and an oil-modified acrylic latex primers to radiata pine exposed outdoors for five to ten days. (Evans et al. 1996) They found that the oil-modified acrylic latex bonded well, whereas the unmodified primer did not. Williams et al. (1999) reported more than 15 years of service life for oil-modified latex paint applied to severely weathered western redcedar and redwood. In this case the oil in the coating stabilized the surface, thereby improving the paint adhesion.

Semitransparent stain on southern pine pressure treated with chromated copper arsenate (CCA) had a longer service life than similar wood treated with brush-applied chromic acid prior to staining. (Feist and Williams 1991) The higher penetration and retention of chromium in the pressure-treated wood gave better protection. Plackett and Cronshaw (1992) reported similar results for artificial weathering of western redcedar and radiata pine pressure treated with various concentrations of CCA and brush-applied chromic acid prior to application of a clear moisture-curing urethane.

Performance of clear films on wood depends primarily on the weathering of the wood at the coating wood interface (Black and Mraz 1974, Feist 1979, Feist and Hon 1984, Williams 1983, Nakano et al. 1980; Lindberg 1986). Unless a coating has a pigment or other suitable additive to either absorb, quench, or block UV radiation, the radiation is absorbed by the wood surface at the coating/wood interface. The wood degrades and the coating debonds. MacLeod et al. (1995) measured the interface degradation of western redcedar coated with a clear acrylic periodically as it weathered outdoors for 50 weeks using ATR FTIR. Cut-off filters (320, 360, 390, 490, and 620 nm) were used to evaluate the degradation at various wavelengths. They reported that lignin degradation at interface increased with decreasing wavelength.

Turkulin et al. (1997) exposed finished and unfinished Scots pine and Norway spruce to natural and QUV weathering. Finishes included both film-forming paints and semitransparent stains. On the basis of detailed SEM evaluations and tensile tests of the finish/wood interface, they reported that pine and spruce have the same modes of failure, but spruce had somewhat less structural change; early evidence of degradation were checks in the aspirated bordered pits; as weathering progresses, tension test specimens showed a brittle failure mode; opaque paints fully protected the interface; and solvent-borne stains failed at the interface.

A properly applied paint system gives the greatest protection to a wood surface against UV radiation. The pigments in the paint block the UV radiation, and the paint film retards water absorption. In the absence of a proper paint film, modification of wood with chromium compounds still stands out as the best treatment to retard photodegradation. Unfortunately, they probably will never be used as a commercial wood treatment because of environmental and health concerns. Chemical modification with organic compounds such as acetic anhydride, epoxides, grafted UV stabilizers, and methyl methylacrylate have shown only moderate efficacy. There is considerable research underway to develop new stabilizers, such as HALS, that can be grafted to wood to give a UV-resistant surface. Use of this technique along with UV absorbers in clear coatings will likely give UV-resistant wood/coating systems.

7.6 SUMMARY AND FUTURE CONSIDERATIONS

The middle lamella between the wood cells has a higher lignin content than the cell wall and degrades faster than the cell wall. The α-carbonyl sites in the lignin have been shown to be quite labile to absorption of UV radiation. ESR analysis confirms the formation of free radicals that initiate a series of oxidative degradation reaction. The cross-link density of lignin decreases, resulting in loss of lignin. Water plays a crucial role in washing the degradation products from the surface, leaching hemicelluloses, abrading the surface, and causing checking.

Much of the early research was done using UV radiation below 300 nm, but it showed, in a qualitative way, the depth of penetration of UV radiation, formation of free radicals, the importance of oxygen, and the reaction products. In many of the studies of photodegradation of wood, the UV radiation sources have not simulated the natural UV radiation. In addition, the sources covered extremely broad ranges. As was shown in Table 7.4, specific wavelengths of radiation can cause specific bond dissociation. To gain a better understanding of these reactions, filters should be used to study the chemical changes that occur at narrow wavelength ranges. This is the approach being used to study coating degradation by J. Martin and others at the National Institute of Standards and Technology. (Martin and Bauer 2002, Bauer and Martin 1999) By using narrow ranges of UV radiation, the chemistry specific to that wavelength can be studied. By using this approach, it might be possible to resolve the broad carbonyl absorptions that are observed in FTIR spectra of wood surfaces following photodegradation. By measuring the chemical changes at specific wavelengths, it might be possible to gain a better understanding of the mechanism of wood weathering.

REFERENCES

Anderson, E.L., Pawlak, Z., Owen, N.L., and Feist, W.C. (1991a). Infrared studies of wood weathering. Part I: Softwoods. *Applied Spectroscopy* 45(4):641–647.

Anderson, E.L., Pawlak, Z., Owen, N.L., and Feist, W.C. (1991b). Infrared studies of wood weathering. Part II: Hardwoods. *Applied Spectroscopy* 45(4):648–652, 1991.

Arnold, M., Feist, W.C., and Williams, R.S. (1992). Effects of weathering of new wood on the subsequent performance of semitransparent stains. *Forest Products Journal* 42(3):10–14

Arnold, M., Sell, J., and Feist, W.C. (1991). Wood weathering in fluorescent ultraviolet and xenon arc chambers. *Forest Products Journal* 41(2):40–44.

Banks, W.B. and Evans, P.D. The degradation of wood surfaces by water. *Fifteenth Annual Meeting of the International Research Group on Wood Preservation*, IRG/WP/3289. May 28–June 1, 1984.

Barta, E., Tolvaj, L., and Papp, G. (1998). Wood degradation caused by UV-laser of 248 nm wavelength. *Holz als Roh- und Werkstoff* 56(5):318.

Bauer, D.R. and Martin, J.W. (1999) (Eds.) Service Life Prediction of Organic Coatings: A Systems Approach. *American Chemical Society Series* 722, Oxford Press, New York, 1999.

Bentum, A.L.K. and Addo–Ashong, F. W. (1977). Weathering performance of some Ghanaian timbers. Technical Note 26, Forest Products Research Institute, Ghana.

Black, J.M. and Mraz, E.A. (1974). Inorganic surface treatments for weather-resistant natural finishes. U.S Forest Service Res. Pap. FPL 232, U.S. Department of Agriculture, Forest Service, Forest Products Laboratory, Madison, WI.

Browne, F.L. and Simonson, H.C. (1957). The penetration of light into wood. *Forest Products Journal* 7(10):308–314.

Chang, S.-T. (1985). Effect of light wavelength on the degradation of wood. *Forest Products Industry* 4(2):118–123.

Chang, S.-T. and Chang, H.-T. (2001). Comparison of the photostability of esterified wood. *Polymer Degradation and Stability* 71(2):261–266.

Chang, S.-T., Hon, N.-S., and Feist, W.C. (1982). Photodegradation and photoprotection of wood surfaces. *Wood and Fiber* 14(2):104–117.

de Lange, P.J., de Kreek, A.K., van Linden, A., and Coenjaarts, N.J. (1992). Weathering of wood and protection by chromium studied by XPS. *Surface and Interface Analysis* 19:397–402.

Denes, A.R. and Young, R.A. (1999). Reduction of weathering degradation of wood through plasma-polymer coating. *Holzforschung* 53(6):632–640.

Deppe, H.J. (1981). A comparison of long-term and accelerated aging tests on coated and uncoated wood-based materials. *Holz-Zentralblatt* 107(63–64):1051–1054.

Derbyshire, H. and Miller, E.R. (1981). The photodegradation of wood during solar irradiation. Part 1: Effects on the structural integrity of thin wood strips. *Holz als Roh- und Werkstoff* 39(8):341–350.

Derbyshire, H., Miller, E.R, and Turkulin, H. (1995a). Investigations into the photodegradation of wood using microtensile testing. Part 1: The application of microtensile testing to measurement of photodegradation rates. *Holz als Roh- und Werkstoff* 53(5):339–345.

Derbyshire, H., Miller, E.R., and Turkulin, H. (1995b). Assessment of wood photodegradation by microtensile testing. *Drvna Industrija* 46(3):123–132.

Derbyshire, H., Miller, E.R., Sell, J., and Turkulin, H. (1995). Assessment of wood photodegradation by microtensile testing. *Drvna Industrija* 46(3):123–132.

Derbyshire, H., Miller, E.R., and Turkulin, H. (1996). Investigations into the photodegradation of wood using microtensile testing. Part 2: An investigation of the changes in tensile strength of different softwood species during natural weathering. *Holz als Roh- und Werkstoff* 54:1–6.

Derbyshire, H., Miller, E.R., and Turkulin, H. (1997): Investigations into the photodegradation of wood using microtensile testing. Part 3: The influence of temperature on photodegradation rates. *Holz als Roh- und Werkstoff* 55(5):287–291.

Dirckx, O., Triboulot-Trouy, M.C., Merlin, M., and Deglise, X. (1992). Modification de la couleur du bois d' *Abies grandis* exposé à la lumière solaire. *Ann. Sci. For.* 49:425–447.

Evans, P.D. (1988). A note on assessing the deterioration of thin wood veneers during weathering. *Wood and Fiber Science* 20(4):487–492.

Evans, P.D. (1989). Structural changes in *Pinus radiata* during weathering. *Journal of Institute of Wood Science* 11(5):172–181.

Evans, P.D. (1996). The influence of season and angle of exposure on the weathering of wood. *Holz als Roh- und Werkstoff* 54(3):200.

Evans, P.D. and Banks, W.B. (1986). Physico-chemical factors affecting the degradation of wood surfaces by *Diplodia natalensis*. *Material und Organismen* 21(3):203–212.

Evans, P.D. and Banks, W.B. (1988). Degradation of wood surfaces by water: changes in mechanical properties of thin wood strips. *Holz als Roh- und Werkstoff* 46(11):427–435.

Evans, P.D. and Banks, W.B. (1990). Degradation of wood surfaces by water: Weight losses and changes in ultrastructural and chemical composition. *Holz als Roh- und Werkstoff*, 48(4):159– 163.

Evans, P.D. and Schmalzl, K.J. (1989). A quantitative weathering study of wood surfaces modified by chromium VI and iron III compounds. *Holzforschung* 43(5):289–292.

Evans, P.D., Michell, A.J., and Schmalzl, K.J. (1992). Studies of the degradation and protection of wood surfaces. *Wood Science and Technology* 26(2):151–163.

Evans, P.D, Owens, N.L., Schmid, S., and Webster, R.D. (2002). Weathering and photostability of benzoylated wood. *Polymer Degradation and Stability* 76(2):291–303, 2002.

Evans, P.D., Pirie, J.D.R., Cunningham, R.B., Donnelly, C.F., and Schmalzl, K.J. (1994). A quantitative weathering study of wood surfaces modified by chromium VI and iron III compounds: Part 2: Image analysis of cell wall pit micro-checking. *Holzforschung* 48(4):331–336.

Evans, P.D., Thay, P.D., and Schmalzl, K.J. (1996). Degradation of wood surfaces during natural weathering. Effects on lignin and cellulose and on the adhesion of acrylic latex primers. *Wood Science and Technology* 30(6):411–422.

Feist, W.C. (1987). Weathering of wood and its control by water-repellent preservatives. In Hamel, M.P. (Ed.), *Wood Protection Techniques and the Use of Treated Wood in Construction*. Proceedings Wood Protection Techniques and the Use of Treated Wood in Construction Conference. Forest Products Society, Memphis, TN.

Feist, W.C. (1979). Protection of wood surfaces with chromium trioxide. U.S. Forest Service Res. Pap. FPL 339. USDA Forest Service, Forest Products Laboratory, Madison, WI.

Feist, W.C. (1990). Outdoor wood weathering and protection. In: Rowell, R.M. and Barbour, J.R. (Eds.), *Archaeological Wood: Properties, Chemistry, and Preservation*. Advances in Chemistry Series 225. Proceedings of 196th meeting, American Chemical Society; 1988 September 25–28, Los Angeles. American Chemical Society, Washington DC.

Feist, W.C. and Hon, D.N.-S. (1984). Chemistry and weathering and protection. In: Rowell, R.M. (Ed.), *The Chemistry of Solid Wood*. Advances in Chemistry Series 207. American Chemical Society, Washington, DC.

Feist, W.C. and Mraz, E.A. (1978). Comparison of outdoor and accelerated weathering of unprotected softwoods. *Forest Products Journal* 28(3):38–43.

Feist, W.C., and Rowell, R.M. (1982). Ultraviolet degradation and accelerated weathering of chemically modified wood. In: Hon, D.N.-S. (Ed.), *Graft Copolymerization of Lignocellulose Fibers*. American Chemical Society Symposium Series 187. American Chemical Society, Washington, DC. pp. 349–370.

Feist, W.C. and Williams, R.S. (1991). Weathering durability of chromium-treated southern pine. *Forest Products Journal* 41(1):8–14.

Feist, W.C., Rowell, R.M., and Ellis, W.D. (1991). Moisture sorption and accelerated weathering of acetylated and methacrylated aspen. *Wood and Fiber Sci.* 23(1):128–136.

Futo, L.P. (1976). Effects of temperature on the photochemical degradation of wood. I. Experimental presentation. *Holz als Roh- und Werkstoff* 34(1):31–36.

Futo, L.P. (1976). Einfluss der Temperatur auf den photochemischen Holzabbau: 2. Mitteilung: Rasterelektronenmikroskopische Darstellung. *Holz als Roh- und Werkstoff* 34(2):49–54.

Fuwape, J.A. and Adeduntan, S.O. (1999). Effect of photodegradation on some properties of *Gmelina arborea* wood. *J. Timb. Dev. Assoc.* (India), XLV(3–4):5–9.

Groves, K.W. and Banana, A.Y. (1986). Weathering characteristics of Australian grown radiata pine. *J. Inst. Wood Sci.* 10(5):210–213.

Hayashi, T., Miyatake, A., and Harada, M. (2002). Outdoor exposure tests of structural laminated veneer lumber. I. Evaluation of physical properties after six years. *The Japanese Wood Research Society, J. Wood Sci.* 48:69–74.

Helinska-Raczkowska, L. and Raczkowski, J. (1978). Creep in pine wood subjected previously to atmospheric corrosion in contact with rusting iron. *Holzforschung und Holzverwertung* 30(3):50–54.

Hill, C.A.S., Cetin, N.S., Quinney, R.F., Derbyshire, H., and Ewen, R.J. (2001). An investigation of the potential for chemical modification and subsequent polymeric grafting as a means of protecting wood against photodegradation. *Polymer Degradation and Stability* 72(1):133–139.

Hon, D.N.-S. (1981a). Photochemical degradation of lignocellulosic materials. In: Grassie, N. (Ed.), *Developments in Polymer Degradation–3*. Essex: Applied Science Ltd. Chapter 8, 229–281.

Hon, D.N.-S. (1981b). Weathering of wood in structural use. In: *Environmental Degradation of Engineering Materials in Aggressive Environments*. Proceedings of Second International Conference on Environmental Degradation of Engineering Materials. September 21–23, 1981, Blacksburg, VA.

Hon, D.N.-S., (1983). Weathering reactions and protection of wood surfaces. *J Appl. Polymer Sci.* 37(1):845–864.
Hon, D.N.-S. (1993). Degradative effects of ultraviolet light and acid rain on wood surface quality. *Wood and Fiber Sci.* 26(2):185–191.
Hon, D.N.-S. (1995). Stabilization of wood color: acetylation blocking effective? *Wood and Fiber Sci.* 27(4):360–367.
Hon, D.N.-S. and Chang, S.-T. (1984). Surface degradation of wood by ultraviolet light. *J. of Polymer Science* 22:2227–2241.
Hon, D.N.-S., Chang, S.-T., and Feist, W.C. (1982). Participation of Singlet Oxygen in the Photodegradation of wood surfaces. *Wood Sci. Technol.* 16(3):193–201, 1982.
Hon, D.N.-S., Chang, S.-T., and Feist, W.C. (1985). Protection of wood surfaces against photooxidation. *J. of Applied Polymer Sci.* 30:1429–1448.
Hon, D.N.-S. and Feist, W.C. (1986). Weathering characteristics of hardwood surfaces. *Wood Sci. Technol.* 20(2):169–183.
Hon, D.N.-S. and Feist, W.C. (1992). Hydroperoxidation in photoirradiated wood surfaces. *Wood and Fiber Science* 23(4):448–455.
Hon, D.N.-S. and Ifju, G. (1978). Measuring penetration of light into wood by detection of photo-induced free radicals. *Wood Sci.* 11(2):118–127.
Hon, D.N.-S., Ifju, G., and Feist, W.C. (1980). Characteristics of free radicals in wood. *Wood and Fiber* 12(2):121–130.
Horn, B.A., Qiu, J., Owen, N.L., and Feist, W.C. (1992). FT-IR Studies of weathering effects in western redcedar and southern pine. In: *Chemical Modification of Cellulose*, FRI Bull. 176. Forest Research Institute, Rotorua, New Zealand, pp. 67–76.
Hussey, B.E., and Nicholas, D.D. (1985). The effect of light stabilizers on the iron and water degradation of wood. Proceeding American Wood Preserver's Association, No. 81, 169–173.
Imamura, Y. (1993). Morphological change in acetylated wood exposed to weathering. *Wood Research*, 79:54–61.
Jin, L., Archer, K., and Preston, A. (1991). Surface characteristics of wood treated with various AACs, ACQ and CCA formulations after weathering, Int. Res. Group on Wood Pres., Doc No. IRG/WP/2369.
Kalnins, M.A. (1966). Surface characteristics of wood as they affect durability of finishes. Part II. Photochemical degradation of wood. U.S. Forest Service Res. Pap. FPL 57, Madison, WI, U.S. Department of Agriculture, Forest Service, Forest Products Laboratory: pp. 23–57.
Kalnins, M.A. (1984). Photodegradation of acetylated, methylated, phenylhydrazine-modified, ACC- treated wood. *J. Applied Polymer Science* 29(1):105–115.
Kalnins, M.A. and Feist, W.C. (1993). Increase in wettability of wood with weathering. *Forest Products J.* 43(2):55–57.
Kamoun, D., Merlin, A., Deglise, X., Urizar, S.H., and Fernandez, A.M. (1999). Electron paramagnetic resonance spectroscopy study of photodegradation of lignins extracted and isolated from *Pinus radiata* wood. *Annals of Forest Sci.* 56(7):563–578.
Kang, H.-Y., Park, S.-J., and Kim, Y.-S. (2002). Moisture sorption and ultrasonic velocity of artificially weathered spruce. *Mokchae Konghak*, 30(1):18–24.
Kataoka, Y. and Kiguchi, M. (2001). Depth profiling of photo-induced degradation in wood by FTIR microspectroscopy. *J. Wood Sci.* 47(4):325–327.
Kiguchi, M. (1990). Chemical modification of wood surfaces by etherification. II. Weathering ability of hot-melted wood surfaces and manufacture of self hot-melt bonded particleboard. *Mokuzai Gakkaishi*, 36(10):867–875.
Kiguchi, M. (1992). Photo-deterioration of chemically modified wood surfaces: preliminary study with ESCA. In; *Chemical Modification of Lignocellulosic*, Rotorua, New Zealand, Forest Research Institute Bulletin No. 176:77–86, 1992.
Kiguchi, M. (1997). Photo-deterioration of chemically modified wood surfaces: acetylated and alkylated wood. *JARQ* 31(2):147–154.
Kleive, K. (1986). Weathered wooden surfaces—their influence on the durability of coating systems. *J. Coatings tech.* 58(740):39–43.
Kollmann, F.F.P. and Côté, W.A. (1968). *Principles of Wood Science and Technology: I Solid Wood.* Springer-Verlag, Berlin.
Košíková, B. and Tolvaj, L. (1998). Characterization of lignin fraction isolated from photodegraded wood. *Drevársky Výskum* 43(2):19–28.

Kubel, H. and Pizzi, A. (1981). Protection of wood surfaces with metallic oxides. *J. of Wood Chemistry and Technology* 1(1):75–92.

Kučera, L.J. and Sell J. (1987). Weathering behavior of beech wood in the ray tissue region. *Holz als Roh- und Werkstoff*, 45(3):89–93.

Kuo, M. and Hu, N. (1991). Ultrastructural changes of photodegradation of wood surfaces exposed to UV. *Holzforschung* 45(5):347–353.

Leukens, von U. and Sell, J. (1972). Bewitterungsversuch der EMPA mit Holz und Anstrichen für Holzfassaden. *Separatabdruck aus der Schweizerischen Maler- und Gipsermeister-Zeitung* Nr. 14.

Li, J. (1988). A study on the weathering resistance of lignocellulosic material. *Scientia Silvae Sinicae* 24(3):266–371.

Li, J., Han, S.J., Xu, Z.C., and Peng, H.Y. (1989). Surface deterioration and protection of lignocellulosic material. *J. of Northeast Forestry University* 17(2):48–56.

Li, R. (2000). Environmental degradation of wood-HDPE composite. *Polymer Degradation and Stability* 70(2):135–145.

Lin, S.Y. (1982). Photochemical reaction mechanisms of lignin. Forest Products Industries, Dept. of Forestry, National Taiwan Univ. 1(2):2–19.

Lin, S.Y. and Kringstad, K.P. (1970). Photosensitive groups in lignin and lignin model compounds. *Tappi* 53(4):658–663.

Lindberg, B. (1986). Improving the outdoor durability of wood by surface treatment. *FATIPEC Congress*, 18(2/B):589–608.

Liu, R., Ruddick, J.N.R., and Jin, L. (1994). The influence of copper (II) chemicals on the weathering of treated wood, Part I. ACQ treatment of wood on weathering. Int. Res. Group on Wood Pres., Doc. No. IRG/WP/94-30040.

Lundin, T., Falk, R.H., and Felton, C. (2002). Accelerated weathering of natural fiber-thermoplastic composites: effect of ultraviolet exposure on bending strength and stiffness. Proceedings: Sixth International Conference on Woodfiber-Plastic Composites, Madison, WI: Forest Products Society, pp. 87–93.

MacLeod, I.T., Scully, A.D., Ghiggino, K.P., Ritchie, P.J.A., Paravagna, O.M., and Leary, B. (1995). Photodegradation at the wood-clearcoat interface. *Wood Science and Technology* 29(3):183–189.

Martin, J.W. and Bauer, D.R. (Eds). (2002). *Service Life Prediction Methodology and Metrology*. American Chemical Society Series. 805 New York: Oxford Press.

Matuana, L.M. and Kamdem, D.P. (2002). Accelerated ultraviolet weathering of PVC/wood-flour composites. *Polymer Engineering and Sci.* 42(8):1657–1666.

Matuana, L.M., Kamdem, D.P., and Zhang, J. (2001). Photoaging and stabilization of rigid PVC/wood-fiber composites. *J. of Applied Polymer Sci.* 80(11):1943–1950.

McKellar, J.F. and Allen, N.S. (1979). *Photochemistry of Man-Made Polymers*. London: Applied Science Publishers, Ltd.

Minemura, N., Umehara, K., and Sato, M. (1983). The surface performance of wood coated with water repellents. *J. of the Hokkaido Forest Products Res. Inst.* No. 380:11–16.

Mitsui, K. (2004). Changes in the properties of light-irradiated wood with heat treatment. Part 2. Effect of light-irradiation time and wavelength. *Holz Roh Werkst* 62(1):23–30.

Mitsui, K., Takada, H., Sugiyama, M., and Hasegawa, R. (2001). Changes in the properties of light-irradiated wood with heat treatment. Part 1. Effect of treatment conditions on the change in color. *Holzforchung* 55(6):601–605.

Murakami, K. and Matsuda, H. (1990). Oligoesterified woods based on anhydride and epoxide VIII. Resistance of oligoesterified woods against weathering and biodeteration. *Mokuzai Gakkaishi*, J. of the Japan Wood Res. Soc. 36(7):538–544.

Nakamura, F. and Sato, M. (1978). Degradation of the surface layer of weathered plywood: Measuring the depth of degradation in a Taber-Abrasion test method. *J. of the Hokkaido Forest Products Res. Inst.*, No. 320:13–18.

Nakano, T., Yamashima, H., and Kawakami, H. (1980). Improvement of wood by impregnation with low content resins: III. The durability in outdoor exposure of the coating films on wood plastic composites. Hokkaido For. Prod. Res. Inst., Hokkaido, Japan. *Rinsan Shikenjo Geppo*, 346:14–18.

Németh, K. and Faix, O. (1994). Observation of the photodegradation of wood by quantitative DRIFT spectroscopy. *Holz als Roh- und Werkstoff*, 52(4):261–266.

Ohkoshi, M., Tsuji, N., and Suzuki, K. (1996). Improvement in weathering performance of transparent-resin-coated wood by acetylation and treatment with PEGMA. *Mokuzai kogyo*, 51(5):203–208.

Onishi, M., Tsujimoto, Y., and Sakuno, T. (1989). Degradation behavior of exterior wood after outdoor exposure for 2 years. Research Bulletin of the Tottori University Forests. No. 26:93–99.

Onoda, K. (1989). Chemical modification of wood. *Mokuzai Kogyo*, 44(511):476–480.

Ostman, B.A.L. (1983). Surface roughness of wood-based panels after aging. *Forest Products Journal*. 33(7,8):35–42.

Panda, R. and Panda, H. Chemical modification of wood; protection of wood surfaces against photooxidation. *Advances in forestry Res. in India* Vol. XIV:220–249.

Pandey, K.K. and Khali, D.P. (1998). Accelerated weathering of wood surfaces modified by chromium trioxide. *Holzforschung*, 52(5):467–471.

Pandey, K.K. and Pitman, A.J. (2002). Weathering characteristics of modified rubberwood (*Hevea brasiliensis*). *J. of Appl. Polm. Sci.* 85(3):622–631.

Papp, G., Tolvaj, L., and Barta, E. (1999). Proceedings: 4th International Conference on the Development of Wood Science 4th M'issenden Abbey, UK, 14–16 July, 1999, pp. 494–501.

Park, B.S., Furuno, T., and Uehara, T. (1996). Histochemical changes of wood surfaces irradiated with ultraviolet light. *Mokuzai Gakkaishi* 42(1):1–9.

Pizzi, A. (1981a). The chemistry and kinetic behavior of Cu-Cr-As/B wood preservatives. I. Fixation of chromium on wood. *J. Polym. Sci. Chem Ed.* 19(12):3093–3121.

Pizzi, A. (1981b). Wood water proofing and lignin crosslinking by means of chromium trioxide/guajacyl units complexes. *J. of Applied Polymer Sci.* 25(11):2547–2553.

Pizzi, A. and Mostert, D. (1980). Model compound kinetics of the polymerization of CrO_3 with lignin guajacyl units. *Holzforschung und Holzverwertung* 32(6):150–152.

Plackett, D.V. and Cronshaw, D.R. (1992). Performance of transparent finishes on chemically pretreated timber surfaces. *Surface Coatings Australia* 29(11):11–15.

Raczkowski, J. (1982). Seasonal effects on the atmospheric corrosion of spruce micro-sections. *Holz als Roh- und Werkstoff*, 38(6):231–234.

Raczkowski, J. (1982). Modification of wood with polystyrene improves its resistance against accelerated weathering in contact with rusting iron. In: Meyer, R.W. and Kellogg, R.M. (Eds.), *Structural Uses of Wood in Adverse Environments*. pp. 150–155.

Rånby, B. and Rabek, J.F. (1975). *Photodegradation, photooxidation and photostabilization of Polymers, Principles and Applications*. London: John Wiley and Sons, pp. 50–76.

Rapp, A.O. and Peek, R.-D. (1999). Melamine resin treated as well as varnish coated and untreated solid wood during two years of natural weathering. *Holz als Roh- und Werkstoff*, 57(5):331–339.

Rowell, R.M., Feist, W.C., and Ellis, W.D. (1981). Weathering of chemically modified southern pine. *Forest Products Journal* 12(4):202–208.

Rowell, R.M., Lange, S.E., and Jacobson, R.E. (2000). Weathering performance of plant-fiber/thermoplastic composites. *Mol. Cryst. and Liq. Cryst.* 353:85–94.

Sandberg, D. (1999). Weathering of radial and tangential wood surfaces of pine and spruce. *Holzforschung*, 53(4):355–264.

Schmalzl, K.J., Forsyth, C.M., and Evans, P.D. (1995). The reaction of guaiacol with iron III and chromium VI compounds as a model for wood surface modification. *Wood Science and Technology*, 29(4):307–319.

Schmalzl, K.J., Forsyth, C.M., and Evans, P.D. (2003). Evidence for the formation of chromium III diphenoquinone complexes during oxidation of guaiacol and 2,6-dimethoxyphenol with chromic acid. *Polymer Degradation and Stability* 82(3):399–407.

Sell, J. and Leukens, U. (1971). Investigations on weathered wood surfaces. Part II. Weathering: phenomenon of unprotected wood species. *Holz als Roh- und Werkstoff* 29(1):23–31.

Stark, N.M. and Matuana, L.M. (2002). Photostabilization of wood flour filled HDPE composites. Annual Technical Conference, Society of Plastics Engineers, 60(2):2209–2213.

Stark, N.M., Matuana, L.M., and Clemons, C.M. (2004). Effect of processing method on surface and weathering characteristics of wood–flour/HDPE composites. *J. Appl. Polymer Sci* 93(3):1021–1030.

Sudiyani, Y., Immura, Y., and Takahashi, M. (1996). Weathering effects on several properties of chemically modified wood. Proceedings: Annual Meeting of the Japan Wood Res. Soc., 46th, Kumamoto, April 1996, *Wood Research* No. 83:55–58.

Sudiyani, Y., Takahashi, M., Immura, Y., and Minato, K. (1999a). Physical and biological properties of chemically modified wood before and after weathering. *Wood Research* No. 86:1–6.

Sudiyani, Y., Tsujiyama, S., Immura, Y., Takahashi, M., Minato, K., and Kajita, H. (1999b). Chemical characterization of surfaces of hardwood and softwood deteriorated by weathering. *J. Wood Sci.* 45(4):348–353.

Sudiyani, Y., Fujita, M., and Immura, Y. (2002). Deposition of PF-resin on lumen surfaces and its effect on wood structure during weathering. Abstracts, 5th Pacific Wood Anatomy Conference, Yogyakarta, *IAWA J.* 23(4):477–478.

Tolvaj, L. and Faix, O. (1995). Artificial aging of wood monitored by DRIFT spectroscopy and CIE L*a*b* color measurements. *Holzforschung*, 49(5):397–404.

Turkulin, H., Arnold, M., Derbyshire, H., and Sell, J. (1997). SEM study of the weathering effects on painted wood. *Drvna Industrija* 48(2):61–78.

Underhaug, Å., Lund, T.J., and Kleive, K. (1983). Wood protection–the interaction between substrate and product and the influence on durability. *J. of the Oil and Colour Chemists' Association* 66(11):345–355.

Voulgardis E.V. and Banks, W.B. (1981). Degradation of wood during weathering in relation to water repellent long-term effectiveness. *J. Inst. Wood Sci.* 9(2):72–83.

Wang, S.Y. (1981). Studies on the properties of wood deterioration. VI. The reduction in strength properties of some Taiwan species after 4 years exposure in outdoor environments. *Quarterly Journal of Chinese Forestry* 14(4):29–39.

Wang, S.Y. (1990). Reduction of mechanical properties of seventeen Taiwan native-wood species subjected to a seven-year exposure in an outdoor environment. *Mokuzai Gakkaishi* 36(1):69–77.

Wang, S.Y., Chiu, C.M., and Chen, Z.C. (1980). Studies on the properties of wood deterioration. 1. The weathering resistance of sixteen different Taiwan native wood species tested by accelerated weathering resistance method. 2. The decay resistance of eighteen different Taiwan native wood species tested by accelerated decay resistance method. *Quarterly Journal of Chinese Forestry* 12(1):21–39 and 13(1):55–93.

Wang, S.Y. and Lin, S.J. (1991). The effect of outdoor environmental exposure on the main components of wood. *Mokuzai Gakkaishi* 37(10):954–963.

Wang, S.Y. and Tsai, M.J. (1988). The improving effects on weathering-resistance of picnic-tables made from Japanese-cedar and China-fir woods treated with CCA, CCB, and paraffin wax. *Quarterly J. Expt. For. NTU* 2(3):17–30.

Williams, R.S. (1983). Effect of grafted UV stabilizers on wood surface erosion and clear coating performance. *J. of Applied Polymer Sci.* 28:2093–2103.

Williams, R.S. (1987). Acid effects on accelerated wood weathering. *Forest Products Journal.* (37(2):37–38.

Williams, R.S. (1988). Effect of dilute acid on the accelerated weathering of wood. *JAPCA* 38:(2)149–151.

Williams, R.S. and Feist, W.C. (1984). Application of ESCA to evaluate wood and cellulose surfaces modified by aqueous chromium trioxide treatment. *Colloids and Surfaces* 9(3):253–271.

Williams, R.S. and Feist, W.C. (1985). Wood modified by inorganic salts: mechanism and properties. I. Weathering rate, water repellency, and dimensional stability of wood modified with chromium (III) nitrate versus chromic acid. *Wood and Fiber Science* 17(2):184–198.

Williams, R.S. and Feist, W.C. (1994). Effects of preweathering, surface roughness, and wood species on the performance of paint and stains. *J. Coatings Tech.* 66(828):109–121.

Williams, R.S. and Feist, W.C. (1993). Durability of paint or solid-color stain applied to preweathered wood. *Forest Products Journal* 43(1):8–14.

Williams, R.S., and Feist, W.C. (2001). Duration of wood preweathering: effect on the service life of subsequently applied paint. *J. Coatings Tech.* 73(930): 65–72.

Williams, R.S., Knaebe, M.T., Evans, J.W., and Feist, W.C. (2001c). Erosion rates of wood during natural weathering: Part III. Effect of exposure angle on erosion rate, *Wood and Fiber Sci.* 33(1):50–57.

Williams, R.S., Knaebe, M.T., and Feist, W.C. (2001b). Erosion rates of wood during natural weathering: Part II. Earlywood and latewood erosion rates, *Wood and Fiber Sci.* 33(1):43–49.

Williams, R.S., Knaebe, M.T., Sotos, P.G., and Feist, W.C. (2001a). Erosion rates of wood during natural weathering: Part I. Effect of grain angle and surface texture, *Wood and Fiber Sci.* 33(1):31–42.

Williams, R.S., Miller, R., and Gangstad, J. (2001d). Characteristics of ten tropical hardwoods from certified forests in Bolivia: Part 1, Weathering characteristics and dimensional change. *Wood and Fiber Sci.* 33(4):618–626.

Williams, R.S., Sotos, P., and Feist, W.C. (1999). Evaluation of several finishes on severely weathered wood. *J. Coatings Tech.* 71(895):97–102.

Williams, R.S., Winandy, J.E., and Feist, W.C. (1987). Paint adhesion to weathered wood. *J. of Coatings Tech.* 59(749):43–49.

Wood Handbook: 1999 Wood as an engineering material. Forest Products Laboratory General Technical Report FPL-GTR-113. Forest Products Laboratory, Madison, WI: 1999.

Yata, S. and Tamura, T. (1995). Histological changes of softwood surfaces during outdoor weathering. *Mokuzai Gakkaishi* 41(11):1035–1042.

Yoshida, H. (1998). Strength properties of weathered tropical wood plywoods. In: Hse, C.Y. (Ed.), *Adhesive Technology and Bonded Tropical Wood Products*. TFRI Extension Series No. 96. Taipei, Taiwan: Taiwan Forestry Research Inst. pp. 445–463.

Yoshida, H. and Taguchi, T. (1977). Bending properties of weathered plywood. I. Analysis of strength loss of exposed plywood. *Mokuzai-Gakkaishi* 23(11):547–551.

Yoshida, H. and Taguchi, T. (1977). Bending properties of weathered plywood. II. Degradation of outermost plies. *Mokuzai-Gakkaishi* 23(11):552–556.

Zhao, B., Han, S., Su, R., Zhao, B.S., Han S.J., and Su, R.Z. (1995) Factors affecting free-radicals on the wood surface. *Scientia Silvea Sinicae*, 31(1):56–59.

8 Surface Characterization

Mandla A. Tshabalala
USDA, Forest Service, Forest Products Laboratory, Madison, WI

CONTENTS

8.1 Overview of Surface Properties 187
8.2 Microscopic Methods for Characterizing Surface Properties 188
    8.2.1 Confocal Laser Scanning Microscope 188
    8.2.2 Scanning Electron Microscopy 190
    8.2.3 Atomic Force Microscopy 192
8.3 Spectroscopic Methods for Characterizing Surface Properties 196
    8.3.1 Molecular Spectroscopy 196
    8.3.2 Electron Spectroscopy 199
    8.3.3 Mass Spectroscopy 202
8.4 Thermodynamic Methods for Characterizing Surface Properties 204
    8.4.1 Contact Angle Analysis 204
    8.4.2 Inverse Gas Chromatography 206
    8.4.3 Total Surface Energy 207
        8.4.3.1 Dispersive Component of the Total Surface Energy 207
        8.4.3.2 Acid-Base Component of the Total Surface Energy 207
8.5 Conclusions and Outlook 209
References 209

8.1 OVERVIEW OF SURFACE PROPERTIES

Surface properties of wood play an important role when wood is used or processed into different commodities such as siding, joinery, textiles, paper, sorption media or wood composites. Thus, for example, the quality and durability of a wood coating are determined by the surface properties of the wood and the coating. The same is true for wood composites, as the efficiency of stress transfer from the wood component to the non-wood component is strongly influenced by the surface properties of both components.

Surface properties of wood can be divided into two major groups: physical and chemical properties. Physical properties include morphology, roughness, smoothness, specific surface area and permeability. Chemical properties include elemental and molecular, or functional, group composition. Together, these two major groups of properties determine the thermodynamic characteristics of the wood surface, such as surface free energy and surface acid-base acceptor and donor numbers.

Wood has a cellular structure, the cell walls of which are composed of three major constituents: cellulose, hemicelluloses, and lignin. In addition to these major constituents, the cell walls also contain pectins, extractives and trace metals. The surface properties of wood are therefore determined by the morphology of the cell wall at the surface of a wood element (particle, fiber, flake, or chip), and the distribution of the major and minor constituents in the cell wall. Hence, to optimize the interaction between a wood surface and a coating, or a matrix in a wood composite, the surface properties of both the wood and the coating, or the matrix in a composite, must be known.

Methods for characterizing surface properties of wood may be divided into three broad categories: microscopic, spectroscopic, and thermodynamic. Microscopic methods provide information about surface morphology; spectroscopic methods provide information about surface chemistry, and thermodynamic methods provide information about the surface energy.

8.2 MICROSCOPIC METHODS FOR CHARACTERIZING SURFACE PROPERTIES

Many types of microscopic methods are available for characterizing the physical properties of various material surfaces, but only a few have been particularly useful for characterizing the physical properties of wood surfaces: Confocal laser scanning microscopy (CLSM), scanning electron microscopy (SEM), and atomic force microscopy (AFM).

8.2.1 CONFOCAL LASER SCANNING MICROSCOPY

A typical confocal laser scanning microscopy (CLSM) instrument consists of a light microscope equipped with scanning mechanisms and a motorized focus, a laser source (He/Ne, krypton or argon), and a computer system with software for instrument control, three-dimensional (3-D) reconstruction, and image processing and analysis.

The basic principle of CLSM is illustrated schematically in Figure 8.1. A collimated, polarized laser beam from an aperture is reflected by a beam splitter (dichroic mirror) into the rear of the objective lens and is focused on the specimen. The reflected, or emitted, longer-wavelength fluorescent light returning from the specimen passes back through the same lens. The light beam is focused by the beam splitter into a small pinhole, the confocal aperture, to eliminate all of the out-of-focus light that comes from regions of the specimen that are above or below the plane of focus. A photo multiplier tube (PMT), positioned behind the confocal aperture, converts the detected in-focus light beam from each specimen point into an analog output signal, which is stored in a computer in digital form. A point-by-point digital image is obtained by scanning the beam over an XY plane. The thickness of such an XY image in the Z direction depends on the diameter of the detector pinhole: the more open the pinhole, the thicker the XY image (Pawley 1990, Boyde 1994, Lichtman 1994, Béland and Mangin 1995, Leica 1999).

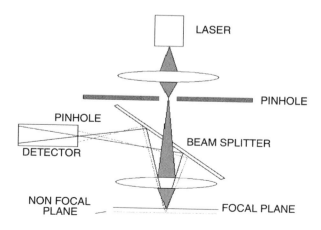

FIGURE 8.1 Principle of CLSM. Out-of-focus light beam from non focal plane is excluded from the detector by the confocal detector pinhole. (Courtesy of Lloyd Donaldson, M.Sc. Hons, D. Sc. Cell Wall Biotechnology Center, Forest Research, New Zealand.)

Surface Characterization

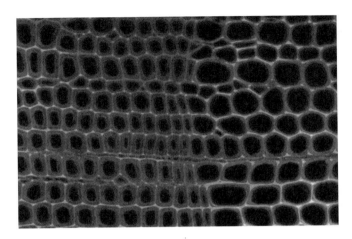

FIGURE 8.2 Transverse view of thick-wall latewood (left) and thin-wall early wood (right) tracheids of *Pinus radiata*.(Micrographs provided by Lloyd Donaldson, M.Sc. Hons, D.Sc. Cell Wall Biotechnology Center, Forest Research, New Zealand.)

Perhaps the greatest advantage of CLSM over other forms of optical microscopy is that it allows 3-D imaging of thick and opaque specimens, such as wood surfaces, without physically slicing the specimen into sections. Figures 8.2–8.4 show 625- × 625-µm micrographs of a wood specimen that were obtained with a Leica TCS/NT confocal microscope. The wood specimen was stained with acriflavin (Donaldson 2003). The micrographs clearly show the anisotropic morphology of wood surfaces. In the transverse view (Figure 8.2) thin-walled earlywood and thick-walled latewood tracheids are clearly distinguishable. In the radial view (Figure 8.3) radial bordered pits and a uniseriate heterogeneous ray are clearly visible. In the tangential view (Figure 8.4) uniseriate rays, small tangential pits in latewood, and larger tangential pits in earlywood are also clearly visible.

CLSM has also been used in the study of resin distribution in medium density fiberboard (Loxton et al. 2003). In another study, CLSM was used to study the effect of simulated acid rain on coatings for exterior wood panels (Lee et al 2003).

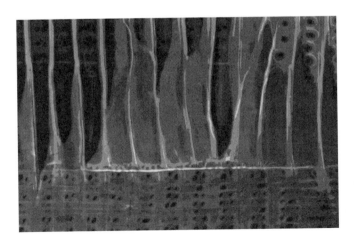

FIGURE 8.3 Radial view of tracheids of *Pinus radiata* showing radial bordered pits (upper left-hand corner) and uniseriate heterogeneous ray (center of micrograph).

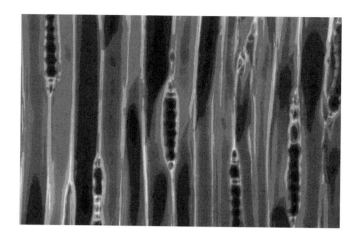

FIGURE 8.4 Tangential view of tracheids of *Pinus radiata* showing fusiform rays, and uniseriate heterogeneous and homogeneous rays.

8.2.2 SCANNING ELECTRON MICROSCOPY

The basic principle of scanning electron microscopy (SEM) is illustrated in Figure 8.5. The primary electron beam, which is produced under high vacuum at the top of the microscope by heating a metallic element, is scanned across the surface of a specimen. When the electrons strike the specimen, a variety of signals is generated, and it is the detection of specific signals that produces an image of the surface, or its elemental composition. The three signals that provide the greatest amount of information in SEM are the secondary electrons, backscattered electrons, and X-rays.

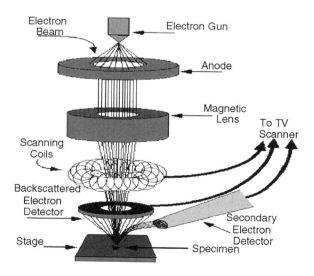

FIGURE 8.5 Principle of SEM. The primary electron beam produced at the top of the microscope by the heating of a metallic filament is focused and scanned across surface of a specimen by means of electromagnetic coils. Secondary and backscattered electron beams ejected from the surface are collected in detectors, which convert them to a signal that is processed into an image. (Courtesy of Prof. Scott Chumbley, Iowa State University, Materials Science and Engineering Department.)

Secondary electrons are emitted from the atoms occupying the top surface layer and produce an image of the surface. The contrast in the image is determined by the surface morphology. A high-resolution image can be obtained because of the small diameter of the primary electron beam.

Backscattered electrons are primary beam electrons, which are reflected by the atoms in the surface layers. The contrast in the image produced is determined by the atomic number of the elements in the surface layers. The image will therefore show the distribution of different chemical phases in the specimen surface. Because backscattered electrons are emitted from a depth in the specimen, the resolution in the image is usually not as good as for secondary electrons.

Interaction of the primary beam with atoms in the specimen causes electron shell transitions, which result in the emission of an X-ray. The emitted X-ray has an energy that is characteristic of its parent element. Detection and measurement of X-ray energy permits elemental analysis and is commonly referred to as Energy Dispersive X-ray Spectroscopy (EDS or EDX or EDXA). EDS can provide rapid qualitative, or with adequate calibration standards, quantitative analysis of elemental composition with a sampling depth of 1–2 microns. X-rays may also be used to form maps or line profiles, showing the elemental distribution in a specimen surface.

One of the latest innovations in SEM is environmental scanning microscopy (ESEM). ESEM differs from conventional SEM in two crucial aspects (Danilatos 1993, Donald 2003). First, ESEM allows the introduction of a gaseous environment in the specimen chamber although the electron gun is kept under the standard SEM high vacuum. Second, the wood specimens do not need to be coated with a metallic layer, as is the case in conventional SEM.

It should be noted that the gaseous environment in the specimen chamber plays a key role in signal detection (Danilatos 1988, Meredith et al. 1996). As the secondary electrons travel toward the positively charged detector they collide with the gas molecules. Each such collision leads to the generation of an additional daughter electron so that an amplified cascade of electrons reaches the detector. Along with additional daughter electrons, positive ions are also produced. These positive ions drift down toward the specimen, and hence serve to compensate charge build-up at the surface of non-conductive specimens such as wood. It is for this reason that non-conductive specimens do not need to be coated with a metallic layer to prevent charging artifacts and consequent loss of image quality.

Figures 8.6–8.8 show ESEM micrographs of specimens prepared from the softwood *Pinus taeda*.

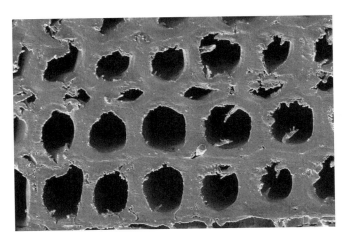

FIGURE 8.6 ESEM micrograph of the softwood *Pinus taeda* showing the transverse surface of thick-walled latewood tracheids. A uniseriate ray is visible above the bottom row of tracheids. (Courtesy T.A. Kuster, USDA Forest Service, Forest Products Laboratory, Madison, WI.)

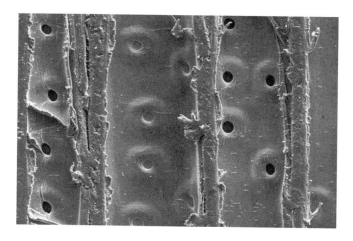

FIGURE 8.7 ESEM micrograph of the softwood *Pinus taeda* showing the radial surface of tracheids. Bordered pits are clearly visible. The warty membrane that lines the lumen is also clearly visible, especially in the left lumen where it has peeled back at some spots. (Courtesy T.A. Kuster, USDA Forest Service, Forest Products Laboratory, Madison, WI.)

8.2.3 Atomic Force Microscopy

The basic principle and application of atomic force microscopy (AFM) has been the subject of a number of excellent reviews (Meyer 1992, Frommer 1992, Hoh and Engel 1993, Frisbie et al. 1994, Hanley and Gray 1994, 1995, Louder and Parkinson 1995, McGhie et al. 1995, Rynders et al. 1995, Schaefer et al 1995, Hansma et al. 1998). In atomic force microscopy (AFM), a probe consisting of a sharp tip (nominal tip radius on the order of 10 nm) located near the end of a cantilever beam is raster scanned across the surface of a specimen using piezoelectric scanners. Changes in the tip-specimen interaction are often monitored using an optical lever detection system, in which a laser is reflected off of the cantilever and onto a position-sensitive photodiode. During scanning, a particular operating parameter is maintained at a constant level, and images are generated through a feedback loop between the optical detection system and the piezoelectric scanners.

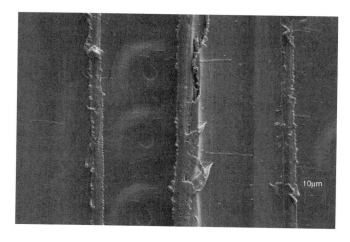

FIGURE 8.8 ESEM micrograph of the softwood *Pinus taeda* showing the tangential surface of thick-walled latewood tracheids. The tangential surface has relatively fewer bordered pits compared to the radial surface. (Courtesy T.A. Kuster, USDA Forest Service, Forest Products Laboratory, Madison, WI.)

Surface Characterization

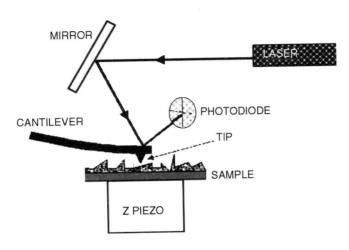

FIGURE 8.9 Schematic diagram of contact mode AFM. The tip makes soft physical contact with the sample. Optical detection of the position of the cantilever leads to a topographic map of the sample surface. (Courtesy of National Physical Laboratory, Teddington, Middlesex, UK, TW11 0LW. © Crown Copyright 2004. Reproduced by permission of the Controller of HMSO.)

Three imaging modes, contact mode, non-contact mode, and intermittent contact or tapping mode can be used to produce a topographic image of the surface. In contact mode (see Figure 8.9), the probe is essentially dragged across the surface of the specimen. During scanning, a constant bend in the cantilever is maintained. A bend in the cantilever corresponds to a displacement of the probe tip, z_t, relative to an undeflected cantilever, and is proportional to the applied normal force, $P = k \cdot z_t$, where k is the cantilever spring constant. As the topography of the surface changes, the z-scanner must move the position of the tip relative to the surface to maintain this constant deflection. Using this feedback mechanism, the topography of the specimen surface is thus mapped during scanning by assuming that the motion of the z-scanner directly corresponds to the surface topography. To minimize the amount of applied force used to scan the sample, low spring constant ($k \leq 1$ N/m) probes are normally used. However, significant deformation and damage of soft samples (e.g., biological and polymeric materials) often occur during contact mode imaging in air because significant force must be applied to overcome the effects of surface roughness or adsorbed moisture as is the case with wood specimens. The combination of a significant normal force, the lateral forces created by the dragging motion of the probe tip across the specimen, and the small contact areas involved results in high contact stresses that can damage either the specimen surface, the tip, or both. To overcome this limitation, contact mode imaging can be performed under a liquid environment, which essentially eliminates problems due to surface moisture so that much lower contact forces can be used. In fact, the ability to image samples under a liquid environment is often a desirable capability of AFM, but in some cases it might not be practical or feasible. Also, working with liquid cells for many commercial AFM systems can be tricky because of the potential for spills and leaks, which can introduce liquid into the scanners.

To mitigate or completely eliminate the damaging forces associated with contact mode, the cantilever can be oscillated near its first bending mode resonance frequency (normally on the order of 100 kHz) as the probe is raster scanned above the specimen surface in either non-contact mode or tapping mode. In non-contact mode (see Figure 8.10), both the tip-specimen separation and the oscillation amplitude are on the order of 1 nm to 10 nm, such that the tip oscillates just above the specimen. The resonance frequency and amplitude of the oscillating probe decrease as the specimen surface is approached as a result of interactions with van der Waals and other long-range forces

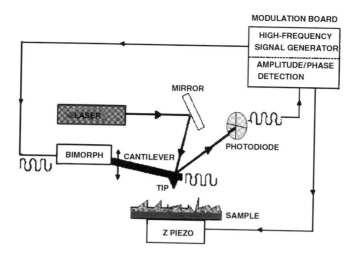

FIGURE 8.10 Schematic diagram of non-contact mode AFM. The cantilever is vibrated near the surface of the sample. Changes in the resonance frequency or vibration amplitude of the cantilever as the tip approaches the surface are used to measure changes in the sample surface topography. (Courtesy of National Physical Laboratory, Teddington, Middlesex, UK, TW11 0LW. © Crown Copyright 2004. Reproduced by permission of the Controller of HMSO.)

extending above the surface. These types of forces tend to be quite small relative to the repulsive forces encountered in contact mode. Both a constant amplitude or constant resonance frequency is maintained through a feedback loop with the scanner, and, similar to contact mode, the motion of the scanner is used to generate the topographic image. To reduce the tendency of the tip to be pulled down to the surface by attractive forces, the cantilever spring constant is normally much higher compared to contact mode cantilevers. The combination of weak forces affecting feedback and large spring constants causes the non-contact AFM signal to be small, which can lead to unstable feedback and requires slower scan speeds than either contact mode or tapping mode. Also, the lateral resolution in non-contact mode is limited by the tip-specimen separation and is normally lower than that in either contact mode or tapping mode.

Tapping mode tends to be more applicable to general imaging in air, particularly for soft samples, as the resolution is similar to contact mode while the forces applied to the specimen are lower and less damaging. In fact, the only real disadvantages of tapping mode relative to contact mode are that the scan speeds are slightly slower and the AFM operation is a bit more complex, but these disadvantages tend to be outweighed by the advantages. In tapping mode, the cantilever oscillates close to its first bending mode resonance frequency, as in non-contact mode. However, the oscillation amplitude of the probe tip is much larger than for non-contact mode, often in the range of 20 nm to 200 nm, and the tip makes contact with the sample for a short time in each oscillation cycle. As the tip approaches the specimen, the tip-specimen interactions alter the amplitude, resonance frequency, and phase angle of the oscillating cantilever. During scanning, the amplitude at the operating frequency is maintained at a constant level, called the *set-point amplitude*, by adjusting the relative position of the tip with respect to the specimen. In general, the amplitude of oscillation during scanning should be large enough for the probe to maintain enough energy to allow the tip to tap through and back out of the surface layer.

One recent development in tapping mode is the use of the changes in phase angle of the cantilever probe to produce a second image, called a phase image or phase contrast image. This image often provides significantly more contrast than the topographic image and has been shown to be sensitive to material surface properties, such as stiffness, viscoelasticity, and chemical composition. In general, changes in phase angle during scanning are related to energy dissipation during tip-sample interaction and can be caused by changes in topography, tip-specimen molecular interactions, deformation at the tip-specimen contact, and even experimental conditions. Depending on the operating conditions, different levels of tapping force might be required to produce an accurate and reproducible image of a surface. Also, the amount of tapping force used will often affect the phase image, particularly with regard to whether local tip-specimen interactions are attractive or repulsive.

Similar contrast images can be constructed concurrently with the topographic image in contact mode. One example is lateral force or friction force imaging, in which torsional rotation of the probe is detected while the probe is dragged across the surface in a direction perpendicular to the long axis of the cantilever. Friction force imaging with a chemically modified probe (i.e., a probe that has been coated with a monolayer of a specific organic group) is often referred to as chemical force microscopy. Another example of contrast imaging is force modulation, which combines contact mode imaging with a small oscillation of the probe tip at a frequency far below resonance frequency. This oscillating force should deform softer regions more than harder regions of a heterogeneous sample so that contrast between these regions is observed. In practice, however, the difference between the elastic modulus of the different regions usually has to be substantial (e.g., rubber particles in a plastic, carbon fibers in an epoxy) for contrast to be realized. Thus far, these types of AFM contrast images are purely qualitative as a result of inaccurate or unknown spring constants, unknown contact geometry, and contributions from different types of surface properties.

Figure 8.11 shows topographic (height) and phase contrast images of a bordered pit on the surface of the cell wall of a softwood specimen. The images were acquired and recorded simultaneously by tapping-mode AFM at ambient conditions, using a Nanoscope MultiMode SPM™.

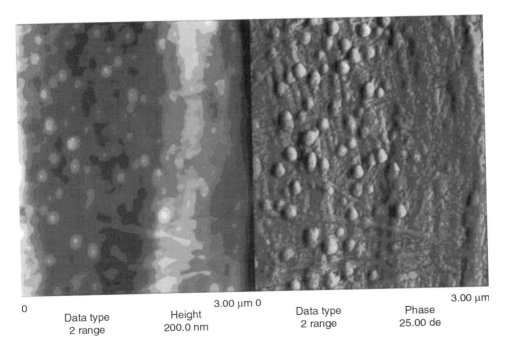

FIGURE 8.11 Topographical (height) and phase contrast images of a bordered pit showing the fibrillar morphology of the margo. The phase contrast image clearly shows the network of nanofibrils that constitute the margo, and nodules located close to the pores that permeate the margo.

8.3 SPECTROSCOPIC METHODS FOR CHARACTERIZING SURFACE PROPERTIES

Although a broad array of spectroscopic methods for surface analysis of materials exists, only a handful of these have been applied in the characterization of wood surfaces. The focus of this section will be on those methods that the author deemed to be of greatest interest.

Spectroscopic methods can be divided into three classes: molecular, electronic, and mass spectroscopies. Molecular spectroscopy includes Fourier transform infrared (FTIR), Fourier transform attenuated total reflectance infrared (FT-ATR), and Fourier Transform Raman (FT-Raman). Electronic spectroscopy includes X-ray photoelectron (XPS) and energy dispersive X-ray spectroscopy (EDXA, EDX or EDS). Mass spectroscopy includes secondary ion mass spectroscopy (SIMS) and static secondary ion mass spectroscopy (SSIMS).

8.3.1 MOLECULAR SPECTROSCOPY

Molecular spectroscopic methods can provide information about the chemical composition of a surface. However, successful characterization of a wood surface by any of the molecular spectroscopic techniques depends upon a judicious choice of sampling techniques and critical interpretation of the data. Detailed descriptions of the fundamentals and applications of molecular spectroscopy can be found elsewhere (Mirabella 1998, Stuart 2004). As significant advances in instrumentation are realized, new applications of infrared spectroscopy in surface characterization of wood have emerged.

TABLE 8.1
Assignment of IR Bands of Solid Wood or Wood Particles Measured by Diffuse Reflectance FTIR

Band Position, cm^{-1}	Assignment
3380 (3418)[a]	O-H stretch vibration (bonded)
2928 (2928)	C-H stretch vibration
1736 (1745)	C=O stretch vibration (unconjugated)
1655 (1660)	H-O-H deformation vibration of adsorbed water and conjugated C=O stretch vibration
1595 (1596)	Aromatic skeletal and C=O stretch vibration
1504 (1507)	Aromatic skeletal vibration
1459 (1465)	C-H deformation (asymmetric) and aromatic vibration in lignin
1423 (1427)	C-H deformation (asymmetric)
1372 (1374)	C-H deformation (symmetric)
1326 (1329)	-CH$_2$ wagging vibration in cellulose
1267 (1270)	C-O stretch vibration in lignin, acetyl and carboxylic vibration in xylan
1236 (1242)	C-O stretch vibration in lignin, acetyl and carboxylic vibration in xylan
1158 (1165)	C-O-C asymmetric stretch vibration in cellulose and hemicellulose
1113 (1128[b])	O-H association band in cellulose and hemicellulose
1055[b] (1083)	C-O stretch in cellulose and hemicellulose
1042 (1036)	C-O stretch
1000 (1003)	C-O stretch in cellulose and hemicellulose
897 (899)	C1 group frequency in cellulose and hemicellulose
667 (670)	C-O-H out-of-plane bending mode in cellulose

[a] Quantities given in parenthesis correspond to band positions obtained from solid wood chips or wood particles undiluted with KBr.
[b] Highest intensity band

Source: Pandey and Theagarajan 1997.

Surface Characterization

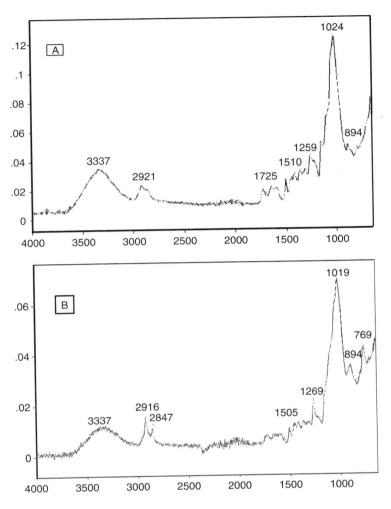

FIGURE 8.12 FTIR-ATR spectra of a wood wafer before (A) and after (B) coating with a thin film of alkoxysilanes.

FTIR and FT-ATR spectra of wood samples have been reported in the literature (Deshmukh and Aydil 1996, Tshabalala et al. 2003). Assignment of infrared absorption bands measured by different FTIR techniques is given in Table 8.1. Figure 8.12 shows examples of infrared spectra of the surface of a softwood wafer obtained by FTIR-ATR.

Infrared analysis of wood surfaces is generally not considered to be sufficiently surface sensitive because the sampling depth of infrared radiation is in the order of 100 µm. Consequently, changes in infrared spectral features caused by changes in surface chemistry are often masked by the spectral features of the underlying bulk chemistry of the wood specimen. In the example shown in Figure 8.12(B), the peaks at 2916, 2847, 1269, 894, and 769 cm^{-1} were attributed to the alkoxysilane thin film that was deposited on the wood wafer (Tshabalala et al. 2003). The peaks at 2916 and 2847 cm^{-1} were assigned to the asymmetric and symmetric C-H stretch modes of the –CH$_3$ and –CH$_2$ groups (Sali et al. 1994, Deshmukh and Aydil 1996, Conley 1996); the peak at 1269 cm^{-1} was assigned to the Si-CH$_3$ bending mode, and the peaks at 894 and 769 cm^{-1} were assigned to Si-C, Si-O and Si-O-CH$_3$ groups (Selamoglu et al. 1989, Fracassi et al. 1992, Libermann and Lichtenberg 1994, Conley 1996).

TABLE 8.2
Raman Band Assignment in the Spectra of Softwood Cellulose and Lignin

Band Position cm⁻¹	Assignment	Attributed to
3065 m[a]	Aromatic stretch	lignin
3007 sh	C-H stretch in OCH$_3$, asymmetric	lignin
2938 m	C-H stretch in OCH$_3$, asymmetric	lignin
2895 vs	C-H and CH$_2$ stretch	cellulose
2886 sh	C-H stretch in R$_3$C-H	lignin
2848 sh	C-H and CH$_2$ stretch	cellulose
2843 m	C-H stretch in OCH$_3$, symmetric	lignin
1658 s	Ring conj. C=C stretch of coniferyl alcohol; C=O stretch of coniferylaldehyde	lignin
1620 sh	Ring conjugated C=C stretch of coniferylaldehyde	lignin
1602 vs	Aryl ring stretch, symmetric	lignin
1508 vw	Aryl ring stretch, asymmetric	lignin
1456 m	H-C-H and H-O-C bending	cellulose
1454 m	O-CH$_3$ deformation; CH$_2$ scissoring; guaiacyl ring vibration	lignin
1428 w	O-CH$_3$ deformation; CH$_2$ scissoring; guaiacyl ring vibration	lignin
1393 sh	Phenolic O-H bend	lignin
1377 m	H-C-C, H-C-O and H-O-C bending	cellulose
1363 sh	C-H bend in R$_3$C-H	lignin
1333 m	Aliphatic O-H bend	lignin
1298 sh	H-C-C and H-C-O bending	cellulose
1297 sh	Aryl-O of aryl-OH and aryl-O-CH$_3$; C=C stretch of coniferyl alcohol	lignin
1271 m	Aryl-O of aryl-OH and aryl-O-CH$_3$; guaiacyl ring (with C=O group)	lignin
1216 vw	Aryl-O of aryl-OH and aryl-O-CH$_3$; guaiacyl ring (with C=O group)	lignin
1191 w	A phenol mode	lignin
1149 sh	C-C and C-O stretch plus H-C-C and H-C-O bending	cellulose
1134 m	A mode of coniferaldehyde	lignin
1123 s	C-C and C-O stretch	cellulose
1102 w	Out of phase C-C-O stretch of phenol	lignin
1095 s	C-C and C-O stretch	cellulose
1073 sh	C-C and C-O stretch	cellulose
1063 sh	C-C and C-O stretch	cellulose
1037 sh	C-C and C-O stretch	cellulose
1033 w	C-O of aryl-O-CH$_3$ and aryl-OH	lignin
1000 vw	C-C and C-O stretch	cellulose
971 vw	C-C and C-O stretch	cellulose
969 vw	-C-C-H and –HC=CH- deformation	lignin
926 vw	-C-C-H wag	lignin
900 vw	Skeletal deformation of aromatic rings, substituent groups and side chains	lignin
899 m	H-C-C and H-C-O bending at C6	cellulose
787 w	Skeletal deformation of aromatic rings, substituent groups and side chains	lignin
731 w	Skeletal deformation of aromatic rings, substituent groups and side chains	lignin
634 vw	Skeletal deformation of aromatic rings, substituent groups and side chains	lignin
591 vw	Skeletal deformation of aromatic rings, substituent groups and side chains	lignin
555 vw	Skeletal deformation of aromatic rings, substituent groups and side chains	lignin
537 vw	Skeletal deformation of aromatic rings, substituent groups and side chains	lignin
520 m	Some heavy atom stretch	cellulose
491 vw	Skeletal deformation of aromatic rings, substituent groups and side chains	lignin
463 vw	Skeletal deformation of aromatic rings, substituent groups and side chains	lignin
458 m	Some heavy atom stretch	cellulose

(*Continued*)

TABLE 8.2
Raman Band Assignment in the Spectra of Softwood Cellulose and Lignin (*Continued*)

Band Position cm⁻¹	Assignment	Attributed to
435 m	Some heavy atom stretch	cellulose
384 w	Skeletal deformation of aromatic rings, substituent groups and side chains	lignin
380 m	Some heavy atom stretch	cellulose
357 w	Skeletal deformation of aromatic rings, substituent groups and side chains	lignin
351 w	Some heavy atom stretch	cellulose

ªNote: vs = very strong; s = strong; m = medium; w = weak; vw = very weak; sh = shoulder. Band intensities are relative to other peaks in the spectrum.

Source: Agarwal 1999.

Raman and infrared spectra give basically the same kind of molecular information, and both methods can be used to supplement or complement each other. Although Raman spectra are theoretically simpler than the corresponding infrared spectra, primarily because overlapping bands are much less common in Raman spectroscopy, its use in the study of wood surfaces is not widespread. The primary reason for this lack of popularity has been the difficulty of obtaining sufficiently intense Raman bands, which are not completely masked by the fluorescence bands emitted by the wood components when irradiated with visible light. However, with the advent of FT-Raman and the introduction of the red laser source (excitation wavelength, 1064 nm), this difficulty has been largely overcome, and application of FT-Raman to the study of wood surfaces is likely to increase.

FT-Raman spectra of wood and wood components have been reported in the literature (Agarwal and Ralph 1997, Agarwal 1999). Table 8.2 gives a summary of band assignment in the FT-Raman spectra of wood components.

8.3.2 ELECTRON SPECTROSCOPY

Electron spectroscopy is concerned with the energy analysis of low energy electrons (generally in the range of 20-2000 eV) liberated from the surface of a specimen when it is irradiated with soft X-rays or bombarded with an electron beam (Watts 1990). Two methods of electron spectroscopy have been developed: X-ray photoelectron spectroscopy (XPS), also known as electron spectroscopy for chemical analysis (ESCA), and Auger electron spectroscopy (AES). Since AES has not been applied to the study of wood surfaces, the rest of this discussion will focus on XPS, which has been widely used for the characterization of the elemental composition of wood surfaces.

In XPS, the surface of a sample maintained in a high vacuum ($10^{-6} - 10^{-11}$ torr) is irradiated with X-rays, usually from a Mg or Al anode, which provide photons of 1253.6 eV and 1486.6 eV, respectively. Electrons are ejected from the core levels of atoms in the surface, and their characteristic binding energies are determined from the kinetic energy of the ejected electrons and the energy of the incident x-ray beam.

The kinetic energy (E_K) of the electron is the experimental quantity measured by the spectrometer. The binding energy of the electron (E_B) of the electron is related to the experimentally measured kinetic energy by the following relationship:

$$E_B = h\nu - E_K - W \quad (8.1)$$

where $h\nu$ is the incident photon energy and W is the spectrometer work function.

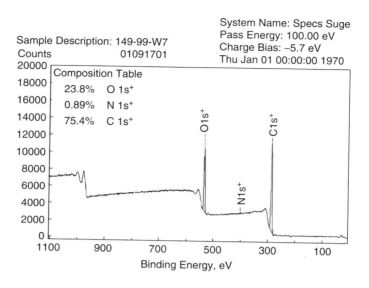

FIGURE 8.13 XPS survey spectrum of a wood specimen of *Pinus taeda* shows the elemental composition of the surface.

The kinetic energy of the electrons ejected from the surface of a specimen is determined by means of an analyzer that gives a photoelectron spectrum. The photoelectron spectrum gives an accurate representation of the electronic structure of an element, as all electrons with a binding energy less than the incident photon energy will be featured in the spectrum.

The characteristic binding energy of a given atom is also influenced by its chemical environment. This dependence allows the assignment of a given atom to a particular chemical functional group on the surface. The sampling depth of XPS is about 5–10 nm.

Examples of survey and high-resolution XPS spectra of a wood specimen are shown in Figures 8.13 and 8.14, respectively. The survey spectrum gives the surface elemental composition

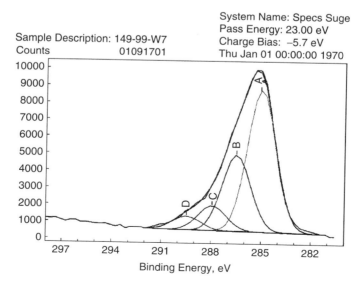

FIGURE 8.14 High-resolution XPS spectrum of the $C1_{s*}$ carbon showing the existence of different states or classes of carbon atoms on the surface of the wood specimen (A = C1; B = C2; C = C3; and D = C4).

of the wood specimen. As expected, the spectrum is dominated by carbon and oxygen peaks, which are the elements that make up the constituents of wood. The high-resolution spectrum of the carbon peak shows the presence of different chemical states, or classes, of carbon on the wood surface.

In early studies on the surface analysis of paper and wood fibers by XPS, C1s and O1s spectra were obtained for samples of Whatman™ filter paper, bleached kraft and bleached sulfite paper, spruce dioxane lignin, stone groundwood pulp, refiner mechanical pulp, and thermomechanical pulp.

The C1s peak was observed to consist of four main components, which were ascribed to four classes of carbon atoms present in wood components. Class I carbon atoms are those bonded to carbon or hydrogen; Class II carbon atoms are bonded to a single non-carbonyl oxygen atom; Class III carbon atoms are bonded to two non-carbonyl or to a single carbonyl oxygen atom, and Class IV carbon atoms are ascribed to the carboxyl carbon. The four components were designated C1, C2, C3, and C4, respectively. The change in the relative magnitude of these components as a function of the oxygen ratio was taken to suggest that both lignin and extractives contribute to the ongoing change in surface composition from pure cellulose to mechanical wood pulps (Dorris and Gray 1978a,b, Gray 1978).

In another pioneer study on the applicability of XPS to the chemical surface analysis of wood fibers, the oxygen-to-carbon (O/C) ratio of wood and wood fibers prepared by different methods was determined. The deviation of the observed O/C ratios from theoretical values was used to provide qualitative characterization of the surface chemical composition. For example, samples of unextracted and extracted pine chips showed O/C ratios of 0.26 and 0.42, respectively. This was interpreted as indicative of surfaces rich in lignin because the observed ratios were very close to the theoretical values of 0.33–0.36 calculated for spruce lignin and considerably lower than the theoretical value of 0.83 for cellulose (Mjörberg 1981).

In a number of studies, the XPS technique has been applied for monitoring the modification of wood surfaces. Wood surfaces treated with aqueous solutions of nitric acid and sodium periodate were analyzed by XPS, and it was observed that the periodate treatment led to a dramatic increase in the relative magnitude of the C2 component. The nitric acid treatment, on the other hand, led to the appearance of a C4 component and a significant increase of the C3 component at the expense of a decrease in both the C1 and C2 components (Young et al. 1982). In another study, XPS data of wood and cellulose surfaces treated with aqueous chromium trioxide showed that at least 75% of Cr(VI) was reduced to Cr(III), and the C1s spectra showed that the surface concentration of hydroxyls decreased, while the hydrocarbon component increased. It was suggested that this change occurred through oxidation of primary alcohols in the cellulose to acids, followed by decarboxylation, and that there were possible cellulose-chromium interactions in addition to the previously proposed chromium-lignin interactions (Williams and Feist 1984).

Changes in the surface chemical composition of solid residues of quaking aspen (*Populus tremuloides*) wood, extracted with supercritical fluid methanol, were monitored by XPS, and it was shown that the C1s peak provided information that allowed for the rapid measurement of the proportion of carbon in polyaromatics. The components of the O1s peak were also tentatively assigned to oxygens in the major wood components and to minor extractives, recondensed material, and strongly adsorbed water (Ahmed et al. 1987, Ahmed et al. 1988). In another XPS study of weathered and UV-irradiated wood surfaces, the observed increase in the O/C ratio was interpreted as an indication of a surface rich in cellulose and poor in lignin (Hon 1984, Hon and Feist 1986).

The surface composition of grafted wood fibers has also been characterized by XPS. By grafting poly(methylmethacrylate) onto wood fibers, this study demonstrated the possibility of tailoring the chemical surface composition of the wood fiber for specific end uses in thermoplastic composites (Kamdem et al. 1991).

The XPS technique has also been applied to the study of the surface composition of wood pulp prepared by the steam explosion pulp (SEP), conventional chemimechanical pulp (CMP) and conventional chemithermomechanical pulp (CTMP) processes (Hua et al. 1991, Hua et al. 1993a, b). Based on the theoretical O/C ratios and C1 contents of the main components of the wood fibers (i.e., carbohydrates, lignin and extractives), a ternary diagram was constructed to illustrate the relative amounts of the three components on the surface. A tentative assignment of two components of the oxygen 1s peak, O1 and O2, was also made. The O1 component, which has a lower binding energy, was assigned to the oxygen in lignin, and O2 was assigned to the oxygen in the carbohydrates. The investigators suggested that the percentage of the O1 peak area could be viewed as a measure of the lignin on the fiber surface.

8.3.3 Mass Spectroscopy

Surface mass spectroscopy techniques consist of measuring the masses of secondary ions that are ejected, in a process known as sputtering, from the surface of a specimen when a primary beam of energetic particles bombards it. The secondary ions can provide unique information about the chemistry of the surface from which they originated. Thus, surface mass spectroscopy and surface electron spectroscopy complement each other. Indeed, it is common practice to use electron spectroscopy in combination with surface mass spectroscopy to characterize the surface chemistry of an unknown specimen.

Three main methods of surface mass spectroscopy are commonly used: secondary ion mass spectrometry (SIMS), laser ionization mass spectrometry (LIMA), and sputtered neutral mass spectrometry (SNMS). However, SIMS is by far the most commonly used surface mass spectroscopic techniques. While SIMS always revolves around the sputtering process, different modes of operation of SIMS exist. The three main modes of operation are static, dynamic, and scanning (or imaging) SIMS (Johnson and Hibbert 1992).

Static SIMS (SSIMS) operates under gentle bombardment conditions and provides information about the chemistry of the upper surface of a specimen, while causing negligible surface damage. Dynamic SIMS (DSIMS) operates under relatively harsher bombardment conditions and provides information about the chemistry of a surface from a few nanometers to several hundred microns in depth. Scanning or imaging SIMS, using highly focused ion beams, provides detailed chemical images with spatial resolution approaching that associated with scanning electron microscopy (less than 100 nm).

In SSIMS, the surface of a sample maintained in a high vacuum (10^{-6}–10^{-11} torr) is bombarded with ions (Ar^+, Xe^+, Cs^+ or Ga^+) at well-defined energies in the range of 1–4 KeV. Secondary ions are sputtered from the surface and are detected in a mass spectrometer. To keep depletion of the surface to a minimum, the primary current is held at approximately 1 nA/cm^2. Under these conditions, only a very small fraction (approximately 1%) of the uppermost monolayer is consumed. The relatively low secondary ion flux is typically detected and analyzed in a time-of-flight mass spectrometer (TOFMS). The TOFMS has high sensitivity, extended mass range, and high mass resolution, and is applicable to virtually all fields of science and technology where solid surfaces and their behavior are important (Wattws 1992, Winogard 1993, Benninghoven et al. 1993, Comyn 1993, Li and Gardella 1994,).

The development of SIMS and its application in the study of paper surfaces is the subject of an excellent review by Detter-Hoskin and Busch (1995). Although SSIMS spectra are complicated and sometimes difficult to interpret, the SSIMS technique has been successfully applied in the study of the surface composition and topochemistry of paper surfaces. Brinnen compared the information content of XPS and SSIMS for characterizing sized paper surfaces. He observed that while XPS readily detected the presence of sizing agents, it provided little structural information. TOF-SIMS, on the other hand, could be used to identify the chemical structures. Brinen also showed that the spatial distribution of the sizing agents on the paper surfaces could be obtained using TOF- SSIMS (Brinen 1993). In a combination of XPS, TOF-SIMS, and paper chromatography, Brinen et al studied the cause of sizing difficulties in sulfite paper. Using SIMS imaging, they observed concentrated islands of pitch, which were implicated in the desizing effect (Brinen and Kulick 1995).

In another study of paper surfaces, Pachuta and Staral demonstrated that TOF-SSIMS, with its extended mass range and high sensitivity, was a useful tool for the nondestructive analysis of

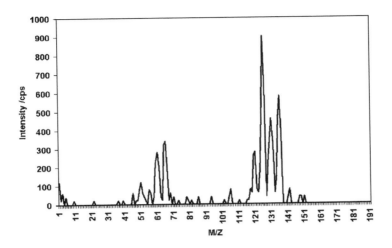

FIGURE 8.15 SSIMS positive ion spectrum of a wood specimen acquired on a MiniSIMS™ desktop chemical microscope (Millbrook Instruments, Ltd.).

colorants on paper. They observed that the use of SSIMS conditions produced no detectable alteration of the paper samples (Pachuta and Staral 1994).

SIMS has also been used in the imaging of fiber surfaces. Tan and Reeve used SIMS imaging to reveal the micro distribution of organochlorine on fully bleached pulp fibers. They observed that pulp-bound chlorine was present over the entire surface and throughout the fiber cross-section. On the fiber's outer surface, the chlorine concentration was observed to be higher in some small areas. In cross-section, chlorine was observed to be concentrated in the middle of the secondary wall. They concluded that, since most of the pulp bound organochlorine is covalently linked to large carbohydrate molecules held deep within the fiber wall, it was not likely to diffuse out of the fiber even after a long time (Tan and Reeve 1992).

SSIMS spectral libraries are still at the early stages of development, and as more spectra of biomaterials, including lignocellulosic materials are added to the libraries, SSIMS will become a very productive technique for characterizing the surface chemistry of wood and wood composites. Figure 8.15 shows a SSIMS positive ion spectrum of the surface a wood specimen, and Figure 8.16

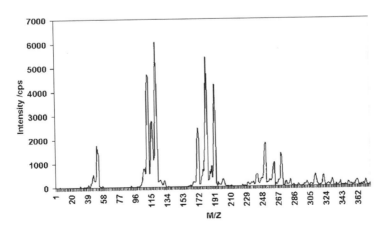

FIGURE 8.16 SSIMS positive ion spectrum of a wood specimen coated with a thin film of alkoxysilanes acquired on a MiniSIMS™ desktop chemical microscope (Millbrook Instruments, Ltd.).

shows a SSIMS positive ion spectrum of the surface of the same wood specimen after coating it with a thin film of alkoxysilanes. Although the peaks were not assigned, the effect of the coating on the surface chemistry of the wood specimen is clearly evident from the appearance of additional peaks (M/Z = 111, 116, 121, 171, 181, 191, 242, 252, 262, and 270).

8.4 THERMODYNAMIC METHODS FOR CHARACTERIZING SURFACE PROPERTIES

Over the years, two methods for determining the surface thermodynamic properties (surface energy and acid-base characteristics) of wood have been developed. The first method, which is known as contact angle analysis (CAA), is based on wetting the wood surface with a liquid. The second technique, which is known as inverse gas chromatography (IGC), is based on the adsorption of organic vapors on the wood surface. Another major difference between the two methods is that solid samples for IGC analysis should be in a finely divided form, while samples for CAA can be in any form, including small coupons or wafers.

8.4.1 Contact Angle Analysis

There are two approaches to contact angle analysis (CAA): static and dynamic contact angle measurements. In static CAA the liquid-solid angle of a sessile drop of liquid on the surface of a wood specimen is observed and measured by means of contact angle meter. The contact angle, θ, of a drop of liquid on the surface of a specimen is shown schematically in Figure 8.17. The cosine of the contact angle (Cos θ) is related to the surface energy of the specimen by the Young equation (Nguyen and Johns 1978).

$$\gamma_{LV} \cos \theta = \gamma_{SV} - \gamma_{SL} = \gamma_S - \gamma_{SL} - \pi_e \tag{8.2}$$

where γ_S is the surface energy of a solid measured in vacuum, γ_{SL} and γ_{SV} are the surface-free energies on the interface between solid and liquid and of the solid and saturated vapor, respectively; γ_{LV} is the surface tension of the liquid against its vapor; and π_e is the equilibrium spreading pressure of the adsorbed vapor of the liquid on solid and is defined as:

$$\pi e = \gamma_S - \gamma_{SV} \tag{8.3}$$

There are only two quantities in Equation 8.2 that can be determined experimentally: γ_{LV} and Cos θ. The surface tension of the liquid, γ_{LV} and Cos θ, can be measured by several methods

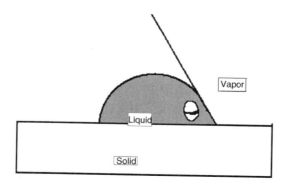

FIGURE 8.17 Schematic diagram of the contact angle, θ at the solid-liquid interface.

Surface Characterization

(Adamson 1967), but contact angle θ can only be obtained when the surface tension of the liquid is higher than the surface-free energy of the solid (assuming that γ_{SL} and π_e in Equation 8.2 are approximately equal to zero). In that case, the surface energy of the solid can be estimated from Equation 8.4:

$$\gamma_S \cong \gamma_{LV} \cos \theta \tag{8.4}$$

Zisman (Adamson 1967) introduced the well-known empirical approach to estimate the surface-free energy of a solid by plotting the cosine of the contact angle θ versus the surface tension of a series of liquids of known surface tension. The point at which the resulting straight line plot intercepts the horizontal line, $\cos \theta = 1$ (zero contact angle), is called the critical surface tension, γ_C. Zisman's plot is generally described by the following equation:

$$\cos \theta = 1 + b(\gamma_C - \gamma_{LV}) \tag{8.5}$$

where b is the slope of the line. Zisman's critical surface tension provides one of the most convenient means of expressing the surface energy of a solid.

In dynamic CAA, the contact angle is measured by means of a dynamic contact angle analyzer, which consists of a precision balance for measuring the wetting force exerted on a specimen as it is dipped and withdrawn from a liquid of known surface tension, γ_{LV}. As shown in the schematic diagram in Figure 8.18, the weight of the solid specimen is recorded by means of a precision balance as it is immersed and withdrawn from the liquid. Immersion and withdrawal of the specimen is accomplished by raising or lowering a cup of liquid placed on motorized elevator. The immersion graph, which is also known as a hysteresis loop, is captured by a data system that is interfaced to the control system of the dynamic contact angle analyzer. The immersion graph is essentially a plot of the measured wetting force versus the immersion depth of the specimen.

The wetting force is related to the contact angle by the following equation:

$$F = \gamma_{LV} P \cos \theta \tag{8.6}$$

where F is the wetting force, γ_{LV} is the surface tension of the liquid, and P is the wetted perimeter of the solid specimen.

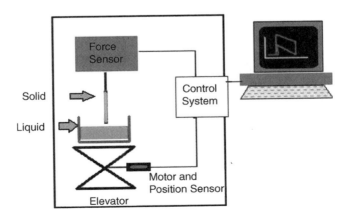

FIGURE 8.18 Schematic diagram of a Dynamic Contact Angle Analyzer.

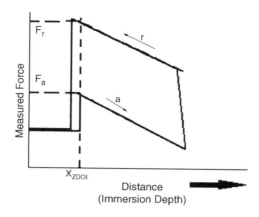

FIGURE 8.19 Schematic diagram of the immersion graph as the specimen is immersed (a) and withdrawn (r) from the liquid. The vertical broken line coincides with the point when the specimen first comes into contact with the liquid at the zero depth of immersion, $x = X_{ZDOI}$.

As shown in Figure 8.19, the advancing and receding contact angles can be calculated from the advancing wetting force F_a and the receding wetting force F_r by the following equations:

$$\cos \theta_a = F_a / \gamma_{LV} P \qquad (8.7)$$

$$\cos \theta_r = F_r / \gamma_{LV} P \qquad (8.8)$$

F_a and F_r are located on the immersion graph at the points of intersection of the vertical line drawn through $x = X_{ZDOI}$ and the advancing line (a) and receding line (r) respectively.

The surface energy of the solid can be estimated by combining Equation 8.4 with Equation 8.7 or Equation 8.8; hence,

$$\gamma_S \cong F_a / P \quad \text{or} \quad \gamma_S \cong F_r / P \qquad (8.9)$$

8.4.2 INVERSE GAS CHROMATOGRAPHY

Inverse gas chromatography (IGC) involves packing the material to be studied, typically of a fibrous or particle form, into the column of a gas chromatograph. The retention behavior of a known vapor probe, which is injected into the carrier gas that passes through the packed column, allows the solid-vapor adsorption characteristics, and thus the surface thermodynamic properties, of the material to be quantified (Shultz and Lavielle 1989, Fowkes 1990, Kamdem and Reidl 1991, Lavielle and Schultz 1991, Felix and Gatenhholm 1993, Kamdem et al. 1993, Farfard et al. 1994, Williams 1994, Chtourou et al. 1995, Belgacem 2000).

The key parameter in IGC measurements is the specific retention volume, $V_g^°$, or the volume of carrier gas required to elute a probe from a column containing one gram of the sample of material. $V_g^°$ is related to the experimental variables by the following equation:

$$V_g^° = (273.15 / T_c) \cdot [(t_r - t_m) / W] \cdot F \cdot J \cdot C$$

where C is the correction factor for vapor pressure of water in the soap bubble flowmeter and J is the correction factor for the pressure drop across the column.

$$C = 1 - (P_w / P_0) \qquad (8.10)$$

Surface Characterization

and

$$J = 1.5\,[(P_i/P_0)^2 - 1]/[(P_i/P_0)^3 - 1] \qquad (8.11)$$

T_c is the column temperature in K; W is the amount of material packed into the column in g; t_r is the retention time of the probe in minutes; t_m is the retention time of a reference probe, such as methane or argon, in minutes; F is the carrier gas flow rate in mL/min.; P_w is the vapor pressure of water at the column temperature in mm Hg; P_0 is the carrier gas pressure at the column outlet (atmospheric pressure) in mm Hg; and P_i is the carrier gas pressure at the column inlet in mm Hg.

8.4.3 TOTAL SURFACE ENERGY

The total surface energy may be regarded as the sum of the contribution of two components: a non-polar and a polar component. The non-polar component is associated with London dispersive forces of interaction, and the polar component is associated with acid-base interactions.

8.4.3.1 Dispersive Component of the Total Surface Energy

The interaction of neutral probes, such as saturated n-alkanes, with the surface of the sample of the material is predominated by London dispersive forces of interaction. Under conditions of infinite dilution of the injected probe vapors, it has been shown that the dispersive component, γ_s^D, of the total surface energy of the material is related to V_g° by the following equation:

$$RT \cdot \ln V_g^\circ = 2N(\gamma_s^D)^{1/2}\,a\,(\gamma_L^D)^{1/2} + C^t \qquad (8.12)$$

where N is Avogadro's number, a is the surface area of the probe molecule, γ_L^D is the dispersive component of the surface energy of the probe, and C^t is a constant that depends on the reference states.

A plot of $RT \cdot \ln V_g^\circ$ versus $2N \cdot a(\gamma_L^D)^{1/2}$ should give a straight line with slope, $(\gamma_s^D)^{1/2}$.

8.4.3.2 Acid-Base Component of the Total Surface Energy

The interaction of polar probes with the surface of the material involves both dispersive and acid-base interactions. The free energy of desorption, ΔG_{AB}°, corresponding to specific acid-base interactions can be related to V_g° by the following equation:

$$RT \cdot \ln\!\left(V_g^\circ / V_g^{\circ *}\right) = \Delta G_{AB}^\circ \qquad (8.13)$$

where V_g° is the specific retention volume of the polar probe and $V_g^{\circ *}$ is the specific retention volume of a reference neutral n-alkan.

This equation suggests that values of $RT \cdot \ln V_g^\circ$, plotted against $2N \cdot a(\gamma_L^D)^{1/2}$ for polar probes, should fall above the straight line obtained by plotting $RT \cdot \ln V_g^{\circ *}$ versus $2N \cdot a(\gamma_L^D)^{1/2}$ for the reference neutral n-alkane probes. The difference of ordinates between the point corresponding to the specific polar probe and the reference line gives the value of the free energy of desorption corresponding to the specific acid-base.

The free energy of desorption corresponding to the specific acid-base interactions may be related to the enthalpy of desorption, ΔH_{AB}°, by the following equation:

$$\Delta G_{AB}^\circ = \Delta H_{AB}^\circ - T\Delta S_{AB}^\circ \qquad (8.14)$$

TABLE 8.3
Values of the Dispersive Component of the Surface Energy of Lignocellulosics Obtained under Infinite Dilution Conditions

Material	Treatment	γ^D_S at (T °C), mJ/m²	Reference
Cellulose	untreated	48.5 (25°C)	Belgacem (2000)
Wood birch	none	43.8 (50°C)	Kamdem et al. (1993)
Bleached softwood kraft pulp	untreated	37.9 (40°C)	Belgacem (2000)
Lignin	kraft	46.6 (50°C)	Belgacem (2000)
Wood eastern white pine	unextracted	37 (40°C)	Tshabalala (1997)
Kenaf powder	unextracted	40 (40°C)	Tshabalala (1997)
Wood eastern white pine	Extracted with toluene/ethanol (2:1 v/v)	49 (40°C)	Tshabalala (1997)
Kenaf powder	Extracted with toluene/ethanol (2:1 v/v)	42 (40°C)	Tshabalala 1997

where T is the temperature in K and $\Delta S°_{AB}$ is the entropy of desorption corresponding to the specific acid-base interactions. A plot of $\Delta G°_{AB}/T$ versus $1/T$ should yield a straight line with slope $\Delta H°_{AB}$.

It has been shown that the enthalpy of desorption can be used to obtain the acidic and basic constants of a substrate. The acidic and basic constants, K_A and K_B, can be regarded as the acceptor and donor numbers, respectively, and are analogous to the acceptor number, AN*, and donor number, DN, of any compound, as defined by the Gutmann theory of acids and bases. The enthalpy of desorption corresponding to the specific acid-base interactions is related to K_A, K_B, AN*, and DN by the following expression:

$$\Delta H°_{AB} = K_A DN + K_B AN^* \qquad (8.15)$$

A plot of $\Delta H°_{AB}/AN^*$ versus DN/AN^* should yield a straight line with slope K_A and intercept K_B. Values of DN and AN* for various solvents are available in the literature (Tshabalala 1997).

IGC has been used to characterize the surfaces of cellulose fibers, birch wood meal, polyethylene and wood pulp fibers, synthetic polymers, and lignocellulosic fibers grafted with poly(methylmethacrylate). The values of the dispersive component of the surface energy of some lignocellulosic materials are summarized in Table 8.3, and Table 8.4 summarizes values of acid-base properties, expressed in terms of the ratio K_A/K_B. For acidic surfaces the value of this ratio is generally greater than 1.0.

TABLE 8.4
Acid-Base Properties of Lignocellulosic Surfaces Obtained under Infinite Dilution Conditions

Material	Treatment	K_A/K_B	Reference
Cellulose powder, 20-μm particle size	none	24	Tshabalala (1997)
Wood birch	none	1.5	Kamdem et al. (1993)
Bleached softwood kraft pulp	untreated	2.2-5.7	Belgacem (2000)
Wood eastern white pine	Extracted with toluene/ethanol (2:1 v/v)	1.4	Tshabalala (1997)
Kenaf powder	unextracted	0	Tshabalala (1997)
Kenaf powder	Extracted with toluene/ethanol (2:1 v/v)	1.6	Tshabalala (1997)

8.5 CONCLUSIONS AND OUTLOOK

Characterization of the surface properties of wood is a very complex and difficult undertaking. No single technique is adequate to completely characterize the surface chemistry of wood and related lignocellulosic materials. Rather, a combination of spectrometric techniques, used together with microscopic and thermodynamic techniques, can provide good insight into the chemical composition of lignocellulosic surfaces, the surface distribution and topography of acid-base sites, and the effect of chemical composition on the reactivity of lignocellulosic surfaces. Hence, as new surface sensitive spectroscopic techniques are applied to the study of lignocellulosic surfaces, it is likely that better data will be developed that should greatly improve our understanding of the surface chemistry of wood. One of the most urgent challenges is the development of a SSIMS spectral database of lignocellulosic materials.

REFERENCES

Adamson, A.W. (1967). *Physical Chemistry of Surfaces* (2nd ed.). Interscience, New York.

Agarwal, U.P. (1999). *An Overview of Raman Spectroscopy as Applied to Lignocellulosic Materials in Advances in Lignocellulosic Characterization* (Dimitris S. Argyropoulos, Ed.). TAPPI Press, Atlanta, GA.

Agarwal, U.P. and Ralph, S.A. (1997). FT-raman spectroscopy of wood: Identifying contributions of lignin and carbohydrate polymers in the spectrum of black spruce (*Picea mariana*). *Appl. Spectrosc.* 51(11):1648–1655.

Ahmed, A., Adnot, A., and Kaliaguine, S. (1987). ESCA study of solid residues of supercritical extraction of *Populus tremuloides* in methanol. *J. Appl. Polym. Sci.* 34:359–375.

Ahmed, A., Adnot, A., and Kaliaguin, S. (1988). ESCA analysis of partially converted lignocellulosic materials. *J. Appl. Polym. Sci.* 35:1909–1919.

Béland, M.-C. and Mangin, P.J. (1995). *Three-Dimensional Evaluation of Paper Surfaces Using Confocal Microscopy in Surface Analysis of Paper* (Conners, T.E. and Banarjee, S., Eds.). CRC Press, Inc., Boca Raton, FL.

Belgacem, M.N. (2000) Characterisation of polysaccharides, lignin and other woody components by inverse gas chromatography. *Cellulose Chem. Technol.* 34:357–383.

Benninghoven, A., Hagenhoff, B., and Niehuis, E. (1993). Surface MS: Probing real-world samples. *Anal. Chem.* 65(14):630A–640A.

Boyde, A. (1994). Bibliography of Confocal microscopy and its applications. *Scanning* 16:33–56.

Brinen. J.S. (1993). The observation and distribution of organic additives on paper surfaces using surface spectroscopic techniques. *Nordic Pulp Pap. Res. J.* 8:123–129.

Brinen, J.S. and Kulick, R.J. (1995) SIMS imaging of paper surfaces, Part 4: The detection of desizing agents on hard-to-size paper surfaces. *Int. J. Mass Spectrom. Ion Proc.* 143:177–190.

Chtourou, H.B., Reidel, B., and Kokta, B.V. (1995). Surface characterization of modified polyethylene pulp and wood pulp fibers using XPS and inverse gas chromatography. *J. Adhesion Sci. Technol.* 9(5): 551–574.

Comyn, J. (1996). Surface analysis and adhesive bonding. *Anal. Proc.* 30:27–28.

Conley, R.T. (1996). *Infrared Spectroscopy*. Allyn and Bacon, Boston.

Danilatos, G.D. (1988). Foundations of environmental scanning electron microscopy. Advances in Electronics and Electron Physics 71:109–250.

Danilatos, G.D. (1993). Introduction to the ESEM instrument. *Microsc. Res. Tech.* 25:354–361.

Deshmukh, S.C. and Aydil, E.S. (1996). Investigation of low temperature SiO_2 plasma enhanced chemical vapor deposition. *J. Vac. Sci. Technol.* B14:738.

Detter-Hoskin, L.D., and Busc, K.L. (1995). SIMS: Secondary ion mass spectrometry in surface analysis of paper (Conners, T.E. and Banerjee, S.S., Eds.). CRC Press, Inc., Boca Raton, FL.

Donald, A.M. (2003). The use of environmental scanning electron microscopy for imaging wet and insulating materials. *Nat. Mater.* 2(8):511–516.

Donaldson, L. (2003). Private communication. Cell Wall Biotechnology Center, Forest Research, New Zealand.

Dorris, G.M. and Gray, D.G. (1978a). The surface analysis of paper and wood fibres by ESCA I. *Cellulose Chem. Technol.* 12:9–23.

Dorris, G.M. and Gray, D.G. (1978a). The surface analysis of paper and wood fibres by ESCA II. *Cellulose Chem. Technol.* 12:721–734.

Farfard, M., El-Kindi, M., Schreiber, H.P., G. Dipaola-Baranyi, G., and Hor, A.M. (1994.) Estimating surface energy variations of solids by inverse gas chromatography. *J. Adhesion Sci. Technol.* 8(12):1383–1394.

Felix, J.M. and Gatenholm, P. (1993). Characterization of cellulose fibres using inverse gas chromatography. *Nordic Pulp Pap. Res. J.* 1:200–203.

Fowkes, F.M. (1990). Quantitative characterization of the acid-base properties of solvents, polymers and inorganic surfaces. *J. Adhesion Sci. Technol.* 4(8):669–691.

Fracassi, F., d'Agostino, R., and Favia, P. (1992). Plasma – Enhanced chemical vapor-deposition of organosilica thin-films from tetraethoxysilane-oxygen feeds. *J. Electrochem. Soc.* 139(9):2936–2944.

Frisbie, C.D., Rozsnyai, L.F., Noy, A., Wrighton, M.S., and Lieber, C.M. (1994). Functional group imaging by chemical force microscopy. *Science* 265:2071–2074.

Frommer, J. (1992). Scanning tunneling microscopy and atomic force microscopy in organic chemistry. *Angew. Chemie* (int. ed., in English) 31:1298–1328.

Gray, D.G. (1978). The surface analysis of paper and wood fibres by ESCA III. *Cellulose Chem. Technol.* 12:735–743.

Hanley, S.J. and Gray, D.G. (1994). Atomic force microscope images of black spruce wood sections and pulp fines. *Holzforschung* 48(1):29–34.

Hanley, S.J. and Gray, D.G. (1995). Atomic Force Microscopy in Surface Analysis of Paper (Conner, T.E. and Banerjee, S.S., Eds.). CRC Press, Inc. Boca Raton, FL.

Hansma, P.K., Ellings, V.B., Marti, O., and Bracker, C.E. (1998). Scanning tunneling microscopy and atomic force microscopy: Application and technology. *Science* 242:209–216.

Hoh, J.H. and Engel, A. (1993). Friction effects on force measurements with an atomic force microscope. *Langmuir* 9:3310–3312.

Hon, D.N.-S. (1984). ESCA study of oxidized wood surfaces. *J. Appl. Polym. Sci.* 29:2777–2784.

Hon, D.N.-S. and Feist, W.C. (1986). Weathering characteristics of hardwood surfaces. *Wood Sci. Technol.* 20:169–183.

Hua, X., Ben,Y., Kokta, B.V., and Kalinguin, S. (1991). Application of ESCA in wood and pulping chemistry. *China Pulp Pap.* 10:52–57.

Hua, X., Kaliaguine, S., Kokta, B.V., and Adnot, A. (1993a). Surface analysis of explosion pulps by ESAC, Part 1: Carbon (1s) spectra and oxygen-to-carbon ratios. *Wood Sci. Technol.* 27:449–459.

Hua, X., Kaliaguine, S., Kokta, B.V., and Adnot, A, (1993b). Surface analysis of explosion pulps by ESAC, Part 2: Oxygen (1s) and sulfur (2p) spectras. *Wood Sci. Technol.* 28:1–8.

Johnson, D. and Hibbert, S. (1992). Applications of secondary ion mass spectrometry (SIMS) for the analysis of electronic materials. *Semicond. Sci. Technol.* 7:A180–A184.

Kamdem, D.P., Bose, S.K. and Luner, P. (1993). Inverse gas chromatography of birch wood meal. *Langmuir* 9:3039–3044.

Kamdem, D.P. and Reidl, B. (1991). IGC characterization of PMMA grafted onto CTMP fiber. *J. Wood Chem. Technol.* 11(1):57–91.

Kamdem, D.P., Reidl, B., Adnot, A., and Kaliaguine, S. (1991). ESCA spectroscopy of poly(methylmethacrylate) grafted onto wood fibers. *J. Appl. Polym. Sci.* 43:1901–1912.

Lavielle, L. and Schultz, J. (1991). Surface properties of carbon fibers determined by inverse gas chromatography. *Langmuir* 7(5):978–981.

Lee, B.H., Kim, H., Lee, J.J., and Park, M.J. (2003). Effects of acid rain on coatings for exterior wooden panels. *J. Ind. Eng. Chem.* 9(5):500–507.

Leica TCS Confocal Systems User Manual, Version 1.0, January, Leica Microsystems Heidelberg GmbH. Edited and written by EDV-Service Dr. Kehrel, Heidelberg, Germany, 1999.

Li, J-X. and Gardella, J.A., Jr. (1994). Quantitative static secondary ion mass spectrometry of pH effects on octadecylamine monolayer Langmuir-Blodgett films. *Anal. Chem.* 66(7):1032–1037.

Libermann, M.A. and Lichtenberg, A.J. (1994). *Principles of Plasma Discharges and Materials Processing.* John Wiley & Sons, New York.

Lichtman, J.W. (1994) Confocal microscopy. *Sci. Am.* 271:40–45.

Louder, D.R. and Parkinson, B.A. (1995). An update on scanning force microscopies. *Anal. Chem.* 67(9):297A–303A.

Loxton, C., Thumm, A., Grisby, W.J., Adams, T.A., and Ede, R.M. (2003). Resin distribution in medium density fiberboard: Quantification of UF resin distribution on blowline- and dry-blended MDF fiber and panels. *Wood Fiber Sci.* 35(3):370–380.

McGhie, A.J., Tang, S.L., and Li, S.F.Y. (1995). Expanding the uses of AFM. *Chemtech* 25(7):20–26.

Meredith, P., Donald, A.M., and Thiel, B. (1996). Electron-gas interactions in the environmental SEM's gaseous detector. *Scanning* 18:467–473.

Meyer, E. (1992). Atomic force microscopy. *Prog. in Surf. Sci.* 41:3–49.

Mirabella, F.M. (Ed.) (1998). *Modern Techniques in Applied Molecular Spectroscopy*. John Wiley & Sons, New York.

Mjörberg, P.J. (1981). Chemical surface analysis of wood fibres by means of ESCA. *Cellulose Chem. Technol.* 15:481–486.

Nguyen, T. and John, W.E. (1978). Polar and dispersion force contributions to the total surface free energy of wood. *Wood Sci. Technol.* 12:63–74.

Pachuta, S.J. and Staral, J.S. (1994). Nondestructive analysis of colorants on paper by time-of-flight secondary ion mass spectrometry. *Anal. Chem.* 66(2):276–284.

Pandey, K.K. and Theagarajan, K.S. (1997). Analysis of wood surfaces and ground wood by diffuse reflectance (DRIFT) and photoacoustic (PAS) Fourier transform spectroscopic techniques. *Holz als Roh- und Werkstoff* 55:383–390.

Pawley, J.B. (1990) *The Handbook of Biological Confocal Microscopy*. Plenum Press, New York.

Perry, S.S. and Somorjai, G.A. (1994). Characterization of organic surfaces. *Anal. Chem.* 66(7):403A–415A.

Rynders, R.M., Hegedus, C.R., and Gilicinski, A.G. (1995). Characterization of particle coalescence in waterborne coatings using atomic force microscopy. *J. Coat. Technol.* 667(845):59–69.

Sahli, S., Segui, Y., Ramdani, R., and Takkouk, Z. (1994). RF plasma deposition from hexamethyldisiloxane oxygen mixtures. *Thin Solid Films* 250(1–2):206–212.

Schaefer, D.M., Carpenter, M., Gady, B., Reifenberger, R., Demejo, L.P., and Rimai, D.S. (1995). Surface roughness and its influence on particle adhesion using atomic force techniques. *J. Adhesion Sci. Technol.* 9(8):1049–1062.

Selamoglu, N., Mucha, J.A., Ibbotson, D.E., and Flamm, D.L. (1989). Silicon oxide deposition from tetraethoxysilane in Radio-Frequency down stream reactor mechanisms and stop coverage. *J. Vac. Sci. Technol.* B7:1345.

Schultz, J. and Lavielle, L. (1989). Interfacial properties of carbon fiber-epoxy matrix composites. In: Lloyd, D.R., Ward, T.C., and Schreiber, H.P. (Eds.), *Inverse Gas Chromatography* [ACS symposium series] 391:168–184.

Stuart, B.H. (2004). *Infrared Spectroscopy – Fundamentals and Applications*. John Wiley & Sons, New York.

Tan, Z. and Reeve, D.W. (1992). Spatial distribution of organochlorine in fully bleached kraft pulp fibres. *Nordic Pulp Pap. Res. J.* 6:30–36.

Tshabalala, M.A. (1997). Determination of acid-base characteristics of lignocellulosic surfaces by inverse gas chromatography. *J. Appl. Polym. Sci.* 65:1013–1020.

Tshabalala, M.A., Kingshott, P., VanLandingham, M.R., and Plackett, D. (2003). Surface chemistry and moisture sorption properties of wood coated with multifunctional alkoxysilanes by sol-gel process. *J. Appl. Polym. Sci.* 88(12):2828–2841.

Watts, J.F. (1990). *An Introduction to Surface Analysis by Electron Spectroscopy*. Oxford University Press, New York.

Watts, J.F. (1992). Investigation of adhesion phenomena using surface analytical techniques. *Anal. Proc.* 29:396–398.

Williams, D. (1994). Inverse gas chromatography. In: Ishida, H. (Ed.), *Characterization of Composite Materials*. Butterworth-Heinemann, Boston, MA.

Williams, R.S. and Feist, W.C. (1984). Application of ESCA to evaluate wood and cellulose surfaces modified by aqueous chromium trioxide treatment. *Colloids Surf.* 9:253–271.

Winograd, N. (1993). Ion beams and laser positionization for molecule-specific imaging. *Anal. Chem.* 65(14):622A–629A.

Young, R.A., Rammon, R.M., Kelley, S.S., and Gillespie, R.H. (1982). Bond formation by wood surface reactions, Part I: Surface analysis by ESCA. *Wood Sci.* 14:100–119.

Part III

Wood Composites

9 Wood Adhesion and Adhesives

Charles R. Frihart
USDA, Forest Service, Forest Products Laboratory, Madison, WI

CONTENTS

9.1	General	216
9.2	Wood Adhesive Uses	217
9.3	Terminology	219
9.4	Application of the Adhesive	220
	9.4.1 Adhesive Application to Wood	220
	9.4.2 Theories of Adhesion	221
	9.4.3 Wood Adhesion	225
	9.4.4 Wood Surface Preparation	225
	9.4.5 Wood Bonding Surface	226
	9.4.6 Spatial Scales of Wood for Adhesive Interaction	229
	9.4.7 Wetting and Penetration in General	230
	9.4.8 Wetting, Flow, and Penetration of Wood	231
9.5	Setting of Adhesive	234
	9.5.1 Loss of Solvents	235
	9.5.2 Polymerization	236
	9.5.3 Solidification by Cooling	237
9.6	Performance of Bonded Products	238
	9.6.1 Behavior under Force	238
	9.6.2 Effect of Variables on the Stress-Strain Behavior of Bonded Assemblies	241
	9.6.3 Bond Strength	242
	9.6.4 Durability Testing	245
9.7	Adhesives	246
	9.7.1 Polymer Formation	246
	9.7.2 Self-Adhesion	248
	9.7.3 Formaldehyde Adhesives	249
	9.7.3.1 Phenol Formaldehyde Adhesives	250
	9.7.3.2 Resorcinol and Phenol-Resorcinol Formaldehyde Adhesives	252
	9.7.3.3 Urea Formaldehyde and Mixed Urea Formaldehyde Adhesives	254
	9.7.3.4 Melamine Formaldehyde Adhesives	255
	9.7.4 Isocyanates in Wood Adhesives	257
	9.7.4.1 Polymeric Diphenylmethane Diisocyanate	258
	9.7.4.2 Emulsion Polymer Isocyanates	260
	9.7.4.3 Polyurethane Adhesives	260
	9.7.5 Epoxy Adhesives	261
	9.7.6 Polyvinyl and Ethylene-Vinyl Acetate Dispersion Adhesives	263
	9.7.7 Biobased Adhesives	265

		9.7.7.1	Protein Glues	265
		9.7.7.2	Tannin Adhesives	266
		9.7.7.3	Lignin Adhesives	267
	9.7.8	Miscellaneous Composite Adhesion		267
	9.7.9	Construction Adhesives		268
	9.7.10	Hot Melts		269
	9.7.11	Pressure Sensitive Adhesives		269
	9.7.12	Other Adhesives		270
	9.7.13	Formulation of Adhesives		270
9.8	Environmental Aspects			272
9.9	Summary			272
References				273

9.1 GENERAL

The recorded history of bonded wood dates back at least 3,000 years to the Egyptians (Skeist and Miron 1990, River 1994a), and adhesive bonding goes back to early mankind (Keimel 2003). Although wood and paper bonding are the largest applications for adhesives, some of the fundamental aspects are not fully understood. Better understanding of the critical aspects in wood adhesion should lead to improved composites. The chemistry of adhesives has been covered in detail; however, how the adhesives hold wood together when under external and internal stresses need to be better understood from the basic scientific principles. This chapter is aimed at more in-depth coverage of those items that are not covered elsewhere. It will touch briefly on topics covered by other writers and the reader should examine the recommended books and articles for more details. Many of the books on adhesives and adhesion are long and complicated, but at least one is briefer, while still being quite thorough (Pocius 2002). Adhesives are designed for specific applications, leading to thousands of products (Rice 1990). Petrie has broken adhesives into 20 groups of synthetic structural, 11 groups of elastomeric, 12 groups of thermoplastic, and six groups of natural adhesives (Petrie 2000). Brief has summarized the vast number of markets for adhesives (Brief 1990).

Understanding how an adhesive works is difficult since adhesive performance is not one science of its own, but the combination of many sciences. Adhesive strength is defined mechanically as the force necessary to pull apart the substrates that are bonded together. Mechanical strength is dependent upon primary and secondary chemical bonds of the polymer chains in the adhesive, wood and adhesive-wood interphase. Thus, one needs to consider both the chemical and mechanical aspects of bond strength, and the interrelation of the two factors. Because adhesive strength is a measurement of failure, the process determines where the localized stress exceeds the bond strength under specific test conditions. One concept is the idea of the bonded assembly being a series of links representing each phase with the failure occurring in the weakest link (Marra 1980). Although the bend is actually more a continuum than discrete links. The localized stress is usually very different from applied stress due to stress distribution and concentration (Dillard 2002). It is generally preferred that the adhesive bond be stronger than the substrate so that the failure mechanism is one of substrate fracture.

There are generally three steps in the process of adhesive bonding. The first is usually the preparation of the surface to provide the best interaction of the adhesive with the substrate. Even though a separate treatment step may not be used in some cases, the knowledge of material science (surface chemistry and morphology) is important for understanding this interaction. Preparation of the surface can involve either mechanical or chemical treatment or a combination of the two. In some cases, the adhesive is modified to deal with problems in wetting of the surface or contamination on the surface. Surface analysis techniques are often more difficult on wood than other materials due to the complex chemistry and morphology of the wood.

The second step is that the adhesive needs to form a molecular-level contact with the surface; thus, it should be a liquid so that it can develop a close contact with the substrates. This process involves both the sciences of rheology and surface energies. Rheology is the science of the deformation and flow of matter. Surface energies are determined by the polar and non-polar components of both the adhesive and the wood. Improving the compatibility by changing one or both of the components can lead to stronger and more durable bonds.

The third step is the setting, which involves the solidification and/or curing of the adhesive. Most adhesives change physical state in the bonding process, with the main exception being pressure sensitive adhesives that are used on tapes and labels. The solidification process depends on the type of adhesive. For hot melt adhesives, the process involves the cooling of the molten adhesive to form a solid, whether this is an organic polymer as in craft glues, or an inorganic material as in the case of solder. Other types of adhesives have polymers dissolved in a liquid, which may be water (e.g., white glues) or an organic (e.g., rubber cement). The loss of the solvent converts these liquids to solids. The third type of adhesive is made up of small molecules that polymerize to form the adhesive, for example, super glues or two-part epoxies. Most wood adhesives involve both the polymerization and solvent loss methods. Understanding the conversion of small molecules into large molecules requires knowledge of organic chemistry and polymer science.

Once the bond is prepared, the critical test is the strength of the bonded assembly under forces existing during the lifetime use of the assembly. This involves internal forces from shrinkage during the curing of the adhesive and differential expansion/contraction of the adhesive and substrate during environmental changes, or externally applied forces. Understanding the performance of a bonded assembly requires knowledge of both chemistry and mechanics. Often the strength of a bonded assembly is discussed in terms of adhesion. Adhesion is the strength of the molecular layer of adhesive that is in contact with the surface layer of the substrate, such as wood. The internal and applied energies may be dissipated at other places in the bonded assemblies than the layer of molecular contact between the adhesive and the substrate. However, failure at the interface between the two is usually considered unacceptable. Understanding the forces and their distribution on a bond requires knowledge of mechanics.

An appreciation of rheology, material science, organic chemistry, polymer science, and mechanics leads to better understanding of the factors controlling the performance of the bonded assemblies; see Table 9.1. Given the complexity of wood as a substrate, it is hard to understand why some wood adhesives work better than other wood adhesives, especially when under the more severe durability tests. In general, wood is easy to bond to compared to most substrates, but it is harder to make a truly durable wood bond. A main trend in the wood industry is increased bonding of wood products as a result of the use of smaller diameter trees and more engineered wood products.

9.2 WOOD ADHESIVE USES

Because adhesives are used in many different applications with wood, a wide variety of types are used (Vick 1999). Given the focus of this book on composites, the emphasis will be more on adhesives used in composite manufacturing than on those used in product assembly. Factors that influence the selection of the adhesive include cost, assembly process, strength of bonded assembly, and durability.

The largest wood market is the manufacturing of panel products, including plywood, oriented strandboard (OSB), fiberboard, and particleboard. Except for plywood, the adhesive in these applications bonds small pieces of wood together to form a wood-adhesive matrix. The strength of the product depends on efficient distribution of applied forces between the adhesive and wood phases. The composites (strandboard, fiberboard, and particleboard) have adhesive applied to the wood (strands, fibers, or particles); then they are formed into mats and pressed under heat into the final product. This type of process requires an adhesive that doesn't react immediately at room temperature (pre-mature cure), but is heat-activated during the pressing operation. Given the weight

TABLE 9.1
Wood Bonding Variables

Resin	Wood	Process	Service
Type	Species	Adhesive amount	Strength
Viscosity	Density	Adhesive distribution	Shear modulus
Molecular weight distribution	Moisture content	Relative humidity	Swell–shrink resistance
Mole ratio of reactants	Plane of cut: radial, tangential, transverse, mix	Temperature	Creep
Cure rate	Heartwood vs. sapwood	Open assembly time	Percentage of wood failure
Total solids	Juvenile vs. mature wood	Closed assembly time	Failure type
Catalyst	Earlywood vs. latewood	Pressure	Dry vs. wet
Mixing	Reaction wood	Adhesive penetration	Modulus of elasticity
Tack	Grain angle	Gas-through	Temperature
Filler	Porosity	Press time	Hydrolysis resistance
Solvent system	Surface roughness	Pretreatments	Heat resistance
Age	Drying damage	Posttreatments	Biological resistance: fungi, bacteria, insects, marine organisms
pH	Machining damage	Adherend temperature	Finishing
Buffering	Dirt, contaminants		Ultraviolet resistance
	Extractives		
	pH		
	Buffering capacity		
	Chemical surface		

Note: Norm Kutscha contributed most of the information for this table.

of adhesive (2–8%) compared to the product weight, cost is an issue. In addition, since the wood surfaces are brought close together, gap filling is not an important issue, but over penetration is. On the other hand, for plywood, the surfaces are not uniformly brought in such close contact, requiring the adhesive to remain more above the surface. Light colored adhesives are important for some applications, but many of these products have their surfaces covered by other materials. Most of the adhesives used in wood bonding have formaldehyde as a co-monomer, generating concern about formaldehyde emissions. Dunky and Pizzi have discussed many of the commercial issues relating to the use of adhesives in manufacture and the use of wood composites (Dunky and Pizzi 2002).

For laminating lumber and bonding finger joints, the adhesive can either be heat or room-temperature cured. The cost of the adhesive has become more critical as the thickness of the wood has decreased from glulam to laminated veneer lumber and parallel strand lumber (Moody et al. 1999). Generally, color is not critical unless it is in a trim application, but moisture and creep resistance are more important because these products are often used structurally.

Adhesives used in construction and furniture assembly usually have long set times and are room-temperature cured. Furniture adhesives are light-colored, low-viscosity, and generally do not need much moisture resistance. On the other hand, construction adhesives generally have a high viscosity and need flexibility, but can be dark-colored.

The movement away from solid wood for construction to engineered wood products has increased the consumption of adhesives. A wooden I-joist can have up to five different adhesives in its construction; see Figure 9.1. The wood laminates that form the top and bottom members may be finger joined with a melamine-formaldehyde adhesive and glued together with a phenol-resorcinol-formaldehyde adhesive. The OSB that forms the middle part is often produced using both phenol-formaldehyde and polymeric diphenylmethane diisocyanate adhesives. This middle section is then

Wood Adhesion and Adhesives

FIGURE 9.1 The importance of adhesives is illustrated by the need for different adhesives to make the flange by the bonding of laminate pieces and the oriented strandboard from the flakes and the final I-joist by attachment of the strandboard to the flange.

joined to the top and bottom members with emulsion-polymer isocyanate. Each of these adhesives has different chemistries, and some are bonded under different conditions of time, temperature, and pressure to a variety of wood surfaces, and is subjected to different forces during use. Thus, it is not surprising that a simple model for satisfactory wood adhesion has been difficult to derive.

9.3 TERMINOLOGY

Confusion can be caused if there is not a clear understanding of the terminology; this chapter generally follows that given in the ASTM Standard D 907-00 (ASTM International 2000a). *Adhesive joint failure* is "the locus of fracture occurring in an adhesively-bonded joint resulting in the loss of load-carrying capability" and is divided into interphase, cohesive, or substrate failures. *Cohesive failure* is within the bulk of the adhesive, while *substrate failure* is within the substrate or adherend (wood). The least clear failure zone is that occurring within the *interphase*, which is "a region of finite dimension extending from a point in the adherend where the local properties (chemical, physical, mechanical, and morphological) begin to change from the bulk properties of the adherend to a point in the adhesive where the local properties are equal to the bulk properties of the adhesive." Figure 9.2 shows the various regions of a bonded assembly. The bulk properties are the properties of one phase unaltered by the other phase.

The *assembly time* is "the time interval between applying adhesive on the substrate and the application of pressure, or heat, or both, to the assembly." This time can be closed with substrates brought into contact or open with the adhesive exposed to the air; these times are important to penetration of the adhesive and evaporation of solvent. *Set* is "to convert an adhesive into a fixed or hardened state by chemical or physical action, such as condensation, polymerization, oxidation, vulcanization, gelation, hydration, or evaporation of volatile constituents." *Cure* is "to change the physical properties of an adhesive by chemical reaction…." Note that cure is only one way in the adhesive setting step. However, because cure is a function of how it is measured, there is no universal value for an adhesive. Separating partial cure from total cure is important because they usually have very different properties, and in most bonded products, total cure is not usually obtained. *Tack*

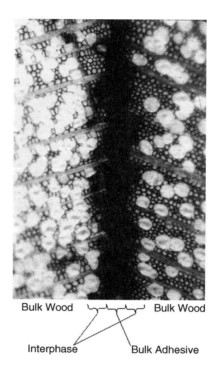

FIGURE 9.2 A transverse scanning electron microscope image of a resorcinol bond of yellow-poplar, showing the zones of bulk wood, interphase region, and bulk adhesive.

is "the property of an adhesive that enables it to form a bond of measurable strength immediately after the adhesive and adherend are brought into contact under low pressure." Tack is important for holding composites together during lay up and pre-pressing.

A *structural adhesive* is "a bonding agent used for transferring required loads between adherends exposed to service environments typical for the structure involved" (ASTM International 2000a). For wood products, structural implies that failure can cause serious damage to the structure, and even loss of life (Vick 1999), while semi-structural adhesives need to carry the structural load, but failure is not as disastrous, and nonstructural adhesives typically support merely the weight of the bonded product.

Other terms are used in different ways that can also cause confusion. The term *adhesive* can refer to either the adhesive as applied or the cured product. On the other hand, a *resin* is often used to refer to the uncured adhesive, although the ASTM defines a resin as "a solid, semisolid or pseudosolid organic material that has an indefinite and often high molecular weight, exhibits a tendency to flow when subject to stress, usually has a softening or melting range, and usually fractures conchoidally" (ASTM International 2000a). Thus, a crosslinked adhesive is not a resin, but the adhesive in the uncrosslinked state may be. Glue was "originally, a hard gelatin obtained from hides, tendons, cartilage, bones, etc. of animals," but is now generally synonymous with the term adhesive.

9.4 APPLICATION OF THE ADHESIVE

9.4.1 Adhesive Application to Wood

The first step in bond formation involves spreading the adhesive over the wood surface. The physical application of the adhesive can involve any one of a number of methods, including using spray, roller coating, doctor blade, curtain coater, and bead application technologies. After the adhesive application, a combination of some open and closed assembly times is used depending on the specific bonding process. Both give the adhesive time to penetrate into the wood prior to bond formation, but the open assembly time will cause loss of solvent or water from the formulation. Long open times can cause the adhesive to dry out on the surface causing poor bonding because flow is needed for bonding to the substrate. In the bonding process, pressure is used to bring the surfaces closer together. In some cases, heat and moisture are used during the bonding process, both of which will make the adhesive more fluid and the wood more deformable (Green et al. 1999).

For any type of bond to form, molecular-level contact is required. Thus, the adhesive has to flow over the bulk surface into the voids caused by the roughness that is present with almost all surfaces. Many factors control the wetting of the surface, including the relative surface energies of the adhesive and the substrate, viscosity of the adhesive, temperature of bonding, pressure on the bondline, etc. Wood is a more complex bonding surface than what is generally encountered in most adhesive applications. Wood is very anisotropic because the cells are greatly elongated in the longitudinal direction, and the growth out from the center of the tree makes the radial properties different from the tangential properties. Wood is further complicated by differences between heartwood and sapwood, and between earlywood and latewood. Adding in tension wood, compression wood, and slope of grain increases the complexity of the wood adhesive interaction. The manner in which the surface is prepared also influences the wetting process. These factors are discussed in later sections of this chapter and in the literature (River et al. 1991), but for now we will assume that the adhesive is formulated and applied in such a manner that it properly wets the surface.

9.4.2 Theories of Adhesion

Adhesion refers to the interaction of the adhesive surface with the substrate surface. It must not be confused with bond strength. Certainly if there is little interaction of the adhesive with the adherent, these surfaces will detach when force is applied. However, bond strength is more complicated because factors such as stress concentration, energy dissipation, and weakness in surface layers often play a more important role than adhesion. Consequentially, the aspects of adhesion are a dominating factor in the bond formation process, but may not be the weak link in the bond breaking process.

It is important to realize that, although some theories of adhesion emphasize mechanical aspects and others put more emphasis on chemical aspects, chemical structure and interactions determine the mechanical properties and the mechanical properties determine the force that is concentrated on individual chemical bonds. Thus, the chemical and mechanical aspects are linked and cannot be treated as completely distinct entities. In addition, some of the theories emphasize macroscopic effects while others are on the molecular level. The discussion of adhesion theories here is brief because they are well covered in the literature (Schultz and Nardin 2003, Pocius 2002), and in reality, most strong bonds are probably due to a combination of the ideas listed in each theory.

In a mechanical interlock, the adhesive provides strength through reaching into the pores of the substrate (Packham 2003). An example of mechanical interlock is Velcro; the intertwining of the hooked spurs into the open fabric holds the pieces together. This type of attachment provides great resistance to the pieces sliding past one another, although the resistance to peel forces is only marginal. In its truest sense, a mechanical interlock does not involve the chemical interaction of the adhesive and the substrate. In reality, there are friction forces preventing detachment, indicating interaction of the surfaces. For adhesives to form interlocks, they have to wet the substrate well enough so that there are some chemical as well as mechanical forces in debonding. For a mechanical interlock to work, the tentacles of adhesive must be strong enough to be load bearing. The size of the mechanical interlock is not defined, although the ability to penetrate pores becomes more difficult and the strength becomes less when the pores are narrower. It should be noted that generally mechanical interlocks provide more resistance to shear forces than to normal forces. Also, many substrates do not have enough roughness to provide sufficient addition to bond strength from the mechanical interlock. Roughing of the substrate surface by abrasion, such as grit blasting or abrasion, normally overcomes this limitation.

If the concept of tentacles of adhesive penetrating into the substrate is transferred from the macro scale to the molecular level, the concept is referred to as the diffusion theory (Wool 2002). If there are also tentacles of substrate penetrating into the adhesive, the concept can be referred to as interdiffusion. This involves the intertwining of substrate and adhesive chains. The interface is strong since the forces are distributed over this intertwined polymer network (Berg 2002). However, the concept can also work if only the adhesive forms tentacles into the substrate. For this to occur, there has to be good compatibility of the adhesive and substrate. This degree of compatibility is not that common for most polymers. When it does occur a strong network is formed from a combination of chemical and mechanical forces.

The other theories are mainly dependent upon chemical interactions rather than truly mechanical aspects. Thus, they take place at the molecular level, and require an intimate contact of the adhesive with the substrate. These chemical interactions will be discussed in order of increasing strength of the interaction (Kinlock 1967). The strengths of various types of bonds are given in Table 9.2, along with examples of some of the bond types in Figure 9.3. It is important to remember that the strength of interaction is for just a single interaction. To make a strong bond these interactions need to be large in number and evenly distributed across the interface.

The weakest interaction is the London dispersion force (Wu 1982a). This force is the dispersive force that exists between any set of molecules and compounds when they are close to each other. The dispersion force is the main means of association of non-polar molecules, such as polyethylene (Figure 9.3). Although this force is weak, where the adhesive and the adherend are in molecular contact, the force exists between all the atoms and can result in appreciable total strength. The ability of the gecko to walk on walls and ceilings has been attributed to this force (Autumn et al. 2002).

The other types of forces are generally related to polar groups (Pocius 2002). The weakest are the dipole-dipole interactions. For polar bonds, there is a separation of charge between the atoms; this process creates a natural, permanent dipole. Two dipoles can interact if positive and negative ends of the dipole match up with the opposite ends of another dipole. The strength of this interaction

TABLE 9.2
Table of Bond Strengths from Literature Bond Types and Typical Bond Energies

Type	Bond Energy (kJ·mol^{-1})
Primary bonds	
Ionic	600–1100
Covalent	60–700
Metallic, coordination	110–350
Donor-acceptor bonds	
Brønsted acid-base interactions	Up to 1000
(i.e., up to a primary ionic bond)	
Lewis acid-base interactions	Up to 80
Secondary bonds	
Hydrogen bonds (excluding fluorines)	1–25
Van der Waals bonds	
Permanent dipole-dipole interactions	4–20
Dipole-induced dipole interactions	Less than 2
Dispersion (London) forces	0.08–40

Source: Data from Fowkes 1983, Good 1966, Kinloch 1987, Pauling 1960,

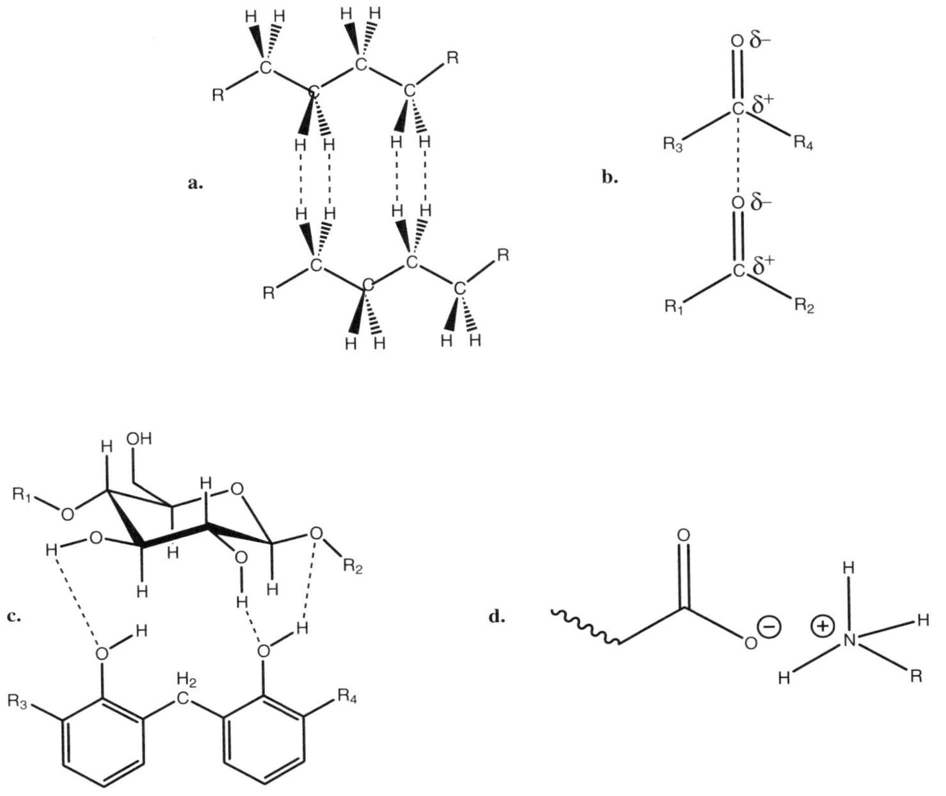

Wood Adhesion and Adhesives

FIGURE 9.3 Examples of various types of bonds, including (a) dispersive bonds between two hydrocarbon chain, such as exist in polyethylene, (b) a dipole bond between two carbonyl group, such as in a polyester, (c) hydrogen bonds between a cellulosic segment and a phenol-formaldehyde polymer, (d) an ionic bond between an ammonium group and a carboxylate group.

depends on proper alignment of the dipoles, which is not difficult for small molecules in solution, but can be very difficult between two chains because they have constrained translation and rotation (Wu 1982a). A variation of this concept is the dipole-induced dipole, but this interaction is usually weaker than the permanent dipole interaction and also suffers from the same alignment problem in polymers.

Strongest of the secondary interactions is the hydrogen bond formation. This type of bond is common with polar compounds, including nitrogen, oxygen, and sulfur groups with attached hydrogens, and carbonyl groups. This type of bond involves sharing a hydrogen atom between two polar groups, and is extremely likely with wood and wood adhesives because both have an abundance of the proper polar groups. Almost all wood components are rich in hydroxyl groups and some contain carboxylic acid and ester groups. Both of these groups form very strong internal hydrogen bonds that give wood its strength, but are also available for external hydrogen bonds. All major wood adhesives have polar groups that can form internal and external hydrogen bonds. The bio-based adhesives depend heavily on hydrogen bonds for their adhesive and cohesive strength. Many synthetic adhesives are less dependent upon the hydrogen bond for their cohesive strength because they have internal crosslinks, but most certainly form hydrogen bonds to wood. One limitation of the hydrogen bond is its ability to be disrupted in the presence of water. Water and other hydrogen bonding groups can insert themselves between the two groups that are present in the hydrogen bond. This process softens the inter-chain bonds so that they are less able to resist applied loads. Thus, a material that adsorbs and absorbs water, like wood, will lose some of its strength when it is wet. The same is true of the adhesion between the wood and the adhesive—it is certainly possible that hydrogen bonds weaken enough to serve as a failure zone.

An interesting aspect of secondary bonds (dispersive, dipolar, and hydrogen bonds) is that after disruption, they can reform while fractured covalent bonds usually do not reform. The reformability of hydrogen bonds has been known about for a long time, but recent work has indicated that it can be an important part of wood's ability to maintain strength even after there is some slippage of the bonds (Keckes et al. 2003), and this process has been referred to as "Velcro" mechanics (Kretschmann 2003). The role of this process in allowing the adhesives to adjust and maintain strength as the wood changes dimensions is not well understood, but could play a significant factor.

Strong bonds can be formed from donor-acceptor interactions. The most common of these interactions with wood-adhesive bonds are the Brønsted acid-base interactions. Some acid-base interactions of cations with anions are possible in adhesion to substrates. Wood contains some carboxylic acids that can form salts with adhesives that contain basic groups, such as the amine groups in melamine-formaldehyde, protein, and amine-cured epoxy adhesives.

Generally, with most materials, the strongest interaction is when a covalent bond forms between the adhesive and the substrate. However, for wood adhesion, this has been an area of great debate, because of the difficulty in determining the presence of this bond type given the complexity of both the adhesive and the wood and the difficulty of generating a good model system. Because wood has hydroxyl groups in its three main components—cellulose, hemicellulose, and lignin— and many of the adhesives can react with hydroxyl groups, it is logical to assume that these reactions might take place. However, others contend that the presence of large amounts of free water would disrupt this reaction (Pizzi 1994a). More sophisticated analytical methods will be needed to answer this issue (Frazier 2003).

It is commonly assumed that the strongest interaction will control the adhesion to the substrate. This overlooks the fact that the adhesion is the sum of the strength of each interaction times the frequency of its occurrence. Thus, covalent bonds that occur only rarely may not be as important to bond strength as the more common hydrogen bonds or dipole-dipole interactions. Hydrogen bonds may be less significant under wet conditions than other bonds if the water disrupts these bonds. It is more important to think about forming stronger adhesion, not by a single type of bond, but by a large number of bonds of different types. Another point to consider is that the adhesive can adhere strongly to a surface and still not form a strong bond overall, due to failure within either the adhesive or the adherend interphases.

One model of adhesion that is generally not related to the bond formation step, but is observed during bond breakage, is the electrostatic model. This model assumes that adhesion is due to the adhesive or the adherend being positive while the other is the opposite charge. It is unlikely that such charges generally exist prior to bond formation, and therefore cannot aid in adhesion; however, they can occur during the debonding process.

Another model that has limited applicability to most cases of adhesion is deep diffusion, which involves polymers from the adhesive and adherend mixing to form a single, commingled phase. Although it is unlikely that the wood will dissolve in the adhesive, it is quite likely that many of the adhesive molecules will be absorbed into the wood cell walls. This diffusion can form one of several types of structures that more strongly lock the adhesive into the wood. This is one type of penetration, and it will be covered in section 9.4.8. In many cases, the strength of this penetration could be as strong as covalent linkages.

Most of these adhesion models play not only a role in bond formation, but also aid the bonded assembly in resisting the debonding forces. The important part to remember is that, depending on the origin of the forces, the stresses can be either concentrated at the interface or dispersed throughout the bonded assembly. If the forces are dispersed, then the force felt at the interface may be quite small.

It is often asked which model of adhesion is correct. This question assumes that there is only a single factor dominating the interaction of the adhesive and the substrate. In reality, there is often a combination of factors that play a role to some degree. The general rule is that the more of each mode of adhesion existing at the interface, the greater the bond durability.

9.4.3 Wood Adhesion

The comprehension of wood adhesive bonds requires both an understanding of the uniqueness of the wood structure for bond formation and an understanding of the modes of energy dissipation and concentration of wood during environmental changes. Because adhesive strength is a mechanical property, the polymer properties of the adhesive, wood, and wood-adhesive interphase regions, are covered in the following sections. Generalizations are difficult in the sense that wood is a nonhomogenous substrate. The adhesive needs to interact with many different types of bonding surfaces. In softwood, large longitudinal tracheids opened by vertical transwall cleavage are the main part of the surface, but parenchyma cells, various ray cells, and resin canals that are also exposed to the adhesive are additional bonding surfaces. In hardwood, small fiber cells and large vessels form the main bonding surface, with rays and other cells also being involved. The wood surface structure is complicated by the presence of thinner-walled earlywood cells and thicker-walled latewood cells. Although generally for bonding studies the sapwood is used, in actual products there can be considerable heartwood, which is less polar. Adding to the complexity, the wood can have juvenile, tension and compression wood. Adhesion studies use samples that are mainly tangential with a small slope of grain, only tiny knots, and no splits, but in commercial wood these factors are less controlled.

Most observations of adhesive interaction with wood are concentrated on scales of millimeter or larger (Marra 1992). However, the wood-adhesive interaction needs to be evaluated in three spatial scales (millimeter, micrometer, and nanometer) (Frazier 2002, Frihart 2003b). The millimeter or larger involves observations by eye or light microscopy. The use of scanning electron microscopy allows observations on the micrometer or cellular level. On the other hand, the size of the cellulose fibrils, hemicellulose domains, and lignin regions are on the nanometer scale. The nanometer level is also the spatial scale in which the adhesive molecules need to interact with the wood for a bond to form. Tools, such as atomic force microscopy, developed for making observations on the nanoscale can be difficult to use with wood because its surface is rough on the micrometer scale.

To understand the adhesive interaction with the wood, we need to consider in more detail the aspects of surface preparation, types of wood surfaces, and spatial scales of wood surfaces. This provides the appropriate background for discussing the adhesion bonding as the steps of wetting the surface and solidification of the adhesive. The wood-adhesive interaction is important in the ultimate strength and durability of the bonded assembly.

9.4.4 Wood Surface Preparation

On the larger scale, wood is a porous, cellular, anisotropic substrate. It is porous in that water and low molecular weight compounds will be rapidly absorbed and move through the wood. The elongated cells of varying size and shape with the differences between the radial and tangential directions lead to the wood being very anisotropic. A simple model cannot be developed because of the large differences between wood species in the chemistry and morphology of the wood surfaces. The cell types and sizes are dramatically different between hardwood and softwood. The individual species in each of these classes vary considerably in their ability for liquids to penetrate, the amount of extractives, as well as the distribution of the various cell types. Even within a species, there is the problem of earlywood versus latewood, sapwood versus heartwood, and juvenile, compression, and tension wood having distorted cell structures. The earlywood cells with the thinner walls should be easier to bond because of a more accessible lumen. The sapwood of a species is generally considered easier to bond than the heartwood due to changes in the extractives. The juvenile, compression, and tension wood all have distorted cell structures that should weaken the wood adhesive interphase region. To simplify the discussion, the emphasis will be placed on the wood that meets the selection criteria for standard testing.

The surface preparation has been shown to have a large effect on the quality of a wood surface (River et al. 1991). One concern is a weak boundary layer, which is a layer between the bulk materials and the true adhesive-adherend interface that is often the weak link and fails cohesively within that layer (Bikerman 1968, Wu 1982b). A classic example was the difficulty in bonding to aluminum, as a result of its weak aluminum oxide surface layer, until the FPL etch was developed (Pocius 2002). Stehr and Johansson have broken down the weak boundaries of wood into those that are chemically weak and those that are mechanically weak (Stehr and Johansson 2000). The distinction is that the chemically weak layer involves extractives coming to the surface, while the mechanically weak layer involves a crushed or fractured cell layer. The role of extractives has been widely considered to be a major factor in poor adhesive strength. Certainly, low-polarity, small molecules coming to the surface can hurt the wetting process. However, it is not clear that they are normally a cause of poor bond strength. Chemically weak boundary layers are certainly an issue in oily woods, such as teak, where wiping the surface with solvent to remove the oils will solve most bonding problems. The issue of extractives should not be confused with the more general phenomena of over-dried wood. The latter case also involves chemical alteration of the wood by excessive heating that leads to poor wetting and weaker bonds (Christiansen 1990, Christiansen 1991, Christiansen 1994).

Wetting is an important issue, especially since most wood adhesives are water-borne. Water has such a high surface energy that wetting of many surfaces is difficult. Although surfactants can lower surface energy, they are often avoided since they can create a chemically weak boundary layer. The monomers and oligomers in the adhesive can lower surface energies, as can added low-molecular-weight alcohols. Wetting should be less of an issue with adhesives that have organic solvents or are 100% solids.

Mechanically weak boundary layers are often an issue with wood (River 1994a). The general problem is from crushing the wood during the surface preparation. Wood cells, especially earlywood, are weak in the radial and tangential direction. Crushed cells are easy to visualize by looking at cross sections microscopically. If the adhesive does not penetrate through the layer of crushed cells, then this layer will generally be the source of fracture under test or use conditions. The best method for preparing a wood surface for bonding is to use sharp planar blades. Unsharpened blades can crush cells and cause a very irregular surface (River and Miniutti 1975). The difference in penetration of an adhesive on well- and poorly-planed wood surfaces is shown in Figure 9.4. Abrasively planed surfaces and saw-cut surfaces also suffer from crushed and fractured surface cells. Hand sanding is generally acceptable because it causes less damage to the cells. For laminates, ASTM prescribes that the wood surfaces be planed with sharp blades and then be bonded within 24 hours to provide the most bondable surface (ASTM International 2000b).

9.4.5 WOOD BONDING SURFACE

The wood-bonding surface varies considerably both chemically and morphologically depending on how the surface is prepared and what type of wood is being used. The morphology is better characterized than the surface chemistry, and will be discussed first. Except for fiber bonding, the desire is to have sufficient open cells on the surface so that the adhesive can flow into the lumen of the cells to provide more area for mechanical interlock. The accessibility of open cells is dependent upon the tree species, types of cells, and method of preparation. When the cell wall is thin in comparison to the diameter of the cell, then there will be more longitudinal transwall fracture.

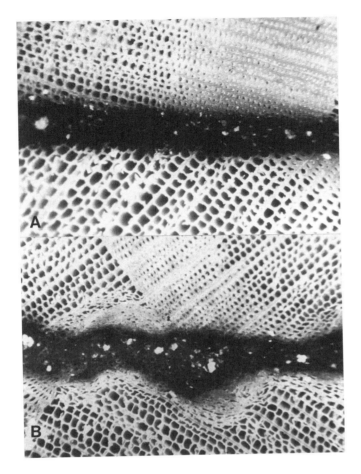

FIGURE 9.4 Bondlines show good adhesive penetration for (a) a sound wood surface, but not for (b) a crushed and matted wood surface.

Hardwood vessels and earlywood cells have thin walls that are easily split to open the lumens to the adhesive for good penetration. On the other hand, hardwood fiber cells and latewood cells have thick walls that are not easy to fracture, so cleavage often occurs more in the middle lamella providing less area for mechanical interlock (River et al. 1991). The open ends of any cells and cracks in the cell walls allow the adhesive to penetrate into the lumens. The differences in the surfaces can be large, by comparing the scanning electron microscopy pictures for southern yellow pine and hard maple (Figure 9.5), with pine having more open cells, while many of the maple's cells are closed.

The chemical composition of the wood-bonding surface is less well understood because the surface is very hard to characterize. The roughness of the surface, the presence of many different surfaces (lumen walls, middle lamella, and fractured cell walls), and the changes of the surfaces with time, heat, and moisture add to the difficulty. The main components of the wood are the cellulose, hemicellulose and lignin fractions. The interactions of phenol-formaldehyde and urea-formaldehyde polymers with cellulose have been modelled (Pizzi 1994b). Although cellulose is the main component of wood, it may not be the main component on the surface. Prior work has indicated that hemicellulose is the main site of interaction with water for hydrogen bonding because of its greater accessibility (River et al. 1991, Salehuddin 1970). The preparation of the

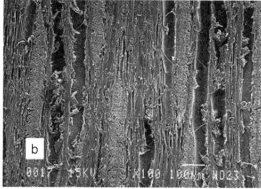

FIGURE 9.5 Scanning electron microscopy pictures of transverse sections of (a) southern yellow pine and (b) hard maple.

wood surface by planing can create many types of surfaces, depending on how the cells fracture, as illustrated in Figure 9.6. If the cell walls are cleaved in longitudinal transwall fashion as desired, then the lumen should be the main bonding surface. The lumen walls are often a large part of the bonding surface, especially for earlywood cells of softwoods, and vessel elements in hardwoods. The lumen walls' compositions can vary from being highly cellulosic, if the S_3 layer is exposed, to highly lignin if they are covered by a warty layer. The middle lamella is also rich in lignin. However, for the most part we do not know when the walls are fractured if the cleavage plane runs through any of the three main fractions or between the lignin-hemicellulose boundary, which may be the weakest link in the wood cellular structure. Complicating this consideration of the bonding surface is that the typical mechanical ways of preparing binding surfaces cause a lot of fragmentation and smearing of the cell wall components. Only by careful microtome sectioning can the clean splitting of the cell walls be observed. Other methods give surfaces that are a lot less intact (Wellons 1983). As can be seen by Figures 9.5 and 9.7, there is a lot of debris on the surface even with sharp planer blades. Hardwood tends to give even more smearing of the surface. Thus, the theory of many open lumens into which the adhesive can flow is not always correct, which may be why the penetration of the adhesive into the lumens is not always that fast.

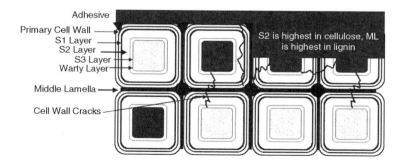

FIGURE 9.6 Illustration of a transverse section of wood showing fracture points of the wood cellular structure and surfaces available with which adhesives can interact, assuming clean fractures are occurring.

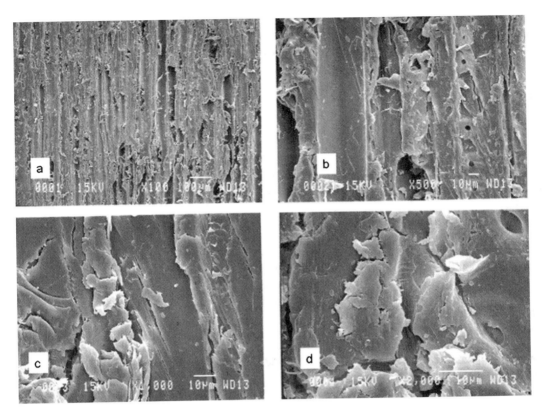

FIGURE 9.7 Scanning electron microscopy of yellow poplar surfaces at four levels of magnification showing the extensive fracturing of the surface and generation of weakly bonded fragments even with sharp planar blades.

9.4.6 Spatial Scales of Wood for Adhesive Interaction

Wood bonds need to be considered on three different spatial scales: millimeter and larger, micrometer, and nanometer (Frazier 2002, Frihart 2004). The millimeter and larger scale is normally used for evaluating the bonding and debonding processes of wood. The micrometer scale relates to the cellular and cell wall dimensions. The nanometer and smaller scale correlates to the sizes of the cellulose, hemicellulose, and lignin domains and the molecular interactions of the adhesive with the wood. Each domain size requires different observation methods and has different implications on bonding and debonding processes.

The millimeter and larger scale is the normal method for dealing with both the bonding and the debonding processes. Usually the naked eye or feel by hand touch is used to judge the smoothness of the surface for bonding. On this scale, measurement of the spread by the adhesive across the surface is typically done by contact angles. Examination of the adhesive bond failure is generally limited to this scale. This information is valuable for understanding bond formation and failure aspects as the first stage in evaluation of adhesive performance. However, it is important to move on to the smaller spatial scales to gain a fuller understanding of wood bonding.

The micrometer scale involves the adhesive interaction with the lumens and cell walls. While the earliest theory on the strength of wood adhesive bonds involves mechanical interlock (McBain and Hopkins 1925), others proposed that there were specific interactions of adhesives of the wood surface (Browne and Brouse 1929). Flow into the lumen of cells is still considered important, leading to many microscopic studies on penetration (Johnson and Kamke 1992). However, there has not been enough consideration of what happens to the adhesive-wood interphase as the cells and the adhesive undergo differential expansion caused by changes in moisture and temperature. These aspects are covered in more detail in the individual bonding and debonding sections (9.4.8 and 9.6.3). The tools for looking at this level of interaction are more complicated because it is at the high end of light microscopy magnification, but it is certainly in the range of scanning electron and transmission electron microscopy (SEM and TEM, respectively).

The nanometer and smaller scale is important because it is the size of the basic domains of wood and of the adhesive-wood interactions (Fengel and Wegener 1984). The size of the cellulose fibrils, the hemicellulose portions, and the lignin networks are in the tens of nanometers. For there to be adhesion, the adhesive needs to interact with the wood at the molecular level; independent of whatever mechanism is involved. The idea of wood adhesion being more than a mechanical interlock was proposed in the 1920's with the concept of specific adhesion as being critical (Browne and Brouse 1929). The problem with understanding this specific adhesion is our lack of understanding what is on the wood surface. Although cellulose is the main component of wood, it may not be the main component on the wood surface. If bonding to lumen walls is important, then adhesion to lignin is important since the warty layer present in many species is high in lignin content (Tsoumis 1991). Cleavage in the middle lamella, as may occur with latewood cells, fiber cells in hardwood or fibers prepared for fiberboard, leads to a surface high in lignin content. Until we can better define how the adhesive has to interact with the wood to form durable bonds, this area is still quite speculative. Although instrumental methods, such as atomic force microscopy, surface force microscopy, and nanoindentation can look at surfaces at this scale, they work best when the surface morphology changes only by nanometers while the roughness of the wood surface varies by micrometers.

9.4.7 WETTING AND PENETRATION IN GENERAL

For a bond to form, the adhesive needs to wet and flow over a surface, and in some cases penetrate into the substrate. It is important to understand that the terms mean different things even though they sound familiar. Wetting is the ability of an adhesive drop to form a low contact angle with the surface. In contrast, flow involves the adhesive spreading over that surface under reasonable time. Flow is important because covering more of the surface allows for a stronger bond. Thus, a very viscous adhesive may wet a surface, but it might not flow to cover the surface in a reasonable time frame. Penetration is the ability of the adhesive to move into the voids on the substrate surface or into the substrate itself. The filling of the lumens has long been one measure of penetration, but penetration can also involve the movement of the adhesive into the cell wall. The difference between flow, penetration, and transfer are illustrated in Figure 9.8.

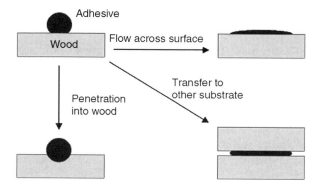

FIGURE 9.8 Adhesive wetting of wood surfaces, showing the difference between flow, penetration, and transfer.

First, we will consider the aspects of wetting, flow, and penetration that are common to most substrates. In the next section, we will discuss how these need to be modified for wood bonding. It is important to understand the general aspects since a number of adhesives are used to bond wood to other substrates. Some laminated structures will have a fiber-reinforced plastic (FRP) layer bonded to the wood; therefore, it is important to understand how the bonding to the FRP may be different from bonding to wood. Other applications could involve the bonding of wood to concrete or metal. One type of lignocellulosic material that is hard to bond to is wheat straw because it has a nonpolar waxy surface that makes it hard for the adhesive to wet and penetrate to the cellular structure.

For a bond to form, the adhesive must intimately encounter most of the substrate surface (Berg 2002). With many plastics having low surface energies, this is a significant problem since the adhesive can find it difficult to wet the substrate. An extreme example is the bonding of Teflon, which has a very low surface energy so that very few adhesives will wet it. In fact, an adhesive applied to the surface forms a bead rather than wets the surface. For bonding to many polyethylene and polypropylene materials, wetting by an adhesive is also a significant problem because of their low surface energies. For laminates with FRP, most wood adhesives have very high surface tensions due to being water-borne, and will not bond well to the FRP. Thus, a great deal of the literature places emphasis on the measurement of contact angles to determine the wetting of the surface. The contact angle is the angle at the edge of a droplet and the plane of that surface upon which it is placed. Therefore, a material with a high contact angle has poor surface wetting ability. The addition of surfactants or less polar solvents reduces the adhesive's surface energy as indicated by a decreased contact angle. With many plastics, surface treatments such as oxidation by flame or corona discharge are used to increase the polarity and surface energy of the plastic surface to improve its bondability. It is important to remember that most contact angle measurements are equilibrium values, and may not reflect the dynamics of the bonding process well. Another very important property that is closely associated with wetting is flow over the surface. Flow is dependent upon not only the contact angle, but also the viscosity of the adhesive. With a lower viscosity, the adhesive flows better and wets more of the surface.

While flow is movement across the surface, penetration is the movement into the substrate. Adhesives will not penetrate into the bulk of many substrates like metals and many plastics, but penetration is important in the sense of movement of the adhesive into the microcrevices on the surface (Berg 2002). Most surfaces have some degree of roughness, which an adhesive must penetrate. Like flow, penetration is dependent upon surface energies and adhesive viscosity, but it also depends on the size of the capillary or void that it is penetrating. For a strong bond, the adhesive must penetrate into all microscale roughness. A typical problem is a displacement of air, water, or oil on the surface. As discussed in the next section, penetration has a very different meaning for wood, due to its structure.

9.4.8 WETTING, FLOW, AND PENETRATION OF WOOD

Wood bonding faces many of the same issues as discussed in the previous section on general aspects of wetting, flow, and penetration, but there are many characteristics that are unusual about wood that require additional consideration. Wood has a relatively polar surface that allows the general use of water-borne adhesives, although some woods are harder to wet. Examples are some tropical woods that have a very oily nature, such as teak, and wood that has been treated with creosote. Wetting of the surface can be improved by removal of the oily components through solvent wiping, mechanical, or oxidation techniques. In Figure 9.9, the effect of sanding on improving the wetting of yellow birch veneer is illustrated. It has been shown that oxidation of wood surfaces by corona treatment can improve wetting and adhesion for some woods (Sakata et al. 1993). A lot of work has been done on examining the wetting of wood; however, it is not clear what this data means. Wetting experiments have been done with water at room temperature; while most bonding is done

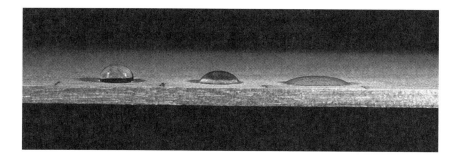

FIGURE 9.9 Water droplets on a yellow birch veneer show the improved wetting by removal of surface contaminants. The photograph was taken 30 seconds after placing three droplets on the surface. The left drop was on an untreated surface, the middle was renewed by two passes of 320 grit sandpaper, and the right drop was renewed with four passes with the sandpaper.

using water solutions of organics, higher temperatures and pressure, all of which improve wetting of surfaces. In addition, the studies to relate extractives with bonding have not found good correlations (Nussbaum 2001).

Understanding flow over the surface is complicated by the fact that the surface has very macroscopic roughness, and penetration is taking place at the same time. As mentioned in the previous section, penetration generally involves wetting of the micro-roughness. On the other hand, wood's cellular nature allows significant penetration of the adhesive into the substrate. A main complication is that different species of woods have different cellular structures, and therefore, adhesives will penetrate them to different degrees. This leads to problems in trying to achieve uniform penetration when bonding different species of wood, as occurs in OSB production. For a more porous wood, an adhesive can over-penetrate into the wood and not be on the surface for bonding, while the same adhesive on a less porous wood sits on the surface and may not give significant bonding. Thus, adhesives are formulated for different applications given the type of wood, the type of application, and the application conditions. An adhesive that is sprayed onto OSB tends to be much lower in viscosity for better spraying than one that is formulated for spreading on plywood that needs to sit more on the surface. Aspects of formulating adhesives are covered in later sections.

In most bonding applications, adhesive penetration into the adherend does not occur to any great degree, but it is very important for wood. The proper degree of penetration influences both the formulation of the adhesive and the bonding conditions. The proper balance needs to be obtained in that poor bonds will result from either under- or over-penetration. In under-penetration, the adhesive is not able to move into the wood enough to give a strong wood-adhesive interaction. In contrast, with over-penetration so much of the adhesive moves into the wood that insufficient adhesive remains in the bondline to bridge between the wood surfaces, resulting in a starved joint. To solve these problems, the viscosity and composition of the adhesive can be adjusted, as well as the temperature and time for the open and closed assemblies. Some species are known to be more porous compared to other species, leading to complications when bonding mixed species. This is a significant issue for composites that usually use a wide mixture of species and a frequently changing mixture. Using mixed species certainly could lead to both over- and under-penetration and to potentially reduced bond strength. Although it is generally known that proper penetration is important to strong bonds, it is not clear whether penetration into the lumens or the cell walls is more critical.

The penetration of adhesives into wood is most often examined at the cellular level. Some lumens have openings on the surface as a result of slope of grain so that the adhesive can flow into the lumen; this is more likely with larger diameter cells in softwood. In hardwoods, most of the filling of lumens is of the larger vessels rather than the smaller fiber cells. Factors that influence the filling of the lumens can be classified into those that are:

- Wood-related, such as diameter of the lumen and exposure on the wood surface.
- Adhesive-related, such as its viscosity and surface energy.
- Process-related, such as assembly time, temperature, pressure, moisture level.

It is normally assumed that the filling of lumens contributes to bond strength. Resin penetration into lumens has been extensively investigated in the wood bonding literature because it is easy to determine by visible light, fluorescence, and scanning electron microscopy. The problem is that these data have not been related to bond strength or level of bond failure. An example, where a filled ray cell contributed to adhesion of a coating after environmental exposure, has been shown by light microscopy (Dawson et al. 2003).

In addition to filling the lumens, an important part of wood adhesion, especially for durable bonds, might be flow of adhesive components into cell walls (Nearn 1965, Gindel et al. 2002b). A significant number of lower molecular weight compounds can go into cell walls due to their ability to swell. This includes both adhesive monomers and oligomers, but not higher molecular weight polymers. Polyethylene glycol molecules of up to 3000 g/mole were shown to penetrate into the transient capillaries or micropores in cell walls (Tarkow et al. 1965). It would be expected that not only the molecular weight, but also the hydrodynamic volumes of the penetrating compound would affect its ability to move through the transient capillaries. An additional factor is the compatibility of the adhesive with the wood structure. Generally, solubility parameters are widely used to determine the compatibility of adhesives and coatings to interact with surfaces (Barton 1991). Limited studies have been done trying to relate the solubility parameters of the components of wood to its ultrastructure (Hansen and Bjorkman 1998), which would then relate to the components' interaction with adhesives.

Do adhesive components enter into cell walls? The observation of adhesive components in cell walls has been shown by a variety of methods. The migration of phenol-formaldehyde resins into cell walls has been shown using fluorescence microscopy (Saiki 1984), audioradiography (Smith 1971), transmission electron microscopy (Nearn 1965), scanning electron microscopy with x-ray dispersive emissions (Smith and Cote 1971), dynamic mechanical analysis (Laborie et al. 2002), and anti-shrink efficiency (Stamm and Seborg 1936). For polymeric diphenylmethane diisocyanate, pMDI, the presence of adhesives in cell walls has been shown by x-ray micrography, and nuclear magnetic resonance spectroscopy (Marcinko et al, 1998, Marcinko et al. 2001). These and other techniques such as UV microscopy (Gindl et al. 2002) and nano-indentation (Gindl and Gupta 2002) have been used to show the presence of urea-formaldehyde, melamine-formaldehyde, and epoxy resins in the wall layers (Bolton et al. 1985, Bolton et al. 1988, Furuno and Goto 1975, Furuno and Saiki 1988). Because both chemical and mechanical data show the presence of adhesives in cell lumens and cell walls, it is likely that the wood portion of the interphase has very different properties than the bulk wood.

Although it has been shown that adhesive components can migrate into cell walls, only in one case has it been claimed to improve bond strength (Nearn 1965). Several models can be proposed as to how these adhesive components may influence bond strength. The simplest is that the oligomers and monomers are simply soluble in the cell walls, but do not react, being too diluted by the cell wall components. In this case, they would maintain the cell walls in the expanded state due to steric constraint (bulking effect); thus, the process would reduce the stresses due to less dimensional change. A second model is that the adhesives react with cell wall components and possibly crosslink some of the components, thereby increasing the strength properties of the surface wood cells, as shown in Figure 9.10. A third model is that the adhesives polymerize to form molecular scale fingers of the adhesive in the wall, providing a nanoscale mechanical interlock. The fourth is that they form an interpenetrating polymer network within the wood, providing improved strength

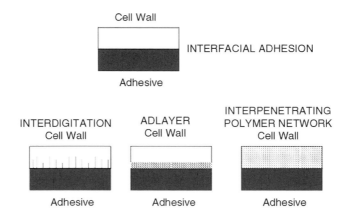

FIGURE 9.10 Modes of adhesive interaction within wood cell walls are depicted for true interfacial adhesion with no cell wall penetration, interdigitation of fingers of adhesive penetrating the microchannels, adlayer of crosslinking in the surface cell wall and interpenetrating polymer network deep in the cell wall.

(Frazier 2002). All of these models have the adhesive reducing the dimensional changes of the surface cells, and therefore reducing the stress gradient between the adhesive and the wood, thereby improving the bond strength.

Knowing that adhesive components do migrate into the cell wall, the next questions is: Are they associated with any specific cell layer or the middle lamella, and are they more in the cellulose, hemicellulose or lignin domains? One study indicates that the isocyanates seem to be more concentrated in the lignin domains (Marcinko et al. 2001). Peeling experiments have shown that an epoxy adhesive gave failure in the S_3 layer while a phenol-formaldehyde adhesive resulted in failure deeper in the S_2 layer (Saiki 1984).

9.5 SETTING OF ADHESIVE

Once an adhesive is applied to wood, the adhesive needs to set to form a product with strength. Set is "to convert an adhesive into a fixed or hardened state by chemical or physical action, such as condensation, polymerization, oxidation, vulcanization, gelation, hydration, or evaporation of volatile solvent." Although the ASTM terminology uses solvent to refer to organic solvents, this chapter uses it in the more general sense of both water and organics because wood adhesives are usually water-borne. Water-borne adhesives often contain some organic solvent to help in the wetting of wood surfaces. For some of the polymeric adhesives, including polyvinyl acetate, casein, blood glue, etc., the loss of solvent sets the adhesive. For many others, including the formaldehyde-cured adhesives, the set involves both the loss of water and polymerization to form the bond. For polymeric diphenylmethane diisocyanate, the set is by polymerization. For hot melt adhesives, cooling to solidify the polymer is sufficient. In wood bonding, all of these mechanisms are applicable, dependent upon the adhesive system that is being used.

The original wood adhesives were either hot-melt or water-borne natural polymers (Keimel 2003). These had several limitations in relation to speed of set, formation of a strong interphase region, and environmental resistance. All of the biomass-based adhesives had poor exterior resistance. The use of composites and laminated wood products has greatly expanded with the development of synthetic adhesives with good moisture resistance. Instead of being mainly polymers with limited and reversible crosslinks, these adhesives have strong covalent crosslinks to provide environmental resistance. In addition, these synthetic adhesives generally cure by both polymerization and solvent loss, leading to a faster setting process. Having multiple modes of set allows both the use of lower viscosity polymers for good wetting and polymers with a higher molecular weight for a faster cure. This combination gives a fast set rate that allows for higher production speeds.

9.5.1 Loss of Solvents

With many adhesive uses, solvents are a problem because of the non-porous nature of the substrate preventing removal of the solvent by migration into and through the substrate. However, wood is quite effective in allowing solvent to migrate away from the bondline, thus allowing adhesives to set. Of course, this property is very dependent upon the wood species and the moisture level of the wood (Tarkow 1979). It is not surprising that wet wood will less rapidly absorb moisture, thus making it harder for water-borne adhesives to move into the wood. The dynamics of water movement have a large effect on the bonding process. The factors involve penetration of the adhesive into the wood, rate of adhesive cure, flow of heat through composites, and premature drying of the adhesive. Most bonding processes require the wood to be within a set range of moisture content to get an acceptable set rate. The desire is to have the bonded product be near the normal in-use moisture condition to reduce internal stress and dimensional changes (Marra 1992).

Penetration of the adhesive into the wood is an important part of the bonding process. Green wood is difficult to bond with most adhesives because there is little volume into which the adhesive can penetrate. (See Figure 9.11 for the generalized effect of bonding parameters on penetration.) At the other extreme, overly dry wood can also be difficult for the adhesive to penetrate because the wood surface is more hydrophobic and harder to wet (Christiansen 1994). Thus, wood with a 4% to10% moisture range is typically good for optimum penetration and set rates. While green wood hinders the adhesive flow into lumens for forming mechanical and chemical bonds, wood can also be too dry so that there is poor absorption of the water and adhesive. For the adhesive to set, the solvent needs to flow away from the adhesive into the adjoining and further removed cell walls. The sorption of the water into the nearby cell walls allows the formation of the solid, cured adhesive. Although most of the studies on uptake of small molecules into wood have naturally concentrated on water, other solvents are also readily absorbed/adsorbed by wood.

For many of adhesives, cure rate is dependent upon the moisture content. Many setting reactions involve condensations that give off water; higher moisture levels can retard the reactions as expected

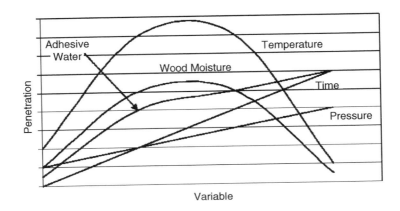

FIGURE 9.11 General effect of conditions on adhesive penetration. The temperature makes the adhesive more fluid until too much causes polymerization. At low wood moisture the water is drawn from the adhesive, while at high wood moisture the water retards the penetration. As the water content of the adhesive increases, the viscosity of the adhesive is lower and penetration increases. Both an increase in bond pressure and a longer time promote adhesive penetration.

by normal chemical equilibrium theory and from limited collisions due to dilution. The amount of water present also alters the mobility of polymer chains during the curing process, which can change the product distribution for the adhesive polymers. On the other hand, many isocyanates depend on a small amount of water to start the curing process.

A very important issue in the rate of setting is the heat flow through composites or laminates to the bond surface, especially since wood is a good insulator. In composites, water in the wood near the surface or added steam helps transfer heat to the core of the composite. Use of core resins that cure at lower temperatures than face resins is important for fast production cycles. Controlling heat transfer and moisture levels is important for fast, reproducible composite production. In laminates, the use of water vapor for heat transfer is not available, thus leading to longer heating cycles. The ability of resorcinol-formaldehyde and phenol-resorcinol-formaldehyde to cure rapidly at room temperature favors them over phenol-formaldehyde resins despite their higher cost. Another way to accelerate cure is to use radiation methods, such as radio frequency curing.

With some adhesives, premature drying can be a problem if the open time is too long. This involves too much loss of solvent so that the adhesive does not flow to wet the other surface. Proper control of moisture level and penetration determines the length of open- and closed-assembly times.

9.5.2 Polymerization

For a strong bond, higher molecular weight and more crosslinked polymers are generally better. In most cases, adhesives consist of monomers and/or oligomers, which are a small number of monomers linked together. Because adhesives need to have stability prior to application, there needs to be some method for activation of polymerization. This activation can include heat, change in pH, catalyst, addition of a second component, or radiation. Sometimes a combination of methods is used for faster acceleration. The acceleration method is closely tied to the process for making the wood product.

Heat is a very common way to accelerate polymerization reactions. Most chemical processes are controlled by the transition state activation energy, using the standard Arrhenius equation. One typical factor is that rates of reaction double for every 10°C increase in temperature, but this does not always apply. This means that if a fast reaction is desired and the normal reaction temperature is not extraordinarily high, there will be appreciable reaction at room temperature limiting storage life of the adhesive. Since wood is a good insulator, uniform heating of the adhesive continues to be a problem for many composites and laminates. Incomplete heating gives poor bond strength as a result of incomplete formation of the polymer. To overcome this problem adhesive producers try to have the adhesive formulation to as advanced a degree of polymerization as is possible while still having good flow and penetration into the wood. Having a more advanced resin means that fewer reactions need to take place to obtain the strength properties needed from the adhesive. This balance between the advancement of the resin for fast curing while still having good bonding properties has been optimized by intense study of reaction mechanisms over the years to allow higher production rates. On the other hand, the understanding of heat and moisture levels within the composites is still being studied to allow further improvement in production rates.

Many of the adhesive polymerization rates are sensitive to pH. This is especially true of the formaldehyde polymers, but the effect varies with the individual type of co-reactant and the different steps in the reaction. For urea-formaldehyde (UF) resins, the initial step of addition of formaldehyde to urea is base catalyzed, while the polymerization of hydroxymethylated urea is acid catalyzed. Thus, UF resins are kept at a more neutral pH for storage stability, but then accelerated by lowering the pH during the bonding process. For phenol-formaldehyde (PF) resins, there is a different pH effect with condensation reactions being faster at high pH's and very low pHs. One issue of concern is how much the pH and neutralization capacity of wood alter the adhesives' polymerization rates near the interface and within the wood. This is complicated by the fact that different woods have different pHs and buffering abilities (Marra 1992).

Another aspect that alters the polymerization rate is the addition of catalysts and accelerators. A true catalyst is one that is not consumed in the process, while an accelerator can be consumed via reaction. A number of accelerators are incorrectly termed catalysts. As mentioned in the previous paragraph, changes in pH can catalyze polymerizations. In some cases, the pH is not changed directly, but compounds are added that can generate acids, such as the ammonium chloride or ammonium sulfate accelerators for UF resins that decomposes upon heat to yield hydrochloric acid or sulfuric acid, respectively (Pizzi 2003e). Certain metal ions are known to be catalysts for phenol-formaldehyde resins. Ortho esters are often described as catalysts for PF resins, but in actuality are consumed in the process, making them accelerators (Conner et al. 2002). A number of compounds have been found to speed up PF curing (Pizzi 1994c). In some cases, co-reactants, such as formaldehyde, have been referred to as accelerators, but in their general use, they serve as hardeners.

Many adhesives are two-part products. Because the components are not mixed together until shortly before the bonding process, each component alone has a long storage life. However, the addition of a second component allows the polymerization to begin. Because the adhesive is applied at ambient temperatures and most of the polymerizations need higher temperatures, setting is slow until the composite or laminate reaches the heated press. Rapid ambient polymerizations are not desirable because they limit the adhesive's ability to wet and penetrate the wood, and to transfer when the wood surfaces are brought into contact. One area of concern is the uniformity of mixing of two components. Off-ratio mixtures do not form as strong a bond as those at optimum ratio because of the poor stoichiometry. The better the compatibility and more equal the viscosity of the two components, the better the uniformity of the product upon mixing. Most application equipment is designed to give good mixing, but this may not be as true in laboratory testing or during upsets in plant operations. A special type of two-component application, where one component is applied to one surface and the other component to the other surface has been called a honeymoon adhesive (Kreibich et al. 1998). The two surfaces need to be brought into the proper contact to allow mixing and the two components need to have good mutual solubility for this system to work well.

Another method of activation of an adhesive is the use of some type of radiation. The use of ultraviolet light and electron beam radiation are common for the curing of coatings, but trying to get light into a wood adhesive bond is more difficult. However, other types of radiation can penetrate wood, including microwaves and radio frequencies, which activate curing by causing heat generation in the bondline to initiate thermal polymerization.

9.5.3 Solidification by Cooling

Although hot melts are a small part of the wood adhesive market, understanding the interaction of molten polymers with wood to form a strong durable interface is important for the wood-plastic composite field. Many wood adhesives used by the early civilizations were hot melts (Keimel 2003). Some hot melt adhesives have been used for bonding plastics to wood and are used in some wood assembly markets, such as cabinet construction, edge banding, window manufacturing, and mobile home construction. Because hot-melt adhesives and plastics used for composites are polymeric, they have a limited ability to flow. Heating the polymers above their softening point will allow them to flow. The lower the molecular weight of the polymer and the higher the temperature, the better the flow. However, both of these aspects can reduce the final strength and lengthen the set time. The formulation of the polymer backbone and additives can have a great effect on the set time. In fact, formulation is often used to control the set time so that the adhesive does not solidify before the two components are in place or take so long that extended clamping times are needed. Unlike other adhesives, high viscosities of hot melts limit their ability to penetrate into the wood lumens and flow across the wood surfaces. As the adhesive cools, its viscosity raises rapidly to further limit the wetting. Although the wetting of the wood is limited, there has still been reported flow into lumens (Smith et al. 2002). Understanding the wood-molten polymer interaction is very critical for making improved wood-plastic composites.

Some of the newer hot-melt adhesives are reactive types to allow for better wetting and greater strength of the adhesive. Normally, hot melts need to be of high molecular weight for strength, but if the adhesive cures after application, then the initial strength is not such a critical issue. The curing also makes the adhesive a thermoset to eliminate remelting of the adhesive or flow (creep) with time. Some of these products are isocyanates so that they cure by reacting with moisture that is readily available in the wood. Thus, the combination of modes of set provides benefits that are not available over adhesives with a single mode of setting.

9.6 PERFORMANCE OF BONDED PRODUCTS

Because an adhesive is used to hold two adherends together under normal use conditions, it is important to comprehend the properties of an adhesive that allow it to perform this function. The definition of an adhesive is mechanical in nature, making it important to understand the internal and external forces on the bondline and the distribution of those forces across the bonded assembly. Mechanical properties are dependent upon the chemical structure; thus, knowing the structure of the adhesive and interphase helps to understand the adhesive's performance. Bonded assemblies are usually weaker in tension than in shear or compression because it is easier to pull the chains apart. To understand the performance of bonded products, the structures of the wood adhesive polymers and the mechanical properties of polymers need to be appreciated. Greater strength in the bulk of the adhesive does not necessarily result in more strongly bonded assemblies because the weakest portion may still be in the interphase region. Another factor is the need to know the forces that the bondline must withstand under normal use conditions. The effect of external forces on the bondline can be analyzed through a variety of standard tests; however, the internal forces are not as clearly evaluated. There are commonly accepted durability tests, but the forces that are exerted on the bondline during these tests are not well understood. The relationship of mechanical properties that are usually observed on the millimeter scale to the chemical structure that is formed under the nanoscale has to be examined.

9.6.1 BEHAVIOR UNDER FORCE

The evaluation of the integrity of a bonded object rests upon understanding the viscoelastic dissipation of energy for each of the components (bulk adherend, bulk adhesive, and adhesive-adherend interphase). A basic test is a stress-strain curve, which shows the response of a material to an applied force, usually in tension. Although the behavior of material can be measured in tension, compression, or shear, tension is usually measured because it is the most likely mode of failure.

Stress-strain data are presented for a variety of materials in Figure 9.12. A very stiff material, such as a non-ductile metal or glass, does not elongate (% strain) much before the material breaks; thus, the applied force accumulates as stress until it exceeds the strength of the material, as indicated by curve A. The stiffness or modulus of A is defined as the stress divided by strain at low strain. Plastics are represented by curve B or C in that at some point the elastic limit (when deformation is no longer reversible) is exceeded at the yield point. The modulus of B is the linear portion prior to the yield point. The applied force is elastically stored in the plastic prior to the yield point, but stretches inelastically after the yield point. For a lower molecular weight plastic B, at some point on this plateau the applied force exceeds what that plastic can take and the sample breaks. However, a higher molecular weight plastic C will have a strain-induced crystallization that causes the curve to bend upward again. The last example D represents a rubber that does not store much energy as stress, but the energy causes the material to elongate. The modulus in this case is much lower and hard to measure since the initial linear section is short. In addition to the stress, strain, and modulus obtained from these tensile tests, another important piece of information is the area under the curve, which is related to toughness.

Wood Adhesion and Adhesives

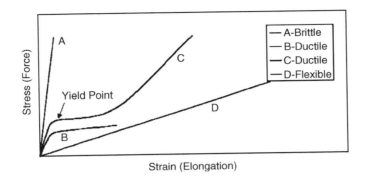

FIGURE 9.12 General stress-strain data for polymers. The rigid polymers resist applied force and build the stress showing a high modulus (stress/strain) until the material breaks. A ductile material will resist initially, but then start to flow at the yield point, with higher molecular polymers showing strain-induced crystallization. The flexible polymer will offer little resistance to the applied force, giving a high elongation.

For wood-bonding applications, a polymer of type D is not acceptable since there is not enough rigidity in the adhesive. However, type D is excellent for caulking and sealant applications since these materials need to be flexible given the expansion and contraction of buildings. Curves B and C have large areas under the stress-strain curve giving these materials good toughness, especially for impact resistance. Curve C represents plastic used in wood-plastic composites. Some wood adhesives represented by curve B are the polyvinylacetate resins, emulsion polymerized isocyanates, polyurethanes, contact cement, and hot-melt adhesives.

Curve A represents structural adhesives that have low creep, the lack of flow under force. This non-flow characteristic under normal conditions means that bonded products will retain their shape. Most wood adhesives fall into this class, including the widely used urea-formaldehyde, phenol-formaldehyde, resorcinol-formaldehyde, and combinations such as melamine-urea-formaldehyde and phenol-resorcinol formaldehyde. The polymeric diphenylmethane diisocyanate and epoxy adhesives also are members of this class.

The data in these graphs represent the materials at a specific temperature. As the temperature of a material increases it softens so that a class A polymer becomes like B. The transition of going from a glassy (hard and brittle) material to a more pliable one involves going through the glass transition temperature, T_g. However, there are limits on softening for curable adhesives because they can continue to cure and become more rigid at elevated temperatures and can begin to degrade at some point, thus changing their physical properties.

Knowing the chemical structure of the adhesives allows the prediction of the general class of polymer properties, but does not allow the calculation of the specific shape of the curve. The class D polymers are generally linear or branched organics that have low crystallinity. They also include a major non-organic adhesive and sealant type, the silicone adhesives that are actually poly(dimethylsiloxanes) and their derivatives and copolymers. These materials will creep, that is, flow under an applied force, unless they are crosslinked. The crosslinks prevent the polymer chains from continuing to flow past one another. As the number of crosslinks increases, the material becomes stiffer, usually resulting in a reduction in the ultimate elongation.

For non-crosslinked polymers, the properties are dependent not only upon the chemical structure, but also upon the conditions to which the material has been exposed. As would be expected, the lower the rotational energy around the bond in the backbone, the more flexible and impact-resistant the product is. Thus, Si-O-Si bonds provide the most flexibility and are class D, with C-O-C next, and then C-C-C bonds being the least flexible. Replacement of a linear structure with a cyclic group increases the stiffness of the backbone, and having an aromatic ring provides even higher stiffness. Interchain interactions, such as hydrogen or ionic bonds between chains and the formation of crystalline regions to act as reversible crosslinks, also alter the properties. These interactions reduce chain mobility, and thus increase the stiffness and T_g of the polymer. However, these interactions will be weakened by heat or water exposure, reducing the strength of the polymer. Additionally, the history of the polymer affects its properties. Plastics generally have a fair degree of crystallinity; this association of the molecules causes a reduction in the mobility of the polymer chains compared to more amorphous polymers. The quantity and structure of the crystalline regions depend very much on how the material solidifies. Fast cooling creates fewer and smaller crystals, resulting in a less stiff product than does slow cooling (annealing). At the interfaces, the type of adjoining surfaces influences the crystallization of the polymer.

The chemical structure and amount of crosslinking play a major role in making an A-type polymer. The backbones usually contain aromatic groups, sometimes cyclic groups, and generally few aliphatic groups, and the polymers tend to be highly crosslinked. Because many wood products are used for structural applications, it is necessary that under applied load most will not exhibit any significant elongation; thus, a high modulus is required. Unfortunately, the same factors that lead to a high modulus generally lead to brittleness in the polymer.

Crosslinking of polymer chains is required to convert a thermoplastic resin to a thermoset resin. The tying of the chains together eliminates the plastic flow of the polymers, which is necessary to eliminate creep over time. Natural rubber was known about for a long time but had little commercial utility because it softened under heat. After much research, vulcanization processes were developed which allowed rubber to retain its deformability, but eliminated the flow. As would be expected at low crosslinking levels, rubber has large segmental mobility, resulting in a very flexible product. As the crosslinking and molecular weight increases, the segments have less mobility, making the product more rigid. Unfortunately at high crosslinking levels, not only does the product become more rigid, it also becomes more brittle.

Figure 9.13 shows some idealized stress-strain curves that demonstrate the effect of increasing polymerization and crosslinking on the properties of different adhesives, and the effect of conditions on the adhesive. For thermoplastics, increasing the molecular weight mainly increases the elongation at break. This means as the adhesive cures, it is able to withstand greater force. The conversion from a thermoplastic to a thermoset will increase the stiffness at some expense of ductility. For both thermoplastics and thermosets, an increase in temperature or moisture will soften the material. Thus, in composite production, both the heat and moisture factors are working against the adhesive as it is trying to hold the material together to resist either a blowout (void in panels caused by steam bubbles) or excessive springback (tendency of compressed or bent materials to return to their original state).

Some classes of adhesives are more amenable to changing their properties by altering their formulations than are others. Polyurethanes and polyamide adhesives can go from very flexible to quite rigid depending on the formulation. Phenol-formaldehyde and polymeric methanediphenyl diisocyanate adhesives do not have a similar formulation flexibility. For some resins, incorporating flexible segments, which are softer than the main backbone and improve the impact resistance and reduce the brittleness of the polymer, can improve the polymer's properties.

However, the adhesive formulator does have a number of tools for varying the stress-strain behavior of these products. It should be noted that many of these additives are added for other purposes, such as lower cost, reduction of over-penetration, increase of resin tack, and improvement of wet out, but our concern here is how they affect the stress-strain behavior. The additives are divided into the classes of fillers, extenders, plasticizers, and tackifiers see section 9.7.13.

Fillers are common additives because they lower the cost, and thus are used at as high a level as possible to make the adhesive more economical. Fillers increase the stiffness of the adhesive, but usually also reduce its elongation and increase its viscosity. At low levels extenders have a small impact on an adhesive's properties, but at high levels they cause decreased elongation and higher viscosity. On the other hand, plasticizers soften an adhesive, resulting in a decreased modulus and T_g, and an increased elongation. For most wood adhesives the desire is to have a rigid bond;

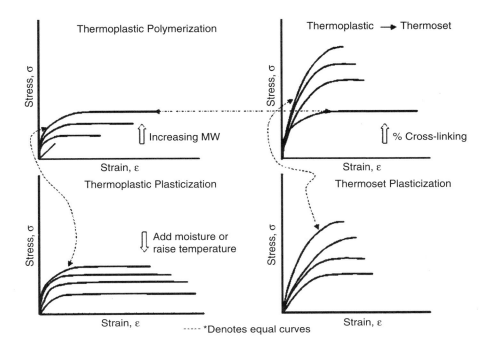

FIGURE 9.13 Effect of polymer changes on physical properties. For a thermoplastic, increasing the molecular weight leads to increases in both the stiffness and the ductility, while the thermoset loses ductility as it becomes stiffer with higher crosslinking. When the polymers are plasticized or the temperature is raised, both the thermoplastic and thermoset lose stiffness.

thus, plasticizers are not generally used. Tackifiers are often confused with plasticizers, but provide very different responses in raising the glass-transition temperature while decreasing the modulus.

9.6.2 Effect of Variables on the Stress-Strain Behavior of Bonded Assemblies

The discussion, so far, has been on the stress-strain behavior of adhesives under one condition and in tension. It is important to understand what happens to the strength properties under other conditions. For wood adhesives, the two most important changes in conditions are changes in temperature and moisture. Additionally, it is important to consider more than just the cohesive strength of the bulk adhesive and bulk wood. Although the properties of the bonded assembly are a continuum, Marra's weakest link concept is useful in understanding failure (Marra 1992). Thus, it is important to understand the properties of the interphase, as well as the bulk adhesive and the wood. Applied forces are not going to be expressed as a uniform force throughout the bonded assembly for several reasons (Dillard 2002). The differences in mechanical properties of the wood, adhesive, and interphase regions imply that stress concentrations are likely to occur in the zone of greatest change, i.e., the interphase zone. Additionally, the interphase has the greatest internal stress caused by volume reduction in the adhesive upon setting. With environmental exposure, the interphase has to accommodate the large dimensional changes between the wood and the adhesive. If the applied stresses can be dispersed over the entire volume of the material, then localized stresses can be reduced and higher total bond strengths obtained. The ability of the applied forces to be dissipated in certain domains without catastrophic failure can lead to higher bond strengths (the shock absorber approach). On the other hand, high natural internal stresses can add to the applied force and cause unexpected failure. The stresses can be concentrated such as at a flaw causing early failure (Liechti 2002).

For a bonded assembly, the overall properties are hard to predict because less is known about the properties of the interphase regions compared to the bulk properties of adhesives and adherends. The bulk mechanical properties of many wood species have been well studied (Green et al. 1999). The bulk properties of many adhesives have also been investigated, but many of the wood adhesives form brittle, inhomogenous films that do not yield good mechanical property measurements. However, the interphase properties change from those of the bulk adhesive to those of the bulk wood. This gradient of property change can be gradual or sharp, and it is expected that a more gradual change should be better, as the stress concentration would be smaller. Large internal forces can be generated when the adhesive and the adherend have different responses to environmental changes, such as moisture and heat. The difference in expansion coefficient between metals and adhesives has been well studied as a cause of adhesive failure. A major issue with wood is the difference in expansion coefficients with moisture changes between adhesives and wood, mainly in the radial and tangential directions. How these expansion differences are handled in the bonded assembly may be very important to its durability. It is important to remember that the internal forces can be as significant as the applied forces for bond strength.

The strength properties of most polymers are sensitive to temperature changes (see Figure 9.13). The increased vibration and therefore mobility of polymers at higher temperatures cause the polymer to be less resistant to applied forces. However, the effect is greatly influenced by the structure of the polymer. Thermoplastic polymers soften at the glass transition temperature and eventually flow at the melt transition temperature, leading to a lower T_g. Crystalline segments will limit the effect of temperature until the melting point of the crystallites is reached. The addition of crosslinks, even non-covalent crosslinks, such as hydrogen bonds, can improve the resistance to softening at elevated temperatures. Covalent crosslinks that exist in many wood adhesives give good resistance to temperature changes in the bulk adhesive. However, there can be significant differences in the thermal expansion coefficients of the wood and the adhesive causing interphase stresses (Pizzo et al. 2002).

An even greater issue is the effect of moisture changes on bonded assemblies, especially in the interphase region. Some adhesives, like polyvinyl acetates, lose much of their strength at high moisture levels, as a result of polymer plasticization. Urea-formaldehyde adhesives are known to depolymerize under high moisture environments, as shown by increased release of formaldehyde (Dunky 2003). On the other hand, wood adhesives, like phenol-formaldehyde and resorcinol-formaldehyde, do not change drastically in their adhesion to wood at higher moisture levels. Wood is known to weaken at higher moisture levels, and to change dimensionally in the radial and tangential directions. When an adhesive does not change dimensionally as the wood swells and shrinks, then stress concentration will occur in the interphase region.

The setting process can generate additional internal forces due to shrinkage of the adhesive. The loss of solvent/water and the polymerization process reduce the volume of the adhesive, while the surface area of the wood stays constant. This difference can cause significant forces that may exceed the strength of the adhesive. Weakness in the bulk of the adhesive urea-formaldehyde has been shown to cause cracks in the adhesive (River et al. 1994c); adding flexible groups to the urea-formaldehyde formulation reduces this deficiency (Ebewele et al. 1991), especially those groups of low to medium molecular weight (Ebewele et al. 1993). In other cases, the forces alone are not sufficient to cause fracture, but may be sufficient to cause fracture when combined with small applied external loads or swelling of the wood as a result of a higher sum of internal and external forces.

9.6.3 BOND STRENGTH

Adhesives are used to hold two materials together; thus, the viscoelastic dissipation of internal and external forces is the most important aspect of adhesive performance. The forces that a bond assembly has to withstand depend very much on the type of product and the use of that product. The effects of internal forces are often not considered, but such forces can be very high in wood.

Wood Adhesion and Adhesives

The most rigorous test for laminated wood is the ASTM D 2559 cyclic delamination test (ASTM International 2000b). Many adhesives that have strong wood bonds under dry conditions show significant delaminations and do not pass this test; however, the extent of the delaminations can be reduced by using a hydroxymethylated resorcinol primer (Figure 9.14). An interesting aspect of D 2559 is that no external force is applied; swelling and shrinking forces alone cause the bond failures. This test involves cycles of vacuum water soaks, followed by oven drying, with a water boil in the second cycle. The fact that dimensional changes, along with some warping of wood, are sufficient to cause substantial bondline failure shows the power of these internal forces. The problem with internal forces is that they are very hard to quantify. However, a test like the D 2559 may exaggerate these forces since the rapid drying provides sufficient force to cause excessive fracture of the wood, while under normal use conditions the moisture change in the wood is more gradual, allowing stress relaxation of the wood.

The forces on bondlines are divided into three modes: I, II, and III (Figure 9.15) (Liechti 2002). The normal force of mode I is perpendicular to the bond and is the direction in which adhesives are the weakest, usually because of stress concentration. On the other hand, the shearing force of mode II is the direction in which the adhesive is the strongest. The torsional forces of mode III are an intermediate test of adhesive strength. All three types of forces are common in bonded wood products. Mode I forces exist in strandboard as it resists springback from its compressed state and in the internal bond test. The mode II force is common in laminated veneer lumber under normal external loading or during swelling under high moisture conditions. Mode III forces exist in plywood as a result of the cross-ply construction.

The performance tests are generally covered by the ASTM and other standards (Vick 1999, River et al. 1991). Normally, the tests tend to be hard to pass to allow safety factors in construction. The general rule with most wood products is to have as much good bonding surface and to have as much of the force in the shear mode as possible. Knowledge of wood bond strength has generally been gained using laminated wood and plywood specimens. Distributing the adhesives as droplets on irregular surfaces of strands or fibers has made understanding the bonding for strandboard and fiberboard more difficult. The problem in relating this work is that laminates and plywood are normally tested in shear, while a primary test for particle board, strandboard, and fiberboard, the

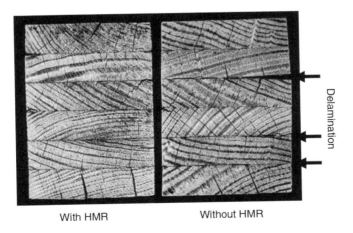

FIGURE 9.14 ASTM D 2559 causes bond failure, as shown by delamination from the shrinking and swelling of the wood. The test is severe enough to cause cracking in the wood, but an acceptable adhesive gives minimal bondline failure. The same adhesive was used in both specimens, but the wood on the left that was first primed with hydroxymethylated resorcinol (HMR) resisted the delamination much better than the untreated wood on the right (Vick et al. 1976), see section 9.7.3.2.

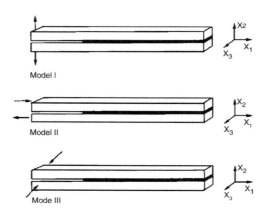

FIGURE 9.15 The force on the bondlines is often a combination of the three modes of force. Mode I is a tensile force in the normal mode and is usually the one in which the adhesive is the weakest. Mode II is the common shear force and is usually the mode in which the adhesive is the strongest. Mode III is the less common torsional force.

internal bond test, involves mode I forces. The moduli of rupture and elasticity are more a property of the wood and the flake orientation rather than the adhesive bond. Summaries of much of the work on performance testing have already been published (River et al. 1991, River 1994a).

Considering the bonded assembly as a series of links in a chain (Marra 1992), the chain will hold unless the force exceeds the strength of one of the links. Thus, the process for making improved adhesives involves understanding what the weak link is, and why it failed. The strength of the links can vary with conditions. For example, if an adhesive, such as uncrosslinked poly(vinyl acetate), softens with heat, then it is likely to become the weak link under hot conditions. Many adhesives give strong bonds to wood under dry conditions so that the wood is the weak link. However, under wet conditions the weak link may be in the interphase because of a greater strength loss in this link than in the bulk wood or adhesive. Epoxies exhibit a high percent wood failure when dry, but low wood failure when wet (Vick 1999); thus the conditions cause the weak link to change. Recent data has indicated that the weak link is the epoxy interphase region (Frihart 2003a), leading to the need to strengthen the epoxy or reduce the stress concentration in the interphase. Using the chain analogy can also aid in understanding why adhesives do not bond as well to dense wood species. If the strength of the bulk adhesive and the adhesive-wood interphase links are enough to hold 2000 psi and the wood strength is only 1000 psi, then the wood breaks first. If the wood strength increases to 3000 psi, then the fracture is not in the wood. This is not to imply that more dense woods may not be harder to bond in some cases, but the data needs to be considered in light of the strength of the wood.

With this concept of the chain links to represent bond strength, it becomes important to understand where failure occurs. Failures within the bulk wood and bulk adhesive are generally easy to see using the naked eye or microscopy. The failure in the interphase is more complicated, especially for wood. In Figure 9.16, the different types of interfacial failure are illustrated. Understanding failure mechanism is important because it leads to better routes for improving the adhesive to solve the problem. One study showed that phenol-formaldehyde gave fracture in the S_2 layer while an epoxy gave failure in the S_3 under peel, suggesting that the PF gave deeper penetration of the cell walls (Saiki 1984).

At this time, there is insufficient knowledge to predict how well a new adhesive will hold wood pieces together without testing the bonding with the same type of wood and a similar bonding process that will be used commercially and then carrying out the performance tests. The current limitations are in not understanding what is necessary about the adhesive-wood interactions to give

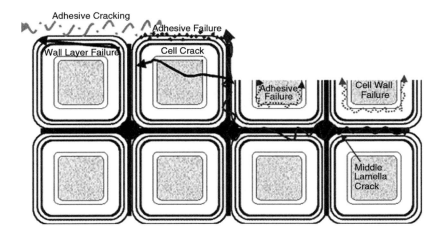

FIGURE 9.16 Failure in the interphase region of wood bonds is complicated. Besides the true interfacial failure that leads to adhesive on one surface and wood on the other, there are a number of other failure zones. The adhesive near the wood may not cure as well, leading to failure in the adhesive near the surface. The adhesive may bond strongly to the wood, but the wood itself may split between layers or within a cell wall layer.

strong durable bonds. This has been hard to examine because of the complex chemistry and morphology of wood. However, improved analysis will help to shed light on this issue.

9.6.4 Durability Testing

ASTM defines durability "as related to adhesive joints, the endurance of joint strength relative to the required service conditions"(ASTM International 2000a). Because wood products are used over many years, accelerated tests are used to estimate the long-term performance. Some studies were carried out using field-testing to understand the performance of adhesives under some conditions (River 1994b, Okkonen and River 1996). In addition, for many adhesives, there is in-use experience over many years. Several accelerated tests have been developed that give similar results on durability with these adhesives (River et al. 1991). The key factor that has often been overlooked is that the failure mode must be the same for both long-term use and accelerated test results; thus, it is of paramount importance to validate the accelerated aging tests. If the failure modes are different, then the accelerated tests are not a reliable predictor of long-term performance.

The most common problem with wood durability is the adhesive's inability to withstand the swelling and shrinking of wood with moisture changes. Most wood products are subjected to moisture changes, but those in uncontrolled environments are subjected to greater changes. The swelling of wood can subject the bond to mode I, II, or III types of forces depending on the joint design. Swelling has normally been considered on the basis of the macroscopic changes; however, it should be considered also on the basis of the cellular (micrometer) scale. The available data indicate that the swelling of cells usually involves thickening of the cell walls rather than shrinking of the lumen diameter (Skaar 1984). Thus, large forces are exerted on the adhesive at the cell wall edges (Frihart et al. 2004). One study indicates that a phenol-resorcinol-formaldehyde adhesive yields more under wet conditions, but the changes were not as large as the dimensional changes of wood during the wetting process (Muszynski et al. 2002). A key question is whether durable adhesives have more compliance with moisture changes, whether they stabilize the cell walls so that there is less swelling and shrinking with moisture changes, or whether they are better able to distribute the interfacial stresses? Understanding these points is key to designing more durable adhesives. Another factor that has to be considered is that accelerated tests involve rapid wetting and drying of the wood. These changes can be so fast that the wood structure does not have a chance to stress relax during the tests; thus, artificially high stresses may be created that would not be observed in normal use.

Wood adhesives have to pass other durability tests, but, for the most part, these have not been as difficult. Certainly, adhesives used in structural and semi-structural applications have to resist creep under load. Given the rigid nature of the polymer backbone and the crosslinking, this has not been a significant issue. Wood adhesives also have to resist decay, and therefore they may be formulated using additives so that fungi do not grow on the surface (ASTM International 2001).

9.7 ADHESIVES

The properties of an adhesive need to not only match the needs of the bonded assembly in its end use, but also need to be compatible with the substrate and the bonding process conditions. Generally for wood bonding, adhesives are of the structural type; that is, they are able to transfer load between adherens. Although these adhesives fall into the general classification of structural type, in wood bonding they can be divided into structural, semi-structural and non-structural—see section 9.3 (River et al. 1991). Although rigidity is often good, adhesives can be too rigid for some applications.

Adhesives need to be compatible with the bonding conditions used commercially. Heat-cured adhesives are compatible with the manufacture of panel products for the following reasons:

1. They cure slowly at room temperature, allowing time for the wood components to be coated with the adhesive and brought together for assembly.
2. The heat and moisture let the wood soften, allowing the adjoining wood surfaces to be brought into close contact.
3. Upon heating, the adhesive cures quickly, reducing springback when the pressure is released.

However, a room temperature cure is better for thick laminates because heating the deep layers is more difficult. For manufacturing bonded products, low-cost and rapid setting of the adhesive are important factors, but for construction adhesives, a longer set time and easy dispensing from cartridges are important because it takes time to bring the surfaces into contact. In many non-wood applications, water-borne adhesives are not used because of poor surface wetting and the inability of the water to move away from the bondline. Neither of these issues is as critical for wood adhesives. However, the penetration of adhesives into wood without over-penetration is important for wood bonding, but not a factor in the bonding of most other materials.

To understand the application, setting, and performance of adhesives, some general polymer chemistry and polymer properties need to be covered. The specific adhesives refer back to this general discussion. The properties of polymers are controlled by the structure of the backbone and the number of crosslinks, if any. In a few cases, such as polyurethanes, domain separation is also an important factor.

9.7.1 POLYMER FORMATION

Knowledge about the structure of polymers leads to a better understanding of their properties; the properties of polymers are important both in the bonding process and in the ultimate end use performance of the bonded material. Aspects of polymers that need to be considered include use, class, type, and size.

Adhesives can be grouped not only by their structural, semi-structural, and non-structural use, but also by their permanence and durability. Permanent is more stable than wood under irreversible environmental conditions, while nonpermanent is less stable than wood under irreversible environmental conditions (River et al. 1991). Durable is stronger, more rigid than wood, and more stable under reversible environmental effects, while nondurable is weaker, less rigid than wood, and less stable under reversible environmental effects.

Different applications require materials of different mechanical properties, with these being greatly influenced by the chemical structure of the polymer. It must be remembered that as discussed in Section 9.6.1, the properties of polymers are greatly influenced by the conditions under which they are measured. For example, most adhesives will tend to soften and therefore are less able to carry a load as the temperature increases. When many adhesives absorb small molecules, including water, they will soften and, in some cases, will develop cracks that will expand and ultimately cause failure. In addition, the properties of many polymers change as they age. If a polymer is susceptible to oxidation, over time this can either make the material stiffer or depolymerize the adhesive, making it weaker. Chemicals, such as ozone, acids, and bases can also alter the performance of many adhesives.

Polymer classes are determined by how the polymer is constructed. Some polymers are homopolymers, such as poly (vinyl acetate). This means that the polymer (AAA…) is made up of individual monomer units (A) that are all the same. Much more common are those polymers made up of two or more components, such as A and B. One way of putting the components together is a random process where two or more monomer units form the copolymer (AAABAABBBB…), but there is no specific order to the adjacency of the components. An example of this class is the styrene-butadiene rubber that is used in many sealants and mastics. Another way of putting the components together is an alternating copolymer (ABABABAB). Two components can also be combined by making block co-polymers where there are long stretches of monomer A that are then attached to sections of monomer B. Often the A and B components are not compatible when polymerized, so materials tend to separate into individual domains, with examples being polyurethanes and styrenated block copolymers. While the random and alternating copolymers exhibit the average properties of the homopolymers, the block copolymers often exhibit properties not obtainable with either of the homopolymers. A fourth way of reacting two monomers is a grafting process, in which monomer B is attached along the sides of a polymer A backbone. An example is the reaction of grafting of acrylate polymers onto a polyolefin backbone.

Polymer types can be used to group adhesives with different topology independent of their grouping, according to class. The same polymer type can be either a homopolymer or copolymer. One type is a linear polymer where all the monomer units go in head to tail fashion with one another to form the polymer chain. Polyethylene and polypropylene are for the most part linear polymers. However, in the case of polyethylene, there are often branches off the linear chain; the properties of the polymer change dramatically as the type and degree of branching changes. In going from the linear high density polyethylene to the slightly branched low density polyethylene and onto the much more branched very low density polyethylene, there are changes in melting point, flexibility, and strength. Another type classifies polymers according to whether they are crosslinked (thermoset) or not crosslinked (thermoplastic). Some wood adhesives are thermoplastic, including uncrosslinked poly(vinyl acetate) and hot melts. The problem with thermoplastics is that at elevated temperatures or moisture levels, they will flow, leading to creep (flow under load over time) problems. For structural and semi-structural applications creep is very undesirable. Thus the great majority of wood adhesives are thermoset. The term thermoset is used to indicate crosslinked polymers even though the setting process may not be caused by heat. Hot press adhesives are certainly thermoset because they need heat activation to develop the crosslink. On the other hand, moisture-cured adhesives, such as some polyurethanes and silicones, are crosslinked not by the heat process but by the presence of moisture, but are also considered themosets.

Another variation in polymer backbone involves whether the structures are linear aliphatics, such as the case with polyethylene, or whether they are cyclic structures, such as cylcohexane or aromatic rings. The cyclical nature of the monomers makes the polymers much stiffer because they have less ability to rotate around the backbone bonds. Aromatic rings make the adhesives even more rigid due to less rotation in the backbone. Many wood adhesives tend to be made from cyclical compounds, including phenol, resorcinol, and melamine, to produce much more rigid polymers with high glass transition temperatures.

Another factor in the properties of polymers is their size or molecular weight. This is an area that illustrates the two competing natures that an adhesive needs to exhibit. For bond formation, the adhesive needs to flow and penetrate into lumens and cell walls very well, favoring low molecular weights. However, once the bond is formed, it is desirable that the product has great resistance to flow, which favors higher molecular weight. A higher molecular weight adhesive will tend to set faster because fewer reactions are needed to form the cured product. The higher molecular weight polymer can lead to solubility and stability problems of the adhesive. Thus, in designing polymers to be used as adhesives, a balance is needed between low molecular weight for a good wetting of the wood and higher molecular weight for more rapid set and to resist flow once the bond is formed.

Aside from these obvious differences in formulations, changes in the curing conditions can have an effect on the properties of the resin. It is well known with epoxies that additional heating causes additional crosslinking reactions. An epoxy cured at room temperature becomes a rigid gel so that the remaining unreacted groups are not physically able to find each other. As the epoxy is heated, the mobility of the polymers increases, allowing additional groups to come into physical contact to add more crosslinks in the matrix, making the product more rigid and usually more brittle. This effect has also been observed with phenolic resins, in that postcure times influenced both the degree of cure and the mechanical properties (Wolfrum and Ehrenstein 1999). This is important in considering the production of composites. For particleboard, strandboard, and fiberboard, the adhesive near the surface is at a higher temperature for longer times and at a lower moisture content compared to the adhesive toward the center of the board. The gradient in the heat and moisture causes less polymerization and crosslinking to occur in the center of the composite. The primary curing problem can be reduced by using a faster reacting resin or a higher molecular weight resin in the core than in the face. However, the gradient in the reaction rates can influence the properties of the board and makes studies on the curing process exceptionally difficult.

9.7.2 Self-Adhesion

Under certain conditions wood can self-adhere, but generally adhesives are needed to give sufficient strength. The forces working against good self-adhesion are the roughness of the surface and the lack of mobility of the wood components. For good adhesion, the two surfaces have to be brought into contact at the molecular level. Obviously, this is difficult with the high surface roughness of a cellular material like wood. It becomes more likely if the surface cells are pressed together under high pressure and if the wood is more compliant, such as when one goes from wood laminates to chips to particles and finally to fibers. In fact, the only product made with little or no added adhesive is high-density fiberboard. The adhesion of the hardboard is dependent upon hydrogen bonding and auto-crosslinking (Back 1987). Of the main wood components, the greatest likelihood for self-adhesion is with lignin components. Both lignin and hemicellulose soften under high moisture and temperature conditions. Hemicellulose more readily forms hydrogen bonds to bond the adjoining fibers, while lignin more readily forms chemical bonds. The process works adequately for hardboard, but other wood products are not bonded under sufficient heat and pressure to obtain high intersurface bonding. Recently, vibrational welding has been demonstrated to cause bond formation (Gfeller et al. 2003). This process uses the heat and cellular distortion generated by friction to bond the wood together.

Chemical modification of wood surfaces has been shown to give improved bond strengths. A base activation of wood was found to give significant improvement in the dry strength of wood bonds, but not the wet strength (Young et al. 1985). Iron salts with hydrogen peroxides will give more durable bonds with wood particles than with unactivated wood (Stofko 1974, Westermark and Karlsson 2003), and surface activation with peracetic acid has also been used in making particleboard (Johns and Ngyuen 1977).

Wood Adhesion and Adhesives

The use of enzyme modification of wood has been shown to increase the strength of bonded wood (Felby et al. 2002, Widsten et al. 2003). Although most studies have been at the laboratory stage, at least one investigated has been done at the pilot plant stage (Kharazipour et al. 1997).

9.7.3 Formaldehyde Adhesives

The most common wood adhesives are based on reactions of formaldehyde with phenol, resorcinol, urea, melamine, or mixtures thereof. The reactions can involve three steps:

1. Formaldehyde reaction with a nucleophilic center of the comonomer to form a hydroxymethyl derivative
2. Condensation of two of these hydroxymethyl groups to form a bismethylene ether group, with loss of a water molecule
3. Elimination of formaldehyde from the bismethylene ether to form a methylene bridge

However, the hydroxymethyl derivative can be directly attacked by a nucleophile to form the methylene-bridged product, making it a two step process. The discussion of the chemical reactions in this section is quite basic and does not involve the details because these have been well covered in other books. The intent is to cover the general concepts; the reader is encouraged to read the more detailed discussions cited for each adhesive type.

The rates of the individual reactions depend very much on the nucleophile that is copolymerized with the electrophilic formaldehyde. All of these reactions are very pH dependent (see Figure 9.17), but the effect of pH varies depending on the nucleophile. For example, under acidic conditions,

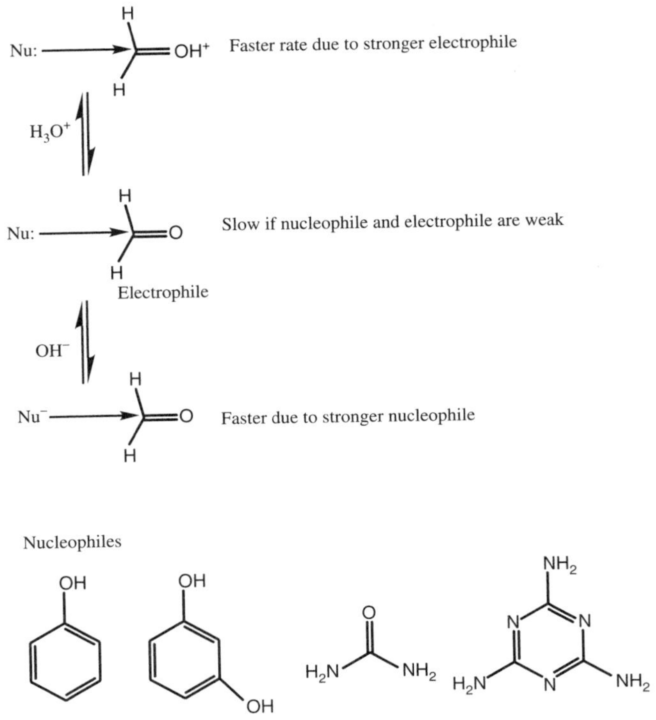

FIGURE 9.17 Reaction of formaldehyde with phenol, resorcinol, urea, and melamine. All of these compounds will copolymerize with formaldehyde, generally in an alternating fashion. The first step is the reaction of a nucleophile with an electrophilic formaldehyde that can be promoted under acidic or basic conditions.

formaldehyde addition to phenol is a slower step than the condensation step to form the methylene-bridged product, while the relative rates of these two reactions is reversed under basic conditions. Thus, control of the pH is very important in controlling the polymerizations, and the pH and buffering capacity of the wood may alter the curing in the interphase region. In addition to the pH, these reactions are also controlled by adjusting the temperature and adding catalysts or retarders. Some specifics are discussed with each adhesive type in the following sections.

The formaldehyde adhesives are usually water-borne resins so that the curing process is not only polymerization, but also the loss of the water solvent. Because the polymerization process evolves water, too much water remaining in the bondline retards the reaction. On the other hand, too little water prior to polymerization not only influences wetting, but also can reduce the mobility of the resins and limit collisions needed for polymerization, in addition to limiting heat transfer. Control of both the open and closed assembly times are important for controlling both the penetration and water content of the bondline.

Most wood bonding applications need an adhesive that does not creep over time, leading to the use of crosslinked or thermoset adhesives. High glass transition temperature polymers could also exhibit low creep, but they have been too expensive and hard to use for wood bonding. The formaldehyde copolymers produce thermoset polymers by crosslinking in the later stages of curing. These reactions occur by formaldehyde bridging the reactive sites on different chains. The comonomers used with the formaldehyde all have three or more reactive sites, leading to plentiful opportunities to crosslink. Having many available reactive sites is important due to the limited mobility of the polymer backbones, which allows close proximity between only a few locations. It is highly unlikely that every site that is converted to a hydroxymethyl group can find another group in close proximity with which to react. Longer cure times at higher temperatures will tend to push the product to a higher degree of cure. Thus, the ultimate performance of the adhesives is going to depend on the processing conditions.

Generally, the molar ratio of formaldehyde needs to be greater than that of the coreactant to accommodate the need for extra formaldehyde to crosslink the chains, to compensate for formation of bismethylene ethers, and to allow for unpolymerized hydroxymethyl groups. Extra formaldehyde was therefore used to produce fast-setting adhesives with a high degree of curing. However, this caused the problem of significant formaldehyde emissions from the bonded product, especially those made using urea. The formulations needed to be adjusted to reduce the formaldehyde levels, but still give good final cures and fast set rates. This has been accomplished through a good understanding of the adhesive chemistry, but then there has been some sacrifice in operability of the bonding process and performance of the bonded assembly.

Formaldehyde copolymer adhesives are used for the production of most laminates, finger joints, and composite products, although the isocyanates are taking over some of the market share as the result of a lower sensitivity to wood moisture and process temperatures. These adhesives provide good wood adhesion and rigid bonds that do not creep because the formaldehyde not only forms the polymeric chain, but also provides the crosslinking group. However, the properties vary depending on the coreactant used with the formaldehyde. Urea-formaldehyde adhesives are the least expensive of all wood adhesives, but they have poor durability under wet conditions. Phenol-formaldehyde adhesives offer a good balance of cost and water resistance. Higher cost melamines are used because they also provide good water resistance, and are light in color compared to the phenol resins. Resorcinol-formaldehyde resins are useful because they cure at room temperature, but are very expensive. The coreactant or combination of coreactants used with formaldehyde is selected depending on the costs, production conditions, and expected performance of the product.

9.7.3.1 Phenol Formaldehyde Adhesives

Phenol formaldehyde (PF) polymers are the oldest class of synthetic polymers, having been developed at the beginning of the 20th century (Detlefsen 2002). These resins are widely used in both laminations and composites because of their outstanding durability, which derives from their good adhesion to wood, the high strength of the polymer, and the excellent stability of the adhesive. In most durability testing, PF adhesives exhibit high wood failure and resist delamination. There are a vast number of possible formulations, and selection of the wrong one can lead to poor bond strength. Among the factors that can lead to poor adhesion are incomplete polymerization due to too little time at temperature; a resin with too high a molecular weight, leading to poor wetting and penetration; not enough assembly time to allow good wood penetration; or too much assembly time or pressure leading to over penetration of the adhesive and a starved bondline. In general, PF adhesives can meet the bonding needs for most wood applications if cost and heat curing times are not an issue.

For all these adhesives, phenol is reacted with formaldehyde or a formaldehyde precursor under the proper conditions to produce a resin that can undergo further polymerization during the setting process. There are two basic types of pre-polymers, novolaks that have a formaldehyde/phenol (F/P) ratio of less than 1 and are generally made under acidic conditions, and resole resins made under basic conditions with F/P ratios of greater than 1. Although at first glance the acid and base processes may seem to be similar, the chemical reactions and the polymer structures are quite different. For most wood adhesive applications, the resole resins are used because they provide a soluble adhesive that has good wood wetting properties and the cure is delayed until activated by heat allowing product assembly time.

The formaldehyde addition depends on the electron-donating hydroxyl group for activation of the aromatic ring, specifically at the positions *ortho* and *para* to the hydroxyl group; these positions are nucleophilic enough to attack the electrophilic formaldehyde. Although all three sites are activated, the reaction conditions control which sites are more reactive toward the initial and subsequent modifications. The availability of three positions for reaction leads to the ability to form a crosslinked polymer that is necessary for good strength and durability. The chemistry described here is general, and more detailed discussions of the reactions have been published (Detlefsen 2002, Pizzi 2003a, and Robins 1986).

Novolak resins are made using acidic conditions with typical formaldehyde-to-phenol ratios of 0.5 to 0.8 at a pH of 4 to 7. The chemistry involves, first, the addition of the acid-activated formaldehyde to the phenol via a nucleophilic reaction to the activated *ortho* or *para* positions of the phenol. This molecule can then lose a water molecule under acidic conditions due to stabilization with the phenol group. The methylene group is then reactive with another phenol group to form the methylene-bridged dimer. Continuation of this process leads to a low molecular-weight linear novolac oligomer. Under acid conditions, the linking step is faster than the addition step, which would lead to mainly polymer if the formaldehyde content were not limited. To form polymers from the oligomer, additional formaldehyde, often in the form of paraformaldehyde, which is usually called the hardener, is added just prior to application. Novolak oligomers are generally not used for wood bonding due to their low water solubility and high acidity.

Resole resins are generally made using alkali hydroxides with a formaldehyde to phenol ratio of 1.0 to 3.0 at a pH of 7 to13. The chemistry involves the reaction of the base-activated phenol with the formaldehyde, as shown in Figure 9.18. In contrast to the reaction under acidic conditions, the addition of formaldehyde to phenol under basic conditions is the rapid step, while the conversion of the hydroxymethyl derivatives to oligomers is the slow step. Thus, higher formaldehyde levels can be used without forming the final polymer until sufficient heating is applied. Some of the hydroxymethylphenols may dimerize to form a bismethylene ether bridge and are then always converted to the methylene-bridged species. This process is used to generate oligomers with sufficient reactive groups to cure under the proper heating conditions. The molecules in Figure 9.18 show the fully functionalized species, but the molar ratio of formaldehyde to phenol is usually less than 3, leading to enough groups to form the polymer backbone and some crosslinking spots. Drawings often depict only one position of reaction, but it should be remembered that all the *ortho* and *para* positions are reactive, with position selectivity due to the reaction conditions. After being applied to the wood, these resins are then converted to the final adhesive by using sufficient heating

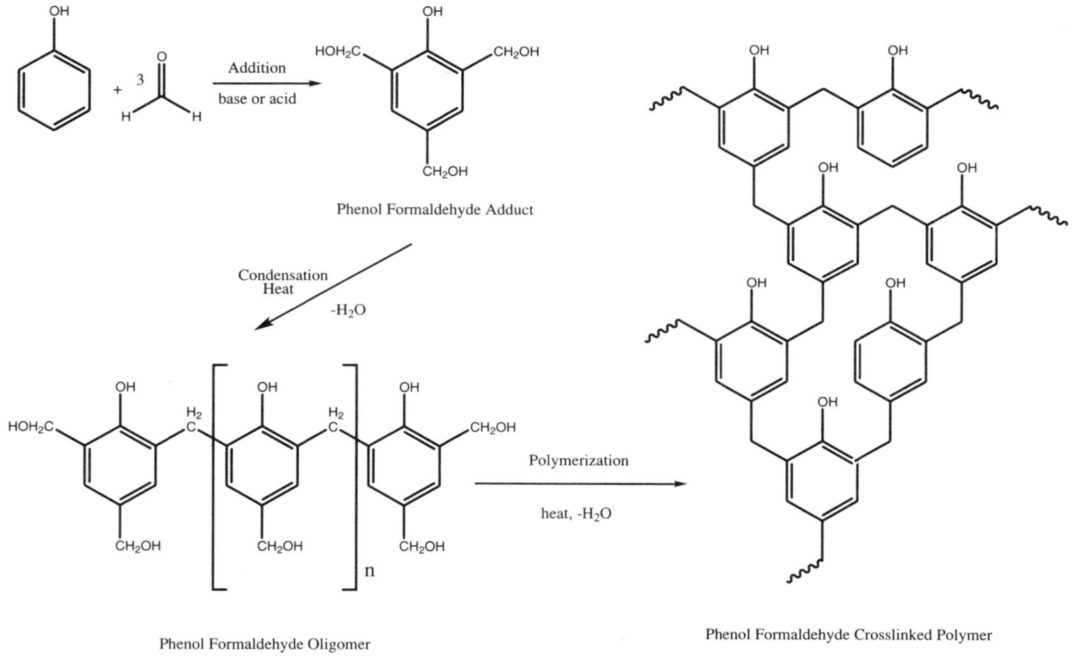

FIGURE 9.18 Phenol-formaldehyde chemistry involves first formation of the hydroxymethyl group, followed by partial polymerization to the oligomer that makes up the adhesive. After applying adhesive to the substrate the polymization is completed to form a crosslinked polymer network.

and water removal conditions. The structure in Figure 9.18 shows the limited mobility of the polymer chain; there are also hydroxylmethyl groups that cannot find a reactive site.

The phenol-formaldehyde adhesives could serve in almost all wood bonding applications, as long as the adhesive in the assembly can be heated; however, in many cases, environmental resistance is not needed so a lower cost urea-formaldehyde adhesive is used. Like most adhesives, the commercial products contain more than just the resin depending on the application. The most common additive is urea to provide improved flow properties, to scavenge free formaldehyde, and to reduce cost. It is generally assumed that the urea does not become part of the polymer backbone due to its low polymerizability under basic conditions. For plywood, fillers and extenders are added to provide holdout on the surface and control the rheology for the specific application method.

9.7.3.2 Resorcinol and Phenol-Resorcinol Formaldehyde Adhesives

Resorcinol-formaldehyde (RF) resins have the advantage over PF resins of being curable at room temperature due to being 10 faster in reaction. Resorcinol is 1,3-dihydroxybenzene, and is very reactive because of the combined effect of the two hydroxyl groups on the aromatic ring in activating the 2-, 4-, and 6-positions toward reaction with formaldehyde for the addition reaction, and with hydroxymethylresorcinol in the condensation step (Pizzi 2003b). The activation of both steps leads to a fast modification and polymerization. Because phenol and resorcinol have three reactive sites, they are able to crosslink to form a thermosetting adhesive. The chemistry of modification and polymerization is illustrated in Figure 9.19. The resorcinol copolymerizes well with formaldehyde at room temperature. Thus, it is important to have a formaldehyde-resorcinol ratio low enough to make a non-crosslinked novolac polymer, but it also requires the addition of a formaldehyde hardener just prior to applying the adhesive to wood for completing the cure.

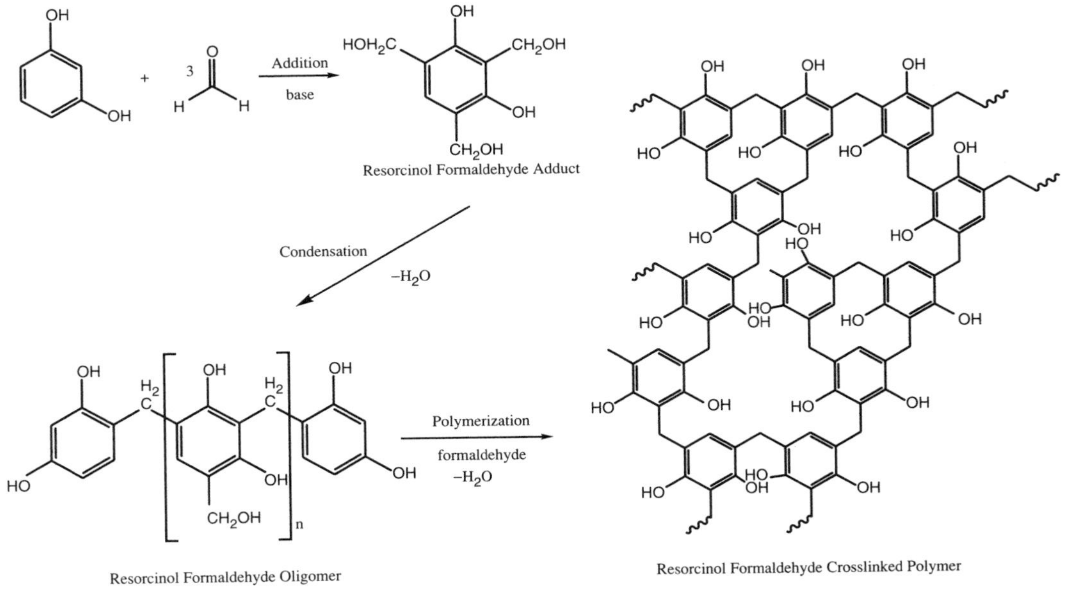

Wood Adhesion and Adhesives

FIGURE 9.19 Resorcinol-formaldehyde chemistry is similar to the phenol-formaldehyde in Figure 9.18, but the reaction rates are fast enough that heat does not need to be applied.

An interesting use of a RF resin is for the production of a low solids primer, called hydroxymethylated resorcinol (HMR). This primer has been found to be very useful in improving the delamination resistance of phenol-resorcinol-formaldehyde adhesive to CCA treated wood (Vick 1995), epoxy bonds to Douglas-fir (Vick et al. 1998), polyurethane and epoxy to yellow birch and Douglas-fir (Vick and Okkonen 1998, Vick 1997), yellow cedar with PRF adhesive (Okkonen and Vick 1998), and epoxy to Sitka spruce (Vick et al. 1996). The original primer had to be manufactured shortly before use and had a short use time, but an improved process has solved these issues (Christiansen and Okkonen 2003).

Like the phenol-formaldehyde resins, these adhesives form very durable bonds. They are resistant to both bond failure and to degradation. The main drawback to resorcinol adhesives has been the cost of the resorcinol. To lower the cost, but to maintain the room temperature curing properties, phenol-resorcinol-formaldehyde (PRF) adhesives were developed. PRF adhesives are widely used in wood lamination and finger jointing. PRFs are covered in this section because they behave more like RFs than PFs in their cure.

Three different phenol-resorcinol-formaldehyde polymers can be prepared, but all depend on the ability of the resorcinol to react at room temperature.

- A phenol-formaldehyde resole is reacted with resorcinol at the hydroxymethyl sites to form a resorcinol-terminated adhesive that is then mixed with a formaldehyde hardener just prior to bonding.
- A phenol-formaldehyde resole is mixed with a resorcinol hardener just prior to bonding.
- A phenol-formaldehyde resole is reacted with resorcinol at the hydroxymethyl sites to form a resorcinol-terminated adhesive that is mixed with a phenol-formaldehyde resole just prior to bonding.

The three methods give different polymer structures, and each has its own advantages and disadvantages depending on the specific application. The PRFs generally have a lengthy assembly time because of the room temperature cure. If the cure were rapid at room temperature, then there would not be enough time to mix the components, spread them on the wood, and press the wood pieces together prior to adhesive curing. The slow cure results in a longer clamping time before the adhesive has sufficient strength to allow handling of the wood pieces. Thus, a room temperature cure is desirable, to avoid heating large laminated pieces, but suffers from the long clamping times.

Another type of PRF is the honeymoon adhesive, developed for finger jointing and laminating; this process circumvents the long clamping times associated with room temperature cures. In this application, the adhesive is placed on one wood surface and the activator or copolymer material is placed on the other, with the mating of these two pieces leading to the faster cures (Pizzi 2003b). One system for fast curing used an amine cure-promoter on one wood piece and a formaldehyde-based adhesive on the other, and showed that this produced rapid curing with good bonds even to green wood (Parker et al. 1997). The use of hydrolyzed soybean flour and a PRF adhesive as the two components has been shown to produce very good finger joints (Kreibich et al. 1998).

9.7.3.3 Urea Formaldehyde and Mixed Urea Formaldehyde Adhesives

Urea-formaldehyde (UF) adhesives have several strong positive aspects: very low cost, non-flammable, very rapid cure rate, and a light color. On the negative side, the bonds are not water-resistant and formaldehyde continues to evolve from the adhesive. UF adhesives are the largest class of amino resins, and are the predominate adhesives for interior grade plywood and particleboard.

The chemistry of the urea-formaldehyde adhesives involves several steps, with the first being the addition of the formaldehyde to the urea under neutral or basic conditions (Pizzi 2003e, Updegaff 1990). Although there are only two nitrogen atoms on which the formaldehyde can add, the literature shows that the N,N,N´-tris(hydroxymethyl)urea, along with the bis- and mono-hydroxymethyl ureas are the primary products. These hydroxymethyl compounds then react under slightly acidic conditions and heat to generate oilgomers, in which the urea molecules are linked by bismethylene ether or methylene bridges see Figure 9.20. After reaching the desired molecular weight for the specific application, the polymerization is slowed by raising the pH and cooling. Usually an additional charge of urea is added to reduce formaldehyde emissions from the resin. The UF resins contain a latent acid catalyst that produces an acid catalyst during the heat cure. Latent catalysts can be salts, such as ammonium sulfate or chloride, that generate ammonia and sulfuric or hydrochloric acid, respectively. These acids and heat cause the UF to cure rapidly, giving the UF adhesive its desirable rapid setting properties. UF resins rapidly develop strength, leading to shorter press times than with other adhesives. The chemistry and formulation are much more complicated than there is space here to describe and understanding the chemistry has led to efficient products that are used commercially (Pizzi 2003e, Updegaff 1990).

Concern about formaldehyde emissions during production and indoor applications has led to lower formaldehyde/urea ratios in current products. However, this has not come about without some sacrifice in ultimate strength and robustness of commercial production, but was needed to meet current environmental standards. The specific UF formulation and bonding conditions are adjusted to meet acceptable formaldehyde emissions for the end product. The classes of products are more rigidly defined in Europe (Dunky 2003) than in the United States. The formaldehyde emissions are high initially, and decrease with time, but do not go to zero even over a long time.

A major drawback of UF adhesives is their poor water resistance; in that they have high bondline failure under accelerated aging tests, restricting them to indoor applications. Another area of concern is the long-term hydrolytic stability of these adhesive polymers, which generally show the least durability of any formaldehyde-copolymer adhesive. UF resins are believed to depolymerize resulting in continuing emission of formaldehyde. The use of some modified ureas can reduce the poor resistance to the mechanical effects of accelerated aging (Ebewele et al. 1993). The poor water-resistance of

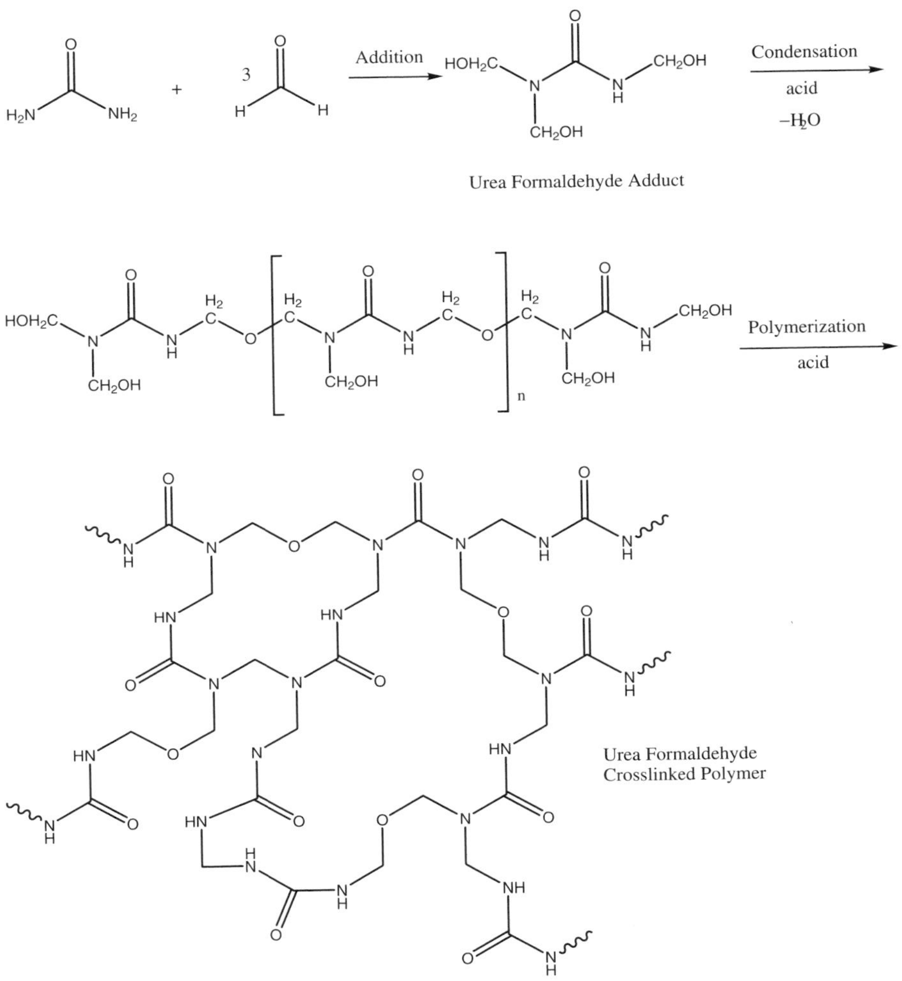

FIGURE 9.20 Urea-formaldehyde polymerization goes through an addition reaction and then condensation to give an oligomer that is applied to the wood. After application, the polymerization is completed to give a crosslinked network.

urea-formaldehyde adhesives has led to the development of melamine-urea-formaldehyde (MUF) adhesives that are covered in the next section.

9.7.3.4 Melamine Formaldehyde Adhesives

Like formaldehyde adhesives made with phenol and resorcinol, melamine-formaldehyde (MF) adhesives have acceptable water resistance, but they are much lighter in color than the others. MF resins are most commonly used for exterior and semi-exterior plywood and particleboard, and for finger joints. Another significant use is for impregnating paper sheets used as the backing in making plastic laminates. The resins for paper impregnation are different in many respects (degree of polymerization, addition of copolymerizing additives, viscosity, etc.), but will not be covered here because they have been discussed in detail elsewhere (Pizzi 2003f). The limitation of the MF adhesives is their high cost due to the cost of the melamine. This has led to the use of melamine-urea-formaldehyde (MUF) resins that have much of the water resistance of MF resins, but at substantially lower cost. The MUF adhesives, depending on the melamine-to-urea ratio, can be considered as a less expensive MF that has lower durability or as a more expensive UF that has better water resistance (Dunky 2003). The MUF adhesives can replace other adhesives that are used for some exterior applications.

Like most formaldehyde curing, the first step in MF curing is the addition of the formaldehyde to the melamine, see Figure 9.21. Because the melamine is a good nucleophile, the addition reaction with the electrophilic formaldehyde occurs under most pH conditions, although the rate is slower at neutral pH. The melamine reacts with up to six formaldehyde groups to form two methylol groups on each exocyclic amine group. The mixture of hydroxymethyl compounds then react by condensation to form the resin. Two types of condensation reactions can occur:

- Bismethylene ether formation by the reaction of two hydroxymethyl groups, $RCH_2OH + R'CH_2OH \Rightarrow RCH_2OCH_2R' + H_2O$
- Methylene bridge formation by reaction of the hydroxymethyl group with an amine group, $RNH_2 + R'CH_2OH \Rightarrow RHNCH_2R' + H_2O$

The chemistry for the addition and condensation reactions is illustrated in Figure 9.21. The addition reaction is reversible, though generally the equilibrium is far to the right side. On the other hand, the condensation reaction to form oligomers and polymers is not very reversible, which is important for the water resistance of the product and makes it different from UF. It is evident from the dimers illustrated that many isomers can be produced. Considering that each melamine has three amine groups, with each amine group having up to two hydroxymethyl groups attached; formation of both methylene and bismethylene ether bridges occur, and formation of dimers, trimers, and higher oligomers take place, the chemistry rapidly becomes very complex (Pizzi 2003f). Sato and Naito have studied the chemistry of some of these reactions (Sato and Naitio 1973). The

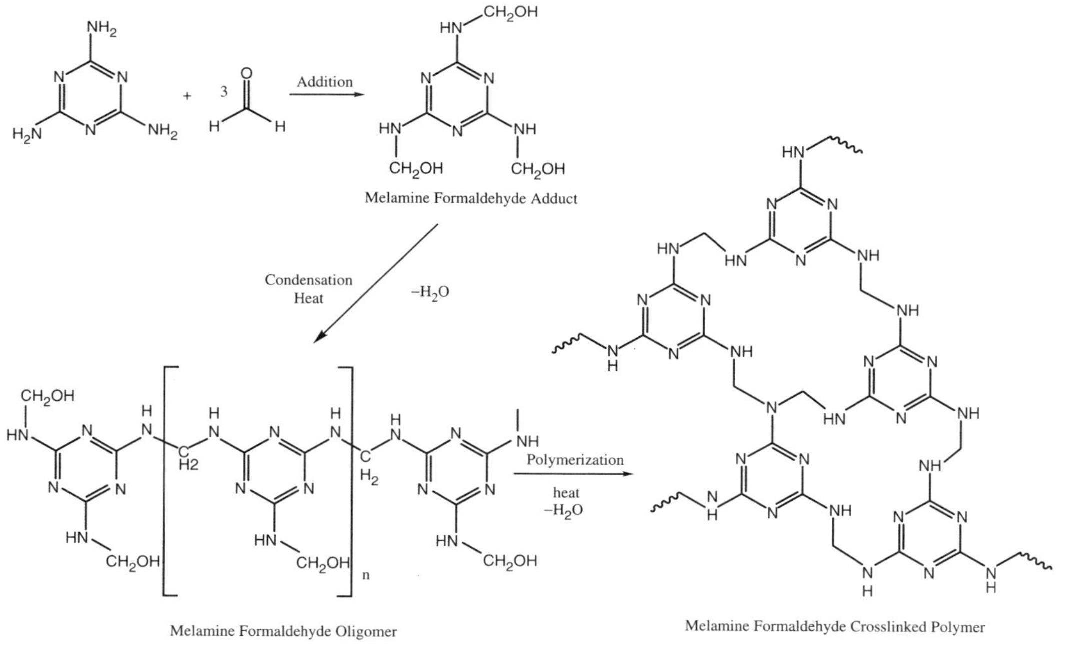

FIGURE 9.21 Melamine-formaldehyde chemistry goes through similar steps as the urea-formaldehyde in Figure 9.20.

reaction conditions of time, temperature, formaldehyde/melamine (F/M) ratio, pH, and catalyst will influence the composition and structure of the resin that makes up the adhesive.

The melamine-formaldehyde adhesive that is sold commercially is a mixture of oligomers made by heat polymerization in standard agitated reactors. In normal applications, the formaldehyde-to-melamine ratio is about 2 to 3. The formulations are altered depending on the specific application. In some cases, the water content of the resin is reduced, and usually some additives (rheology modifiers, fillers, extenders, etc.) are added. A typical wood bonding MF resin is of 53–55% solids with a pH of 9.9 to 10.3 (pH is raised at the end of the reaction to slow down the polymerization for stabilization of the resin after manufacturing).

The MF adhesive needs to be activated to give good polymerization to the final product. Similar to UF, this usually involves lowering the pH and raising the temperature. The catalysts added to the MF resin are either acids or acid precursors that liberate acid upon heating. Often hardener, such as ammonium chloride or sulfate, is added that will generate either hydrogen chloride or hydrogen sulfate, plus ammonia, which migrates from the adhesive. In most applications, the products are heat-cured. Although the bonded products show respectable water resistance, phenol-containing resins are preferred for exterior uses in the United States. Unlike the urea-formaldehyde adhesives, the melamine-formaldehyde resins do not show degradation during water boiling (Pizzi 2003f). They do show some loss of bond strength during accelerated and exterior exposure tests (Selbo 1965). Care often needs to be taken in comparing the performance of different classes of adhesives, for usually only one of many commercial products is tested for the evaluation. In most countries, the adhesive manufacturer has to show data that its product passes the accelerated aging test as required by specific standards.

The chemistry of the MUF adhesives is similar to the MF and UF adhesives, but more variations exist due to the ratio of melamine to urea, the sequence for addition of the components, temperature, pH and time factors. In summary, the MUFs are a good compromise between the good performance of melamine adhesives and low cost of the urea adhesives.

9.7.4 Isocyanates in Wood Adhesives

Several classes of adhesives used in wood bonding involve the use of isocyanates. Isocyanates are widely used because of their reactivity with groups that contain reactive hydrogens, such as amine and alcohol groups at room temperature. This allows great flexibility in the types of products produced because they can self-polymerize or react with many other monomers. Isocyanates are most often used to produce polyurethanes by reacting with liquid diols.

The high reactivity of isocyanates is both an advantage and a disadvantage. The advantage is that polymerization proceeds rapidly and usually to high conversion. One disadvantage is that isocyanates will react rapidly with water that is present in most wood products. This water can reduce the effective molecular weight by altering the stoichiometry and can compete with desired reactions with the wood, such as the hydroxyl groups in the cellulose and hemicellulose fractions as well as the phenols and hydroxyl groups in the lignin domains. Another disadvantage is they can react rapidly with many compounds present in human bodies. These reactions are rapid under physiological conditions and are not readily reversible which means that safety of handling isocyanates is a concern. The concern occurs mainly during the manufacturing stage when low molecular weight and volatile isocyanates are still present; once these react, the resulting ureas and esters are quite safe. An exception is that the heat of combustion causes the formation of free isocyanate groups. Isocyanates used in wood bonding are not as hazardous as some other isocyanates in that they are generally higher molecular weight so their volatility and the number of free isocyanate groups is diminished.

The most common wood adhesive is a self-curing isocyanate, which reacts with water in the wood to initiate the curing process. These are used in both the production of composites and in the gluing of laminated wood products. The polymeric diphenylmethane diisocyanate (pMDI) is used mainly for production of the core of oriented strandboard and is the only known adhesive that works well for the bonding of strawboard. pMDI can be used in the face of OSB, but mold release is needed to reduce sticking to the platens. Although the costs are higher the pMDI is taking market share away from the phenol-formaldehyde adhesives due to its rapid cure and ability to wok at lower application rates.

Another common wood adhesive is the emulsion-polymer isocyanate, which is a two-component adhesive. These adhesives are used in bonding of oriented strandboard in engineered wood products. Non-waterborne two-component isocyanates are very common in coating applications, but are not used much as wood adhesives due to their rapid reaction rate. Mixing of the diisocyanate with a diol starts the curing process to form primarily a linear polymer with, usually, a moderate degree of crosslinking to provide more flexible products.

The most common reactive hot melt adhesives contain isocyanate groups, attached to the polymer backbone. Hot-melt adhesives are very desirable in product assembly because they develop their bond strength as the molten polymer cools and transforms from the melt to a solid. Unfortunately, the bond generally has poor resistance to heat, long-term stresses, and in some cases, moisture. Moisture-cured hot-melt isocyanates behave like typical hot-melts with good initial strength, but also crosslink to yield a thermoset that can resist the effect of heat, long-term loads, and moisture. These are being widely used in product assembly areas, many of which involve wood.

The other class of isocyanate adhesives is polyurethanes, which are being used in more specialty wood-bonding applications. Their advantage is wide formulation ability, given the great variety of raw materials that can be used. Polyurethanes have shown good potential for bonding green wood (Lange et al. 2001).

9.7.4.1 Polymeric Diphenylmethane Diisocyanate

Isocyanate adhesives have shown increasing use at the expense of other adhesives due to their high reactivity and efficiency in bonding. Polymeric diphenylmethane diisocyanates (pMDI) are commonly used in wood bonding and are a mixture of the monomeric diphenylmethane diisocyanate and methylene-bridged polyaromatic polyisocyanates, illustrated in Figure 9.22 (Frazier 2003). The higher cost of the adhesive is offset by its fast reaction rate, its efficiency of use, and its ability to adhere to difficult-to-bond surfaces.

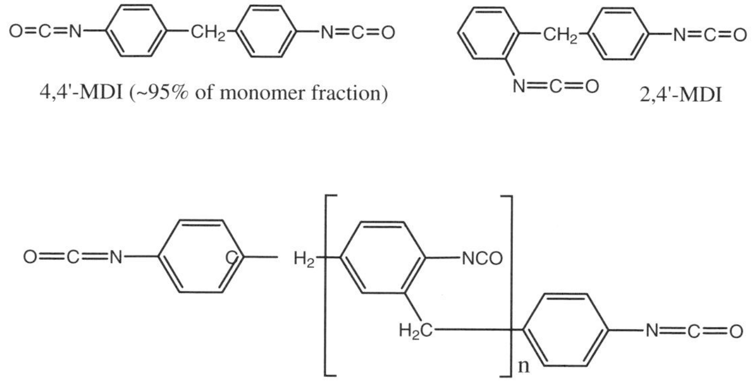

Polymeric MDI is an isomeric mixture of methylene-bridged polyisocyanates

FIGURE 9.22 The polymeric diphenylmethane diisocyanates is a mixture of the monomeric and polyfunctional isocyanates.

Wood Adhesion and Adhesives

The pMDI forms a homoploymer, but needs water for the activation, which is not a problem with wood, but may be for bonding to other substrates. The chemistry involves several steps:

- The isocyanate first reacts with water to form a carbamic acid:
  R-NCO + H_2O => R-NHCOOH.
- The unstable carbamic acid gives off carbon dioxide to form an amine:
  R-NHCOOH => R-NH_2 + CO_2.
- The amine then reacts with another isocyanate group to form a urea:
  R-NH_2 + OCN-R => R-NHCONH-R.
- Some of the urea molecules react further with isocyanate to form a biuret:
  R-NHCONH-R + R-NHCON(CON –R)-R.

As shown by these mechanisms, once the isocyanate reacts with the water the rest of the process proceeds rapidly as long as there is enough isocyanate for reaction in comparison to groups with other reactive hydrogens, see Figure 9.23. Sufficient water is not a problem with wood given its high water content, but some other substrates need to be wetted for proper bonding. However, high water levels could inhibit polymer formation by producing too many amine groups, but this has not been found to be the case in wood bonding. The carbon dioxide off gas can be a problem since it creates voids in the adhesive that can reduce the strength. Generally, these reactions are not reversible under normal conditions, leading to good bond integrity of the isocyanate-bonded wood.

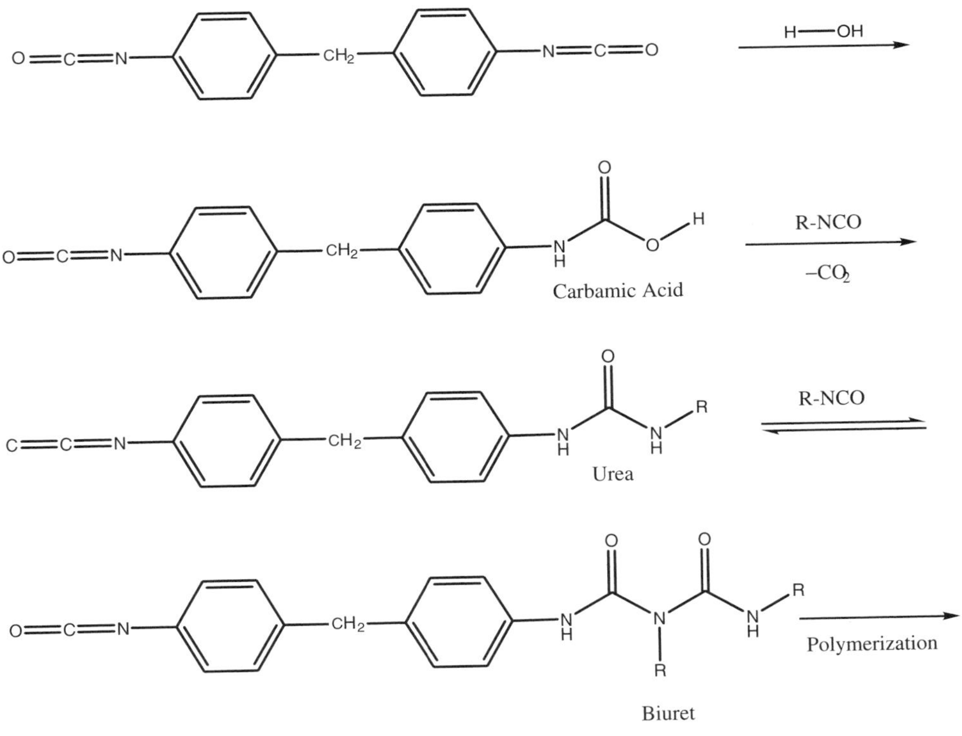

FIGURE 9.23 The isocyanate needs water to start the polymerization process. This reaction ends up forming carbon dioxide that can cause bubbles in the adhesive, but once the amine forms, self-polymerization takes place rapidly.

The interaction of pMDI with the composite surface is quite different from other wood adhesives, such as phenol-formaldehyde. pMDI's low polarity and low viscosity compared to other wood adhesives leads to very rapid penetration into the wood (Frazier 2003). Normally this can lead to a starved bondline and poor strength, but this does not happen with the pMDI-bonded wood. It may be that the strength derives from the strong bridge created at the point where the wood is brought into close contact with the isocyanate adhesive. Some consider the isocyanate to be the most likely of all wood adhesives to form covalent bonds to wood due to the ease of isocyanate reacting with the hydroxyl groups of the wood to form urethane bonds (Frazier 2003). On the other hand, others believe that the fast reaction of the isocyanate with water and the large number of water groups present, especially on the wood surface makes the urethane formation unlikely (Pizzi 1994a). In any case, the ureas that form should be strong hydrogen bonders with the polar wood surface.

The unique properties of pMDI adhesives give it advantages in several markets. Its rapid polymerization and ability to form bonds in the presence of high water levels has led to its use as a core resin for OSB. The higher water content and lower temperatures of the OSB core section can make sufficient cure of a phenol-formaldehyde resin in the core more difficult, leading to increased use of pMDI. This ability to form bonds with high-moisture-content wood has also led to pMDI's use in bonding green or wet lumber. Low polarity allows pMDI to find cracks in the waxy coating of straw, leading to its use in strawboard, for which PF resins are unsatisfactory.

There are several disadvantages to pMDI besides cost. Unlike many wood adhesives that are poor bonders to substrates other than wood, pMDI bonds very well to other materials, including metal caul plates or press platens. Thus, this adhesive is less likely to readily displace PF from the face layer of OSB. Isocyanate's hazards have limited the use of pMDI due to the extra cost of maintaining safe operations in the plants. However, the safety issues can be addressed and no hazard exists in the bonded product due to the reaction of the isocyanate groups. pMDI is promoted as a formaldehyde-free adhesive.

9.7.4.2 Emulsion Polymer Isocyanates

Emulsion polymer isocyanates are generally two-part adhesives that are mixed prior to use and have been used for panel bonding, bonding of plastics to wood surfaces, and for bonding OSB web into the flange to make I-joists. The components are a water-emulsifiable isocyanate and a hydroxy-functionalized emulsion latex. The emulsion allows higher molecular weight polymers to be used while keeping a low solution viscosity for ease of application. Because the isocyanate is emulsifiable it readily disperses when mixed with the latex, coming into contact with the hydroxy groups as the water separates into the wood. Like all two-component systems, adequate mixing is important. As the adhesive cures, polyurethane groups are formed between the two components. The hydroxy-functionalized pre-polymer can be varied in both its backbone structure and the number of hydroxy groups to control the crosslinking. The variations in the latex portion allow products to be made with a wide range of stress-strain behaviors for different applications.

These adhesives can form fairly durable bonds depending on the formulation; some are known to give good water-resistance. The ability to bond plastics and other non-wood substrates is an advantage of these resins over many other wood adhesives. The higher cost and the need to mix the two components prior to use are disadvantages.

9.7.4.3 Polyurethane Adhesives

Polyurethanes are widely used in coatings and adhesives, but less common in wood bonding. Polyurethanes can be either one- or two-component systems, with the selection depending on the specific application. To obtain good wetting, the components need to be low molecular weight or a solvent needs to be added to reduce the viscosity for good wetting. Low molecular weight of the isocyanate components are not desirable because it leads to excessive volatility and health problems. The one-component system is an isocyanate-functionalized polymer that has remaining isocyanate groups. These groups will react with moisture causing the generation of amines that react with other isocyanate groups to form the backbone and crosslinking connections. The two-component adhesive has an isocyanate portion and an isocyanate-reactive portion. Good mixing of these two components just prior to bonding is critical.

The market for these products has been somewhat limited because of their marginal levels of wood failure in some applications. They are widely used in many other bonding markets due to their good strength, flexibility, impact resistance, and ability to bond many substrates.

9.7.5 Epoxy Adhesives

Epoxy adhesives and coatings are widely used because of their good environmental resistance and the ability to bond to a wide variety of surfaces, including wood, metals, plastics, ceramics, and concrete. They are less commonly used in wood bonding because they cost more than most wood adhesives, and in some cases, their durability is limited. On the other hand, they are structural adhesives that cure at ambient temperatures, have good gap filling ability, and do bond to many other surfaces, while most wood adhesives require heat cure, are not gap filling and do not bond well to other substrates. Thus, epoxies continue to be examined for their use in bonding wood to other materials and for in-place repair of damaged wood structural members. Besides cost, a main limitation of epoxies is their lack of acceptance for applications that require durable bonds (American Institute of Timber Construction 1990).

Although there are some cases of self-polymerization under the influence of acid or tertiary amine catalysts, most epoxies have an alternating ABABAB backbone that is highly crosslinked, usually using the multi-functionality of the hardener. The standard terminology is for the epoxy to be called the resin and the other component that crosslinks the epoxy to be called the hardener. The formulation is expressed as parts per hundred resin (phr) with the weight of the epoxy as 100 and the rest of the components given relative to the epoxy weight. The hardener is anything that will react with the epoxy groups, including amines, thiols, hydroxides, and acid groups, but amines are the most common hardeners.

The most common epoxy resin is the diglycidly ether of a bisphenol A (DGEBA), although other multi-functional epoxies can be used. The starting material is made by the condensation of phenol with acetone to give the bisphenol A (bisA) as illustrated in Figure 9.24. This is then reacted with epichlorohydrin under basic conditions to yield the DGEBA molecule and sodium chloride. The removal of the salt is especially important in electronic applications to minimize metal corrosion by the chloride. The DGEBA epoxies vary in molecular weight due to polymerization through the epoxy groups. Another important class of the epoxies is made from the novalak resins via the condensation of phenol with formaldehyde, as discussed in section 9.7.3.1. Bis-F resins have also been used; these are similar to the bis-A resins except that formaldehyde is used in place of acetone for the condensation. More flexible resins can be made using epoxidized oils and other non-aromatic epoxides. Brominated epoxies are often used for fire resistance.

In contrast to the limited types of epoxies, the hardeners or curatives have a wide variety of chemical structures. The hardeners have an active hydrogen attached to a nucleophile, which essentially adds across the epoxy group. The process involves the nucleophile attacking the terminal carbon of the epoxy as illustrated in Figure 9.25, with the hydrogen then migrating to the hydroxy anion. For less nucleophlic groups, addition of a tertiary amine that interacts with the oxygen atom in the epoxy makes the epoxy ring easier to open. This continues until all the active hydrogens are reacted with epoxide or the epoxide is used up. Thus, for amine with two reactive hydrogens on a nitrogen, two epoxy groups can react, but the second addition is much slower. For formulating the ratio of hardener, the equivalent weight of the hardener is calculated by dividing its molecular weight by the number of active hydrogens.

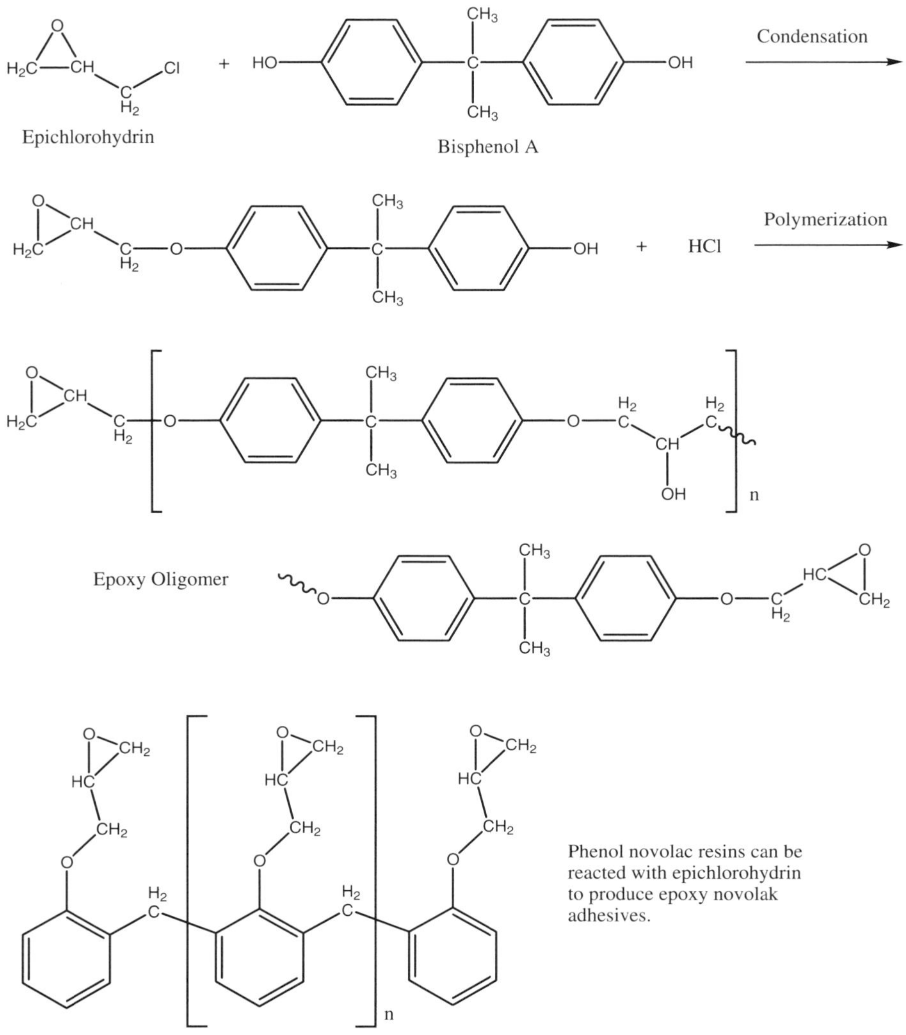

FIGURE 9.24 The epoxies are made by reaction of epi-chlorhydrin with phenols to produce glycidyl ethers. The bis-A epoxy is much more common than the novolak epoxies, but the latter can provide better heat resistance.

The amine hardeners are the most common room-temperature curable epoxies; they can be divided into three different classes. The main class comprises polyamines, such as those made from the reaction of ethylene oxide, to give products such as diethylene triamine, triethylene tetraamine, and tetraethylene pentaamine. These low molecular amines are hazardous due to their corrosivity and their ability to chelate metals. Their reasonable cost and high reactivity make them the most common curatives. Other amines used to make products more flexible are the amine-capped polypropylene oxide polymers and branched six carbon diamines. The other amine-containing curatives contain fatty acids that make the epoxies somewhat more flexible and hydrophobic. Polyethylene polyamines can be reacted with fatty acids to make amidoamines. They can also be reacted with dimer acid, made via the dimerization of unsaturated fatty acids, reacted with the polyethylene polyamines, resulting in polyamide curatives.

The other curing agents are not used much in wood bonding. While the standard epoxies in home stores involve amine hardeners, those that exhibit a five-minute cure time use mercaptans

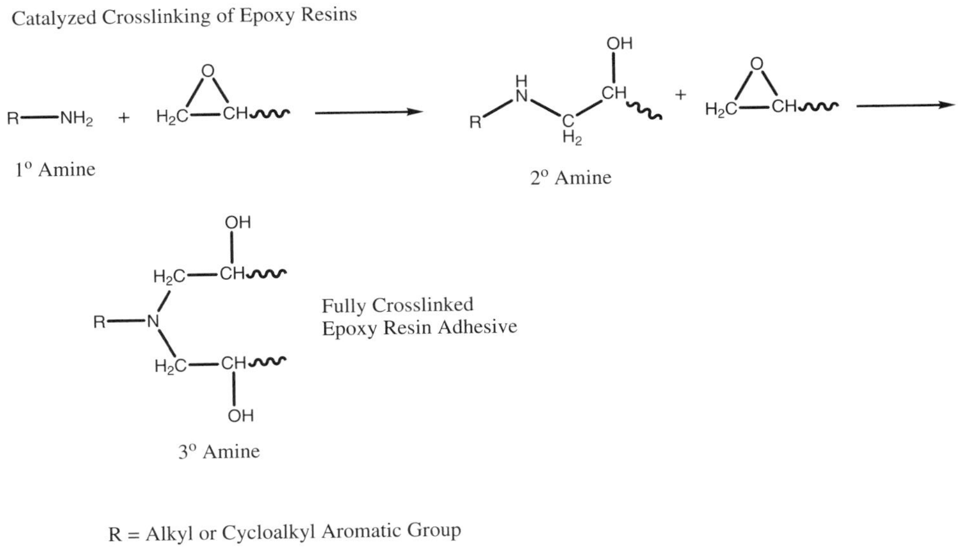

Wood Adhesion and Adhesives

Catalyzed Crosslinking of Epoxy Resins

FIGURE 9.25 Epoxies react readily with nucleophiles, such as amines. Most amines are primary allowing multiple additions to provide a crosslinked network.

hardeners. Epoxies can also be cured using anhydride hardeners or tertiary amine catalysts, but the high cure temperatures limit their use in wood products.

Although epoxies give strong and durable bonds to many substrates, they do not result in durable bonds to wood under all conditions. There is some disagreement on the durability of epoxy bonds under wet conditions, but most standards limit epoxies for load bearing applications (American Institute of Timber Construction 1990). Considerable work has been done on supporting the use of epoxies for restoration work, but examination under rigorous testing shows that commercial epoxies do not pass the test requirements. Examination of the failure indicates that much of the failure is in the epoxy interphase region (Frihart 2003a). Higher wood failure under wet conditions has been found using acetylated wood (Frihart et al. 2004), while using a hydroxymethylated resorcinol primer allowed the bonded assemblies to pass the durability tests (Vick 1997).

9.7.6 POLYVINYL AND ETHYLENE-VINYL ACETATE DISPERSION ADHESIVES

These water-borne adhesives, poly(vinyl acetate), PVA, and poly(ethylene-vinyl acetate), EVA, find wide utility in the assembly of wood and paper products into finished goods. Common white glue (PVA) and EVA dispersions are mainly used for wood bonding, such as furniture construction. The pre-formed polymers do not require any heat to cure, are cost effective and are easy to use. Water-borne adhesives are set by the water being absorbed into the wood or paper product (and eventually released into the atmosphere), leading to wide use in manufacturing and construction operations involving wood. Because these products are not cured, they will lose much of their strength at high moisture levels.

The processes for making PVA and EVA dispersions are similar in many respects (Geddes 2003, Jaffe et al. 1990). The monomers are dissolved in water containing poly(vinyl alcohol), making an emulsion; the monomers in the droplets of the emulsion are polymerized to form a dispersion of an organic polymer in water. Making a stable product requires having an emulsion with small droplet sizes. Addition of the monomers is controlled to prevent overheating caused by the exothermic polymerization. If there is not enough solvency, the phases will separate without

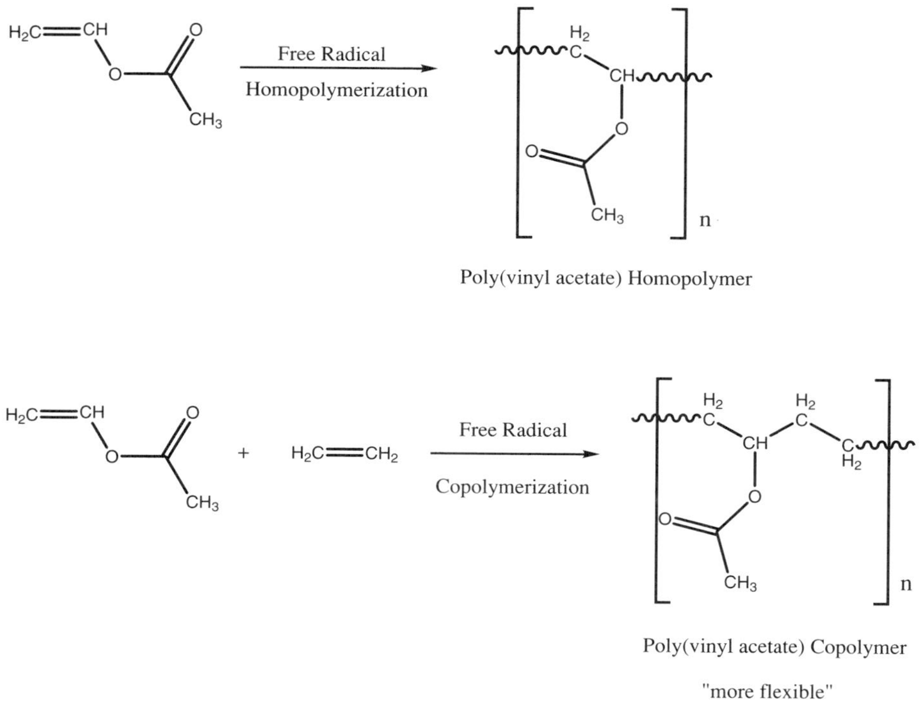

FIGURE 9.26 Polyvinyl acetate is made by the self-polymerization of vinyl acetate usually under free radical conditions. The chains can be altered by adding ethylene to form a copolymer.

forming the necessary fine dispersion. If there is too much surfactant, the product can have poor adhesion due to a weak boundary layer. After application, the water evaporates and the beads of adhesive coalesce to form a film, but the coalescence needs to take place on the wood surface. The polarity of the adhesive can be reduced by incorporation of ethylene in the polymerization to produce ethylenevinyl acetate polymers for bonding to less polar surfaces.

PVAs are linear polymers with an aliphatic backbone; thus, they are very flexible adhesives as opposed to the rigid nature of formaldehyde copolymers normally used as wood adhesives. These PVA adhesives being water-borne generally exhibit good flow into the exposed cell lumens, but, given their high molecular weight, they most likely do not penetrate cell walls. PVAs, with their high content of acetate groups and a flexible backbone, can form many hydrogen bonds with the various fractions of the wood for good interfacial adhesion. They maintain much of their bond strength as the wood expands and contracts due to dissipation of the energy into flexing of the polymer backbone. With this dissipation of the stress into the polymer chains, there is limited stress concentration at the interface. However, PVAs do not work well at high moisture levels due to loss of strength or high constant stress levels because of a lack of creep resistance. A solution to these problems is to convert the thermoplastic PVA into a thermoset. This is accomplished by crosslinking the linear thermoplastic, using covalent bond formation, such as reaction with glyoxal, formaldehyde resins, or isocyanates, or using ionic bond formation, such as reaction with organic titanates, chromium nitrates, aluminum chloride, or aluminum nitrates. Crosslinked PVAs have improved resistance at high moisture levels, higher temperatures, and under load, but they are not as convenient because the crosslinker needs to be added just prior to application. However, they can be used in other applications, such as windows and door construction, for which regular PVA cannot be used.

Poly(vinyl acetate) is converted to poly(vinyl alcohol) (PVOH) by hydrolysis (Jaffe and Rosenblum 1990). PVOH products vary in degree of hydrolysis and molecular weight depending on the specific application, the main uses being textile and paper sizing, adhesives, and emulsion polymerization of vinyl acetate. Given the water solubility, these adhesives need to be either crosslinked or gelled to give some permanence under moist conditions. This conversion is done by crosslinking using a covalent bond formation, such as reactions with glyoxal, urea-formaldehyde, and melamine-formaldehyde or using ionic bonds formations with metal salts, such as cupric ammonium complexes, organic titanates, and dichromates. The gelation is usually accomplished using boric acid or borax.

Poly(vinyl alcohol) is converted to poly(vinyl butyral) that is used as the adhesive interlayer in laminated safety glass. This adhesive needs to bond to glass tightly so that when the glass fractures it does not fly away causing injury; moreover, it must have very good resistance to light degradation.

9.7.7 Biobased Adhesives

The most common bio-based adhesives are protein-based. In contrast to the other wood adhesives, protein glues have been used for thousands of years. Many of the early civilizations learned how to make adhesives from plants and animals (River et al. 1991, Keimel 2003). Although the original bonded wood products were made using natural protein adhesives, these bonds were durable only at low moisture levels and generally softened at high moisture levels. This leads to delamination, in addition to biological degradation when wet. Many sources have been used for protein-based adhesives, including animal bones and hides, milk (casein), blood, fish skins, and soybeans. The bonded wood industry expanded greatly due to the use of these adhesives in the early 1900s, as processes were developed to make more effective adhesives. The biggest advancement was the development of soybean flour adhesives that allowed interior plywood to become a cost effective replacement of solid wood. Today, most of these bio-based adhesives have been replaced by synthetic adhesives due to cost, durability, and availability factors. Research has been done on incorporating soybean flour or protein into phenol-formaldehyde resins, but more success has been obtained by using the phenol and formaldehyde to crosslink the denatured soy flour protein (Wescott and Frihart 2004).

Trees and bushes, themselves, provide many adhesive materials, some of which have been used in wood bonding. Pitch from trees was one of the earliest adhesives because of its availability, usefulness without processing and ability to bond many materials (Regert 2004). From this grew the naval stores industry, with the name indicating its importance to ship construction for sealing and bonding applications. Naval stores research led to the development of rosin resins that are important additives in both ethylenevinyl acetate hot-melt and pressure sensitive adhesives, and to the development of fatty acid derivatives that are converted to epoxy hardeners and polyamide hot-melt adhesives. Tannins have been used as phenol replacement in some formaldehyde copolymers. Due to their phenolic nature, lignins have been examined extensively as phenol replacement in PF resins. Both tannins and lignin adhesives tend to have good moisture resistance and are not readily attacked by microorganisms.

Carbohydrates have been used as adhesives, but have not found much utility in wood bonding (Conner and Baumann 2003). Starch is widely used in paper bonding, especially in the construction of corrugated board used in many packaging applications, but generally lacks the strength and water resistance needed for use in wood bonding. The cellulosic adhesives not only suffer from loss of strength under wet conditions, but also support the growth of microorganisms.

9.7.7.1 Protein Glues

Protein glues, once the dominant wood bonding adhesive, are now used in specialty applications where they provide some distinctive properties, and continue to be examined for use in wood bonding where limited durability is needed. Like most biomass materials, proteins are not uniform in composition as the source varies; thus, the processes for using these proteins and the properties of the adhesives vary as the protein source changes. To make a useful adhesive, the native protein structure has to be denatured to expose the polar groups for solubilization and bonding. The primary structure involves a polyamide backbone made from the condensation of amino acids, while the secondary and tertiary structures are based on intrachain and interchain interactions, respectively, which are hydrogen bonds, disulfide linkages, or coordination around metallic sites. The main denaturization involves breaking the hydrogen bonds, while breaking other secondary and tertiary bonds depends on the denaturization conditions. Once the protein has been denatured, then it has the ability to flow onto and into the wood, and to form hydrogen bonds with the wood structure. The setting step involves reformation of the hydrogen bonds between the protein chains to establish bond strength. The main method of denaturization for adhesive applications is hot aqueous conditions (Lambuth 2003), but some other processes are also used. The aqueous process is often done under caustic conditions and may also involve adding other chemicals to either stabilize the denatured glue or add strength to the final bonds.

Of the protein-based adhesives, soybean flour was used in the largest volume; the flour is ground soybean meal, the residue after the soybeans have had the traditionally more valuable oil removed. The flour is finely ground and is processed through a number of steps to disperse the meal and denature the protein (Lambuth 2003). In many cases, the denatured protein has to be used within eight hours, before the adhesive starts to degrade. Soybean protein adhesives allowed the development of the interior plywood industry in the early 1900s. The adhesives were improved to give better water resistance, but never achieved sufficient moisture resistance to make exterior grade plywood. Phenol-formaldehyde (PF) resins were slow to displace soybean adhesives due to cost and marginal performance. The need for more durable plywood adhesives during World War II led to improved and lower cost PF resins and the ultimate demise of the soybean adhesives. Today, soybean resins are used in some isolated cases, but are more often used in small amounts as an additive to synthetic resins. The upsurge in soybean use during the 1950s shows the potential for soybean adhesives on a cost basis if the water resistance, short storage stability, and inconsistency of properties can be overcome. With the rising cost of petroleum based adhesives, soy flour based adhesives are again being studied (Wescott and Frihart 2004).

None of the other protein sources are available with sufficiently low cost, large supply, and consistent composition as soybean flour, but they still have advantages because of their special properties. Blood protein from beef and hogs has the best water resistance of any of the commercial protein adhesives but has great inconsistency (Lambuth 2003). To retard spoilage, the blood is spray dried. It is mixed with phenol-formaldehyde adhesives for plywood bonding. Animal bone and hide glues are used in fine furniture manufacturing because they provide flexible bonds for good durability with indoor humidity changes (Pearson 2003). They have many other uses but are being replaced by synthetics, such as ethylene vinyl acetate polymers, due to cost and the synthetics' greater ability to be formulated for specific applications. Casein, like many of the protein adhesives, provides good fire resistance and is therefore used in fire doors. Each of these adhesives has its own process for denaturization and use (Lambuth 2003).

9.7.7.2 Tannin Adhesives

Tannins are polyhydroxypolyphenolics that occur in many plant species, but have a high enough concentration in only a few species to make it worthwhile to isolate them. The commercial supplies of tannins are those isolated mainly from plants in a few countries. Tannins are used because they are more reactive than phenol, but they are also more expensive than phenol. Extraction of the plant material and subsequent purification of the isolates, followed by spray drying, yield powdered tannins (Pizzi 2003c). The purified isolates behave in many ways like a natural form of resorcinol, with their high reactivity and the resulting water-resistant bonds when copolymerized with formaldehyde. Although tannin's reaction rate with formaldehyde is quite similar to that of resorcinol, the polymer structure is quite different. Instead of multiple additions of formaldehyde to a single aromatic ring, formaldehyde adds mainly as single additions and some double additions to the connected rings of the resorcinol, pyrogallol, phloroglucinol and catacheol structures in tannins. Thus, the final polymer structure is very different and will have different properties than the resorcinol product due to its lower crosslink density, despite the similarity in the chemical reactions.

Three limitations of tannins compared to synthetic adhesives are their high viscosity, limited availability, and inconsistent source and therefore reactivity. Their polycyclic structure that leads to fast cure speed also makes solutions of tannins high in viscosity; using more dilute solutions to reduce viscosity leads to additional steam in the hot pressing of the composite. Tannins exist in high enough concentrations to be commercially viable in a few species, but are not available in the large quantities to compete with synthetic adhesives. Like many natural products, the composition of tannins varies depending on growing conditions; thus, making consistently performing adhesives difficult.

Tannins have been used as adhesives in South Africa, Australia, Zimbabwe, Chile, Argentina, Brazil, and New Zealand (Dunky and Pizzi 2002). The use is in composite (particleboard and medium density fiberboard) production, laminate and finger joint bonding, and for damp-ply-resistant corrugated cardboard (Pizzi 2003c). The tannin market will always be of limited volume due to supply limitations.

9.7.7.3 Lignin Adhesives

Although lignins are phenolic derivatives, they are very different from tannins – they are available in large quantities at low cost, but they are much slower in their reaction with formaldehyde. The supply of lignin is large, being the by-product of pulping processes for papermaking; they constitute 24–33% of the woody substance in softwoods and 16–24% in hardwoods. Native lignin is a crosslinked polymer, but the polymers have to be partially degraded to allow them to be separated from the cellulosics. For adhesive purposes, these degraded lignins need to be further polymerized to obtain useful adhesive properties. Despite being almost completely aromatic, lignins have only a few phenolic rings and no polyhydroxy phenyl rings, leading to low reactivity with formaldehyde.

The low value of lignins has led to much research in finding ways to convert the lignin into useful thermoset adhesives. Lignin from the predominant Kraft pulping process does not lead to a useful product because of the cost of separating the lignin from the pulping chemicals and the inconsistency of the lignin product. However, lignosulfonates contained in the spent sulfate liquor (SSL) from sulfite pulping of wood have been found to be a more useful feed for the production of reactive lignins (Pizzi 2003d). Because of lignin's slow reactivity with formaldehyde, other curing mechanisms have been investigated, including thermal cure with acids and oxidative coupling using hydrogen peroxide and catalysts. Three methods of using SSL as the main adhesive with particleboard are to use long press times with a postheating step, to heat with sulfuric acid during bonding, and to heat with hydrogen peroxide (Pizzi 2003d). SSL has also been used as phenol-formaldehyde and urea-formaldehyde extenders. The poor reactivity of lignin can be altered by pre-methyolation with formaldehyde, and this pre-methyolated lignin has been used with PF resins in plywood bonding.

9.7.8 MISCELLANEOUS COMPOSITE ADHESION

Understanding adhesion to wood is as important for composites where wood is the minor component as it is for composites where wood is the main component. Three product areas for wood as a minor component are wood-fiber cement board, wood plastics, and wood filler for plastics. In all three cases, the non-wood component is the main phase holding the material together, but the better the adhesion of the main phase to the wood fiber, the stronger and more durable the product.

Wood-fiber reinforced cement board competes with traditional cement board that uses other reinforcing materials such as fiberglass cloths. The reinforcement serves to reduce the fracture of these preformed panels. Making improved products involves knowing the interaction of wood with the inorganic cement. Plant fiber reinforcement is still being studied, but the market is dominated by fiberglass reinforcement.

Developing good interaction of wood with low polarity plastics is of growing importance. Polyethylene (PE) and polypropylene (PP) being the most significant for wood plastic composites. Both are involved in wood-plastic composites that are used as wood replacements for exterior decking. The small amount of interaction between wood and PE or PP is not surprising, given the large difference in polarity and the difficulty in obtaining good molecular contact between a solid and high molecular weight polymer. The most common method of addressing the polarity difference and the rheological issue is to use medium molecular weight, maleic anhydride-modified polyethylene or polypropylene that can serve as a bridging species. These bridging compounds have the polar maleic anhydride that can either react with hydroxyl groups to form esters or react with water to form organic acid groups that will form polar bonds with the hydroxyl groups in wood; thus, making the plastic more compatible with the wood. Better interaction between the hydrocarbon polymer network and the wood fiber will lead to stronger and more durable products. These products are mainly aimed at replacement of wood in decking (Morton et al. 2003), while retaining the appearance of wood.

Other areas for understanding wood-plastic interactions involve plastics filled with wood for applications such as automotive (Suddell and Evans 2003). These products are made to look like normal plastics, but wood filler is used as a partial replacement of the inorganic filler to reduce the weight of the product. The main polymer network is selected from a wide variety of polymers and more of the main fibers are agricultural (non-wood) than is the case in the previous paragraph. However, the fiber-polymer interactions are still very important, and worthy of further investigation. Poor interaction between the fiber and polymer network can cause early failure due to stress concentration. Although the plastic slows the migration of water to the fiber, under wet conditions, the fiber will eventually become saturated with water and begin to swell, putting additional stress on the interface.

9.7.9 CONSTRUCTION ADHESIVES

Construction adhesives are a large segment of the adhesive market, with their use for attachment of floor and wall coverings, and in assembly of buildings (Miller 1990). Most building construction still uses nails or screws for attachment of wood pieces to each other. However, the use of an adhesive can give extra rigidity to the structure if the panel products are also bonded to the frame. Because the nail or screw holds the wood together, the adhesive does not need to set rapidly. Construction adhesives are normally made to be flexible to provide lateral "give" as the various house components expand and contract with changes in moisture and temperature (Blomquist and Vick 1977). A typical application uses an adhesive that is non-curing, high in molecular weight and with a small amount of solvent to provide some flow. The adhesive is applied at room temperature from a gun to one surface as a bead. Then, the nailing or screwing provides the force necessary for transfer, spreading, and penetration of the adhesive to both surfaces. Because the surfaces are not uniformly brought into close contact, the adhesive has to have gap-filling capabilities. Most standard wood adhesives are not able to be gap-filling due to void formation as the water escapes during the setting process (River et al. 1991). However, Vick made a gap-filling phenol-resorcinol resin (Vick 1973), but it would not have the flexibility needed for a construction adhesive.

Construction adhesives are usually elastomers, which provide the deformability needed for short-range movement to prevent fracture of the bondline as the wood expands and contracts.

However, the adhesive is high in molecular weight to prevent long-range movement that would lead to separation of the bondline. These adhesives provide good strength for many years, but it is unlikely that many will last more than the lifetime of the building because most elastomers will react with oxygen and ozone, leading to embrittlement and fracture over such a long time.

9.7.10 Hot Melts

Hot-melt adhesives are used mainly in specialty applications. The main applications in wood bonding are related to furniture and cabinetry assembly, although they have also been used in window construction and edge banding of decorative laminates due to their ability to form bonds quickly. Rapid bond formation is valuable for manufacturing operations because minimal clamping time is needed for assembly. Hot melts generally set by cooling that turns the molten polymer into a solid, although some hot melts can acquire additional strength by crosslinking. Hot-melt adhesives are fully formed polymers that are molten for application, but they have such high viscosities that their ability to wet wood surfaces is limited. Upon cooling they recover their strength as the polymer melt solidifies.

The hot-melt version of ethylene vinyl acetate (EVA) copolymer sets by cooling to room temperature and is used more with paper products than wood products, although there is some use in wood furniture assembly. These EVAs are made by gas phase polymerization to yield non-solvated polymers that have a range of properties. Variation in the ethylene to vinyl acetate ratio and molecular weight of the final polymer creates the various properties needed for individual applications. These products are formulated with tackifiers and waxes (Eastman and Fullhart 1990). Although EVAs are relatively inexpensive, they often have problems with creep as the temperature increases because they contain large amounts of lower molecular weight compounds (tackifiers and waxes) in the adhesive formulations and there is limited attraction between the polymer chains.

Polyamide hot-melt adhesives are more widely used in wood bonding than EVAs because of their stronger interactions between chains, leading to better creep resistance. These polyamides are made by the reaction of various diamines with "dimer acid," a diacid that is made from the coupling of unsaturated fatty acids at their olefinic sites (Rossitto 1990). These polyamides offer good creep and heat resistance for a thermoplastic polymer due to the strong hydrogen bonds between the chains. These interchain hydrogen bonds resist flow until enough heat is applied to break these bonds, rapidly turning the solid into a fluid. After application to the substrate, cooling then converts the melt into a strong solid, with good adhesive strength. These properties have made the "dimer acid" polyamides useful for edge banding of laminates, cabinet construction, and window assembly. The higher cost of these adhesives limits their use to high-value products that need more durable bonds.

The moisture-cured isocyanates that were discussed in section 9.7.4 and polyesters are other hot melts that are also used in wood products. The polyesters are made by reacting aromatic diacids with aliphatic diols, where the aromatic rings provide rigidity to the polymers (Rossitto 1977).

9.7.11 Pressure Sensitive Adhesives

Pressure sensitive adhesives (PSAs) have been a high growth area not only for tapes and labels, but also for application of decorative laminates. PSAs are different from other adhesives in that there is no setting step in their end use. PSAs readily deform to match the topography of the surface to which they are being bonded. Because PSAs are high molecular weight polymers, and in some cases crosslinked polymers, they have limited ability to flow, though their low modulus allows enough deformation to wet the surface. Although these adhesives may not have high interfacial adhesion, most of the applied force is not concentrated at the interface, because the force is mainly expended in deformation of the elastomeric adhesive (Rohn 1999). Because rheological properties are time and temperature dependent, the development of PSAs has been strongly dependent upon dynamic mechanical analysis (DMA) measurements. DMA provides useful information about a formulation's effect on the glass transition temperature and modulus (inverse of compliance) with the small dimensional changes that occur during bonding (Satas 1999a). However, the debonding process occurs over large dimensional changes and is more dependent upon the stress-strain properties of the formulation. PSAs offer a wide range of properties from easily removed tape or Post-It™ notes to high peel and shear strength tapes, by alterations of both the bonding ability and the energy dissipation ability in debonding.

Given that bonding involves deformation of the adhesive to conform to the substrate surface, PSAs give satisfactory bonds to most surfaces because almost all surfaces are rough on the nanometer scale. Elastomeric polymers provide the strength to the PSA, but the formulations usually contain low molecular weight materials that are used to tackify and plasticize the polymer. Many types of homopolymers and copolymers (random and blocked) are used in PSAs; Satas' book is an excellent reference source for PSAs (Satas 1999b).

Pressure sensitive adhesives are often used for bonding plastics (usually having information or decorations printed on them) to wood products for informational or decorative purposes. Applications for this technology range from indoor office furniture to outdoor signs.

9.7.12 OTHER ADHESIVES

Contact adhesives have been used for bonding of plastic laminates to wood. A contact adhesive is a preformed polymer dissolved in a solvent that is applied to both surfaces, that are brought into contact after the solvent has evaporated. Thus, a countertop is produced by first coating both the particleboard base and the plastic laminate with a contact adhesive, usually neoprene dissolved in a solvent or emulsified in water; then, after the volatiles evaporate, the coated surfaces are pressed together. It is interesting to note that plastic laminates are primarily paper that has been impregnated with resin and then surface coated. Contact adhesives are mainly used in bonding plastic laminates to particleboard for countertops and furniture.

Polymerizable acrylic and acrylate adhesives are not used often for wood because of their high cost. The most common products of this type are super glues that can bond to wood, but these generally require smooth surfaces. These adhesives are more often used in electronics assembly with radiation (light) curing rather than in wood bonding. They do provide rapid cure rates and high strength bonds. Light curing of adhesives does not work with an opaque substrate like wood, but the acrylates can be used for a tough finish over the paper decorative layer on paneling.

Film adhesives involve either partially cured adhesives or adhesives applied onto a carrier such as a fiberglass mat or tissue paper. They are used where applying a liquid adhesive may be difficult, such as in bonding of very thin wood veneers.

9.7.13 FORMULATION OF ADHESIVES

Adhesives are composed of several different components in addition to the base polymer. Although the other components are added for a specific purpose, they often will alter several properties of the adhesive, as applied or after setting.

Base is the polymer, either synthetic, biobased, or a combination, that provides the adhesive the strength to hold the two substrates together. This is the material from which the adhesive usually takes its name, such as phenol-formaldehyde, epoxy or casein. The base material is a solid substance that provides the "backbone" of the adhesive film, controlling its application, setting, and curing.

Solvents are liquids often used to dissolve or disperse the base material and additives in order to provide a liquid system for application to the adherends, but are removed from the adhesive in the setting step. The most common solvent for wood adhesives is water. Water is not used in most other adhesive applications due to poor wetting, low volatility, and corrosion of surfaces, but for wood it is an ideal and low cost solvent. In some cases, the base material of the adhesive is a liquid itself and can be applied in this form without the need for solvents; for example, epoxies or pMDI. These are often referred to as "100-percent solid" adhesives. Such systems shrink little on hardening, thus reducing internal stresses in the film.

Diluents or *thinners* are liquids added to reduce the viscosity of the adhesive systems, and make them suitable for spraying or other special methods of application. However, unlike solvents, they are usually not volatile. A reactive diluent not only reduces the viscosity of the adhesive for application, but also becomes part of the final polymeric chain.

Catalysts or *accelerators* are chemicals added in small amounts to increase the rate of chemical reaction in the curing or hardening process. True catalysts are not consumed in the reaction, while accelerators may be consumed in the reaction. An example of a catalyst is the acid catalyst generated from ammonium salts for curing urea-formaldehyde resins, while an example of an accelerator is an ortho ester used to speed up the cure of phenol-formaldehyde resins.

Curing agents or *hardeners* are chemicals that actually undergo chemical reaction in stoichiometric proportions with the base resin and are combined in the final cured polymer structure. A good example is the amine component that reacts with an epoxy resin to form the final adhesive.

Fillers are solids that are added primarily to lower the cost and to give body to liquid adhesives, reducing undesired flow or over-penetration into wood. Fillers usually increase the rigidity of the cured adhesive. They may also modify the thermal expansion coefficient of the film to more nearly approximate that of the adjacent adherends, thus reducing thermal stresses in the joint formed during the cooling, following heat-curing conditions or when thermally cycled in service. Two examples of such fillers include walnut shell flour, incorporated in urea or resorcinol adhesives to improve spreading or reduce penetration into open wood pores, and china clay that is sometimes added to epoxy resin systems primarily for thickening or to modify thermal coefficients.

Extenders' primary purpose is reducing the adhesive costs while also providing some adhesive properties. At times they can also alter other properties, such as increasing the tack of the adhesive. A good example of an extender is wheat flour added to urea-formaldehyde resins in making hardwood plywood for interior applications where high moisture resistance is not required.

Stabilizers or *preservatives* are chemicals added to the adhesive to protect one or more of the components and/or the final adhesive against some type of deterioration. Preservatives are usually used for preventing biological deterioration, while a stabilizer can protect against either biological or chemical degradation. Prevention of biological deterioration can involve the use of fungicides or biocides, while chemical degradation prevention may involve the use of antioxidants or antiozonates.

Fortifiers are generally other base materials added to modify or improve the durability of the adhesive system under some specific type of service. A good example is the addition of more durable melamine resins to urea-formaldehyde resins in wood bonding to provide greater resistance to deterioration under hot and moist conditions.

Carriers are sometimes used to produce film-type adhesives. The carrier is usually a very thin, rather porous fabric or paper on which the liquid adhesive is applied and then dried. Examples include the use of thin tissue paper as a carrier for phenolic film adhesives in making thin hardwood plywood, where spreading the liquid adhesive on a conventional roller spreader might tear or break the thin veneers.

Adhesive formulating is an important skill, often requiring a mixture of empirical and scientific knowledge. Because there are no universal adhesives, systems must be formulated for different specific applications, e.g., for a given type of joint or even for a given type of production bonding operation. While billions of pounds of phenolic adhesives are used each year in wood bonding, the actual adhesive formulation used in one plant is often quite different from that used in another. Additionally the adhesive formulation used within the same plant may vary with the season. The moisture content or surface roughness of the veneers or in the time sequence between one operation and the next influences the actual types and proportions of additives, solvents, and resins used to make a cost effective adhesive.

9.8 ENVIRONMENTAL ASPECTS

Although wood is a natural material, bonded wood products have caused some environmental concern. There are a number of problem areas, but the foremost area of concern has been formaldehyde emissions from the bonded products, mainly using UF resins. Formaldehyde can react with biological systems in reactions similar to those that are used for curing of adhesives. The problem can arise from both unreacted and generated formaldehyde. Unreacted formaldehyde is also a problem during the manufacturing operation and in freshly produced composites. Formaldehyde emissions from composites decrease with time after production. The rate is high initially, but slowly decreases due to diffusion limitations. On the other hand, formaldehyde can be generated by the decomposition of some formaldehyde copolymer adhesives, in particular the urea-formaldehyde adhesives. These adhesive bonds are more prone to hydrolysis, generating free formaldehyde. The biggest concern is with particleboard, due to the large volume of indoor usage and the high level of adhesive in the product. The formulations of formaldehyde adhesives were altered to reduce the amount of formaldehyde used and in some cases formaldehyde scavengers were added. The reduction in formaldehyde altered the curing rate and the strength of the product; thus, the process required much research. Although many of the formaldehyde concerns have been addressed through adhesive reformulation, products still need to meet the commercial standards (Dunky and Pizzi 2002, Dunky 2003). The science of formaldehyde in wood products has been extensively reviewed (Marutzky 1989). The formaldehyde issue is still of concern, especially in some indoor environments.

The main concern, emissions, has focused on formaldehyde, but this is not the only compound emitted by bonded wood products. Other volatile compounds in the adhesive formulation have also been detected. In addition, a number of other volatiles are present in wood and additional ones can be generated by the heat and moisture in the production of the composite (Wang et al. 2003). Careful analysis has revealed the presence of formaldehyde, other aldehydes, methanol, and pinenes, many of which come from the wood itself rather than from the adhesive (Baumann et. al. 2000).

During the use of the adhesives, volatiles from the monomers that are used to produce the polymers generate additional health concerns. Thus, free formaldehyde, phenol, methanediphenyl diisocyanate, polyethylene polyamines, etc., are all of concern depending on the type of adhesive used. Heating certainly increases the problem because it raises the vapor pressure of these reactive chemicals. In addition, many hot pressing methods cause other chemicals to be entrained in the steam from the presses.

9.9 SUMMARY

Although wood bonding is one of the oldest adhesive applications, it is likely the most complicated to understand. Many modes are possible for both bond formation and failure. Wood structure has so many variables in the different species, cell structures within a species, and spatial scales of examination that it is hard to model the process. Despite these problems, many adhesives have been developed that are stronger and some are even more durable than the wood itself. In addition, many functional adhesives have been developed that allow a wide variety of woods and wood pieces to be glued together in a useful and cost-effective manner.

The area that is best understood is the chemistry of the adhesives, even though there are aspects, such as the effects of the composite processing dynamics, that need to be more thoroughly researched. The development of the physical properties during the setting process and the interaction of the adhesive with the wood needs to be better understood to allow for a more cost effective development of new adhesives.

REFERENCES

American Institute of Timber Construction. (1990). Use of epoxies in repair of structural glued laminated timber. AITC Technical Note 14. AITC, Englewood, CO.

ASTM International, (2000a). D 907-00. *Standard Terminology of Adhesives*. ASTM International, West Conshohocken, PA.

ASTM International, (2000b). D 2559-00. *Standard Specification for Adhesives for Structural Laminated Wood Products for Use Under Exterior (Wet Use) Exposure Conditions*. ASTM International, West Conshohocken, PA.

ASTM International, (2001). D 4300-01. *Standard Test Methods for Ability of Adhesive Films to Support or Resist the Growth of Fungi*. ASTM International, West Conshohocken, PA.

Autumn, K., Sitti, M., Liang, Y.A., Peattie, A., Hansen, W., Sponberg, S., Kenny, T., Fearing, R., Israelachvili, J., and Full, R.J. (2002). Evidence for van der Waals adhesion in gecko setae. *Proc. Natl. Acad. Sci.* 99:12252–12256.

Back, E.L. (1987). The bonding mechanism in hardboard manufacture, *Holzforschung* 41:247–258.

Barton A.F.M. (1991). *CRC Handbook of Solubility Parameters and Other Cohesion Parameters* (2nd ed.). CRC Press, Boca Raton, FL.

Baumann, M.G.D., Lorenz, L.F., Batterman, S.A., and Zhang, G-Z. (2000). Aldehyde emissions from particleboard and medium density fiberboard products. *Forest Products J.*, 50(9):75–82.

Berg, J.C. (2002). Semi-empirical strategies for predicting adhesion. In: Chaudhury, M. and Pocius, A.V. (Eds.), *Adhesive Science and Engineering – 2: Surfaces, Chemistry and Applications*. Elsevier, Amsterdam, pp. 1–73.

Bikerman, J.J. (1968). *The Science of Adhesive Joints* (2nd ed.). Academic Press, New York.

Blomquist, R.F. and Vick, C.B. (1977). Adhesives for building construction. In: Skeist, I. (Ed.), *Handbook of Adhesives* (2nd ed.). Van Nostrand Reinhold, New York, chap. 49.

Bolton, A.J., Dinwoodie, J.M., and Beele, P.M. (1985). The microdistribution of UF resins in particleboard. In: *Proceedings of IUFRO Symposium, Forest Products Research International: Achievements and the Future*. Vol. 6. Pretoria, South Africa, 17.12-1–17.12-19.

Bolton, A.J., Dinwoodie, J.M., and Davies, D.A. (1988). The validity of the use of SEM/EDAX as a tool for the detection of the UF resin penetration into wood cell walls in particleboard. *Wood Sci. Technol.* 22:345–356.

Brief, A. (1990). The role of adhesives in the economy. In: Skeist, I. (Ed.), *Handbook of Adhesives* (3rd ed.). Van Nostrand Reinhold, New York, chap. 2.

Browne, F.L. and Brouse, D. (1929). Nature of adhesion between glue and wood. *Ind. and Chem. Eng.* 21(1):80–84.

Christiansen, A.W. (1990). How overdrying wood reduces its bonding to phenol-formaldehyde adhesives: A critical review of the literature, Part I: Physical responses. *Wood and Fiber Sci.* 22(4):441–459.

Christiansen, A.W. (1991). How overdrying wood reduces its bonding to phenol-formaldehyde adhesives: A critical review of the literature, Part II: Chemical reactions. *Wood and Fiber Sci.* 23(1):69–84.

Christiansen, A.W. (1994). Effect of overdrying of yellow-poplar veneer on physical properties and bonding. *Holz als Roh und Werkstoff* 52:139–149.

Christiansen, A.W. and Okkonen, E.A. (2003). Improvements to hydroxymethylated resorcinol coupling agent for durable wood bonding. *For. Prod. J.* 53(4):81–84.

Conner, A.H. and Baumann, M.G.D. (2003). Carbohydrate polymers as adhesives. In: Pizzi, A. and Mittal, K.L. (Eds.), *Handbook of Adhesive Technology* (2nd ed.). Marcel Dekker, New York, chap. 22.

Conner, A.H., Lorenz, L.F., and Hirth, K.C. (2002). Accelerated cure of phenol-formaldehyde resins: Studies with model compounds. *J. Appl. Polym. Sci.* 86:3256–3263.

Dawson, B., Gallager, S., and Singh, A. (2003). *Microscopic View of Wood and Coating Interaction, and Coating Performance on Wood*. Forest Research Bulletin 228. Forest Research, Rotorua, New Zealand.

Detlefsen, W.D. (2002). Phenolic resins: Some chemistry, technology and history. In: Chadhury, M, and Pocius, A.V. (Eds.), *Adhesive Science and Engineering – 2: Surfaces, Chemistry and Applications*. Elsevier, Amsterdam, chap. 20.

Dillard, D.A. (2002). Fundamentals of stress transfer in bonded systems. In: Dillard, D.A. and Pocius, A.V. (Eds.), *Adhesive Science and Engineering – 1: The Mechanics of Adhesion*. Elsevier, Amsterdam, chap. 1.

Dunky, M. (2003). Adhesives in the wood industry. In: Pizzi, A. and Mittal, K.L. (Eds.), *Handbook of Adhesive Technology* (2nd ed.). Marcel Dekker, New York, chap. 47.

Dunky, M. and Pizzi A. (2002). Wood adhesives. In: Chaudhury, M. and Pocius, A.V. (Eds.), *Adhesive Science and Engineering – 2: Surfaces, Chemistry and Applications.* Elsevier, Amsterdam, chap. 23, pp. 1039–1103.

Eastman, E.F. and Fullhart, L., Jr. (1990). Polyolefin and ethylene copolymer-based adhesives. In: Skeist, I. (Ed.), *Handbook of Adhesives* (3rd ed.). Van Nostrand Reinhold, New York, chap. 23.

Ebewele, R.O., River, B.H., Myers, G.E., and Koutsky, J.A. (1991). Polyamine-modified urea-formaldehyde resins, II: Resistance to stress induced by moisture cycling of solid wood joints and particleboard. *J. Appl. Polymer Sci.* 43:1483–1490.

Ebewele, R.O., River, B.H., and Myers, G.E. (1993). Polyamine-modified urea-formaldehyde-bonded wood joints, III: Fracture toughness and cyclic stress and hydrolysis resistance. *J. Appl. Polymer Sci.* 49:229–245.

Felby, C., Hassingboe, J., and Lund, M. (2002). Pilot-scale production of fiberboards made by laccase oxidized wood fibers: Board properties and evidence for cross-linking of lignin. *Enzyme Microb. Technol.* 31:736–741.

Fengel D. and Wegener, G. (1984). *Wood: Chemistry, Ultrastructure, Reactions.* Walter de Gruyter, Berlin.

Fowkes, F.M., (1983). Acid-Base Interactions in Polymer Adhesion. In: Mittal, K.L. (Ed.), *Physicochemical Aspects of Polymer Surfaces.* Vol. 2. Plenum Press, New York.

Frazier, C.E. (2002). The interphase in bio-based composites: What is it, what should it be? In: *Proc. 6th Pacific Rim Bio-Based Composites Symposium & Workshop on the Chemical Modification of Cellulosics.* Portland, OR, Oregon State University, Corvallis, p. 206.

Frazier, C.E., (2003). Isocyanate wood binders. In: Pizzi, A. and Mittal, K.L. (Eds.), *Handbook of Adhesive Technology.* (2nd ed.). Marcel Dekker, New York, chap. 33.

Frihart, C. R. (2003a). Durable wood bonding with epoxy adhesives. In: *Proceedings, 26th Annual Meeting, Adhesion Society, Inc.* Myrtle Beach, SC, Feb. 23–26, p. 476–478.

Frihart, C.R. (2003b). Wood structure and adhesive strength. In: *Characterization of the Cellulosic Cell Wall—The Proceedings of the SWST International Workshop on the Assessment and Impact of the Cellulosic Cell Wall.* Colorado, Aug. 25–27, Blackwell Publishing, Oxford, United Kingdom (planned 2005).

Frihart, C.R. (2004). Adhesive interactions with wood. General Technical Report FPL-GTR-149. U.S. Department of Agriculture, Forest Service, Forest Products Laboratory, Madison, WI.

Frihart, C.R., Brandon, R., and Ibach, R.E. (2004). Selectivity of bonding for modified wood. In: *Proceedings, 27th Annual Meeting of The Adhesion Society, Inc.* Wilmington, NC, Feb. 15–18, p. 329–331.

Furuno, T. and Goto, T. (1975). Structure of the interface between wood and synthetic polymer, VII: Fluorescence microscopic observation of glue line of wood glued with epoxy resin adhesive. *Mokuzai Gakkaishi* 21(5):289–296.

Furuno, T. and Saiki, H. (1988). Comparative observations with fluorescence and scanning microscopy of cell walls adhering to the glue on fractured surfaces of wood-glue joints. *Mokuzai Gakkaishi* 34(5):409–416.

Geddes, K. (2003). Polyvinyl and ethylene-vinyl acetates. In: Pizzi, A. and Mittal, K.L. (Eds.), *Handbook of Adhesive Technology* (2nd ed.). Marcel Dekker, New York, chap. 35.

Gfeller, B., Properzi, M., Zanetti, M., Pizzi, A., Pichelin, F., Lehmann, M., and Delmotte, L.(2003). Wood bonding by mechanically induced *in situ* welding of polymeric structural wood constituents. *J. Appl. Polymer Sci.* 92(1):243–251.

Gindl, W., Dessipri, E. and Wimmer, R. (2002). Using UV-microscopy to study the diffusion of melamine-urea-formaldehyde resin in cell walls of spruce wood. *Holzforschung* 56:103–107.

Gindl, W. and Gupta, H.S. (2002). Cell wall hardness and Young's modulus of melamine-modified spruce wood by nano-indentation. *Composites: Part A* 33:1141–1145.

Good, R.J. (1966). Intermolecular and interatomic forces. In: Ptrick, R.L. (Ed.), *Treatise on Adhesion and Adhesives, Volume 1: Theory.* Marcel Dekker, New York, chap. 2.

Green, D.W., Winandy, J.E., and Kretschmann, D.E. (1999). Mechanical properties of wood. In: *Wood Handbook: Wood as an Engineering Material.* U.S. Department of Agriculture Forest Service, Forest Products Laboratory, Madison, WI, chap. 4.

Hansen, C.M. and Björkman, A. (1998). The ultrastructure of wood from a solubility parameter point of view. *Holzforschung* 52(4):335–344.

Jaffe, H.L., and Rosenblum, F.M. (1990). Poly(vinyl alcohol) for adhesives. In: Skeist, I. (Ed.), *Handbook of Adhesives* (3rd ed.). Van Nostrand Reinhold, New York, chap. 22.

Jaffe, H. L., Rosenblum, F.M. and Daniels, W. (1990). Poly(vinyl acetate) emulsions for adhesives. In: Skeist, I. (Ed.), *Handbook of Adhesives* (3rd ed.). Van Nostrand Reinhold, New York, chap. 21.

Johns, W.E and Nguyen, T. (1977). Peroxyacetic acid bonding of wood. *Forest Products J.* 27(1):17–23.

Johnson, S.E. and Kamke, F.A. (1992). Quantitative analysis of gross adhesive penetration in wood using fluorescence microscopy. *J. Adhesion* 40:47–61.

Keckes, J., Burgert, I., Frühmann, K., Müller, M., Kölln, K., Hamilton, M., Burghammer, M., Roth, S.V., Stanzl-Tschegg, S., and Fratzl, P. (2003). Cell-wall recovery after irreversible deformation of wood. *Nat. Mater.* 2:810–814.

Keimel, F.A. (2003). Historical development of adhesives and adhesive bonding. In: Pizzi, A. and Mittal, K.L. (Eds.), *Handbook of Adhesive Technology* (2nd ed.). Marcel Dekker, New York, chap. 1, pp. 1–12.

Kharazipour, A., Hüttermann, A., and Lüdemann, H.D. (1997). Enzymatic activiation of wood fibres as a means for the production of wood composites. *J. Adhesion Sci. Technol.* 11(3):419–427.

Kinlock, A.J. (1987). *Adhesion and Adhesives Science and Technology*. Chapman & Hall, London.

Kreibich, R.E., Steynberg, P.J., and Hemingway, R.W. (1998). End jointing green lumber with SoyBond. In: *Wood Residues into Revenue: Residual Wood Conference Proceedings*. Nov. 4–5, 1997, Richmond, British Columbia, MCTI Communications Inc., pp. 28–36.

Kretschmann, D. (2003). Velcro mechanics in wood. *Nat. Mater.* 2:775–776.

Laborie M.P.G., Salmén, L., and Frazier, C.E. (2002). The cooperativity analysis of segmental motion in the wood adhesive interphase: a probe of nanoscale morphology. In: *Proceedings, 6th Pacific Rim Bio-Based Composites Symposium & Workshop on the Chemical Modification of Cellulosics,* Portland, OR, 2002, Oregon State University, Corvallis, pp. 18–25.

Lambuth, A.L. (2003). Protein adhesives for wood. In: Pizzi, A. and Mittal, K.L. (Eds.), *Handbook of Adhesive Technology* (2nd ed.), Marcel Dekker, New York, chap. 20.

Lange, D.A., Fields, J.T., and Stirn, S.A. (2001). Finger joint application potentials for one-part polyurethanes. *Wood Adhesives 2000*. Forest Products Society, Madison, WI, pp. 81–90.

Liechti, K.M. (2002). Fracture mechanics and singularities in bonded systems. In: Dillard, D.A. and Pocius, A.V. (Eds.), *Adhesive Science and Engineering – 1: The Mechanics of Adhesion*, Elsevier, Amsterdam, chap. 2.

Marcinko, J., Phanopoulos, C., and Teachey, P. (2001). Why does chewing gum stick to hair and what does this have to do with lignocellulosic structural composite adhesion. In: *Wood Adhesives 2000,* South Lake Tahoe, NV, June 22–23, 2000. Forest Products Society, Madison, WI, pp. 111–121.

Marcinko, J.J., Devathala, S., Rinaldi, P.L., and Bao, S. (1998). Investigating the molecular and bulk dynamics of pMDI/wood and UF/wood composites. *Forest Products J.* 48(6):81–84.

Marra, A.A. (1980). Applications in wood bonding. In: Blomquist, R.F., Christiansen, A.W., Gillespie, R.H., and Myers, G.E. (Eds.), *Adhesive Bonding of Wood and Other Structural Materials*. Educational Modules for Materials Science and Engineering (EMMSE) Project, Pennsylvania State University, University Park, PA, chap. 9.

Marra, A.A. (1992). *Technology of Wood Bonding: Principles in Practice*. Van Nostrand Reinhold, New York.

Marutzky, R. (1989). Release of formaldehyde by wood products. In: Pizzi, A. (Ed.), *Wood Adhesives Chemistry and Technology*. Vol. 2. Marcel Dekker, New York, chap. 10.

McBain, J.W. and Hopkins, D.G. (1925). On adhesives and adhesive action. *J. Phys. Chem.* 29:188–204.

Miller, R.S. (1990). Adhesives for building construction. In: Skeist, I. (Ed.), *Handbook of Adhesives* (3rd ed.). Van Nostrand Reinhold, New York, chap. 41.

Moody, R.C., Hernandez, R., and Liu, J Y. (1999). Glued structural members. In: *Wood Handbook: Wood as an Engineering Material*. U.S. Department of Agriculture, Forest Service, Forest Products Laboratory, Madison, WI, chap. 11.

Morton, J., Quarmley, J. and Rossi, L. (2003). Current and emerging applications for natural and woodfiber-plastic composites. *Seventh International Conference on Woodfiber-Plastic Composites*. Forest Products Society, Madison, WI, pp. 3–6.

Muszyński, L., Wang, F., and Shaler, S.M. (2002). Short-term creep tests on phenol-resorcinol-formaldehyde (PRF) resin undergoing moisture content changes. *Wood and Fiber Sci.* 34(4):612–624.

Nearn, W.T. (1965). Wood-adhesive interface relations. *Off. Dig., Fed. Soc. Paint Technol.* 37 (June): 720–733.

Nussbaum, R. (2001). Surface interactions of wood with adhesives and coatings. Doctoral thesis, KTH – Royal Institute of Technology, Stockholm, Sweden.

Okkonen, E.A. and Vick, C. B. (1998). Bondability of salvaged yellow-cedar with phenol-resorcinol adhesive and hydroxymethylated resorcinol coupling agent. *Forest Products J.* 48(11/12):81–85.

Okkonen, E.A., and River, B.H. (1996). Outdoor aging of wood-based panels and correlation with laboratory aging: Part 2. *Forest Products J.* 46(3):68–74.

Packham, D.E. (2003). The mechanical theory of adhesion. In: Pizzi, A. and Mittal, K.L. (Eds.), *Handbook of Adhesive Technology* (2nd ed.). Marcel Dekker, New York, chap. 4.

Parker, J.R., Taylor, J.B.M., Plackett, D.V., and Lomax, T.D. (1997). Method of joining wood, U.S. Patent 5,674,338.

Pauling, L. (1960). *The Nature of the Chemical Bond.* Cornell University Press, Ithaca, NY.

Pearson, C.L. (2003). Animal glues and adhesives. In: Pizzi, A. and Mittal, K.L. (Eds.), *Handbook of Adhesive Technology* (2nd ed.). Marcel Dekker, New York, chap. 21.

Petrie, E. M. (2000). *Handbook of Adhesives and Sealants.* McGraw-Hill, New York, chap. 8.

Pizzi, A. (1994a). *Advanced Wood Adhesives Technology.* Marcel Dekker, New York, pp. 275–276.

Pizzi, A. (1994b). *Advanced Wood Adhesives Technology,* Marcel Dekker, New York, pp. 31–39, 92–98.

Pizzi, A. (1994c). *Advanced Wood Adhesives Technology,* Marcel Dekker, New York, pp. 119–126.

Pizzi, A. (2003a). Phenolic resin adhesives. In: Pizzi, A. and Mittal, K.L. (Eds.), *Handbook of Adhesive Technology* (2nd ed.). Marcel Dekker, New York, chap. 26.

Pizzi, A. (2003b) Resorcinol adhesives. In: Pizzi, A. and Mittal, K.L. (Eds.), *Handbook of Adhesive Technology* (2nd ed.), Marcel Dekker, New York, chap. 29.

Pizzi, A. (2003c). Natural phenolic adhesives I: Tannin, In: Pizzi, A. and Mittal, K.L. (Eds,) *Handbook of Adhesive Technology* (2nd ed.).., Marcel Dekker, New York, chap. 27.

Pizzi, A., (2003d). Natural phenolic adhesives II: Lignin. In: Pizzi, A. and Mittal, K.L. (Eds.), *Handbook of Adhesive Technology* (2nd ed.), Marcel Dekker, New York, chap. 28.

Pizzi, A. (2003e). Urea-formaldehyde adhesives. In: Pizzi, A. and Mittal, K.L. (Eds.), *Handbook of Adhesive Technology* (2nd ed.) Marcel Dekker, New York, chap. 31.

Pizzi, A. (2003f). Melamine-formaldehyde resins. In Pizzi, A. and Mittal, K.L. (Eds.), *Handbook of Adhesive Technology* (2nd ed.). Marcel Dekker, New York, chap. 32.

Pizzo, B., Rizzo, G., Lavisci, P. Megna, B., and Berti, S. (2002). Comparison of thermal expansion of wood and epoxy adhesives. *Hoz als Roh- und Werkstoff* 60: 285–290.

Pocius, A.V. (2002). *Adhesion and Adhesives Technology: an Introduction* (2nd ed.,). Hanser, Munich.

Regert, M. (2004). Investigating the history of prehistoric glues by gas chromatography-mass spectroscopy. *J. Sep. Sci.* 27:244–254.

Rice, J. T. (1990). Adhesive selection and screening testing. In: Skeist, I., (Ed.) *Handbook of Adhesives* (3rd ed.). Van Nostrand Reinhold, New York, chap. 5.

River, B.H. (1994a). Fracture of adhesively-bonded wood joints. In: Pizzi, A. and Mittal, K.L. (Eds.), *Handbook of Adhesive Technology.* Marcel Dekker, New York, chap. 9.

River, B.H. (1994b). Outdoor aging of wood-based panels and correlation with laboratory aging. *Forest Products J.* 44(11/12):55–65.

River, B.H., Ebewele, R.O., and Myers, G.E. (1994c). Failure mechanisms in wood joints bonded with urea-formaldehyde adhesives. *Holz als Roh- und Werkstoff* 52:179–184.

River, B.H. and Miniutti, V.P. (1975). Surface damage before gluing – weak joints. *Wood Wood Prod.* 80(7):35–36.

River, B.H., Vick, C.B., and Gillespie, R.H. (1991). Wood as an adherend. In: Minford, J.D. (Ed.), *Treatise on Adhesion and Adhesives.* Vol. 7. Marcel Dekker, New York.

Robins, J. (1986). Phenolic resins. In: Hartshorn, S.R. (Ed.), *Structural Adhesives, Chemistry and Technology.* Plenum Press, New York, chap. 2.

Rohn, C.L. (1999). Rheology of pressure sensitive adhesives. In: Satas, D. (Ed.), *Handbook of Pressure Sensitive Adhesive Technology* (3rd ed.). Satas & Associates, Warwick, RI, chap. 9.

Rossitto, C. (1990). Polyester and polyamide high performance hot melt adhesives. In: Skeist, I. (Ed.), *Handbook of Adhesives* (3rd ed.), Van Nostrand Reinhold, New York, chap. 28.

Saiki, H. (1984). The effect of the penetration of adhesives into cell walls on the failure of wood bonding. *Mokuzai Gakkaishi* 30(1):88–92.

Sakata, I., Morita, M., Tsurata, N., and Morita, K. (1993). Activation of wood surface by corona treatment to improve adhesive bonding. *J. Appl. Polymer Sci.* 49:1251–1258.

Salehuddin, A. (1970). A unifying physico-chemical theory for cellulose and wood and its application in gluing. Ph.D. thesis, North Carolina State University at Raleigh.

Satas, D. (1999a). Dynamic mechanical analysis and adhesive performance. In: Satas, D. (Ed.), *Handbook of Pressure Sensitive Adhesive Technology* (3rd ed.), Satas & Associates, Warwick, RI, chap. 10.

Satas, D. (Ed.). (1999b). *Handbook of Pressure Sensitive Adhesive Technology* (3rd ed.). Satas & Associates, Warwick, RI.

Sato, K. and Naitio, T. (1973). Studies on melamine resin .7. kinetics of acid catalyzed condensation of dimethylolmelamine and trimethylolmelamine. *Polymer. J.* 5(2):144–157.

Schultz, J. and Nardin, M. (2003). Theories and mechanisms of adhesion. In: Pizzi, A. and Mittal, K.L. (Eds.), *Handbook of Adhesive Technology* (2nd ed.). Marcel Dekker, New York, chap. 3.

Selbo, M. L. (1965). Performance of melamine resin adhesives in various exposures. *Forest Products J.* 15(12):475–483.

Skaar, C. (1984). Wood-water relationships. In: Rowell, R. (Ed.), *The Chemistry of Solid Wood,* American Chemical Society, Washington, DC, chap. 3.

Skeist, I. and Miron, J. (1990). Introduction to adhesives. In: Skeist, I. (Ed.), *Handbook of Adhesives.* (3rd) Van Nostrand Reinhold, New York, chap 1.

Smith, L.A. (1971). Resin penetration of wood cell walls – implications for adhesion of polymers to wood. Ph.D. thesis, Syracuse University, Syracuse, NY.

Smith, L.A. and Cote, W.A. (1971). Studies on the penetration of phenol-formaldehyde resin into wood cell walls with SEM and energy-dispersive x-ray analyzer. *Wood Fiber J.* 56–57.

Smith, M.J., Dai, H., and Ramani, K. (2002). Wood-thermoplastic adhesive interface – method of characterization and results. *Int. J. Adhesion Adhesives* 22:197–204.

Stamm, A.J. and Seborg, R.M. (1936). Minimizing wood shrinkage and swelling. Treating with synthetic resin-forming materials. *Ind. Eng. Chem.* 28(10):1164–1169.

Stehr, M. and Johansson, I. (2000). Weak boundary layers on wood surfaces. *J. Adhesion Sci. Technol.* 14:1211–1224.

Stofko, J. (1974). The autoadhesion of wood. Ph.D. thesis, University of California, Berkeley.

Sudell, B.C. and Evans, W.J. (2003). The increasing use and application of natural fiber composite materials within the automotive industry. *Seventh International Conference on Woodfiber-Plastic Composites.* Forest Products Society, Madison, WI, pp. 7–14.

Tarkow, H. (1979). Wood and moisture. In: Wangaard, F.F. (Ed.), *Wood: Its Structure and Properties*, Educational Modules for Materials Science and Engineering (EMMSE) Project, Pennsylvania State University, University Park, PA, chap. 4.

Tarkow, H., Feist, W.C., and Southerland, C.F. (1965). Interpenetration of wood and polymeric materials, II: Penetration versus molecular size. *Forest Products J.* 16(10):61–65.

Tsoumis, G. (1991). *Science and Technology of Wood: Structure, Properties and Utilization.* Van Nostrand Reinhold, New York.

Updegaff, I.V. (1990). Amino resin adhesives. In: Skeist, I. (Ed.), *Handbook of Adhesives.* (3rd) Van Nostrand Reinhold, New York, chap. 18.

Vick, C.B. (1973). Gap-filling phenol-resorcinol resin adhesives for construction. *Forest Products J.* 23(11):33–41.

Vick, C.B. (1995). Coupling agent improves durability of PRF bonds to CCA-treated southern pine. *Forest Products J.* 45(3):78–84.

Vick, C.B. (1997). More durable epoxy bonds to wood with hydroxymethylated resorcinol coupling agent. *Adhesives Age* 40(8):24–29.

Vick, C.B. (1999). Adhesive bonding of wood materials. In: *Wood Handbook: Wood as an Engineering Material.* U.S. Department of Agriculture, Forest Service, Forest Products Laboratory, Madison, WI, chap. 9.

Vick, C.B., Christiansen, A.W., and Okkonen, E.A. (1998). Reactivity of hydroxymethylated resorcinol coupling agent as it affects durability of epoxy bonds to Douglas-fir. *Wood Fiber Sci.* 30(3):312–322.

Vick, C.B. and Okkonen, E.A. (1998). Strength and durability of one-part polyurethane adhesive bonds to wood. *Forest Products J.* 48(11/12):71–76.

Vick, C.B., Richter, K.H., and River, B.H. (1996). Hydroxymethylated resorcinol coupling agent and method for bonding wood. U.S. Patent 5,543,487.

Wang, W., Gardner, D.J., and Baumann, M.G.D. (2003). Factors affecting volatile organic compound emissions during hot-pressing of southern pine particleboard. *Forest Products J.* 53(3):65–72.

Wellons, J.D. (1980). The adherends and their preparation for bonding. In: Blomquist, R.F., Christiansen, A.W., Gillespie, R.H., and Myers, G.E. (Eds.), *Adhesive Bonding of Wood and Other Structural Materials.* Educational Modules for Materials Science and Engineering (EMMSE) Project, Pennsylvania State University, University Park, PA, chap. 3.

Westermark, U. and Karlsson, O. (2003). Auto-adhesive bonding by oxidative treatment of wood. In: *Proceedings, 12th Intl. Symp. Wood and Pulping Chem. Vol. 1*. Madison, WI, June 9–12, pp. 365–368.

Wescott, J. and Frihart, C.R. (2004). Competitive crosslinked soybean flour phenol formaldehyde adhesive for OSB. *38th International Wood Composites Symposium*. Washington State University, Pullman, WA, April 5–8.

Widsten, P., Laine, J.E., Tuominen, S., and Qvintus-Leino, P. (2003). Effect of high defibration temperature on the properties of medium-density fiberboard (MDF) made from laccase-treated hardwood fibers. *J. Adhesion Sci. Technol.* 17(11):67–78.

Wolfrum, J. and Ehrenstein, G.W. (1999). Interdependence between the curing, structure, and the mechanical properties of phenolic resins. *J. Appl. Polymer Sci.* 74:3173–3185.

Wool, R.P. (2002). Diffusion and autoadhesion. In: Chaudhury, M. and Pocius, A.V. (Eds.), *Adhesive Science and Engineering—2: Surfaces, Chemistry and Applications*. Elsevier, Amsterdam, chap. 8, pp. 351–401.

Wu, S. (1982a). *Polymer Interface and Adhesion*. Marcel Dekker, New York, chap. 2.

Wu, S. (1982b). *Polymer Interface and Adhesion*. Marcel Dekker, New York, chap. 12.

Young, R.A., Fujita, M. and River, B.H. (1985). New approaches to wood bonding: A base-activated lignin system. *Wood Sci. Technol.* 19:363–381.

10 Wood Composites

Lars Berglund[1] *and Roger M. Rowell*[2]
[1]Lightweight Structures Division, Royal Institute of Technology, Stockholm, Sweden
[2]USDA, Forest Service, Forest Products Laboratory, and Biological Systems Department, University of Wisconsin, Madison, WI

CONTENTS

- 10.1 Types of Composites and Applications 281
  - 10.1.1 Laminated Timbers 281
  - 10.1.2 Plywood 282
  - 10.1.3 Structural Composite Lumber 282
  - 10.1.4 Composite Beams 283
  - 10.1.5 Wafer- and Flakeboard 283
  - 10.1.6 Particleboard 284
  - 10.1.7 Fiberboard 285
    - 10.1.7.1 Isolation of Fibers 285
    - 10.1.7.2 Low-Density Fiberboard (LDF) 286
    - 10.1.7.3 Medium-Density Fiberboard (MDF) 286
    - 10.1.7.4 High-Density Fiberboard (HDF) 287
  - 10.1.8 Other Types of Composites 287
  - 10.1.9 Nanocomposites 287
- 10.2 Adhesives 288
- 10.3 Production, Properties, Performance, and Applications 289
  - 10.3.1 Glued Laminated Timber 290
  - 10.3.2 Structural Composite Lumber (LVL, PSL, LSL) 290
  - 10.3.3 Plywood 291
  - 10.3.4 Particleboard 293
  - 10.3.5 Flakeboard 293
  - 10.3.6 Fiberboard 295
    - 10.3.6.1 Low-Density Fiberboard (Insulation Board) 296
    - 10.3.6.2 Medium-Density Fiberboard (MDF) 297
    - 10.3.6.3 High-Density Fiberboard (HDF, Hardboard) 298
  - 10.3.7 Nanocomposites 298
- 10.4 Conclusions 300
- References 301

A composite can be defined as two or more elements held together by a matrix. By this definition, what we call "solid wood" is a composite. Solid wood is a three-dimensional composite composed of cellulose, hemicelluloses and lignin (with smaller amounts of inorganics and extractives), held together by a lignin matrix. (See Chapter 3.)

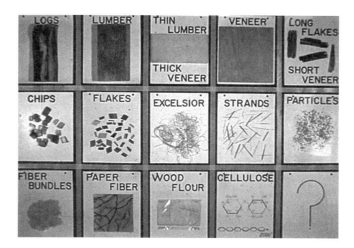

FIGURE 10.1 Basic wood elements, from largest to smallest (Mara 1979).

The advantages of developing wood composites are (1) to use smaller trees, (2) to use waste wood from other processing, (3) to remove defects, (4) to create more uniform components, (5) to develop composites that are stronger than the original solid wood, and (6) to be able to make composites of different shapes.

Historically, wood was used only in its solid form as large timbers or lumber. As the availability of large-diameter trees decreased (and the price increased) the wood industry looked to replace large-timber products and solid lumber with reconstituted wood products made using smaller-diameter trees and saw and pulp mill wastes. There has been a trend away from solid wood for some traditional applications toward smaller element sizes. Marra (1979) put together what he called "a periodic table of wood elements" (Figure 10.1), showing the breakdown of solid wood into composite components. With some modifications, this table still represents most of the types of elements used today to produce composite materials.

New composite products started with very thick laminates for glued laminated beams, to thin veneers for plywood, to strands for strandboard, to flakes for flakeboard, to particles for particleboard, and, finally, to fibers for fiberboard. As the size of the composite element gets smaller, it is possible to either remove defects (knots, cracks, checks, etc) or redistribute them to reduce their effect on product properties. Also, as the element size becomes smaller, the composite becomes more like a true material, i.e. consistent, uniform, continuous, predictable, and reproducible (Marra 1979). For many new fiber-based composite products, the use of fibers will become more common, and these fibers can come from many different agricultural sources.

Glued laminated beams were introduced in an auditorium using a casein adhesive in 1893 in Basel, Switzerland (Wood Handbook 1999). These early laminated beams created a new dimension in design away from the solid wood beam that had been used in construction for hundreds of years. Now it was possible to create a structure from solid wood with graceful lines, and a new structural element that was aesthetic as well as functional was introduced. This is a design element still very much in use today.

The modern plywood industry began around 1910, but the furniture industry had used veneers over solid wood for several hundred years beforethat. Overlaying thin sheets of wood or paper over another material created the "wood look" without actually using very much wood, if any, at all. Furniture designs using plywood were created using rather complex designs but were still limited to the bending properties of thin wood veneers. Today, very thin veneers are made, backed with a thermoplastic sheet that can be overlaid onto many different materials. The best known example of this technology is in the manufacture of business cards using these thin wood/thermoplastic laminated sheets.

The particleboard industry started in the 1940's, the hardboard industry around 1950, and the flakeboard and medium density fiberboard (MDF) industries in the early 1960s (Maloney 1996). In general, all of these products are produced in flat sheets and used in two-dimensional designs. It is possible, however, to produce all of these composites in three-dimensional products. Flakes and particles have been formed into pallets and packing materials using an adhesive and a rather simple mold.

10.1 TYPES OF COMPOSITES AND APPLICATIONS

The earliest composite structures were made of solid beams. Figure 10.2 shows an example of solid beam construction that was done about 200 years ago. Solid beam structures are still being built today but it is far more common to see smaller wood elements that are glued together in composite structures.

10.1.1 LAMINATED TIMBERS

Structural glued-laminated beams (glulam) can be made using thick, wide wood members and are used as structural elements in large, open buildings. Glulam is a structural product that consists of two or more layers of lumber glued together with the grain all going parallel to the length. Figure 10.3 shows a laminated beam being fitted into a steel plate that joins the beam to the ground. It can be formed straight or curved, depending on the desired application. Typically the laminates are 25 to 50 mm in thickness. Douglas fir, southern pine, hem-fir and spruce are common wood species used to make glulam in the United States (Wood Handbook 1999). Solid wood and glulam have a specific gravity of 0.4 to 0.8.

The biggest advantage of using glulam is that large beams can be made using small trees. In addition, lower quality wood can be used, thinner lumber can be dried much faster than large, thick beams, and a variety of curved shapes can be produced.

FIGURE 10.2 Old cabin built using solid wooden beams.

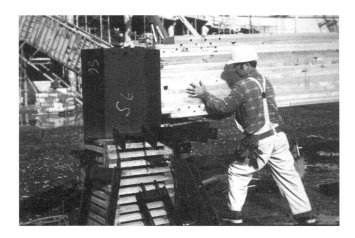

FIGURE 10.3 Large composite beam made from laminated lumber.

10.1.2 PLYWOOD

Thin veneers can be glued together for plywood, a material that is used as a structural underlayment in floors and roofs and in furniture manufacturing. There are two basic types of plywood: construction and decorative. Construction-grade plywood has traditionally been produced using softwoods such as Douglas-fir, southern pines, white fir, larch, and western hemlock, and comes in several grades based on the quality of each layer and the adhesive used. Decorative plywood is usually produced using softwoods for the back and inner layers, with a hardwood layer on the outer surface. Figure 10.4 shows a three-ply composite in which each layer is perpendicular to the layer above and below. The veneers can be produced either by peeling or slicing, and the grade of plywood depends on the defects on the two faces. Different thicknesses of plywood can be produced using multi-layers of veneers. Usually an odd number of layers is used, and the products have a specific density from about 0.4 to 0.8.

10.1.3 STRUCTURAL COMPOSITE LUMBER

Structural composite lumber (SCL) is manufactured by laminating strips of veneers or strands of wood glued parallel to the length. Figure 10.5 shows three types of SCL products: oriented strand

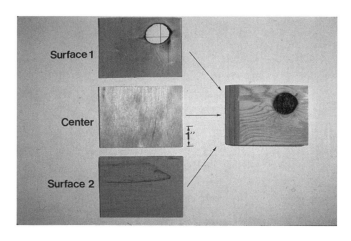

FIGURE 10.4 Thin veneers used to make plywood.

Wood Composites

FIGURE 10.5 Three types of structural composite lumber (SCL). Left—Oriented strand lumber (OSL), center—Parallel strand lumber (PSL), right—Laminated veneer lumber (LVL).

lumber (OSL), parallel strand lumber (PSL) and laminated veneer lumber (LVL). Laminated strand lumber (LSL), oriented strand board (OSB) and OSL are produced using different lengths and sizes of strands. LSL uses strands that are about 0.3 m in length while OSB is produced from shorter strands. PSL is made from strands that are 3 mm thick, approximately 19 mm wide, and 0.6 m in length. Usually Douglas-fir, southern pines, western hemlock, and yellow-poplar are used, but other species are also used. LVL is produced from veneers that are approximately 2.5 to 3.2 mm thick and varying lengths. The major adhesives used to produce SCL products are phenol-formaldehyde or isocyanates. All of these SCL products are used as replacements for solid wood and have a specific gravity of 0.5 to 0.8.

10.1.4 Composite Beams

By combining several elements, composite structural beams can be produced. Figures 10.6, 10.7, and 10.8 show composite beams made from a variety of elements. Figure 10.6 shows an I-beam made of curved plywood sides and laminated plywood top and bottom. Figure 10.7 shows a comply beam made of a flakeboard center with plywood top and bottom. Figure 10.8 shows beams made of plywood, hardboard, flakeboard and oriented strandboard. Prefabricated I-beams are used by builders because they are lightweight, uniform, and easy to use; have increased dimensional stability; and meet codes and standards.

10.1.5 Wafer- and Flakeboard

Large, thin wafers or smaller flakes can be produced by several methods and used to produce a composite board. Wafers are almost as wide as they are long while flakes are much longer than they are wide. Wafers are also thicker than flakes. Figure 10.9 shows the flakes used to produce a construction grade flakeboard. These are used as the structural skin over wall and floor joists. Wafer- and flakeboard are made with a waterproof adhesive, such as phenol formaldehyde or an isocyanate, and usually have a specific gravity of between 0.6 and 0.8 (Wood Handbook 1999).

FIGURE 10.6 Composite laminated beam.

10.1.6 Particleboard

Wood can be broken down into particles of various size and glued together to produce particleboard. Figure 10.10 shows representative particles and an example of an industrial particleboard. Particleboard has a specific gravity of between 0.6 and 0.8 and is usually produced from softwoods such as Douglas-fir, southern pines or other low-value wood sources (Maloney 1993).

FIGURE 10.7 Comply beam.

Wood Composites

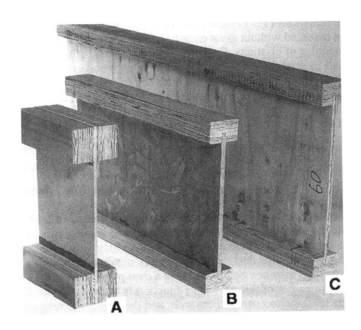

FIGURE 10.8 Composite I beams. A—plywood and hardboard, B—flakeboard and plywood, C—plywood and oriented strandboard.

10.1.7 FIBERBOARD

10.1.7.1 Isolation of Fibers

Wood can be broken down into fiber bundles and single fibers by grinding or refining. In the grinding process, the wood is mechanically broken down into fibers. In the refining process, wood chips are placed between one or two rotating plates in a wet environment and broken down into fibers. If the refining is done at high temperatures, the fibers tend to slip apart as a result of the softening of the lignin matrix between the fibers, and, consequently, the fibers will have a lignin-rich surface. If the refining is done at lower temperatures, the fibers tend to break apart and the surface is rich in carbohydrate polymers. Fiberboards can be formed using a wet-forming or a

FIGURE 10.9 Flakes used to make flakeboard.

FIGURE 10.10 Particles used to make particleboard.

dry-forming process. In a wet-forming process, water is used to distribute the fibers into a mat, which is then pressed into a board. In many cases an adhesive is not used, and the lignin in the fibers serves as the adhesive. In the dry process, fibers from the refiner go through a dryer and a blowline, where the adhesive is applied, and then formed into a web, which is pressed into a board.

10.1.7.2 Low-Density Fiberboard (LDF)

Low-density fiberboards have a specific gravity of between 0.15 and 0.45, and are used for insulation and for light-weight cores for furniture. They are usually produced by a dry process that uses a ground wood fiber.

10.1.7.3 Medium-Density Fiberboard (MDF)

Medium-density fiberboard has a specific gravity of between 0.6 and 0.8 and is mainly used as a core for furniture. Figure 10.11 shows a MDF board with a melamine-paper overlay.

FIGURE 10.11 Thin veneer over medium-density fiberboard.

Wood Composites

FIGURE 10.12 High-density hardboard.

10.1.7.4 High-Density Fiberboard (HDF)

High-density fiberboard has a specific gravity of between 0.85 and 1.2 and is used as an overlay on workbenches and floors, and for siding. It is produced both with and without wax and sizing agents. The wax is added to give the board water resistance. Figure 10.12 shows a few of the different types of high-density fiberboards.

10.1.8 OTHER TYPES OF COMPOSITES

There are a wide variety of composite products that can be made from wood. Figure 10.13 shows a few two- and three-dimensional composites made from a wide variety of wood elements. Sawdust, planer shavings, and bark have been used to produce composite boards. There are also a lot of wood-based composites that are a combination of wood and non-wood elements. Combinations of wood and inorganics, thermoplastics, fiberglass, metals, and other synthetic polymers have been produced; some are commercial, and some are still in the research phase. One commercially used product is a cement-bonded insulation panel for sound damping, which is based on thin wood shavings.

10.1.9 NANOCOMPOSITES

It has been mentioned that the small size of the wood element often has desirable consequences. This is particularly true for fibrous wood elements. The high stiffness and strength of wood is caused by the cellulose reinforcement in the wood cell wall. Thin cellulose micro-fibrils are the major constituent of wood and are aligned close to the axial direction of the fiber. The modulus of

FIGURE 10.13 Wide variety of flat and three-dimensional shaped composites.

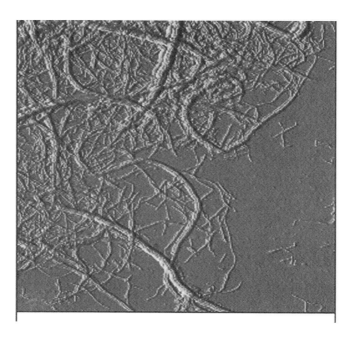

FIGURE 10.14 Atomic force microscopy image of cellulose microfibrils from wood pulp; the horizontal length of the complete image is 5 μm (image by Dr Shannon Notley, Royal Institute of Technology).

these micro-fibrils is in the order of 130 GPa, which is the same as for Kevlar® fibers. A vision for future wood composites is therefore high-performance materials and structures based on stiff and strong cellulose micro-fibrils. Research efforts are focusing on disintegration of very small micro-fibrils from the wood cell wall (see Figure 10.14). Another current research topic is the processing of micro-fibrils and polymers into new composite materials.

10.2 ADHESIVES

Adhesives are thoroughly covered in Chapter 9. In the context of wood composites, adhesive development is driven by adhesive cost-reduction, faster processing time, and specialized products

TABLE 10.1
Typical Choices of Structural Adhesives in Different Service Environments

Service Environment	Adhesive Type
Fully exterior (withstands long-term soaking and drying)	Phenol-formaldehyde Resorcinol-formaldehyde Phenol-resorcinol-formaldehyde Emulsion polymer/isocyanate Melamine-formaldehyde
Limited exterior (withstands short-term water soaking)	Melamine-urea-formaldehyde Isocyanate Epoxy
Interior (withstands short-term high humidity)	Urea-formaldehyde Casein

Source: Data from USDA, 1999.

where complex adhesive formulations are motivated. The basic chemicals most commonly used for wood adhesives and resins are formaldehyde, urea, melamine, phenol, resorcinol, and isocyanate. However, despite the apparent simplicity, in terms of families of chemicals, the formulations are highly complex mixtures of chemicals and additives. Various wood adhesives, along with their typical applications, are listed in Table 10.1.

Although the requirements for cheaper raw materials and reduced press times are the same in Europe and the United States, the emphasis on environmental issues appears to be stronger in Europe. This includes the effect of adhesives on wastewater and on gas emission during panel production. Formaldehyde emission is of significant importance. It is caused by residual un-reacted formaldehyde and by slow adhesive hydrolysis under hot/humid conditions. Modern adhesives show very low formaldehyde emission rates, in compliance with the strict E1 emission class (Pizzi and Mittal, 2003).

10.3 PRODUCTION, PROPERTIES, PERFORMANCE, AND APPLICATIONS

This section first describes laminated timbers (such as glulam) and structural composite lumber. These materials are primarily used in load-bearing building applications in the form of beams. The section then continues with wood composition boards such as plywood, flakeboard, particleboard and fiberboard. Finally, nanocomposites are presented as a concept for new materials with great potential for the future. The wood composition boards can be classified according to density, raw material form, and process type (see Figure 10.15).

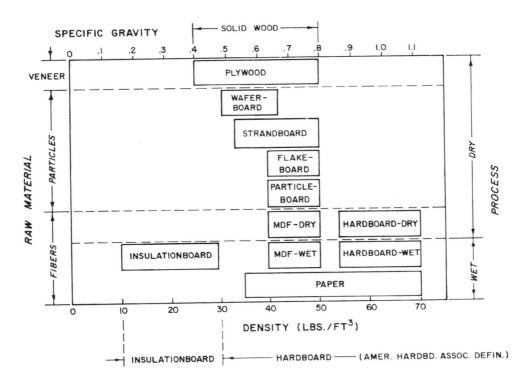

FIGURE 10.15 Classification of wood composite board materials by particle size, density and processing principle (Suchsland and Woodson 1986).

TABLE 10.2
Design Values for Douglas Fir

Design Values	Bending (MPa)	Transverse Compression (MPa)	Shear Stress (MPa)	Modulus of Elasticity (GPa)
Glulam	16.6	4.5	1.1	12.4
Sawn timber	9.3	4.3	0.6	11.0

Source: Adapted from Schniewind 1989.

10.3.1 GLUED LAMINATED TIMBER

Glued laminated timber (glulam) was first used about a hundred years ago with casein as an adhesive system. During the second World War, improved-durability adhesives were developed (e.g., phenol-resorcinol), and glulam was used in aircraft. Later it was also used as framing members for buildings (churches, schools, sports centers, airports, etc.), bridges, truck beds, and marine construction.

Glued laminated timber is currently used for construction members and typically consists of at least four laminations of wood, bonded together with adhesives. The grain direction is parallel to the direction of the construction member. The laminations are typically 45 mm (1.5–2 in.) in thickness and can be end-joined by finger-joints or bonded edge-to-edge to increase the width of the member. In the case of curved beams, laminations are often thinner. The center part of the member may be of lower quality wood. Dried wood is used, typically at 12% moisture content. For indoor applications, the customers thus experience considerable dimensional stability, and the extent of drying cracks is lower as compared to solid wood elements. In addition, the higher strength values for the dry state apply. Selective placement of wood laminations leads to distributed locations of knots and other weak spots. Design values for glulam and sawn timber based on Douglas fir are compared in Table 10.2. The design values for failure stress in bending and in shear are much higher for glulam than for sawn timber.

Phenol resorcinol adhesives are commonly used in the manufacture of glued laminated timber. Melamine-formaldehyde is used where light-colored joints are desired. Very large glulam beams are fabricated industrially. Beams of 60 m in length and 2000 × 215 mm in cross-section have been made. The fire resistance is significant due to the carbonization rate of approximately 40 mm/hr (see Chapter 6).

Glulam members are produced industrially under strict control procedures. The manufacturing requires great care with respect to the milling of finger joints, adhesive preparation and application, pressing conditions, etc. Random extraction of test pieces for strength and durability testing is used to maintain product quality.

Glulam is produced from strength-graded quality timber such as spruce. The timber is dried, planed, and finger-jointed (see Figure 10.16). Adhesive bonding is carried out under pressure. Two different types of bonding are used industrially: high-frequency bonding and hot press bonding.

10.3.2 STRUCTURAL COMPOSITE LUMBER (LVL, PSL, LSL)

Structural composite lumber is often referred to as SCL. Product examples include laminated veneer lumber (LVL), parallel strand lumber (PSL), and laminated strand lumber (LSL). PSL is made from 3-mm thickness veneers cut to 100–300 mm in length and 20 mm in width. Adhesive is applied, and blocks are pressed under high pressure in a continuous process. Beams of desired dimensions are cut from the blocks. LSL is similar to PSL; however, long and slender strands are cut directly from whole logs, in special machines equipped with rotating knives.

LVL is closely related to plywood and is produced in larger quantities than PSL and LSL. Veneers 3 mm in thickness are adhesively bonded together under pressure into thick boards. Typical

Wood Composites

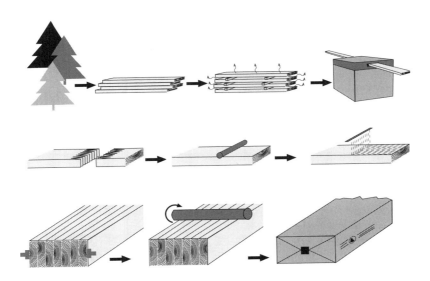

FIGURE 10.16 Sketch of procedure for glulam fabrication. The wood is cut, dried, sorted, fingerjointed, and planed; adhesive is applied; the glulam is pressed and the adhesive cured; final planing is carried out; and the product is packaged (courtesy of the Swedish Glulam Manufacturers Association, www.limtra.se).

thicknesses are 21–75 mm. In contrast with plywood, all veneer is typically oriented in one direction, although certain qualities have some cross-wise oriented layers. Construction members in the form of beams are cut from the LVL-boards. A typical beam width is 45 mm, and beam height is normally 200–400 mm, although up to 900 mm exists. Applications include roof and floor joists, lintels, framework studs, etc. LVL with crosswise veneer is often used for load-bearing panels. There are five main steps in the LVL manufacturing process (in some companies, green or dried veneer is purchased, thereby skipping the first two or three steps):

1. Logs are debarked and then cut to length.
2. Logs are peeled to veneer sheets.
3. Veneer is dried and graded.
4. Veneer sheets are glued, laid up, and pressed.
5. Billets are sawn to required size.

The basic idea behind SCL is similar to that for glulam. Thin layer constituents reduce the probability of large, localized defects. For this reason, average strength is increased and scatter in mechanical properties is reduced. Veneer or strands are dried so that the products have a moisture content of around 12% or lower to provide dimensional stability. An important argument in favor of SCL is that construction elements remain straight after equilibrium moisture conditions have been reached. Mechanical properties of LVL based on Scandinavian pine or spruce are summarized in Table 10.3.

10.3.3 PLYWOOD

Plywood consists of thin veneer wood layers (plies) bonded together by an adhesive into sheet form. Typically, each ply is placed perpendicular to the preceding ply to increase in-plane dimensional stability. The outside plies are called faces, and the inner plies are cores. The core may be lumber, particleboard, or veneer. The panel thickness range is typically 1.6–76 mm. Its history is very old, even Greek and Roman civilizations used plywood-like materials. Hardwood plywood is primarily used for decorative purposes whereas softwood plywood finds applications in the building industry. Competition from oriented strand-board (OSB) is strong. Compared with solid wood, its mechanical

TABLE 10.3
Design Values for LVL (Scandinavian Pine or Spruce) According to Euronorm (EC5)

Calculations of Stress Values	MPa
Bending edgewise	51
Tension parallel to grain	42
Tension transverse to grain	0.6
Compression parallel to grain	42
Compression transverse to grain, edgewise	9
Shear edgewise	6
Calculations of Deformation	
Modulus of elasticity	14000
Shear modulus	960

properties are more isotropic in the plane, and the resistance to splitting is greater. Plywood can cover large areas with a minimum amount of wood since thinner sections are possible. Plywood properties depend on the adhesive used (durability), ply stacking, and quality of veneer. Since plywood is often subjected to bending, the face and back plies are particularly critical to performance.

After debarking, logs are often prepared for the veneer cutting procedure by softening in hot water. The veneer is produced from rotary cutting of blocks on a lathe. The veneer of highest quality is produced initially, then the extent of knots increases as the block diameter decreases. The veneer side pressed against the knife is the tight side, whereas lathe checks occur on the knife side creating a "loose" side. The veneer goes into a dryer to reduce the moisture content and produce flat and pliable veneer. Dried veneer is graded and stacked according to width and grade. Grade A veneers have the fewest defects and Grade D the most. Phenol-formaldehyde adhesives dominate the plywood industry although urea and melamine adhesives are also used. The adhesive is sprayed, roller-coated, curtain-coated, or foamed. Today, the plies are laid up by machine, although core materials may be placed by hand. Cold pressing is often used to flatten the sheet. Hot pressing takes place in multi-opening presses. A typical pressure is 1.2–1.4 MPa and the temperature is 100–165°C, depending on the adhesive. Complete curing is essential to ensure chemical stability and low emissions. Grading of the finished panels then takes place. Typical properties for sheathing-grade plywood are presented in Table 10.4.

TABLE 10.4
Property Values for Sheathing-Grade Plywood

Property	Value
Linear hygroscopic expansion (30%–90%RH)	0.15%
Linear thermal expansion	6.1×10^{-6} m/mK (3.4×10^{-6} in/in°F)
Flexure	
Strength	21–48 MPa (3000–7000 lb/in^2)
Modulus of elasticity	6.9–13 GPa (1–1.9 $\times 10^6$ lb/in^2)
Tensile strength	10–28 MPa (1500–4000 lb/in^2)
Compressive strength	21–35 MPa (3000–5000 lb/in^2)
Edgewise shear	
Shear strength	4.1–7.6 MPa (600–1100 lb/in^2)
Shear modulus	470–760 MPa (68–110 $\times 10^3$ lb/in^2)

Source: Adapted from Youngquist, 1999.

Wood Composites

In principle, the lay-up of plywood (grain direction and stacking sequence of veneer layers) can be tailored for specific applications. If the requirements, in terms of maximum deformation due to moisture and mechanical loads, are known, the plywood lay-up can be optimized. Lay-up optimization can be performed by the use of laminate plate theory, which has been described in the context of wood composites (Bodig and Jayne 1993).

10.3.4 Particleboard

Particleboard is a result of the need to utilize large quantities of sawdust at sawmills. It is primarily used in panel form, although it is possible to produce I-beams, corrugations or even compression–molded, three-dimensional objects. Wood particles are bonded using synthetic adhesives and pressed into sheets. Although the mechanical performance is limited as a result of the inherent weakness of materials composed of particles, high performance flakeboard materials based on oriented fibrous strands were developed from the particleboard concept.

Typically, particleboard consists of a lower density core of coarse particles and outer, higher-density layers of finer particles. This distribution of density and particle size is important with respect to board performance. Many applications involve bending loads, where a high-density skin and a low-density core are advantageous. The particleboard panel functions as a sandwich structure and the ratio of bending stiffness to weight becomes high. Particleboard is mainly used for furniture, in flooring, and as panels.

In a modern particleboard plant, production is by continuous pressing (see outline in Figure 10.17). Such a line may have an annual production capacity of 400,000 m^3. The raw material is typically sawdust from sawmills, although sawmill chips may also be ingredients. Saw dust is used increasingly, and recycled wood is in some countries becoming more common. Where roundwood is used, mobile chippers are convenient to use. Hammer-milling ensures proper particle size and can also be used to process over-sized particles from screening. Oscillating screens allow dry particle screening into core and surface fractions and oversized particles. Drying is carried out in a single-pass drum dryer, where the temperature of the drying gases declines with decreasing moisture content of particles. Flakes are then gravity-fed to resin blenders from dosing bins. Urea-formaldehyde resins are most common, although melamine enhanced resins are often used in applications where exposure to moisture is common.

The forming unit is a key machine, since the structure (and cost) of the final board can be controlled at this stage. Proportions between surface and core flakes can be adjusted. Also, the amount of fine particles in the core can be increased to improve internal bond strength (out-of-plane tensile strength). A conveyor brings the flakes from the forming to the prepress unit. A fine spray of water is added to facilitate heat transfer in the pressing operation. A gentle initial press is applied to the loose fiber mat. The prepressing operation provides increased density and strength and reduces spring-back.

Modern continuous presses can have a length of 35 m and are equipped with cooling zones. This reduces temperature and steam pressure of entrapped moisture, provides a more even moisture distribution, and reduces the risk for board blisters. The final moisture content can be higher; a typical value is 7–8%. This is close to the equilibrium value for many applications, and the risk for warping is thus reduced. Board thicknesses are typically 3–38 mm.

10.3.5 Flakeboard

Flakeboard is a term that also includes waferboard (WB) and oriented strandboard (OSB). They are structural panels produced from wafers obtained from logs. The first waferboard plant was opened in 1963 by MacMillan Bloedel in Saskatchewan, Canada. Aspen was the raw material, and the wafers were randomly oriented. In the late 1980s, most wafers were oriented, resulting in oriented waferboard (OWB). Long and narrow strands are now used in OSB, which typically have 3 or 5 layers. The orientation distribution may be tailored to the application. OWB and OSB compete

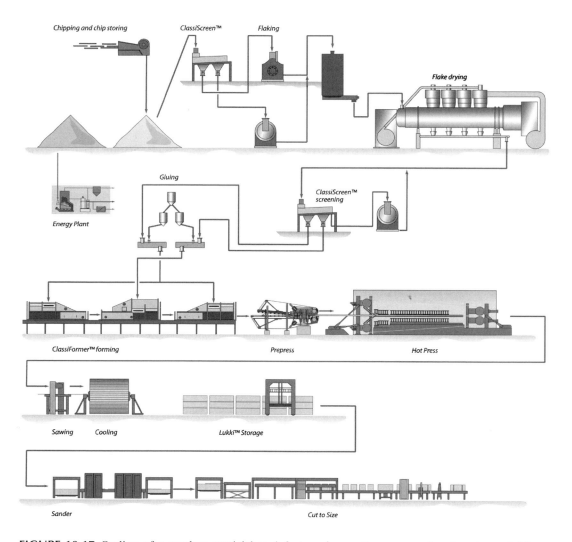

FIGURE 10.17 Outline of a modern particleboard factory for continuous pressing (courtesy of Metso Panelboard).

with plywood in applications such as single-layer flooring, sheathing, and underlayment in lightweight structures. Pines, firs and spruce are used, as well as aspen.

Manufacturing begins with logs, which are debarked and cut to 2.5-m lengths. The waferizer slices the logs into wafers typically 38 mm wide by 76 to 150 mm long by 7 mm thick. The wafers are dried in rotary drum dryers to moisture levels below 10 percent. Screening and separation takes place, and the wafers are stored. Resin, wax, and additives are mechanically mixed in a blender. Phenol-formaldehyde and isocyanates dominate as resins. In the forming step, the wafers are metered out on a moving screen system. In WB production, wafers fall randomly whereas in OSB production, wafers are mechanically oriented in one direction. Forming heads can form distinct layers oriented perpendicularly with respect to each other. The mat is trimmed and sent to either a multi-opening press or a continuous belt press. Curing takes place at elevated temperatures. In some newly developed processes, boards are cooled prior to trimming and packaging.

Typical properties of OSB are presented in Table 10.5. Plywood and OSB are competing materials for wood structural panels, a term used in the building codes. The structural performance

TABLE 10.5
Property Values for Sheathing-Grade Oriented Strandboard

Property	Value
Linear hygroscopic expansion (30%–90%RH)	0.15%
Linear thermal expansion	6.1×10^{-6} m/mK (3.4×10^{-6} in/in°F)
Flexure	
Strength	21–28 MPa (3000–4000 lb/in²)
Modulus of elasticity	4.8–8.3 GPa (0.7–1.2 × 10⁶ lb/in²)
Tensile strength	6.9–10.3 MPa (1000–1500 lb/in²)
Compressive strength	10–17 MPa (1500–2500 lb/in²)
Edgewise shear	
Shear strength	6.9–10.3 MPa (1000–1500 lb/in²)
Shear modulus	1.2–2.0 GPa (180–290 × 10³ lb/in²)

Source: Data from Forest Products Lab at www.fpl.fs.fed.us.

of the two materials is equal in general terms. They share the same exposure durability classifications, performance standards, and span ratings. Both are applied on roofs, walls, and flooring.

One limitation with OSB is edge swelling. Although the edges are coated during transport, subsequent cutting causes limitations in high humidity applications. One advantage, however, is that OSB is more consistent than plywood. Soft spots caused by overlapping knots do not exist, nor do knot holes at edges, and delamination does not take place. Since one OSB sheet may consist of 50 strands, properties are homogeneous, and there are only minor stiffness variations with location in the panel. Through-thickness shear strength is approximately twice as high with OSB. OSB is increasing its market share as a result of its lower cost as compared with plywood. Since OSB is a newer product than OSB, there is also a steady improvement in processing methods, quality control, and performance.

10.3.6 FIBERBOARD

Fiberboards are based on wood or other lignocellulosic fibers held together by an adhesive bond, either by using the inherent adhesive properties of the wood polymers or by adding an adhesive. The boards are in the form of sheet materials, typically less than 25 mm in thickness. Two steps are required in the manufacture of fiberboard. The first is the disintegration of larger wood elements into fibers, and the second is the formation of a sheet or board structure. The disintegration step essentially consists of two major substeps: the reduction of logs to chips and then the conversion of chips to pulp fibers in a refiner or defibrator. Chips are produced in chippers, such as rotating disk chippers, equipped with highly specialized knives which determine the geometry of the chips.

The pulping step is a mechanical process that takes place at elevated temperatures. Increased temperature is critical since it significantly reduces power consumption. There are three basic methods, the Masonite steam explosion process, and the atmospheric or pressurized disk refining processes. Although the board production can take place in both wet and dry processes, environmental issues warrant the dry process. A simple sketch of wet and dry fiberboard process steps is shown in Figure 10.18.

In the wet process, a dilute water suspension of fibers is used to form the mat on Fourdrinier machines (Low-Density Fiberboard or softboard). In the dry process, air-felting machines are used to form the board. Chemical additives, such as binders for adhesion or sizing for reduced water absorption, are, in the wet process, added to the water suspension and precipitated onto the fibers by pH-reduction. In the dry process, spraying and/or mechanical mixing is used for the dispersion of adhesives and additives.

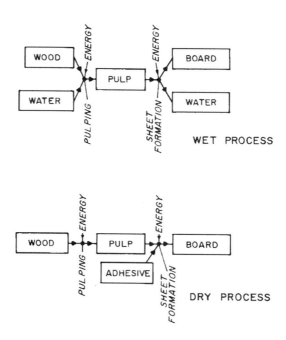

FIGURE 10.18 Sketch of principal steps in wet and dry fiberboard processes (Suchsland and Woodson 1986).

Low-density fiberboards are dried from the wet state, and the integrity of the board is ensured by fiber-fiber bonding caused primarily by hydrogen bonding. Hot pressing is used for the other types of fiberboards. During pressing, the water content is reduced, the mat is densified and fiber-fiber bonding develops by binder curing or solidification of plasticized lignin. With S1S boards (smooth on one side), a screen is placed between the mat bottom and the press in order to facilitate removal of water and steam. The screen pattern is embossed on the finished board. S2S boards (smooth on two sides) are produced from dry-formed mats or dried wet-formed mats.

10.3.6.1 Low-Density Fiberboard (Insulation Board)

Low-density fiberboards originate from the need to utilize paper byproducts. Oversize fiber bundles (screenings) were removed from groundwood pulp and used in board production about a hundred years ago. Agricultural byproducts are also used (sisal, flax, bagasse). A typical density is around 300 kg/m^3. The major application areas are as insulating layers in exterior products (sheathing, roof insulation), interior products (building boards, ceiling tiles, and sound absorption boards) and industrial products (mobile homes, and in the automotive and furniture industries).

Waxes may be added to low-density fiberboards to improve water resistance. Strength is often somewhat reduced with wax addition as a result of weakened fiber-fiber bonding. For the same purpose, asphalt is used in structural insulation board. Asphalt somewhat increases strength but obviously darkens the board color. In the wet process, wax or asphalt is added in emulsion form and is precipitated using alum. Starch can be added to increase strength but also tends to attract rodents.

The typical properties of low-density insulation boards are presented in Table 10.6. In-plane tensile strength is typically 10% of hardboard values, and out-of plane compression strength may be only 1% of hardboard values, because of the low density. Mechanical properties are often not critical for insulation boards. However, insulation sheathing may require a tensile strength of 2 MPa. Typical values for thermal conductivity are in the range 0.055 to 0.069 Wm^{-1}K^{-1}.

TABLE 10.6
Property Values for Low-Density Fiberboard (Insulation-Board)

Property	Values
Density	0.17–0.28 g/cm^3
Flexural strength	1.0–2.7 MPa (142–384 lb/in^2)
Modulus of elasticity (flexure)	80–400 MPa (11400–56900 lb/in^2)
Tensile strength parallel to surface	0.5–1.6 MPa (71–228 lb/in^2)
Out-of-plane tensile strength	0.07–0.17 MPa (10–26 lb/in^2)
Water absorption (24-hr immersion at 20°C)	30–100 w/o
Maximum thickness swelling	12–20%
Linear expansion, from 50–97% RH, 20°C	0.5%
Sound absorption of acoustical board (522 Hz)	50–85%

Source: Adapted from Kollman, 1975.

10.3.6.2 Medium-Density Fiberboard (MDF)

Today, MDF board production is predominantly by the dry process. In many countries, the MDF board market is expanding at the expense of particleboard and high-density fiberboard. Fibers are primarily produced in a pressurized disk refiner. Typical conditions are a pressure of 7–8 bars and a temperature of 170°C. The objective is to disintegrate the chips into individual fibers. The adhesives technology is based on a lignin-rich fiber surface obtained by fiber separation at the middle lamellae. After the refiner step, the furnish is dried to around 4% moisture content. Typically, a urea-formaldehyde adhesive in water solution is then mechanically blended with the fibers. Another alternative, used commercially, is to mix the adhesive with fibers in the transportation line leading from the refiner. A small amount of wax (0.25–1.5%) is often used to improve water resistance. The furnish is then formed into a loose mat, which may be as thick as 300 mm, depending on the final thickness of the board. Recently, continuous pressing has been increasingly used because of the higher productivity as compared with multi-opening presses. Typical press temperatures are in the range 160–210°C.

MDF boards show a significant density gradient through their thickness. The modulus of elasticity in bending correlates strongly with surface layer density (Suchsland et al 1979). Table 10.7 shows property requirements of MDF for interior use. In Table 10.8, the range of data found in eight commercial fiberboards is presented. Considerable variation is apparent. Note also that the density range for MDF extends down toward 500 kg/m^2.

In Europe, MDF tends to be used in indoor applications. In the United States, outdoor applications are also numerous. PF adhesives are often used when requirements for moisture resistance are higher. Another alternative, intermediate in moisture stability, is hybrid adhesives based on urea,

TABLE 10.7
Requirements for Medium-Density Fiberboard for Interior Use

Thickness (in)	Flexural Strength (MPa)	Modulus of Elasticity (MPa)	Internal Bond (Out-of-Plane Strength) (MPa)	Linear Expansion (%)
13/16 and below	20	2000	0.62	0.3
7/8 and above	19	1720	0.55	0.3

Source: Adapted from Schniewind, 1989.

TABLE 10.8
Property Data for Eight Commercial Medium-Density Fiberboards

Mill No	Density	Flexural Strength	Modulus of Elasticity	Internal Bond (Out-of-Plane Strength)	Residual Thickness Swelling
1	0.73 (g/cm^3)	4.84 × 10^3 lb/in^2	466 × 10^3 lb/in^2	125 × 10^3 lb/in^2	4.36%
2	0.90	4.93	576	136	1.61
3	0.79	3.37	432	282	2.48
4	0.82	5.70	635	121	2.83
5	0.95	3.57	517	133	5.45
6	0.80	5.28	578	103	4.52
7	0.77	5.42	572	179	3.18
8	0.71	5.11	858	158	3.03

Source: Adapted from Suchsland, 1979.

melamine and formaldehyde. MDF applications include furniture, cupboards, doors, flooring etc. MDF boards for indoor use are often painted and light fiber color is desirable, with correspondingly low press temperatures. Higher post-treatment temperatures improve moisture stability, due to hemicellulose degradation, but fiber and board colors become darker.

10.3.6.3 High-Density Fiberboard (HDF, Hardboard)

Wet processes for fiberboard production are decreasing in importance because of environmental issues. Instead, many HDF fiberboard materials are produced from the dry MDF process. The board density is higher, typically above 800–900 kg/m^3. Applications include exterior siding, interior wall paneling, household and commercial furniture, and industrial and commercial products. Hardboard is increasingly used in the expanding laminate flooring market where a hardboard sheet is used as a backing for the laminate top layer.

Hardboard dimensional and quality requirements are described in ANSI Standards ANSI/AHA A135.4-1995 Basic Hardboard, ANSI/AHA A135.5-1995 Prefinished Hardboard Panelling, and ANSI/AHA A135.6-1998 Hardboard Siding. The Basic Hardboard standard uses thickness and physical properties for classification. The objective is to facilitate the selection of the product for the customer. The other two standards focus on two particular applications, paneling and siding. The products are classified according to specific properties.

Hardboard properties vary considerably for different types of boards, both in average value and in the distribution of data. Important parameters are raw materials, tempering, additives, and density. Internal bond (strength perpendicular to the surface) correlates strongly with specific gravity, see Figure 10.19. Higher density also increases modulus (Suchsland and Woodson 1986). The correlation between density and modulus is not so strong, probably due to the through-thickness distribution of density. In a bending test, the apparent modulus depends strongly on density distribution rather than on average density. Typical property data for hardboards are presented in Table 10.9.

10.3.7 NANOCOMPOSITES

An important principle of engineered wood products is to reduce variability in the material structure by the creation of new materials where particles, fibers, flakes, strands, veneer or thin boards are adhesively bonded. The probability for concentrated regions of knots, cracks, oriented spiral grain, compression wood, juvenile wood and other defects is therefore very small. The smaller the repeating wood element is, the better the homogeneity and the smaller the variability in properties. Porosity and wood element shape complicates the picture, but as a general principle the statement is valid.

Wood Composites

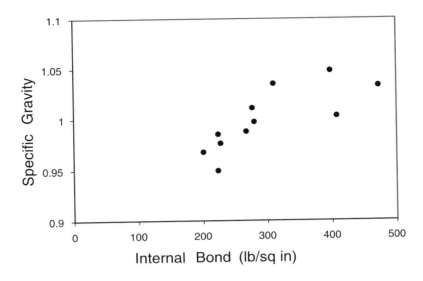

FIGURE 10.19 Effect of density on internal bond strength of fiberboard (data from Suchsland and Woodson 1986).

The smallest load-bearing element in wood is the cellulose fibril. Wood derives its strength and stiffness from cellulose. Lignin provides moisture stability, and hemicelluloses bond cellulose fibrils together, but the intrinsic mechanical performance comes from the cellulose fibrils. They are strongly aligned at an angle close to the axial direction of the wood cells. The Young's modulus of the cellulose unit cell is 134 GPa, a value similar to that of Kevlar® fibers. The reason for this high value is that the cellulose chains are densely packed, and, as they are deformed, strong covalent bonds of high energy take the load.

Cellulose microfibrils of 30 nm in diameter and a few microns in length can be disintegrated from wood pulp fibers using a mechanical milling procedure. It is also possible to take pulp fibers, subject them to acid hydrolysis, and then disintegration will take place in conventional mixing equipment used in plastics processing. The performance of injection molded cellulose fiber thermoplastics can be significantly improved by the use of microfibrils (Boldizar et al. 1987). Rubber polyurethanes have been reinforced with cellulose microfibrils, and strength, as well as strain to failure, increased substantially. The reason is the inherently high performance of the

TABLE 10.9
Property Data for Hardboard

Class	Thickness (in)	Water Absorption (w/o)	Thickness Swelling (%)	Flexural Strength (MPa)	Tensile Strength (MPa)	Internal Bond (out-of-Plane Strength) (MPa)
Tempered	1/8	25	20			
	3/16			41	21	0.9
	3/8	10	9			
Standard	1/8	35	25			
	3/16			31	15	0.6
	3/8	15	10			

Source: Adapted from Schniewind, 1989.

cellulose but also the small scale of the reinforcement. As with other wood composites, mechanisms of failure, such as microcracking and debonding, are delayed to higher stresses and strains. The reason is that small-scale reinforcements also cause damage at a very small scale, which is less detrimental to material performance. Cellulose microfibrils also show interesting properties as a material of their own. If a water suspension of cellulose microfibrils is dried, it forms a hard and tough material, similar to ivory in character. The reason is millions of strong hydrogen bonds forming between the cellulose entities.

The ultimate wood composite would have a high content of cellulose microfibrils oriented in the main direction of loading. Strength and stiffness would be competing with high performance composites used in the aerospace industry. Lightweight sandwich structures could be produced by introducing porosity, for instance through foam cores where the material primarily consists of cellulose.

The wood nanocomposite scheme presented by Yano et al. (2001) is an example of the potential. Veneer layers are subjected to chemical treatment so that significant parts of lignins and hemicelluloses are removed. The veneer is then impregnated with phenol-formaldehyde, compressed, and cured. The resulting wood composite has a Young's modulus of 40 GPa. This is more than twice the value for any commercial wood material or wood composite.

10.4 CONCLUSIONS

Wood composites constitute a wide variety of materials, such as glulam and laminated veneer lumber used in beams, as well as board materials such as particleboard, fiberboard and oriented strand board. Waste-wood and smaller trees can be used in wood composites, strength is increased since defects are removed, and structural shape can be designed. The scale of the wood element, its orientation, and material density are important factors controlling performance.

In laminated timber, plywood, and structural composite lumber, the wood element is cut, for instance by rotary cutting in the case of veneer. Adhesives are applied and the material is pressed and cured. Large standardized elements and continuous production can provide large-scale advantages. Dimensional stability and small variability in properties are selling points in the competition with other materials. Properties may be tailored by the choice of adhesive and, for example, by the design of grain direction in veneer lay-up.

In wafer- and flakeboard, particleboard, and fiberboard, economy depends strongly on the cost of raw materials and their disintegration. Sawdust and recycled wood are therefore increasingly used. Larger scale processing is also a trend, as well as dry processing. Board materials have layered structure. Particleboard and fiberboard typically have higher density in surface layers. This produces an advantageous sandwich structure so that mechanical performance per weight is improved. In addition, wood elements may be oriented, such as in oriented strandboard, so that performance is improved in given directions. The adhesive is an important and sometimes costly constituent, which often controls moisture sensitivity and ultimate properties.

For future developments, there is a need for new methods of breaking wood down into uniform furnish. Such processes must consume less energy, provide high yields, allow mixed species, and give improved performance or reduced cost. In this context, biotechnology is of interest, since enzymatic degradation takes place in water and at ambient temperature.

Optimization of structural shape, in combination with variations in material and material compositions, has great potential for components in building systems. I-beams with different materials in the flange and in the web are simple examples already in existance, although the concept may be brought much further. For example, lightweight structures subjected to bending can be produced through the use of sandwich structures with low-density cores and high-density skins. Such structures are of interest in a more industrialized production of housing, where large sections of the building are prefabricated in factories.

REFERENCES

Bodig, J. and Jayne, B.A. (1993). *Mechanics of Wood and Wood Composites*. Krieger Publishing Co., Melbourne, FL.

Boldizar A., Klason C., Kubat J., Näslund P., and Saha P. (1987). Prehydrolyzed cellulose as a reinforcing filler for thermoplastics. *Int. J. Polym. Mater.* (11):229.

Pizzi, A. and Mittal K.L. (2003). *Handbook of Adhesives Technology* (2nd ed.). Marcel Dekker, New York.

Kollman, F.F. (1975). *Principles of Wood Science and Technology: Wood Base Materials Manufacture and Properties*. Springer-Verlag, New York.

Maloney, T.M. (1996). The family of wood composite materials. *For. Prod. J.* 46(2):19–26.

Maloney, T.M. (1993). *Modern Particleboard and Dry-Processed Fiberboard Manufacture*. Miller-Freeman, Inc., San Francisco, CA.

Marra, G. (1979). Overview of wood as a material. *J. Educ. Modules for Mater. Sci. Eng.* 1(4):699–710.

Schniewind, A.P. (1989). *Concise Encyclopedia of Wood and Wood-Based Materials*. The MIT Press, Cambridge, MA.

Suchsland, O. and Woodson, G.E. (1986). Fiberboard manufacturing practices in the United States. *Agricultural Handbook 640*. Department of Agriculture. Washington, D.C.

Suchsland, O., Lyon, D.E., and Short, P.E. (1979). Selected properties of commercial medium-density fiberboards. *Fo. Prod. J.* 29(9):45–49.

USDA, Forest Service. (1999). Wood as an engineering material. General Technical Report FPL-GTR-113. *Wood Handbook*. USDA, Forest Service, Forest Products Laboratory.

Yano H., Hirose A., Collins P.J., and Yazaki Y. (2001). Effects of the removal of matrix substances as a pretreatment in the production of high strength resin impregnated wood based materials. *J. Mater. Sci. Lett.* (20):1125.

Youngquist, J.A. (1999). Wood-based composites and panel products. Chap. 10 in USDA, Forest Service. Wood as an engineering material. General Technical Report FPL-GTR-113. *Wood Handbook*. USDA, Forest Service, Forest Products Laboratory.

11 Chemistry of Wood Strength

Jerrold E. Winandy[1,2] *and Roger M. Rowell*[1,3]
[1]USDA, Forest Service, Forest Products Laboratory, Madison, WI
[2]Department of Bio-Based Products, University of Minnesota, St. Paul, MN
[3]Department of Biological Systems Engineering, University of Wisconsin, Madison, WI

CONTENTS

11.1 Mechanical Properties 305
11.2 Factors Affecting Strength 307
    11.2.1 Material Factors 307
        11.2.1.1 Specific Gravity 307
        11.2.1.2 Growth Characteristics 307
    11.2.2 Environmental Factors 308
        11.2.2.1 Moisture 308
        11.2.2.2 Temperature 309
    11.2.3 Load Factors 311
        11.2.3.1 Duration of Load 311
        11.2.3.2 Fatigue 311
        11.2.3.3 Mechanical Properties 312
    11.2.4 Flexural Loading Properties 312
        11.2.4.1 Modulus of Rupture 312
        11.2.4.2 Fiber Stress at Proportional Limit 313
        11.2.4.3 Modulus of Elasticity 313
        11.2.4.4 Work to Proportional Limit 314
        11.2.4.5 Work to Maximum Load 314
    11.2.5 Axial Loading Properties 314
        11.2.5.1 Compression Parallel to the Grain 314
        11.2.5.2 Compression Perpendicular to the Grain 314
        11.2.5.3 Tension Parallel to the Grain 315
        11.2.5.4 Tension Perpendicular to Grain 315
    11.2.6 Other Mechanical Properties 316
        11.2.6.1 Shear 316
        11.2.6.2 Hardness 316
        11.2.6.3 Shock Resistance 316
11.3 Chemical Components of Strength 316
    11.3.1 Relationship of Structure to Chemical Composition 316
        11.3.1.1 Macroscopic Level 317
        11.3.1.2 Microscopic Level 317
        11.3.1.3 Composition 317
        11.3.1.4 Microfibril Orientation 319
        11.3.1.5 Molecular Level 319
11.4 Relationship of Chemical Composition to Strength 322

	11.4.1	Below Proportional Limit (Elastic Strength)	322
	11.4.2	Beyond Proportional Limit (Plastic Strength)	325
11.5	Relationship of Structure to Strength		325
	11.5.1	Molecular Level	325
	11.5.2	Microscopic Level	326
	11.5.3	Macroscopic Level	326
11.6	Environmental Effects		326
	11.6.1	Acids and Bases	328
	11.6.2	Adsorption of Elements	328
	11.6.3	Swelling Solvents	329
	11.6.4	Ultra Violet Degradation	329
	11.6.5	Thermal Degradation	330
	11.6.6	Microbial Degradation	332
	11.6.7	Naturally Occurring Chemicals	336
11.7	Treatment Effects		337
11.8	Summary		343
References			343

The source of strength in solid wood is the wood fiber. Wood is basically a series of tubular fibers or cells cemented together. Each fiber wall is composed of various quantities of three polymers: cellulose, hemicelluloses, and lignin. Cellulose is the strongest polymer in wood and, thus, is highly responsible for strength in the wood fiber because of its high degree of polymerization and linear orientation. The hemicelluloses act as a matrix for the cellulose and increase the packing density of the cell wall; hemicelluloses and lignin are also closely associated. The actual role of hemicelluloses in wood strength has recently been shown to be far more critical toward the overall engineering performance of wood than had previously been assumed. We suspect the primary role of hemicelluloses is to act as a highly specific coupling agent capable of associating both with the more random areas (i.e., noncrystalline) of hydrophilic cellulose and the more amorphous hydrophobic lignin. Lignin not only holds wood fibers themselves together but also helps bind carbohydrate molecules together within the cell wall of the wood fiber. The chemical components of wood that are responsible for mechanical properties can be viewed from three levels: macroscopic (cellular), microscopic (cell wall), and molecular (polymeric) (Winandy and Rowell 1984). Mechanical properties change with changes in the thermal, chemical, and/or biochemical environment. Changes in temperature, pressure, humidity, pH, chemical adsorption from the environment, UV radiation, fire, or biological degradation can have significant effects on the strength of wood.

Cellulose has long been thought to be primarily responsible for strength in the wood fiber because of its high degree of polymerization and linear orientation. Hemicellulose may act as a link between the fibrous cellulose and the amorphous lignin. Hemicellulose definitely acts as a matrix for the cellulose and to increase the packing density of the cell wall. Lignin, a phenolic compound, not only holds the fibers together but also acts as a stiffening agent for the cellulose molecules within the fiber cell wall. All three cell wall components contribute in different degrees to the strength of wood. Together the tubular structure and the polymeric construction are responsible for most of the physical and chemical properties exhibited by wood.

The strength of wood can be altered by environmental agents. The changes in pH, moisture, and temperature; the influence of decay, fire, and UV radiation; and the adsorption of chemicals from the environment can have a significant effect on strength properties. Environmentally induced changes must be considered in any discussion on the strength of treated or untreated wood. This susceptibility of wood to strength loss, and the magnitude of that degrade, is directly related to the severity of its thermal/chemical/biochemical exposure.

The strength of wood can also be altered by preservative and fire-retardant compounds used to prevent environmental degradation. In some cases, the loss in mechanical properties caused by these treatments may be large enough that the treated material can no longer be considered the same as the untreated material. The treated wood may now resist environmental degradation but may be structurally inferior to the untreated material. A long-term study of this problem has helped engineers account for these potential alterations from untreated wood in the structural design process. The approach is based on a cumulative-damage approach relating thermal and chemical degradation of the polymers responsible for wood strength to kinetic- or mechanical-based models (Lebow and Winandy 1999a, Winandy and Lebow 2001). With preservative-treated wood, a large amount of work was undertaken in the late 1980s and early 1990s to address treatment-related concerns in the structural design process. This work developed an understanding of the thermochemical issues that primarily control preservative-related strength loss (Winandy 1996a) and was then used to limit treatment-processing levels in standards, especially post-treatment kiln-drying temperatures (Winandy 1996b). With fire-retardant treated (FRT) wood, a large amount of work has recently addressed past problems with in-service thermal degradation of FRT wood exposed to elevated in-service temperatures and that work has been summarized (Winandy 2001).

This chapter presents a theoretical model to explain the relationship between the mechanical properties and the chemical components of wood. This model is then used to describe the effects of altered composition on those mechanical properties. Many of the theories presented are only partially proven and just beginning to be understood. These theories should be considered as a starting point for dialogue between chemists and engineers that will eventually lead to a better understanding of the chemistry of wood strength.

11.1 MECHANICAL PROPERTIES

Even wood that has no discernible defects has extremely variable properties as a result of its heterogeneous composition and natural growth patterns. Wood is an anisotropic material in that the mechanical properties vary with respect to the three mutually perpendicular axes of the material (radial, tangential, and longitudinal). These natural characteristics are compounded further by the environmental influences encountered during the growth of the living tree. Yet wood is a viable construction material because workable estimates of the mechanical properties have been developed.

Mechanical properties relate a material's resistance to imposed loads (i.e., forces). Mechanical properties include the following: (1) measures of resistance to deformations and distortions (elastic properties), (2) measures of failure-related (strength) properties, and (3) measures of other performance-related issues. To preface any discussion concerning mechanical properties, two concepts need to be explained: stress (σ) and strain (ε).

Stress is a measure of the internal forces exerted in a material as a result of an application of an external force (i.e., load). Three types of primary stress exist: tensile stress, which pulls or elongates an object (Figure 11.1a); compressive stress, which pushes or compresses an object (Figure 11.1b); and shear stress, which causes two contiguous segments (i.e., internal planes) of a body to rotate (i.e., slide) within the object (Figure 11.1c). Bending stress (Figure 11.1d) is a combination of all three of the primary stresses and causes rotational distortion or flexure in an object.

Strain is the measure of a material's ability to deform—that is, elongate, compress, or rotate—while under stress. Over the elastic range of a material, stress and strain are related to each other in a linear manner. In elastic materials, a unit of stress (σ) will cause a corresponding unit of strain (ε). This elastic theory yields one of the most critical engineering properties of a material, the elastic modulus (E). The theory is commonly known as Hooke's law (Larson and Cox 1938):

$$E = \sigma/\varepsilon \qquad (11.1)$$

It applies to all elastic materials at points below their elastic or proportional limits.

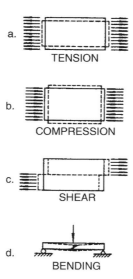

FIGURE 11.1 Examples of the three axil and one flexural types of stress.

Elastic theory relates a material's ability to be deformed by a stress to its ability to regain its original dimensions when the stress is removed. The criterion for elasticity is not the amount of deformation but the ability of a material to completely regain its original dimensions when the stress is removed. The opposite quality is viscosity, which can also be thought of as plasticity. A perfectly plastic body is one that makes no recovery of its original dimensions upon the removal of a stress. Wood is not ideally elastic; it will not completely recover deformation immediately on unloading, but in time, residual deformations tend to be recoverable. Wood is considered a viscoelastic material. This viscoelasticity explains the creep phenomenon in which a given load will induce an immediate deformation, and if that load is allowed to remain on that piece, additional secondary deformation (i.e., creep) will continue to occur over long time periods. However, for simplicity's sake and because the engineering community often also assumes such, wood will be considered as an elastic material in this chapter.

The two main elastic moduli are modulus of elasticity, which describes the relationship of load (stress) to axial deformation (strain), and modulus of rigidity or shear modulus, which describes the internal distribution of shearing stress to shear strain or, more precisely, angular (i.e., rotational) displacement within a material.

Strength values are numerical estimates of the material's ultimate ability to resist applied forces. The major strength properties are limit values for the stress–strain relationship within a material. Strength, in these terms, is the quality that determines the greatest unit stress a material can withstand without fracture or excessive distortion. In many cases the unqualified term strength is somewhat vague. It is sometimes more useful to think of specific strengths, such as compressive, tensile, shear, or ultimate bending strengths.

The American Society for Testing and Materials (ASTM) is the ISO-accredited organization in North America that standardizes testing procedures to provide reliable and universally comparable estimates of wood strength. Several ASTM Standards for wood (see ASTM) outline procedures for determining basic mechanical properties and deriving allowable design stresses. In performing a test, a load is applied to a specimen in a particular manner and the resulting deformation is monitored. The load information allows the internal forces within the specimen (stress) to be calculated. The deformation information allows the internal distortion (strain) to be calculated when accepting specific assumptions. When stress and strain are plotted against each other on a graph, a stress–strain diagram is developed (Figure 11.2).

Chemistry of Wood Strength

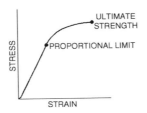

FIGURE 11.2 A typical stress–strain diagram for wood.

The unit stress corresponding to the upper limit of the linear segment of the stress–strain diagram is known as the proportional limit (Figure 11.2, Point A). This proportional or elastic limit measures the boundary of a material's completely recoverable strength. At stress levels below the proportional limit, a perfectly elastic material will regain its original dimensions and form. At stress levels in excess of the proportional limit, an elastic material will not regain its original shape; it will be permanently distorted.

The unit stress represented by the maximum ordinate is the ultimate (maximum) strength (Figure 11.2, Point B). This point estimates the maximum stress at the time of failure. Many of the mechanical properties of interest to the engineer, such as maximum crushing strength or ultimate bending strength, describe this point of maximum stress.

11.2 FACTORS AFFECTING STRENGTH

11.2.1 Material Factors

11.2.1.1 Specific Gravity

Specific gravity is the ratio of the weight of a given volume of wood to that of an equal volume of water. As specific gravity increases, strength properties increase (USDA 1999) because internal stresses are distributed among more molecular material. Mathematical approximations of the relationship between specific gravity and various mechanical properties are shown in Table 11.1.

11.2.1.2 Growth Characteristics

As a fibrous product from living trees, wood is subjected to many environmental influences as it is formed and during its lifetime. These environmental influences can increase the variability of the wood material and, thus, increase the variability of the mechanical properties. To reduce the effect of this inherent variability, standardized testing procedures using small, clear specimens of wood are often used. Small, clear specimens do not have knots, checks, splits, or reaction wood. However, the wood products used and of economic importance in the real world have these defects. Strength estimates derived from small clear specimens are reported because most chemical treatment data have been generated from small clear specimens. Further, comparative analyses of chemical treatment-related effects has clearly shown that clear wood material is affected more than material containing knots and voids (Winandy 1996a), such that it is now commonly assumed within the wood-engineering community that treatment effects are greater the more defect-free the wood material and the straighter its grain.

Because strength is affected by material factors, such as specific gravity and growth characteristics, it is important to always consider property variability. The coefficient of variation is the statistical parameter used to approximate the variability associated with each strength property. The estimated coefficient of variation of various strength properties can be found in Table 11.2.

TABLE 11.1
Relationship between Specific Gravity and Mechanical Properties

Property	Green Wood		Wood at 12% Moisture Content	
	Softwoods	Hardwoods	Softwoods	Hardwoods
Static Bending				
MOR (KPa)	109,600	118,700	170,700	171,300
MOE (MPa)	16,100	13,900	20,500	16,500
WML (KJ/m^3)	147	229	179	219
Impact Bending (N)	353	422	346	423
Compression parallel (KPa)	49,700	49,000	93,700	76,000
Compression perpendicular (KPa)	8,800	18,500	16,500	21,600
Shear parallel (KPa)	11,000	17,800	16,600	21,900
Tension perpendicular (KPa)	3,800	10,500	6,000	10,100
Side hardness (N)	6,230	16,550	85,900	15,300

Source: U.S. Department of Agriculture, Forest Service, 1999.

11.2.2 ENVIRONMENTAL FACTORS

11.2.2.1 Moisture

Wood, which is a hygroscopic material, gains or loses moisture to equilibrate with its immediate environment. The equilibrium moisture content (EMC) is the steady-state level that wood achieves when subjected to a particular relative humidity and temperature. The eventual EMC of two similar

TABLE 11.2
Average Coefficients of Variation for Some Mechanical Properties of Clear Wood

Property	Coefficient of Variation
Static bending	
Modulus of rupture	16
Modulus of elasticity	22
Work to maximum load	34
Impact bending	25
Compression parallel to grain	18
Compression perpendicular to grain	28
Shear parallel to grain	14
Maximum shearing strength	14
Tension parallel to grain	25
Side hardness	20
Toughness	34
Specific gravity	10

Source: U.S. Department of Agriculture, Forest Service, 1999.

Chemistry of Wood Strength

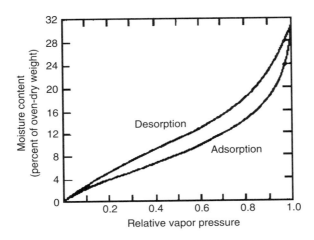

FIGURE 11.3 Adsorption–desorption isotherms for water vapor by spruce at 25°C.

specimens will differ if one approaches EMC under adsorbing conditions and the other approaches EMC under desorbing conditions. For example, if the relative vapor pressure of the environment is 0.65 (i.e., relative humidity of 65%), two similar specimens exposed under either adsorbing or desorbing conditions will equilibrate at moisture contents of approximately 11% and 13%, respectively (USDA 1999) (Figure 11.3).

Wood strength is related to the amount of water in the wood fiber cell wall (USDA 1999, Wilson 1932, Gerhards 1882). At moisture contents from oven-dry (OD) to the fiber-saturation point, water accumulates in the wood cell wall (bound water). Above the fiber-saturation point, water accumulates in the wood cell cavity (free water) and there are no tangible strength effects associated with a changing moisture content. However, at moisture contents between OD and the fiber-saturation point, water does affect strength. Increased amounts of bound water interfere with and reduce hydrogen bonding between the organic polymers of the cell wall (USDA 1999, Rowell 1980), which decreases the strength of wood. The approximate relationships are shown in Tables 11.3 and 11.4.

Not all mechanical properties change with moisture content. The performance of wood under dynamic loading conditions is a dual function of the strength of the material, which is decreased with increased moisture content, and the pliability of the material, which is increased with increased moisture content. Changes in strength and pliability somewhat offset one another, and therefore, mechanical properties that deal with dynamic loading conditions are not nearly as affected by a changing moisture content as are static mechanical properties.

11.2.2.2 Temperature

Strength is related to the temperature of the working environment (USDA 1999, Gerhards 1982, Barrett et al. 1989). At constant moisture content, the immediate effect of temperature on strength is linear (Figure 11.4) and usually recoverable when the temperature returns to normal. In general, the immediate strength of wood is higher in cooler temperatures and lower in warmer temperatures. However, permanent (nonrecoverable) effects can occur. This relationship of permanent strength loss during extended high-temperature exposure can be dramatically influenced by higher moisture contents.

The immediate effects of increased temperature are an increase in the plasticity of the lignin and an increase in spatial size, which reduces intermolecular contact and is, thus, recoverable. Permanent effects manifest themselves as an actual reduction in wood substance or weight loss via

TABLE 11.3
Approximate Change in Mechanical Properties of Clear Wood with Each One Percent Change in Moisture Content

Property		Change per 1% Change in Moisture Content
Static bending		
	Fiber stress at proportional limit	5
	Modulus of rupture	4
	Modulus of elasticity	2
	Work to proportional limit	8
	Work to maximum load	0.5
Impact bending		
	Height of drop causing complete failure	0.5
Compression parallel to grain		
	Fiber stress at proportional limit	5
	Maximum crushing strength	6
Compression perpendicular to grain		
	Fiber stress at proportional limit	5.5
Shear parallel to grain		
	Maximum shearing strength	3
Hardness		
	End	4
	Side	2.5

Source: Winandy and Rowell, 1984.

TABLE 11.4
Relationship between Some Mechanical Properties and Moisture Content

Property	Moisture Content*				
	Green	19%	12%	8%	Oven-Dry
Douglas-Fir					
Modulus of rupture	62	76	100	117	161
Compression parallel to grain	52	68	100	124	192
Modulus of elasticity	80	88	100	108	125
Loblolly Pine					
Modulus of rupture	57	72	100	121	175
Compression parallel to grain	49	66	100	127	203
Modulus of elasticity	78	87	100	109	128
Aspen					
Modulus of rupture	61	75	100	118	165
Compression parallel to grain	50	67	100	126	199
Modulus of elasticity	73	83	100	111	137

Property at 12% moisture content set at 100%.
Source: Winandy and Rowell, 1984.

Chemistry of Wood Strength

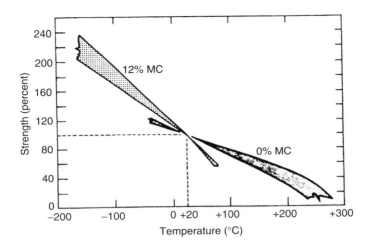

FIGURE 11.4 Immediate effect of temperature on strength properties expressed as a percent of value at 20°C. Trends illustrated are composites from several studies on three strength properties (MOR, $T_\perp$, and $C_\parallel$).

degradative mechanisms and are thereby nonrecoverable. This permanent thermal effect on wood strength has been extensively studied (LeVan et al. 1990, Winandy 2001, Green et al. 2003) and predictive kinetic-based models have been developed (Woo 1981, Lebow and Winandy 1999a, Green et al. 2003). The reasons for these permanent thermal effects on strength relate to changes in the wood polymeric substance and structure, and predictive models have been developed (Winandy and Lebow 2001).

11.2.3 Load Factors

11.2.3.1 Duration of Load

The ability of wood to resist load is dependent upon the length of time the load is applied (USDA 1999, Gerhards 1977, Wood 1951). The load required to cause failure over a long period of time is much less than the load required to cause failure over a very short period of time. Wood under impact loading (duration of load > I_s) can sustain nearly twice as great a load as wood subjected to long-term loading (duration of load > 10 years). This time-dependent relationship (Wood 1951) can be seen graphically in Figure 11.5. Hydrolytic chemical treatments have long been known to incur brashness in wood (Wangaard 1950). More recently, the chemistry of this effect on treated wood was shown to be empirically related to loss in the ability of the hydrolyzed wood material to dissipate strain energy away from localized stress concentrations under impact-type, rapid loadings (ultimate in <1–2 s) (Winandy 1995b).

11.2.3.2 Fatigue

Cyclic or repeated loadings often induce fatigue failures. Fatigue resistance is a measure of a material's ability to resist repeating, vibrating, or fluctuating loads without failure. Fatigue failures often result from stress levels far lower than those required to cause static failure. Repeated or fatigue-type stresses usually result in a slow thermal buildup within the material and initiate and propagate tiny micro checks that eventually grow to a terminal size. When wood is subjected to repeated stress (e.g., 5.0×10^7 cycles), fatigue-related failures may be induced by stress levels as low as 25–30% of the anticipated ultimate stress under static conditions (USDA 1999, Wangaard 1950).

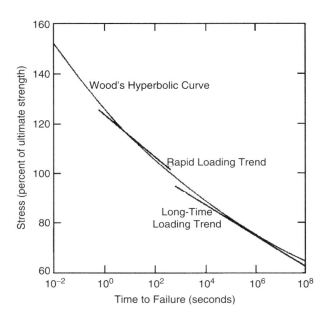

FIGURE 11.5 Hyperbolic load-duration curve with rapid loading and longtime loading trends for bending.

11.2.3.3 Mechanical Properties

To design with any material, mechanical property estimates need to be developed. ASTM standard test methods detail the procedures required to determine mechanical properties via stress–strain relationships (see ASTM).

11.2.4 FLEXURAL LOADING PROPERTIES

Flexural (bending) properties are important in wood design. Many structural designs recognize either bending strength or some function of bending, such as deflection, as the limiting design criterion. Structural examples in which bending-type stresses are often the limiting consideration are bridges or bookshelves. Five mechanical properties are derived from the stress-strain relationship of a standard bending test: modulus of rupture (MOR), fiber stress at proportional limit (FSPL), modulus of elasticity (MOE), work to proportional limit (WPL), and work to maximum load (WML).

11.2.4.1 Modulus of Rupture

The MOR is the ultimate bending strength of a material. Thus, MOR describes the load required to cause a wood beam to fail and can be thought of as the ultimate resistance or strength that can be expected (Figure 11.2, Point B; Figure 11.6, Point B) from a wood beam exposed to bending-type stress. MOR is derived by using the flexure formula:

$$MOR = Mc/I \qquad (11.2)$$

which assumes an elastic response, although that assumption is not exactly true, where M is the maximum bending moment, c is a measure of distance from the highly-stressed flanges of the beam to its neutral axis (i.e., the plane where internal stress becomes zero as those stresses change from tension on one flange to compression on the opposite), and I is the moment of inertia, which relates

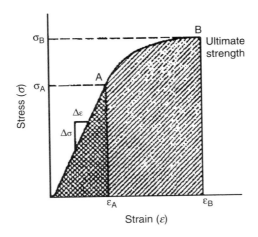

FIGURE 11.6 Examples of the relationship between a typical stress–strain diagram and some mechanical properties. Key A, proportional limit; B, ultimate strength; σ_B, MOR, σ_A, FSPL; $\Delta\sigma/\Delta\varepsilon$ (from origin to A), MOE; $\int_\phi^A \sigma d\varepsilon$, WPL; and $\int_\phi^B \sigma d\varepsilon$, WML.

the bending moment to the geometric shape of the beam. Engineers often simplify these geometric factors (c/I) to a parameter known as S, the section modulus:

$$S = c/I \quad (11.3)$$

For a rectangular or square beam in bending under center point loading, the flexure formula is varied to reflect loading conditions and beam geometry:

$$MOR = 1.5*P*L^2/b*h^2 \quad (11.4)$$

where P is the ultimate load, L is the span of beam, b is the width of beam, and h is the height of beam.

11.2.4.2 Fiber Stress at Proportional Limit

The FSPL is the maximum bending stress a material can sustain under static conditions and still exhibit no permanent set or distortion. It is by definition the amount of unit stress on the y-coordinate at the proportional limit of the material (Figure 11.2, Point A; Figure 11.6, Point A). FSPL is also derived using the flexure formula, where M is the bending moment at the proportional limit and S is the section modulus.

11.2.4.3 Modulus of Elasticity

The MOE quantifies a material's elastic (i.e., recoverable) resistance to deformation under load. The MOE corresponds to the slope of the linear portion of the stress–strain relationship from zero to the proportional limit (Figure 11.6). Stiffness (MOE*I) is often incorrectly thought to be synonymous with MOE. However, MOE is solely a material property and stiffness depends both on the material and the size of the beam. Large and small beams of similar material would have similar MOEs but different stiffnesses. The MOE can be calculated from the stress–strain curve as the change in stress causing a corresponding change in strain.

11.2.4.4 Work to Proportional Limit

The WPL is the measure of work performed, i.e., energy used, in going from an unloaded state to the elastic or proportional limit of a material (Figure 11.6). For a beam of rectangular cross section under center point loading, WPL is calculated as the area under the stress–strain curve from zero to the proportional limit.

11.2.4.5 Work to Maximum Load

WML is the amount of work needed to actually fracture or fail a material. It is a measure of the amount of energy required to fracture the material. Toughness and work to total load are analogous properties, but their final limit state also includes energy absorbed beyond the ultimate failure. WML is calculated as the area under the stress–strain curve from zero to the ultimate strength of the material (Figure 11.6). Because WML is a measure of work both below and beyond the proportional limit, it is derived by either graphical approximations or by means of calculus.

11.2.5 AXIAL LOADING PROPERTIES

Because of the anisotropic and heterogeneous nature of wood, there can be profound differences in the strength in various directions. Wood is stronger along the grain (parallel to the longitudinal axis of the log or longitudinal axis of the wood cell) than perpendicular to the grain (at right angles to the longitudinal axis). Axial loads describe forces that have the same line of action and are, thus, both parallel and concurrent. Because there is no eccentricity in the application of these forces, they do not induce flexure or bending moments.

Mechanical properties dealing with axial loading conditions are maximum crushing strength (compression parallel to the grain), compression perpendicular to the grain, tension parallel to the grain, and tension perpendicular to the grain.

11.2.5.1 Compression Parallel to the Grain

If wood is considered a bundle of straws bound together, then a compression parallel to the grain ($C_\parallel$) can be thought of as a force trying to compress the straws from end to end. The distance through which compressive stress is transmitted does not increase or magnify the stress, but the length over which the stresses are carried is important. If the length of the column is far greater than the width, the specimen may buckle. This stress is analogous to bending-type failure rather than axial-type failure. As long as specimen width is great enough to preclude buckling, $C_\parallel$ is solely an axial property.

Examples of wood in compression parallel to the grain are wooden columns or the top chord of a roof truss. Compression-parallel-to-the-grain strength or the maximum crushing strength is derived at the ultimate limit value of a standard stress–strain curve. The strength of wood in $C_\parallel$ is derived by the following:

$$C_\parallel = P/A \qquad (11.5)$$

where $C_\parallel$ is the stress in compression parallel to grain, P is the maximum axial compressive load, and A is the area over which load is applied.

11.2.5.2 Compression Perpendicular to the Grain

Compression perpendicular to the grain ($C_\perp$) can be thought of as stress applied perpendicular to the length of the wood cell. Therefore, in our straw example, the straws (or wood cells) are being

Chemistry of Wood Strength

crushed at right angles to their length. Until the cell cavities are completely collapsed, wood is not as strong perpendicular to the grain as it is parallel to the grain. However, once the wood cell cavities collapse, wood can sustain a nearly immeasurable load in $C_\perp$. Because a true ultimate stress is nearly impossible to achieve, maximum $C_\perp$ in the sense of ultimate load-carrying capacity is undefined and discussions of $C_\perp$ are usually confined to stress at some predetermined limit state such as the proportional limit or 4% deflection.

Compression-perpendicular-to-the-grain stresses are found whenever one member is supported upon another member at right angles to the grain. Examples of compression perpendicular to the grain are the bearing areas of a beam, truss, or joist.

The $C_\perp$ strength is derived by:

$$C_\perp = P/A \tag{11.6}$$

where $C_\perp$ is the stress in compression perpendicular to the grain, P is the proportional limit load, and A is the area.

11.2.5.3 Tension Parallel to the Grain

A tension parallel to the grain ($T_\parallel$) stress is a force trying to elongate the wood cells, or straws in our straw example. Wood is extremely strong in $T_\parallel$. The distance through which tensile stress is transmitted does not increase the stress. The $T_\parallel$ is difficult to measure because of the difficulty in securely gripping the tensile specimen in the testing machine, especially with clear straight-grained wood. Often $T_\parallel$ of clear straight-grain wood is conservatively estimated by the MOR (ultimate strength in bending). This conversion is accepted because bending failure of clear wood often occurs on the lower face of a bending specimen where the lower face fibers are under tensile-type stresses. An example of tension parallel to the grain would be the bottom chord of a truss that is under tensile stress. The $T_\parallel$ strength of wood is derived by the following formula:

$$T_\parallel = P/A \tag{11.7}$$

where $T_\parallel$ is the stress in tension parallel to the grain, P is the maximum load, and A is the area.

11.2.5.4 Tension Perpendicular to Grain

Tension perpendicular to the grain ($T_\perp$) is induced by a tensile force applied perpendicular to the longitudinal axis of the wood cell. In this case, the straws (or wood cells) are being pulled apart at right angles to their length. The $T_\perp$ is extremely variable and is often avoided in discussions on wood mechanics. However, $T_\perp$ stresses often cause cleavage or splitting failures along the grain, which can dramatically reduce the structural integrity of large beams. Failures from $T_\perp$ are sometimes found in large beams that dry while in service. For example, if a beam is secured by a top and a bottom bolt at one end, shrinkage may eventually cause cleavage or splitting failures between the top and bottom boltholes. Wood can be cleaved by $T_\perp$ forces at a relatively light load. It is this weakness that is often exploited in karate and other demonstrations of human strength. The $T_\perp$ strength of wood is derived by the following:

$$T_\perp = P/A \tag{11.8}$$

where $T_\perp$ is the stress in tension perpendicular to the grain, P is the maximum load, and A is the area.

11.2.6 OTHER MECHANICAL PROPERTIES

11.2.6.1 Shear

Shear parallel to the grain (γ) measures the ability of wood to resist the slipping or sliding of one plane past another parallel to the grain. Shear strength is derived in a manner similar to axial properties, by using the following equation:

$$\gamma = P/A \qquad (11.9)$$

where γ is the shear stress parallel to grain, P is the shearing load, and A is the area of the shear plane through the material.

11.2.6.2 Hardness

Hardness is used to represent the resistance to indentation and/or marring. Hardness (ASTM 2003a) is measured by the load required to embed a 1.128-cm steel ball one-half its diameter into the wood. While a material may be softer or harder in common vernacular, hardness, in engineering terms, is a material property that is measured using specified methods detailing sizes, sources, and test speeds. Beyond those specific test conditions, the term hardness when used in common language may have widely differing meanings to different people.

11.2.6.3 Shock Resistance

Shock resistance or energy absorption is a function of a material's ability to quickly absorb and then dissipate energy via deformation. This is an important property for baseball bats, tool handles, and other articles that are subjected to frequent shock loadings. High shock resistance on energy absorption properties requires both the ability to sustain high ultimate stress and the ability to deform greatly before failing.

Shock resistance can be measured by several methods. With wood, three of the most often used methods are work tests (to maximum load), impact bending tests, and toughness tests. Both of the later two test methods yield measures of strength and pliability, mutually referred to as energy absorption. These two measures of shock resistance are similar but are not particularly relative to one another. Impact bending is tested by dropping a weight onto a beam from successively increasing heights (ASTM 2003a). All that is recorded is the height of drop causing complete failure in a beam such that for a different sized beam or a different mass of weight the measured value would most certainly change. Toughness is the ability of a material to resist a single impact-type load from a pendulum device (ASTM 2003a). Thus, toughness is similar to impact bending in that both are measures of energy absorption or shock resistance. Yet critical differences exist. Toughness uses a single ultimate load and impact-type bending, whereas impact bending uses a series of progressively increasing, multiple loads in which the earlier load history can certainly alter the eventual result. Although each test method defines a material characteristic, each measured property should only be compared within the limited definitions of that method. They should not be compared on a method-to-method basis nor compared if tested on differing sized or conditioned materials.

11.3 CHEMICAL COMPONENTS OF STRENGTH

11.3.1 RELATIONSHIP OF STRUCTURE TO CHEMICAL COMPOSITION

The chemical components responsible for the strength properties of wood can be theoretically viewed from three distinct levels: the macroscopic (cellular) level, the microscopic (cell wall) level, and the molecular (polymeric) level.

Chemistry of Wood Strength

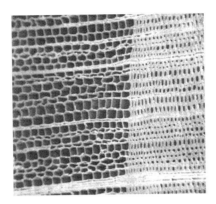

FIGURE 11.7 Scanning electron micrograph showing cellular structure with thin-walled earlywood (left) and thicker-walled latewood (right) at ×120 (A) and ×400 (B).

11.3.1.1 Macroscopic Level

Wood with its inherent strength is a product of growing trees. The primary function of the woody trunk of the living tree is to provide support for the phototropic energy factory (i.e., leaves or needles) at the top and to provide a conduit for moving water and nutrients moving up to those leaves or needles. The phototropic sugars produced by the leaves or needles mostly move down the stem via the bark tissues. Woody tissues, interior to the bark, exist as concentric bands of cells oriented for specific functions (Figure 11.7). Thin-walled earlywood cells act both as conductive tissue and support; thick-walled latewood cells provide support. Each of these cells is a single fiber. Softwood fibers average about 3.5 mm in length and 0.035 mm in diameter. Hardwood fibers are generally shorter (1–1.5 mm) and smaller in diameter (0.015 mm). The earlywood and latewood fibers comprise large composite bands, bonded together by a phenolic adhesive, lignin. Each band is anisotropic in character but is reinforced in two of the three axial directions by longitudinal parenchyma and ray parenchyma cells. These parenchyma cells function as a means of either longitudinal or radial nutrient conduction and as a means of providing lateral support by increased stress distribution (Figure 11.8a,b). Because wood is a reinforced composite material, its structural performance at the cellular level has been likened to reinforced concrete (Freudenberg 1932, Mark 1967).

The macroscopic level of consideration takes into account fiber length and differences in cell growth, such as earlywood, latewood, reaction wood, sapwood, heartwood, mineral content, extractive chemicals, resin content, etc. Differences in cellular anatomy, environmental-controlled growth patterns, and chemistry can cause significant differences in the strength of wood.

11.3.1.2 Microscopic Level

At the microscopic level, wood has been compared to multipart systems, such as filament-wound fiber products (Mark 1967). Each component complements the other in such a manner that, when considering the overall range of physical performance, the components together outperform the components separately.

11.3.1.3 Composition

Within the cell wall are distinct regions (see Figure 3.12), each of which has distinct composition and attributes. For typical softwood the middle lamella and primary wall are mostly lignin (8.4% of the total weight) and hemicellulose (1.4%), with very little cellulose (0.7%). The S_1 layer consists of cellulose (6.1%), hemicellulose (3.7%), and lignin (10.5%). The S_2 layer is the thickest layer

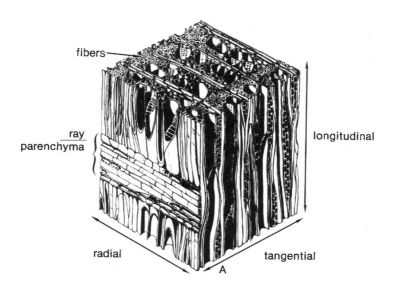

FIGURE 11.8A Cube of hardwood.

and has the highest carbohydrate content; it is mostly cellulose (32.7%) with lesser quantities of hemicelluloses (18.4%) and lignin (9.1%). The S_3 layer, the innermost layer, consists of cellulose (0.8%), hemicelluloses (5.2%), and very little lignin. One interesting caveat is that although the relative lignin ratio is low within the S_2 layer, the largest amount of lignin exists within this layer because of its large overall mass.

The large number of hydrogen bonds existing between cellulose molecules results in such strong lateral associations that certain areas of the cellulose chains are considered crystalline. More than 60% of the cellulose (Stamm 1964) exists in this crystalline form, which is stiffer and stronger than the less crystalline or amorphous regions. The crystalline areas are approximately 60 nm long (Thomas 1981) and are distributed throughout the cell wall.

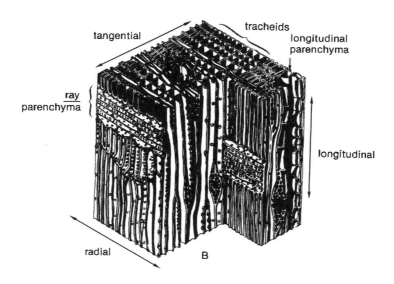

FIGURE 11.8B Cube of softwood.

FIGURE 11.9 Scanning electron micrograph of softwood fibers embedded in lignin (×400).

11.3.1.4 Microfibril Orientation

Microfibrils are highly ordered groupings of cellulose that may also contain small quantities of hemicellulose and lignin. The exact composition of the microfibril and its relative niche between the polymeric chain and the layered cell wall are subjects of great discussion (Mark 1967). The microfibril orientation (fibril angle) is different and distinct for each cell wall layer (see Figure 3.12). The entire microfibril system is a grouping of rigid cellulose chains analogous to the steel reinforcing bars in reinforced concrete or the glass or graphite fibers in filament-wound reinforced plastics. Most composite materials use an adhesive of some type to bond the entire material into a system. In wood, lignin fulfills the function of a matrix material. Yet, it is not truly or solely an adhesive and by itself adds little to strength (Lagergren et al. 1957). Lignin is a hydrophobic phenolic material that surrounds and encrusts the carbohydrate complexes (Figures 11.9 and 11.10). It aids in holding the cell components together at the microscopic level. Lignin also seems to be responsible for part of the stiffness of wood and it most certainly is primarily responsible for the exclusion of water from the moisture-sensitive carbohydrates. Rubbery wood, a viral disease of certain varieties of apple (*Malus* spp.), is characterized by extremely flexible wood. The affected wood has been shown (Prentice 1949) to have cells rich in cellulose but low in lignin.

11.3.1.5 Molecular Level

At the molecular level the relationship of strength and chemical composition deals with the individual polymeric components that make up the cell wall. The physical and chemical properties of cellulose, hemicelluloses, and lignin play a major role in the chemistry of strength. However, our perceptions of wood polymeric properties are based on isolated polymers that have been removed from the wood system and, therefore, possibly altered. The three individual polymeric components (cellulose, hemicelluloses, and lignin) may be far more closely associated and interspersed with one another than has heretofore been believed (Attalla 2002). Recent theories speculate

FIGURE 11.10 Scanning electron micrograph of delignified softwood fiber wall (×16,000).

that the crystalline and amorphous regions of the cellulose chains are more diffuse and less segregated between themselves than may have earlier been believed (Attalla 2002).

Cellulose is anhydro-D-glucopyranose ring units bonded together by β-1-4-glycosidic linkages. The greater the length of the polymeric chain and the higher the degree of polymerization, the greater the strength of the unit cell (Mark 1967, Ifju 1964) and, thus, the greater the strength of the wood. The cellulose chain may be 5000–10,000 units long. Cellulose is extremely resistant to tensile stress because of the covalent bonding within the pyranose ring and between the individual units. Hydrogen bonds within the cellulose provide rigidity to the cellulose molecule via stress transfer and allow the molecule to absorb shock by subsequently breaking and reforming. Past theories have speculated that the cellulose is the predominate factor in wood strength, but recent work has clearly shown that the hydrolytic or enzymatic action upon the hemicelluloses seem to always manifest themselves in the earliest levels of strength loss in woody materials.

The hemicelluloses are a series of carbohydrate molecules that consist of various elementary sugar units, primarily the six-carbon sugars, D-glucose, D-galactose, and D-mannose, and the five-carbon sugars, L-arabinose and D-xylose. Hemicelluloses have linear chain backbones (primarily glucomannans and xylan chains) that are highly branched and have a lower degree of polymerization than cellulose. The sugars in the hemicellulose structure exhibit hydrogen bonding both within the hemicellulose chain as well as between other hemicellulose and amorphous cellulose regions. Most hemicelluloses are found interspersed within or on the boundaries of the amorphous regions of the cellulose chains and in close association with the lignin. Hemicellulose may be the connecting material between cellulose and lignin. The precise role of hemicellulose as a contributor to strength has long been a subject of conjecture. Recent work on hydrolytic chemical agents and enzymatic decay has indicated that early degradation of hemicellulose(s), especially degradation of the shorter branched monomers of D-galactose and L-arabinose along the hemicellulose main chains, seem primarily responsible for the earliest portions of strength loss in wood exposed to severe thermal, chemical, or biological exposures (Winandy and Lebow 2001, Curling et al. 2002, Winandy and Morrell 1993). Other work has speculated that the actions of hydrolytic agents at the aforementioned sheath of hemicellulose along the boundary areas of the cellulose microfibrils (Figure 11.11) may

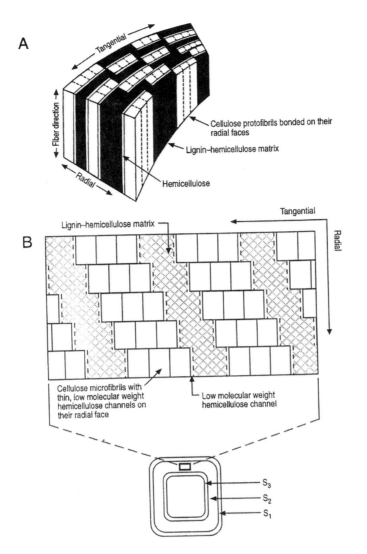

FIGURE 11.11 Representation of proposed ultrastructural models of the arrangement of lignin, cellulose, and hemicellulose in S_2 layer of the wood cell wall. Models proposed by Kerr and Goring 1975 (A) and Larsen et al. 1995 (B).

account for the radial degradation fissures (i.e., interwall cracks of the S2 layer often seen during chemical or biological decay (Larsen et al. 1995)).

Lignin is often considered nature's adhesive. It is the least understood and most chemically complex polymer of the wood-structure triad. Its composition is based on highly organized three-dimensional phenolic polymers rather than linear or branched carbohydrate chains. Lignin is the most hydrophobic (water-repelling) component of the wood cell. Its ability to act as an encrusting agent on and around the carbohydrate fraction, and thereby limit water's influence on that carbohydrate fraction, is the cornerstone of wood's ability to retain its strength and stiffness as moisture is introduced to the system. Dry delignified wood has nearly the same strength as normal dry wood, but wet delignified wood has only approximately 10% of the strength of wet normal wood (Lagergren et al. 1957). Thus, wood strength is due in part to lignin's ability to limit the access of water to the carbohydrate moiety and thereby lessen the influence of water on wood's hydrogen-bonded structure.

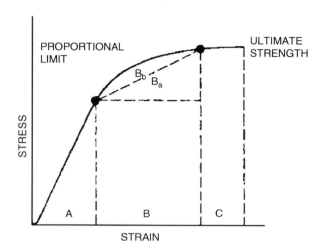

FIGURE 11.12 Typical stress-strain curve showing the three theoretically identifiable regions of mechanical behavior. Key: A, elastic region; B, viscoelastic, partially viscoelastic B_A and partially viscoplastic B_b; and C, plastic region.

11.4 RELATIONSHIP OF CHEMICAL COMPOSITION TO STRENGTH

To relate chemical composition to strength properties, to work and toughness properties, and eventually to elastic parameters, a preliminary model or hypothesis must be developed to aid in conceptualizing the relationship between strength and wood composition. The theoretical relationship of stress to strain can be graphically represented by a diagram (Figure 11.12). If wood is assumed to be an elastic material, a linear region, A, represents the constant relationship below the proportional limit, and the nonlinear regions, B and C, represent nonconstant relationship beyond the proportional limit. If each region of the stress–strain relationship is examined at each of the three distinct levels of wood structure (macroscopic, microscopic, and molecular), the relationship between strength and composition may be hypothetically explained.

As loads are applied to a wood system, stresses are immediately introduced and distributed throughout the material. The stresses cause two types of strain or distortion: immediate, under which wood can be described as an elastic material, and time dependent. Even at low stress levels, permanent set or distortion will eventually be induced in a wood member. This phenomenon, known as creep, explains why wood is considered viscoelastic. But for purposes of simplification, the ensuing discussion will be confined to immediate strain or distortion and will consider wood as an elastic material. Immediate strain or distortion can be conceptualized at each of the three distinct levels of wood structure.

11.4.1 BELOW PROPORTIONAL LIMIT (ELASTIC STRENGTH)

When a load is applied to a piece of wood, at the molecular level, hydrogen bonds between and within individual polymer chains are breaking, sliding (uncoiling), and subsequently reforming (Figure 11.13a, b, c, respectively); C–C and C–O bonds are distorting within the ring structures (Figure 11.14).

At the microscopic level, hydrogen bonds between adjacent microfibrils are breaking and reforming (Figure 11.13a, b, c) to allow the microfibrils to slide by one another with only the disruption of the hydrogen bonds that are subsequently reformed. Additionally, the individual cell wall layers are distorting in relation to each other, but no permanent set or distortion is occurring between these individual cell wall layers.

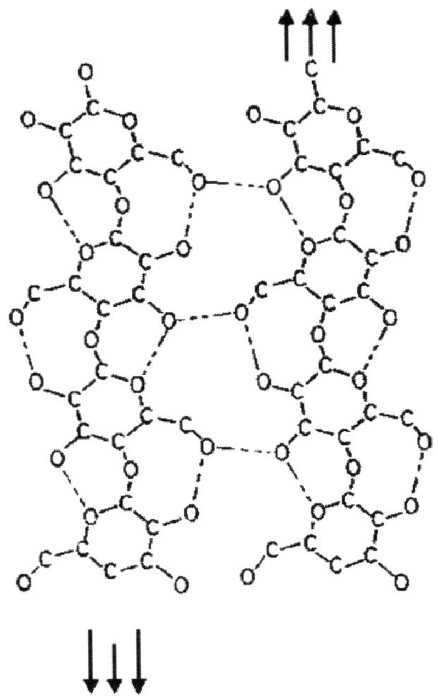

Chemistry of Wood Strength

FIGURE 11.13A Hydrogen bonding (bonded) between polysaccharide chains under shear forces.

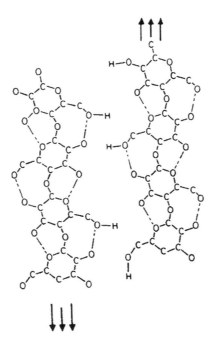

FIGURE 11.13B Hydrogen bonding (sliding, unbounded) between polysaccharide chains under shear forces.

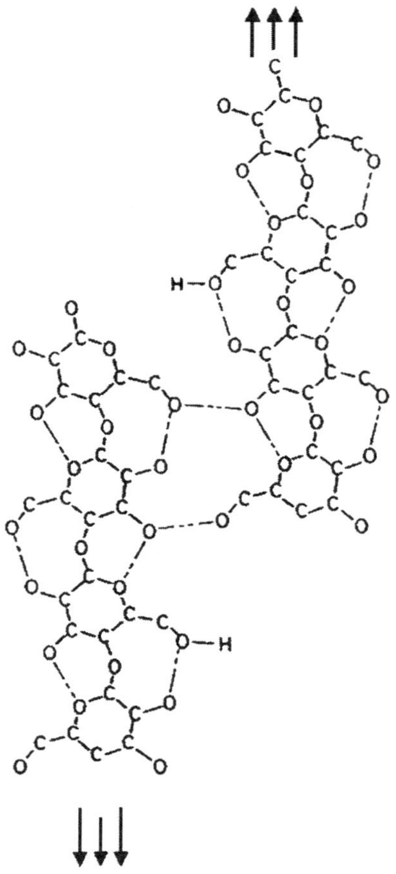

FIGURE 11.13C Hydrogen bonding (rebounded) between polysaccharide chains under shear forces.

At the macroscopic level, there is distortion between the individual cells, but it is not permanent because the stresses are being distributed between the individual cells such that no permanent translocation or set is introduced.

Within the limits of the elastic model, all strain or distortion resulting from the accumulation of stress in this material has been recoverable up to this point. As the proportional limit is approached, the wood material can no longer distribute the stress in a linear elastic manner.

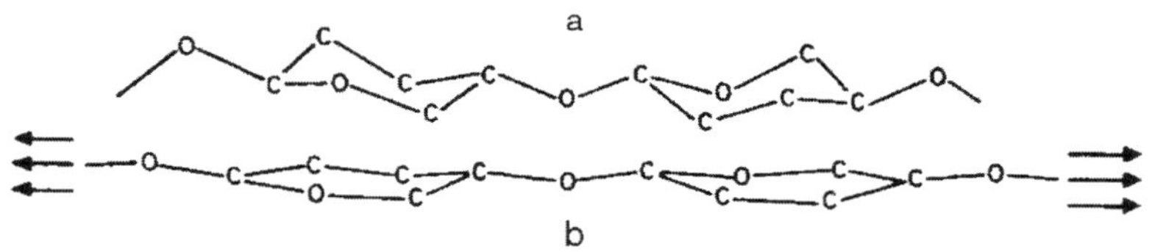

FIGURE 11.14 Flexing and elongation of polysaccharide molecules under tensile force. Key: a, no tensile force, no elongation; and b, tensile force, elongation.

Chemistry of Wood Strength

11.4.2 BEYOND PROPORTIONAL LIMIT (PLASTIC STRENGTH)

As the proportional limit is exceeded (Figure 11.12, Region B), the stress–strain relationship is no longer linear. Stresses are now great enough to induce covalent bond rupture and permanent distortion at all three structural levels.

At the molecular level, the limit of reversible or recoverable hydrogen bonding has been exceeded. Covalent C–C and C–O bonds are breaking, thus reducing larger molecules to smaller ones. This reduction in degree of polymerization by covalent bond scission is nonrecoverable.

At the microscopic level, stresses develop within the crystalline region of the carbohydrate microfibrils. Failure of the microfibril from stress overload causes actual covalent bond rupture and excessive microfibril disorientation. Additionally, the cell wall layers distort such that permanent micro cracks occur between the various cell wall layers. Separation of the cell wall layers is soon noticeable.

At the macroscopic level, entire fibers actually distort in relation to one another, such that recovery of the original position is now impossible. The wood cells or wood fibers are actually failing either by scission of the cell, in which the cell actually fails by tearing into two parts to give a brash type of failure, or by cell-to-cell withdrawal (middle lamella failure), where the cells actually pull away from one another to give a splintering type of failure.

Permanent set is now evident at all levels of consideration, and eventual failure is imminent. In approaching the ultimate strength (Figure 11.12, Region C), molecular level failures occur by C–C and C–O bond cleavage. Stress redistribution within the individual polymers is now impossible. At the microscopic level, the cell walls are distorting without additional stress. These walls are actually deforming at such an exaggerated rate that they can be thought of as being completely viscous or plastic, and they continue to deform and absorb strain energy but they can no longer handle additional stress. The cell wall is being sheared or torn apart. At the macroscopic level, failure is related to cell wall scission or cell-to-cell withdrawal.

11.5 RELATIONSHIP OF STRUCTURE TO STRENGTH

The mechanism of strength, as it relates to wood composition, has been discussed as a theoretical elastic model. To better understand this proposed model, we will look at what may be happening at each of the three structural levels.

11.5.1 MOLECULAR LEVEL

At the molecular level, strength is elastic or recoverable because the polymeric structure can flex and, thus, absorb energy without fracturing the important covalent bonds (Figure 11.12, Region A). The second region of the stress–strain diagram (Figure 11.12, Region B) consists of Region B_a, which is indicative of residual elastic strength such as represented in Region A, and Region B_b, which represents plastic strength or, more appropriately, strength associated with initial permanent set or distortion.

Section B_b is representative of C–C and C–O cleavage at the intrapolymer level, which cannot be recovered. Examples of C–C bond breakage are lignin-hemicellulose copolymer separation, hemi-cellulose depolymerization, and amorphous cellulose depolymerization.

In Region C (Figure 11.12), elastic deformation essentially ends; there is now nearly pure plastic flow in the stress–strain relationship. Strain is continuing with little additional increase in stress and ultimate failure is imminent. This region is characterized by all the same mechanisms as in B_b, but a new and terminal intrapolymeric factor is introduced—the crystalline cellulose failure. As crystalline cellulose failure occurs, the main framework of the wood material at the molecular level is disintegrating.

11.5.2 MICROSCOPIC LEVEL

At the microscopic level, the strength of the phenolic matrix is usually great enough that the cell wall stress reaches failure level in the carbohydrate framework. The S_1 layer microfibrils are oriented in both a right-hand (S helix) and a left-hand (Z helix) arrangement, whereas the S_2 and S_3 have only the S-helix arrangement (Figure 11.9). The S_3 layer can be bihelical or monohelical, but, for the purpose of simplification, it has been assumed to be monohelical in this example. Because of the different linear elongation of the bihelical S_1 layer as compared to the monohelical arrangement of the S_2 and S_3 layers, the cell wall initially fails by S_1–S_2 separation (Siau 1969). As S_1–S_2 separate, the S_2–S_3 layers assume the transferred stresses, and sustained stress increases, which will eventually cause either a brash-type failure (carbohydrate covalent bond failure) or a slow buildup to ultimate stress yielding a fibrous-type failure (phenolic covalent bond failure).

Below the proportional limit (Figure 11.12, Region A), there is elastic transfer of stresses between the S_1–S_2–S_3 cell wall layers. As Region B is entered, stress is still transferred between the S_1–S_2–S_3 cell wall layers as characterized by Section B_a. But S_1–S_2 separation is initiating, causing a sizeable transfer of stresses to the S_2–S_3 layers characterized by Section B_b. In Region C, ultimate strength is now dictated by the S_2–S_3 cell wall layer's ability to sustain additional stress until eventual failure of the substantial S_2 layer.

11.5.3 MACROSCOPIC LEVEL

At the macroscopic level, it is necessary to consider wood a viscoelastic material. As stress is applied to a wooden member, minute cracks initiate, propagate, and terminate throughout the collective cellular system in all directions. They develop in all regions of the stress–strain relationship at the macroscopic level, but only in the elastic region (Figure 11.12, Regions A and B_a) is crack propagation controlled and eventually terminated. In the tangential direction, the concentric ring structure of thin-walled earlywood and thick-walled latewood in softwoods, and porous early-season vessels and dense late-season fibers in hardwoods act as the elements of elastic stress transfer. In the radial direction, the ray structures and the linear arrangement of fibers and vessels are the elements of elastic stress transfer. Every cell in the radial direction is aligned closely with the next cell because each cell in the radial direction has originated from the same cambial mother cell. Thus, the material can transfer stress elastically until an induced crack or a natural growth defect interrupts this orderly cellular arrangement. As stresses are built up within the material, cracks are initiated in the areas where elastic stress transfer is interrupted. These cracks continue to propagate until they are terminated either via dispersion of the energy away from the crack by the structural elements of stress transfer, or by eventual terminal failures as graphically characterized by Regions B_b, and C (Figure 11.12).

11.6 ENVIRONMENTAL EFFECTS

When wood is exposed to environmental agents of deterioration, such as chemical treatments or elevated temperatures, each mechanical property reacts differently. Most commonly, ultimate strength properties are reduced and properties dealing with the proportional limit show little or no effect. However, the strain-to-failure (strain rate) is often considerably reduced, which, due to embrittlement of the fibers, is reflected as a reduction in pliability and energy-related properties such as work, toughness, etc.

As individual wood components are altered in size, stature, or composition, the strength of the wood material is dramatically affected. Hypothetically, when ultimate stress is reduced 5% (Figure 11.15, U_1–U_2) and the proportional limit is not affected, the properties dealing with proportional limit (FSPL, MOE, WPL) reflect this in that they too are unaffected. The mechanical properties dealing with the point of ultimate stress (MOR in bending tests, C in axial-type compression tests,

Chemistry of Wood Strength

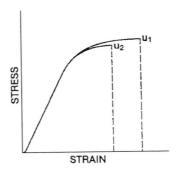

FIGURE 11.15 Hypothetical example of the effect of no change in proportional limit and a 5% reduction in ultimate bending strength on a few mechanical properties: MOE is not affected, MOR is reduced 5%, but WML is reduced by 33% because it is a dual function of stress and strain.

and T and $T_\perp$ in axial-type tensile tests) are reduced 5%. Work to maximum load can be reduced 33% because it is a function of both stress and strain. If the stress level at the proportional limit is reduced, and both the ultimate stress and strain levels are significantly reduced (Figure 11.16), larger decreases will occur in proportional limit properties (MOE, and FSPL, WPL), ultimate strength properties (MOR, C, T, $T_\perp$), and work to maximum load. The examples in both Figures 11.15 and 11.16 show evidence that WML and related properties, such as toughness and impact bending, are usually affected long before the other properties dealing with ultimate strength and the proportional limit are significantly affected.

What causes the phenomenon of stress and strain reduction and why is the reduction in impact and work properties so visible at small or negligible changes in elastic modulus and ultimate strengths? As discussed previously, mechanical properties deal with stress and strain relationships that are simply functions of chemical bond strength. At the molecular level, strength is related to both covalent and hydrogen intrapolymer bonds. At the microscopic level, strength is related to both covalent and hydrogen interpolymer bonds and cell wall layer bonds (S_1–S_2 and S_2–S_3). At the macroscopic level, strength is related to fiber-to-fiber bonding with the middle lamella acting as the adhesive. Thus, any chemical or environmental agent that affects those bonds also affects strength.

In considering the structural performance of the polysaccharide and phenolic polymers in wood fiber, the chemical environment of the fiber is of great importance. Chemicals can swell, hydrolyze, pyrolyze, oxidize, and, in general, depolymerize wood polymers, causing a loss in strength properties

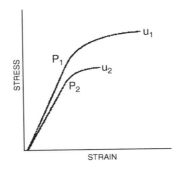

FIGURE 11.16 Hypothetical example of the effect of a 20% reduction in proportional limit and a 30% reduction in ultimate bending strength on a few mechanical properties: MOE is reduced 20%; MOR is reduced 30%, and WML is reduced 50%, because it is a dual function of stress and strain.

due to wood fiber network degradation. Other environmental agents, such as UV light, heat, and biological organisms, have a similar influence in changing strength properties.

11.6.1 ACIDS AND BASES

The average pH of wood is between 3 and 5.5 (Stamm 1964) due to the acetyl content, the presence of acid extractives, and the adsorption of cations that comprise the ash. Even after several hundred years, this naturally mild acidic state does not induce any appreciable strength losses as long as the wood is protected from biological attack (USDA 1999).

If the pH of the environment substantially changes or if temperatures increases to a point where the pH of the environment changes, strength properties can be reduced. These effects are further compounded by time and moisture. In general, the longer the time or the higher the temperature at which wood is exposed to an acid or a base, the greater is the degradative effect on strength (USDA 1999, Wangaard 1950, Stamm 1964). Heartwood is generally more resistant to acid than is sapwood, probably because of heartwood's lower permeability and higher extractives content. Hardwoods are usually more susceptible to degradation by either acids or alkali's than are softwoods. This may be due to hardwood's lower lignin content and higher proportion of pentosan hemicelluloses. Oxidizing acids, such as HNO_3, have a greater degradative action on wood fiber than do nonoxidizing acids. Alkaline solutions are more destructive to wood fibers than are acidic solutions because wood adsorbs alkaline solutions more readily than acidic solutions. Acids with pH values above 2 and bases with pH values below 10 do not degrade the wood fiber greatly over short periods of time at low temperatures (Kollman and Cote 1968). Mild acids, such as acetic acid, have little effect on strength, whereas strong acids, such as H_2SO_4, cause extensive strength losses (Alliot 1926).

11.6.2 ADSORPTION OF ELEMENTS

Chemicals other than acids and bases can also be adsorbed and can cause degradation of the wood fiber. For example, fibers of southern pine exposed to the ocean air can be degraded badly (Figure 11.17). Salt crystals deposited in the void structure (Figure 11.18) can cause extensive

FIGURE 11.17 Scanning electron micrograph of sodium chloride in southern pine cell (×40).

Chemistry of Wood Strength

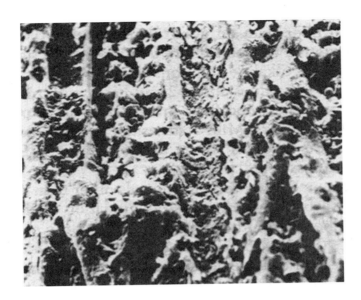

FIGURE 11.18 Scanning electron micrograph of sodium chloride in southern pine cell (×400).

chemical and physical damage. This chemical damage is due, in part, to the salt catalyzing hydrolysis reactions, whereas mechanical damage is related to the hydroscopic salts promoting greater shrinkage and swelling.

Other materials can be adsorbed from the environment if a hydrolytic solvent (e.g., water) is available. When water is available, wood will adsorb iron from oxidized metal (rust) and cause decomposition of the cellulose (Koenigs 1974). This is also true for copper, chromium, tin, zinc, and other similar reactive metals.

11.6.3 Swelling Solvents

Solutions that swell wood tend to plasticize it and reduce its strength properties. Water, for example, swells the intrapolymeric spaces, reduces cross-linking and, thus, reduces strength. In general, the greater the swelling, the greater the strength loss. Nonswelling liquids generally do not decrease strength properties. For example, oven-dry wood and wood saturated with water-free benzene have virtually the same strength (Erickson and Rees 1940, Siau 1969).

11.6.4 Ultra Violet Degradation

Wood exposed to the outdoors undergoes chemical reactions due to UV radiation. UV radiation causes photochemical degradation primarily in the lignin component, which gives rise to characteristic color changes (see Chapter 7). Southern pine, for example, changes from a light-yellow natural color to brown and evenly to gray. As the lignin degrades, the wood surface becomes richer in cellulose content. Although the cellulose is much less susceptible to UV degradation (Kalnins 1966), it is eventually washed off the surface with water during rain, which exposes new lignin-rich surfaces that then start to degrade. As this process continues, the wood surface is said to weather.

Because UV radiation does not penetrate wood more than a few cells deep, weathering is considered a surface phenomenon. Over time it can account for a significant loss in surface fiber (see Figure 7.1). As the degradative process continues, the loss in fiber may eventually cause a reduction in the material's load-carrying capacity.

11.6.5 THERMAL DEGRADATION

Wood strength is inversely related to temperature (see Chapter 6). A nearly linear decrease in strength is observed on increasing the temperature from −200 to +160°C (Figure 11.4; a corresponding loss in strength from two- to threefold (Gerhards 1982, Kollman and Cote 1968, Green 1999). Heat has two types of effects on wood, immediate effects that occur only as long as the increased temperature is maintained and permanent effects that result from thermal degradation of wood polymers. The immediate effects of heat are recoverable, but permanent effects are not. The combination of immediate and permanent effects is multiplicative rather than additive.

In an environment without adequate humidity, the initial effect of heating wood is dehydration. As temperatures approach 55–65°C for extended periods (2–3 months), hemicellulose and cellulose depolymerization slowly begins (Feist et al. 1973, LeVan et al. 1990). This progressively escalates to pyrolysis and volatilization of cell wall polymers, which rapidly occurs at about 250°C, followed by char formation in the absence of air and combustion in the presence of air.

Heating dry Douglas fir in an oven at 102°C for 335 days reduced MOE by 17%, MOR by 45%, and fiber stress at proportional limit by 33% (MacLean 1945,1953, Millett and Gerhards 1972). The same losses might be observed in 7 days at 160°C. In the absence of air, heating softwood at 210°C for 10 min reduced MOR by 2%, hardness by 5%, and toughness by 5% (Stamm et al. 1946). Under the same conditions at 280°C MOR was reduced 17%, hardness was reduced 21%, and toughness was reduced 40%. Both examples illustrate the compound effect of heat, air, and time.

Comparison of photomicrographs of southern pine at 25°C (Figure 11.19 top) and the same sample after heating from 20 to 295°C under nitrogen over a period of 15 min (Figure 11.19 bottom) shows the cell structure still intact, but the cell wall components have been darkened by pyrolysis. LeVan et al. (1990) noted an ongoing darkening of pinewood exposed at 82°C, which corresponded to a loss in arabinose and to a lesser degree xylose. They then attributed the darkening brown color at 82°C to hydrolysis of furan-ringed arabinose and xylose to chocolate-brown colored furfural. Over the last 20 years, the permanent effect of extended high-temperature and cyclic exposure on wood strength has been extensively studied (LeVan et al. 1991, Winandy et al. 1991, Winandy 1995a, LeVan et al. 1996, Green et al. 2003) and fairly thoroughly reviewed (Winandy 2001). Predictive kinetic-based models have also been developed (Woo 1981, Pasek and McIntyre 1990, Winandy and Lebow 1996, Lebow and Winandy 1999, Green et al. 2003). The reasons for these permanent thermal effects on strength relate to changes in the wood polymeric substance and structure and predictive models have been developed (Winandy and Lebow 2001). The comprehensive analyses of almost 10,000 specimens systematically exposed to various high-temperature regimes, followed by their development of kinetic models, debunked one long-held misconception that a thermal threshold existed below which permanent effects did not occur (Lebow and Winandy 1999). That work concluded that thermal degradation of wood was a continuum, but at most ambient temperature exposures below 40–50°C, the rate of degrade was so slow as to be negligible.

Tables 11.5–11.9 show loss in mechanical properties as a function of sugar analysis when southern pine is heated at different temperatures for different times either untreated or treated with various fire retardant chemicals. Table 11.5 shows the effects of these variables on untreated wood, and it can be seen that heat alone results in a decrease in MOR and WML as the time and temperature increase. Along with the loss of mechanical properties, there is an accompanying loss of xylose, galactose, and arabinose. The greatest loss is in arabinose, which may be the causative event leading to initial strength losses.

Table 11.6 shows the effects of these variables on wood that has been treated with phosphoric acid. MOE, MOR, and WML all decrease with increasing temperature and time. Significant losses in glucose, xylose, galactose, arabinose, and mannose also occur as the time and temperature increase. Table 11.7 shows a similar trend for southern pine treated with monoammonium phosphate, Table 11.8 for guanylurea phosphate/boric acid, and Table 11.9 for borax/boric acid. The most sensitive and consistent hemicellulose sugar lost in untreated and fire retardant treated wood is

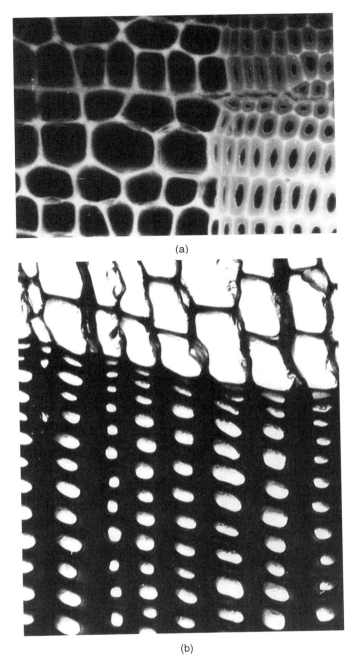

FIGURE 11.19 Southern pine at 25°C (×400) (top) and southern pine after heating in a nitrogen environment from 20 to 295°C over 15 minutes (×400).

arabinose. The loss of arabinose can be used as an approximate indication of strength loss without having to do a bending test to determine strength properties.

These results indicate that strength losses from external exposure of untreated or FRT wood to elevated temperatures are a result of loss of hemicelluloses or cell wall matrix structure, not a loss in the degree of polymerization of cellulose (Sweet and Winandy 1999).

TABLE 11.5
Relationship between Loss of Strength and Sugar Analysis of Fire Retardant Treated Southern Pine—Control

Temp (°C)	Time (days)	MOE (Gpa)	MOR (Mpa)	WML (kj/m³)	Lig (%)	Glu (%)	Xyl (%)	Gal (%)	Arab (%)	Mann (%)
23	3	13.9	117.4	101.2	29.4	48.1	8.0	3.0	1.3	12.9
23	16	14.5	127.9	116.6	27.6	44.9	6.1	2.1	1.2	11.2
54	3	13.9	124.1	110.2	29.6	47.0	7.5	3.4	1.2	12.6
54	7	13.8	116.9	98.5	28.4	48.0	7.0	2.4	1.3	13.2
54	21	13.5	119.5	101.2	29.6	47.1	7.1	2.6	1.2	12.4
54	60	13.8	121.3	101.5	29.4	45.9	6.6	2.5	1.0	12.0
54	160	13.9	120.1	99.3	27.8	52.9	7.1	2.2	1.3	13.6
54	7	14.6	115.9	95.8	28.6	42.9	6.1	1.9	1.0	10.8
66	21	14.0	113.8	96.7	28.5	42.7	5.9	1.8	1.0	11.1
66	60	14.8	122.2	97.3	28.6	42.8	5.9	1.9	0.9	11.1
66	160	13.5	116.0	79.7	28.2	43.1	5.9	2.0	0.8	11.0
66	290	14.3	113.4	74.5	29.3	43.4	5.9	2.2	0.5	10.8
66	560	14.7	96.5	51.3	29.4	42.9	5.7	2.2	0.4	11.0
66	1095	14.6	81.3	32.7	29.7	43.0	5.4	1.7	0.2	10.9
66	1460	11.2	42.5	10.8	32.8	43.0	3.8	0.7	0.1	8.3
82	3	13.5	119.0	112.0	29.2	46.7	7.3	2.6	1.2	12.4
82	7	14.0	125.0	105.3	29.3	46.5	6.9	2.5	1.2	12.3
82	21	14.7	124.2	101.4	29.7	48.0	6.8	2.4	0.9	12.7
82	60	14.0	118.7	84.4	29.1	46.9	6.8	2.6	0.7	12.7
82	160	13.8	104.7	58.0	27.6	53.1	6.9	2.3	0.6	13.4

Lig, Klasson lignin; Glu, glucose; Xyl, xylose; Gal; galactose; Arab, arabinose; Mann, mannose.

11.6.6 Microbial Degradation

When certain organisms come into contact with wood, several types of degradation occur (see Chapter 5). The mechanical damage caused by metabolic action can result in significant losses in strength. Microbial activity via enzymatic pathways induces wood fiber degradation by chemical reactions such as hydrolysis, dehydration, and oxidation. Brown-rot fungi preferentially metabolize holocellulose, especially the strength-critical cellulose fraction (Cowling 1961). White-rot fungi metabolize nearly all fractions of the wood, but strength is not affected to the same degree as with a brown-rot fungus. An initial 10% weight loss for sweetgum (*Liquidambar styraciflua L.*) attacked by a brown-rot fungus reduced the degree of polymerization of the holocellulose from 1500 to 300 (Figure 11.20) (Cowling 1961). As the side-chains of the hemicellulose (e.g., arabinose, galactose) in the wood fiber are degraded, mechanical properties begin to decrease (Winandy and Morrell 1993, Curling et al. 2002). This work has shown that the initial 5–20% of strength loss seems to be directly related to this initial hemicellulose degradation. Later, as the main-chain backbones of the hemicellulose (e.g., xylose, mannose) are degraded, additional strength loss occurs. Finally, after strength has been reduced about 40–60% from the virgin strength level, noticeable degradation of glucose and lignin becomes evident (Winandy and Lebow 2001).

Using isolated polymers derived from wood, previous research has indicated a large drop in degree of polymerization, which was then related to the large drop in wood strength properties without corresponding weight loss (Ifju 1964, Cowling 1961, Kennedy 1958). This work indicated that, in the initial stages of biological attack, hydrolytic chemical reactions play an important part. Yet, more recent work that carefully approached this same subject using less invasive isolation techniques to study the polymers, found that hydrolytic hemicellulose depolymerization preceded

TABLE 11.6
Relationship between Loss of Strength and Sugar Analysis of Fire Retardant Treated Southern Pine—Phosphoric Acid 58.2 kg/m³

Temp (°C)	Time (days)	MOE (Gpa)	MOR (Mpa)	WML (kj/m³)	Lig (%)	Glu (%)	Xyl (%)	Gal (%)	Arab (%)	Mann (%)
23	3	11.5	58.3	22.5	26.7	40.9	5.9	2.6	0.7	10.8
23	160	11.4	56.9	21.6	24.8	44.3	5.8	1.8	1.2	11.3
54	3	11.8	55.8	18.5	27.1	41.5	5.9	2.4	0.8	11.9
54	7	11.2	52.9	17.7	27.0	41.6	5.9	2.7	1.1	10.7
54	21	13.2	55.4	19.1	26.8	43.6	6.1	2.7	1.1	12.2
54	60	10.7	50.8	19.0	27.4	43.2	5.7	2.8	0.8	10.8
54	160	11.3	52.4	17.7	26.7	44.1	6.4	1.5	0.6	10.3
66	7	11.3	49.1	17.2	26.2	41.8	5.6	1.8	0.9	10.9
66	21	10.2	41.1	13.9	27.2	40.2	4.7	1.7	0.7	9.6
66	60	10.5	43.8	13.0	29.3	41.2	3.5	0.9	0.3	8.4
66	160	8.2	27.5	5.5	33.0	40.5	2.2	0.2	0.3	5.3
66	290	7.0	19.2	3.2	36.6	36.6	2.1	0.4	0.1	4.1
66	560	3.3	6.9	0.8	44.4	32.2	1.0	0.2	0.0	1.8
66	1095	0.0	0.0	0.0	0.0	0.0	0.0	0.0	0.0	0.0
66	1460	0.0	0.0	0.0	0.0	0.0	0.0	0.0	0.0	0.0
82	3	10.6	41.0	12.6	28.4	5.4	5.4	2.5	0.7	10.2
82	7	10.4	43.6	12.5	29.3	4.4	4.4	1.8	0.6	10.1
82	21	8.6	30.9	6.5	34.0	2.6	2.6	1.7	0.2	6.2
82	60	7.6	20.4	3.4	37.9	2.0	2.0	0.8	0.0	4.2

Lig, Klasson lignin; Glu, glucose; Xyl, xylose; Gal, galactose; Arab, arabinose; Mann, mannose.

cellulose depolymerization (Sweet and Winandy 1999). In either case, in the initial phase(s) of degradation, large polymers are broken into smaller, more digestible units, but the initial degradation products are not actually consumed by the organisms. Hydrogen peroxide and iron have been proposed as being involved in initial depolymerization (Koenigs 1974). The initial enzymatic attack by microorganisms is probably not only hydrolytic but also oxidative in nature.

Curling et al. (2002) recently evaluated the effect of brown-rot decay on Southern pine (Pinus spp.) exposed in a laboratory test for periods ranging from 3 days to 12 weeks. The relative effects

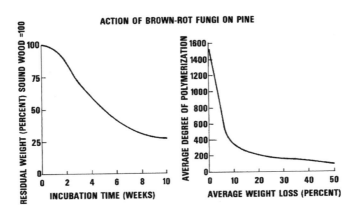

FIGURE 11.20 Action of brown rot fungus on holocellulose (Cowling 1961).

TABLE 11.7
Relationship between Loss of Strength and Sugar Analysis of Fire Retardant Treated Southern Pine—Monoammonium Phosphate 55.5 kg/m³

Temp (°C)	Time (days)	MOE (Gpa)	MOR (Mpa)	WML (kj/m³)	Lig (%)	Glu (%)	Xyl (%)	Gal (%)	Arab (%)	Mann (%)
23	3	12.5	101.8	81.3	27.3	43.8	5.9	2.1	0.7	11.3
23	160	12.7	103.5	79.3	23.6	42.5	6.2	2.1	1.0	11.8
54	3	12.5	96.1	76.6	27.5	43.6	6.6	2.4	1.2	11.7
54	7	12.2	99.2	77.7	27.2	41.8	6.4	2.2	0.9	11.0
54	21	15.6	98.6	62.6	26.6	42.6	6.1	1.7	0.7	11.3
54	60	12.6	95.9	65.1	27.0	41.5	6.2	2.6	0.9	11.1
54	160	13.2	90.9	44.3	26.3	44.2	6.4	2.0	0.6	11.6
66	7	13.4	99.5	69.5	26.6	40.8	5.6	1.8	0.8	10.6
66	21	12.3	90.7	55.1	26.5	40.5	5.7	1.7	0.3	10.2
66	60	12.8	79.2	34.7	27.6	42.1	5.8	2.2	0.4	11.0
66	160	11.7	66.1	25.6	27.8	41.2	5.0	1.6	0.1	10.0
66	290	11.7	57.5	19.0	29.3	40.6	4.0	0.8	0.0	9.1
66	560	10.5	32.7	6.7	33.6	43.0	2.8	0.3	0.0	6.6
66	1095	7.6	18.8	2.6	39.7	41.2	2.2	0.2	0.0	3.7
66	1460	2.2	5.8	0.6	43.4	41.3	1.9	0.1	0.0	3.3
82	3	13.0	89.2	45.7	27.1	41.8	6.2	2.1	0.6	10.9
82	7	12.3	77.5	33.5	27.3	41.1	5.7	2.0	0.5	10.6
82	21	13.3	76.0	29.0	27.7	42.7	6.0	1.7	0.6	10.5
82	60	12.5	61.1	21.1	30.2	43.5	4.9	1.2	0.0	9.8
82	160	10.5	42.7	11.2	26.3	44.7	2.6	0.2	0.0	5.6

Lig, Klasson lignin; Glu, glucose; Xyl, xylose; Gal, galactose; Arab, arabinose; Mann, mannose.

of decay on strength, weight, and chemical composition can be seen by comparing the loss in bending strength (MOR), loss in work to maximum load (WML), and loss in stiffness (MOE) to loss in dry weight (Figure 11.21).

These results demonstrate that considerable bending strength loss occurs before measurable weight loss. At 10% weight loss, strength loss was approximately 40% and the loss in the energy properties, such as work to maximum load, was reduced by 70–80%. The decay–strength relationship

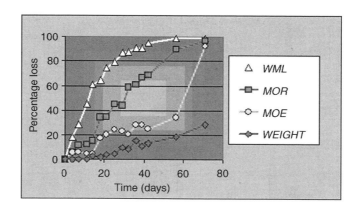

FIGURE 11.21 Effect of decay by *G. trabeum* on weight loss and mechanical properties (Curling et al. 2002).

TABLE 11.8
Relationship between Loss of Strength and Sugar Analysis of Fire Retardant Treated Southern Pine—Guanylurea Phosphate/Boric Acid 55.5 kg/m³

Temp (°C)	Time (days)	MOE (Gpa)	MOR (Mpa)	WML (kj/m³)	Lig (%)	Glu (%)	Xyl (%)	Gal (%)	Arab (%)	Mann (%)
23	3	12.2	108.1	82.8	28.3	41.4	5.7	2.3	1.2	11.2
23	160	12.6	107.8	80.7	25.9	44.1	6.0	2.4	1.1	12.4
54	3	12.6	107.4	84.4	27.9	43.0	6.2	1.9	1.1	11.4
54	7	13.4	108.5	77.6	27.5	43.1	5.9	1.8	1.1	11.4
54	21	12.5	107.6	75.6	28.1	42.6	6.3	2.3	1.2	11.6
54	60	12.9	104.2	72.0	27.9	42.4	5.8	1.9	1.3	11.3
54	160	13.0	104.6	65.1	25.6	46.2	6.2	2.3	0.8	12.0
66	7	14.1	105.9	73.8	26.9	40.5	5.9	1.7	1.0	10.6
66	21	13.3	104.9	72.5	28.1	40.3	5.4	1.8	0.8	10.3
66	60	13.3	98.8	57.5	28.2	40.8	5.7	1.6	0.6	10.4
66	160	12.2	78.3	33.6	29.1	—	5.7	1.7	0.4	10.7
66	290	13.3	69.8	24.6	28.9	41.7	5.6	1.4	0.1	10.4
66	560	12.7	56.2	16.8	30.3	42.8	5.2	1.1	0.1	9.9
66	1095	11.7	38.5	7.6	32.8	41.0	3.7	0.7	0.1	7.3
66	1460	8.0	17.9	3.3	37.6	42.1	2.8	0.4	0.1	5.8
82	3	13.5	111.1	70.1	27.9	42.2	5.7	2.1	1.5	10.9
82	7	12.8	110.3	69.3	28.1	42.3	6.2	2.3	1.4	10.9
82	21	13.3	105.0	55.1	29.5	44.2	6.6	2.4	1.2	11.1
82	60	13.4	83.7	37.2	29.3	44.2	6.5	2.6	0.9	10.4
82	160	12.5	66.3	25.2	26.9	48.4	5.3	1.6	0.3	9.8

Lig, Klasson lignin; Glu, glucose; Xyl, xylose; Gal, galactose; Arab, arabinose; Mann, mannose.

appears very consistent as decay initiates and progresses. It seems not only qualitative but may also be quantitative.

Curling et al. (2002) also found that changes in chemical composition were directly related to strength loss (Figure 11.22). As decay progressed it sequentially affected different chemical components. A virtually similar effect was previously noted by Winandy and Morrell (1993). Initial loss in the mannan and xylan components correlated with the start of measurable weight loss; both

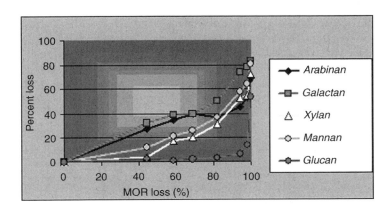

FIGURE 11.22 Comparison of loss of carbohydrate components with loss in bending strength (MOR) caused by *G. trabeum* (Curling et al. 2002).

TABLE 11.9
Relationship between Loss of Strength and Sugar Analysis of Fire Retardant Treated Southern Pine—Borax/Boric Acid 56.3 kg/m³

Temp (°C)	Time (days)	MOE (Gpa)	MOR (Mpa)	WML (kj/m³)	Lig (%)	Glu (%)	Xyl (%)	Gal (%)	Arab (%)	Mann (%)
23	3	12.4	115.7	69.3	27.8	42.2	6.7	2.6	1.21	14.8
23	160	13.3	115.3	65.4	26.3	39.3	5.9	2.4	1.2	10.5
54	3	12.9	106.1	57.6	27.1	42.8	6.5	2.3	1.0	14.7
54	7	13.9	115.4	62.2	26.8	44.4	7.0	2.7	1.4	15.5
54	21	16.0	117.1	59.8	26.7	43.9	6.5	1.9	1.0	15.7
54	60	13.5	126.5	86.3	27.0	43.5	6.8	1.9	1.2	14.9
54	160	13.7	122.7	70.9	25.7	44.9	6.5	2.0	1.0	12.4
66	7	13.8	116.5	69.0	27.4	42.2	6.1	1.9	1.0	11.2
66	21	13.1	116.0	74.7	27.5	41.8	6.4	2.0	1.0	11.1
66	60	13.1	108.5	66.1	28.2	42.0	6.2	1.8	1.0	10.7
66	160	13.4	115.0	77.8	28.9	43.5	6.0	2.0	0.8	11.2
66	290	13.3	108.7	76.3	28.6	44.0	6.0	1.7	0.7	11.1
66	560	13.4	99.2	65.8	29.3	44.1	6.3	1.9	0.8	10.9
66	1095	12.8	73.5	32.3	27.4	42.7	5.4	1.4	0.3	9.7
66	1460	11.7	50.3	16.2	31.6	45.2	4.3	0.8	0.1	8.9
82	3	14.6	123.1	66.9	27.0	43.1	6.7	2.7	1.2	15.7
82	7	14.5	131.1	76.9	27.1	43.5	6.3	1.4	0.6	14.8
82	21	14.4	132.7	73.9	27.0	46.0	6.8	2.3	0.9	15.4
82	60	14.8	137.9	89.1	27.0	42.3	6.5	2.5	0.9	14.5
82	160	14.3	124.0	69.1	26.3	43.8	6.1	2.3	0.7	11.4

Lig, Klasson lignin; Glu, glucose; Xyl, xylose; Gal, galactose; Arab, arabinose; Mann, mannose.

occurred at about 40% strength loss. Because mannan/xylan components make up approximately 18% of the wood, compared to 8% for galactan and arabinan, it is understandable why measurable weight loss is not evident in incipient decay. Final stage of brown-rot decay occurs once the glucan-rich cellulose is broken down, at a strength loss of about 80%. It is also at this stage (80% loss in MOR) that loss in stiffness (MOE) increases rapidly suggesting that the stiffness of the wood is related to the cellulose rather than the hemicellulose composition.

Table 11.10 shows the loss in MOE and maximum compression strength (MCS) in wood exposed to the brown-rot fungus *Postia placenta* in standard ASTM soil block and in ground tests. In the wood blocks, at 2 weeks into the standard 12-week test, there is only a 3% weight loss but already a 13% loss in MOE and a 21% loss in MCS. At five weeks, there is a 26% weight loss but a 53% loss in MOE and a 71% loss in MCS. In the wood stakes, the losses are less since the specimen size is larger and requires more time for the fungus to degrade the wood. Even so, the wood stakes at 5 weeks show a weight loss of 7% and a loss of 12% in MOE and 22% in MCS.

Table 11.11 shows the correlation between time in test and changes in sugar analysis that has occurred during that time. The greatest loss is in the arabinose sugar, which was also the case in FRT wood.

11.6.7 Naturally Occurring Chemicals

Some woods have a higher acidic extractives content that can cause greater strength loss due to hydrolysis. This may be a problem in some of the tropical species coming into the market. These

TABLE 11.10
Strength Loss in Decayed Wood

	Wood Blocks			Wood Stakes		
Time (weeks)	Weight Loss (%)	MOE Reduction (%)	MCS Reduction (%)	Weight Loss (%)	MOE Reduction (%)	MCS Reduction (%)
1	0	3	6	0	5	2
2	3	13	21	0	8	2
3	9	22	37	1	7	8
4	18	34	53	3	9	19
5	26	53	71	7	12	22

Fungus used in test, *Postia placenta*; MOE, modulus of elasticity; MCS, maximum compression strength.

Source: Clausen and Kartal, 2003.

more acidic woods will not only be more susceptible to strength loss but will also increase the corrosion potential of iron fasteners used with the wood. It is not uncommon to find silica and calcium salt crystals in wood fiber (Figure 11.23), and they can lead to strength losses by abrasion of the fibers during machining. Naturally occurring chemicals may also cause increased hygroscopicity and hydrolysis when salts dissolve in water.

11.7 TREATMENT EFFECTS

From a quantitative viewpoint, the effects of preservatives have been investigated by many researchers. A recent review (Winandy 1996a) identified and then classified several treatment-related issues that controlled the treatment effect with waterborne preservatives. In that review, waterborne-preservative treatments were shown to generally reduce the mechanical properties

TABLE 11.11
Correlation between Strength Loss in Decayed Wood and Carbohydrate Analysis

		Carbohydrate Composition					
Sample	Time (weeks)	Arabinan (%)	Galactan (%)	Rhamnan (%)	Glucan (%)	Xylan (%)	Mannan (%)
Blocks	0	1.01	1.37	0.07	42.82	5.88	11.32
	1	0.87	1.40	0.07	42.56	5.05	12.15
	2	0.78	1.10	0.06	42.46	5.60	10.32
	3	0.68	1.02	0.05	42.23	4.82	9.65
	4	0.55	1.07	0.04	41.89	3.91	10.73
	5	0.61	0.86	0.04	39.13	4.36	7.87
Stakes	0	1.02	2.63	0.09	41.97	6.44	9.84
	1	0.93	2.20	0.09	42.80	6.39	10.43
	2	0.94	2.13	0.09	42.07	6.61	10.53
	3	0.86	2.05	0.08	42.23	6.21	10.62
	4	0.82	1.69	0.07	41.25	6.13	9.84
	5	0.77	2.04	0.07	40.48	5.93	8.98

Source: Clausen and Kartal, 2003.

FIGURE 11.23 Scanning electron micrograph of calcium carbonate crystals in lumens of southern pine (×800).

of wood. The effects of waterborne preservatives on mechanical properties are related to several key wood material factors and pretreatment, treatment, and post-treatment processing factors. These key factors include preservative chemistry or chemical type (Wood et al. 1980, Bendtsen et al. 1983, Winandy et al. 1983), retention (Winandy et al. 1985, Winandy et al. 1989), post-treatment drying temperature (Winandy and Boone 1988, Winandy 1989), initial kiln-drying temperature (Winandy and Barnes 1991), grade of material (Winandy et al. 1988, Winandy 1989), and incising (if required) (Lam and Morris 1989, Morrell et al. 1998, Winandy and Morrell 1998). The effects of treatments on time-dependent relationships previously shown in Figure 11.5 have also been studied, but whereas they seemed to have little practical significance related to long-term duration-of-load (Soltis and Winandy 1989) they did have significant effects on short-term impact loading (Winandy 1995b).

In summary, each factor has the potential to influence the effects of the preservative treatment on mechanical properties. As might be expected, the interactive nature of the processing and chemical factors is critical, and these factors must always be considered when attempting to define the effects of waterborne preservatives on mechanical properties. A cumulative-damage model to predict these effects has also been developed (Winandy 1996b).

From a qualitative viewpoint, metallic salts such as chromated copper arsenate (CCA), amine- or ammoniacal-copper quat (ACQ), Copper Azol (CA), and other metallic salts of copper, arsenic, zinc, and tin are used to increase the service life of wood in use against biological attack. One major advantage of many waterborne preservatives is their resistance to leaching by water. Some organo-metallic salt preservatives (e.g., ACQ, CA) become insoluble as the treated wood dries, which dehydrates the preservative complex within the wood. In this process, known as immobilization, the cell wall is not directly affected and subsequently strength is virtually unaffected. Other metal salts (e.g., CCA) undergo hydrolytic reduction upon contact with the reducing sugars found in wood. In this process, known as fixation, the cell wall is oxidized and subsequently strength is affected. At most usable concentrations these preservatives are sufficiently acidic or alkaline to cause some cell wall hydrolysis (Betts and Newlin 1915, Hatt 1906, Wood et al. 1980). The effects on strength are greater in kiln-dried waterborne preservative-treated wood than in air-dried salt-treated wood (Bendtsen et al. 1983, Winandy et al. 1985). Most water-borne preservative salts increase the hygroscopicity of the wood (Bendtsen 1966, Bendtsen et al. 1983). This causes an increased EMC, which further influences strength.

After treatment and either immobilization or fixation, precipitated preservative salts can be seen in the wood structure. Comparison of photomicrographs of untreated southern pine (Figure 11.24 top) and southern pine treated with 2.5 lb CCA/ft^3 (Figure 11.24 bottom) reveals that the salts in

FIGURE 11.24 Scanning electron micrograph of CCA in southern pine.

the treated wood form a rough coating on the lumen walls. The exact final location of these fixed salts is still subject to discussion. Undoubtedly, most of these injected hydrolytic preservatives remain in the cell lumen, but the extent to which these preservatives enter the cell wall and react with the cell wall substance dictates the preservative's effect on strength.

Preservative formulations that contain copper and chromium salts reportedly promote afterglow in treated wood subjected to fire. Once the treated wood starts to burn or glow, the wood may continue to glow until the entire member is consumed, even when no flame is present (Dale 1966, McCarthy et al. 1972). This characteristic can cause serious problems for utility poles, fence posts, and highway signs, structures that might be subjected to accidental fires or controlled ground fire, which is used as a forest or agricultural management tool.

Petroleum-based chemicals such as creosote and pentachlorophenol in oil are also used as wood preservatives. These organic preservatives are inert toward the cell wall substance and do not seem to cause any appreciable strength losses (MacLean 1951, Koukal et al. 1960, Gillwald 1961, USDOD 1979, Barnes and Winandy 1986).

Salts such as sodium tetraborate, diammonium phosphate, trisodium phosphate, diammonium sulfate, and salts of boric acid have long been used as fire retardants. Hygroscopicity, corrosion of fasteners, and increased acidity are also problems with these salts. Like the preservative salts, these

FIGURE 11.25 Scanning electron micrograph of southern pine with FRT crystal deposition on surface.

salts also precipitate in the cell cavity and the cell wall (Figure 11.25). They appear on the fiber surface of pine treated with 4.2 lb of ammonium dihydrogen phosphate ($NH_4H_2PO_4$) per cubic foot of wood. The initial effects of FR treatments on strength are greater in kiln-dried salt-treated wood than in air-dried salt-treated wood (Gerhards 1970, Brazier and Laidlaw 1974, Adams et al. 1979, Johnson 1967). Like the waterborne preservatives, this is due to hydrolytic degradation of the fiber, caused by the combination of high moisture content, high temperature, and acid salts or alkalis during the post-treatment redrying process.

Many fire-retardant salt treatments are very hygroscopic salts causing increased moisture sorption and irreversible swelling (Bendtsen 1966). Most interior FR formulations are not recommended for use where relative humidity is over 92% (AWPA 2003).

FRT plywood roof sheathing is often required by U.S. Building Codes in roof systems for multifamily dwellings having common property walls. In the mid-1980s, users of some FRT materials, especially plywood roof sheathing exposed to elevated in-service temperatures, were experiencing significant thermal degrade (Figure 11.26). From the late 1980s until about 2001, a major +10-year research program was carried on at the U.S. Forest Products Laboratory to identify the factors and mechanisms of FRT-induced strength loss. Qualitatively, the mechanism of thermal degrade in FRT plywood was acid hydrolysis. The magnitude of strength loss could be cumulatively related to FR chemistry, thermal exposure during pretreatment, treatment, and post-treatment processing and in-service exposure. The effects of FR chemistry could be mitigated by use of pH buffers. The strength effects were similar for many quality levels of plywood. Quantitatively, a kinetics-based approach could be used to predict strength loss based on its time-temperature history.

From a mechanistic approach to FR thermal degradation, the effects of FR treatments on strength properties were shown to depend on FR chemistry and thermal processing. The research confirmed that field problems with FRT plywood roof sheathing and roof-truss lumber resulted from thermal-induced acid degradation. In one comprehensive study, over 6000 specimens of density matched southern pine (16- × 35- × 250-mm) were systematically exposed at one of four temperatures: 25°, 54°, 66°, and 82°C for exposures up to 4 years (LeVan et al. 1990, Winandy 1995a, Lebow and Winandy 1999a). Data on the rates and magnitudes of thermal-induced strength loss at 27°, 54°, 66°, and 82°C were obtained for specimens treated with one of six FR-model chemicals or untreated.

Chemistry of Wood Strength

FIGURE 11.26 Example of fully serviceable untreated plywood roof sheathing (top left) adjacent to thermally degraded FR-treated plywood sheathing (center) used next to a gypsum-sheathed 60-minute-rated fire wall (lower right).

The influence of temperature and treatment pH was progressive, as shown by this example at 66°C in which each treatment (going from left to right on z-axis) has a progressively higher pH (Figure 11.27).

Kinetic-based models for predicting strength loss as a function of exposure temperature and duration of exposure were then developed from this data obtained at four temperatures. These kinetic models can be used predict thermal degradation at other temperatures (Lebow and Winandy 1999a). A single-stage approach quantitatively based on time-temperature superposition was used to model reaction rates, such that

$$Y_{ij} = b_j * \exp(-X * A * (H_i/H_o) * e^{(-E_a/RT_{ij})}) \qquad (11.10)$$

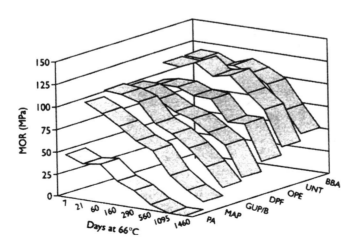

FIGURE 11.27 Strength loss over time of exposed to 66°C. Key: PA, phosphoric acid; MAP, monoammonium phosphate; GUP/B, guanylurea phosphate/boric acid; DPF, dicyandiamide-PA-formaldehyde; OPE, organophosphonate ester; UNT, untreated; and BBA, borax/boric acid.

where
- i = temperature of exposure,
- j = FR chemical,
- Y_{ij} = bending strength (MPa) at Temperature (T_i) for FR_j,
- X = time (days) at Temperature (T_i) for FR_j,
- b_{ij} = initial bending strength (MPa) at time ($X_i = 0$),
- H_i = relative humidity at test,
- H_o = normalized relative humidity (67% R.H. (per ASTM D5516),
- A = pre-exponential factor,
- E_a = activation energy,
- R = gas constant (J/°K*mole), and
- T_{ij} = temperature (°K) and for FR_j.

This research also proved that the use of pH buffers in FR chemicals, such as borates, can partially mitigate the initial effect of the FR treatment on strength and then significantly enhance resistance to subsequent thermal degradation. For modulus of elasticity there appeared to be few real benefits derived from adding borate-based pH buffers to the FR-mixture on the subsequent thermal degrade of FRT plywood exposed to high-temperatures (Winandy 1997). However, after 290 days of exposure at 66°C there were significant ($P < 0.05$) benefits derived with respect to limiting strength loss and loss in energy-related properties, such as work to maximum load, by the addition of borate to the FR-chemical mixture (Figure 11.28). Remedial treatments based on surface application of pH-buffered borate/glycol solutions were also developed to protect against additional in-service strength loss (Winandy and Schmidt 1995).

Other work also found that the rate of strength loss was largely independent of plywood quality or grade (Lebow and Winandy 1998) and could be directly related to changes in treated wood pH (Lebow and Winandy 1999b). Further, variation in redrying temperatures from 49°C to 88°C had little differential effect on the subsequent rate of thermal degradation when the treated plywood was exposed at 66°C for up to 290 days (Winandy 1997, Figure 11.29). This was related to the shorter kiln-residence times required at higher temperatures yielding similar states of entropy via differing, but thermodynamically comparable, temperature-duration histories.

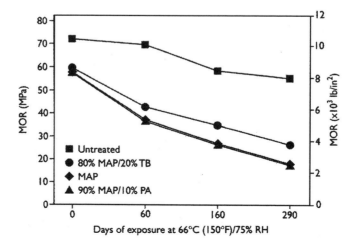

FIGURE 11.28 Effect of pH buffers on the rate of thermal degrade (Winandy 1997). Key: MAP, monoammonium phosphate; PA, phosphoric acid; and TB, sodium tetraoctaborate.

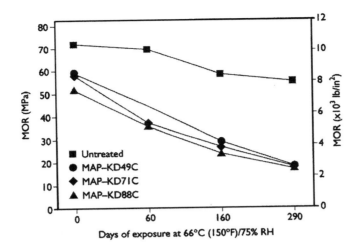

FIGURE 11.29 Effects of kiln drying after MAP treatment at 49, 71, and 88°C on thermal degrade (Winandy 1997).

11.8 SUMMARY

Over the years the strength of wood has been, for the most part, studied by physical chemists and engineers, and the chemistry of wood has been studied by organic chemists and biochemists. In a materials science approach to wood research, these disciplines must work together to relate the physical properties to the chemistry of the wood material.

In this chapter an explanation is presented of certain engineering aspects that are important in understanding the mechanical properties of wood. Individual factors such as growth, environment, chemicals, and use can greatly affect the physical and mechanical properties of the wood material. A theoretical model is presented to explain the relationship between physical properties and chemistry of wood at three distinct levels: macroscopic or cellular, microscopic or cell wall, and molecular or polymeric. These three levels and their implications on material properties must be understood to relate both wood chemistry and wood engineering from a materials science standpoint. When this is accomplished, the treatment and processing of wood and wood products can be controlled to yield more desirable and uniform properties.

While considerable work has been undertaken in the last 20–30 years, many of the underlying governing principles are yet unknown. The theories presented in this paper are offered as a starting point for consideration and discussion. Hopefully they may some day become a point from which, through mutual cooperation between the fields of engineering and chemistry, the chemistry of wood strength may be truly explained.

REFERENCES

Adams, E.H., Moore, G.L., and Brazier, J.D. (1979). The effect of flame-retardant treatments on some mechanical properties of wood. BRE Infor. Paper IP 24/79 Princes Risborough Lab., Aylesbury, Bucks, UK.

Alliott, E.A. (1926). Effect of acids on the mechanical strength of timber. A preliminary study. *J. Soc. Chem. Ind.* 45:463–466T.

American Wood Preservers Association. (2003). Standard C20: Structural lumber: Fire-retardant treatment by pressure processes. In: *AWPA Annual Book of Standards*. Selma, AL.

American Society for Testing and Materials. (2003a) Standard test method for small clear specimens of timber, ASTM Stand. Desig. D 143-94. In: *ASTM Annual Book of Standards.* West Conshohocken, PA.

American Society for Testing and Materials. (2003b). Standard test method for static tests of lumber in structural sizes. ASTM Stand. Desig. D 198-02e1. In: *ASTM Annual Book of Standards.* West Conshohocken, PA.

American Society for Testing and Materials. (2003c). Standard practice for establishing structural grades and related allowed properties for visually graded lumber. ASTM Stand. Desig. D 245-00. In: *ASTM Annual Book of Standards.* West Conshohocken, PA.

American Society for Testing and Materials. (2003d). Standard test methods for establishing clear wood strength values. ASTM Stand. Desig. D 2555-98e1. In: *ASTM Annual Book of Standards.* West Conshohocken, PA.

American Society for Testing and Materials. (2003e). Standard test methods for mechanical properties of lumber and wood based structural material. ASTM Stand. Desig. D 4761-02a. *ASTM Annual Book of Standards.* West Conshohocken, PA.

Atalla, R. (2002). Personal communication.

Barnes, H. M., and Winandy, J.E. (1986). Effects of seasoning and preservatives on the properties of treated wood. *Amer. Wood Pres. Assoc. Proc.* 82:95–104.

Barrett, J.D., Green, D.W., and Evans, J.W. (1989). Temperature adjustments for the North American in-grade testing program. In: *Proceedings of the Workshop on In-Grade Testing of Structural Lumber.* Forest Products Research Society, Madison, WI. 27–38.

Bendtsen, B.A. (1966). Sorption and swelling characteristics of salt-treated wood. USDA, Forest Service, Res. Paper FPL-RP-60. Madison, WI.

Bendtsen, B.A., Gjovik, L.R., and Verrill, S.P. (1983). Mechanical properties of longleaf pine treated with waterborne salt preservatives. USDA, Forest Service, Res. Paper, FPL 434.

Betts, H.S., and Newlin, J.A. (1915). Strength tests of structural timbers. USDA Bull. No. 286. Washington, DC.

Brazier, J.D., and Laidlaw, R.A. (1974). The implications of using inorganic salt flame-retardant treatments with timber. BRE Infor. Supp. IS 13/74 Princes Risborough Lab.: Aylesbury, Bucks, UK.

Clausen, C.A., and Kartal, S.N. (2003). Accelerated detection of brown-rot decay: Comparison of soil block test, chemical analysis, mechanical properties and immunodetection. *Forest Products Journal* 53(11/12):90–94.

Cowling, E.B. (1961). Comparative biochemistry of the decay of sweetgum sapwood by white-rot and brown-rot fungi. USDA, Forest Service, Tech. Bull. No. 1258.

Curling, S.F, Clausen, C.M, and Winandy, J.E. (2002). Solid wood products relationships between mechanical properties, weight loss, and chemical composition of wood during incipient brown-rot decay. *Forest Products J.* 52(7/8):34–39.

Dale, F.A. (1966). Fence posts and fire, *Forest Products Newsletter* 328:1–4.

Erickson, H.D., and Rees, L.W. (1940). Effect of several chemicals on the swelling and crushing strength of wood. *Agric. Res.* 60(1):593–604.

Feist, W.C. Hajny, G.J., and Springer, E.L. (1973). Effect of storing green wood chips at elevated temperatures. *Tappi* 56(8):91–95.

Freudenberg, K. (1932). The relation of cellulose to lignin in wood. *Chem. Ed.* 9(7):1171–1180.

Gerhards, C.C. (1977). Effect of duration and rate of loading on strength of wood and wood-based materials. USDA, Forest Service, Forest Products Laboratory, Res. Paper, FPL 283.

Gerhards, C.C. (1982). Effect of moisture and temperature on the mechanical properties of wood: An analysis of immediate effects. *Wood Fiber* 14(1):4–36.

Gillwald, W. (1961). Influence of different impregnation means on the physical and strength properties of wood. *Holztechnologie* 2:4–16.

Gerhards, C.C. (1970). Effect of fire-retardant treatment on bending strength of wood. USDA, Forest Service, Res. Paper FPL 145, Madison, WI.

Green, D.W., Evans, J.W., and Craig, B. (2003). Durability of structural lumber products at high temperature. Part 1: 66°C at 75% RH and 82° C at 30% RH. *Wood and Fiber Science* 35(4): 499–523.

Green, D.W. (1999). Adjusting Modulus of Elasticity of Lumber for Changes in Temperature. *Forest Products Journal* 49(10):82–94.

Hatt, W.K. (1906). Experiments on the strength of treated timber. USDA, Forest Service, Circ. 39. Washington, DC.

Ifju, G. (1964). Tensile strength behavior as a function of cellulose in wood. *Forest Products Journal* 74(8):366–372.
Johnson, J.W. (1967). Bending strength for small joists of Douglas-Fir treated with fire retardants. Oregon State Univ. Rep. T-23. Forest Res. Laboratory, School of Forestry, Oregon State University, Corvallis, OR.
Kalnins, M.A. (1966). Surface characteristics of wood as they affect curability of finishes. Part II. Photochemical degradation of wood. USDA, Forest Service Res. Paper FPL 57.
Kennedy, R.W. (1958). Strength retention in wood decayed to small weight losses. *Forest Products Journal* 8:308–14.
Koenigs, J.W. (1974). Hydrogen peroxide and iron: a proposed system for decomposition of wood by brown-rot basidiomycetes. *Wood Fiber* 6(1):66–80.
Kollmann, F., and Cote, W.A., Jr. (1968). *Principles of Wood Science and Technology.* Vol. 1. Solid Wood; Springer-Verlag, Berlin.
Koukal, M., Bednarcik, V. and Medricka, S. (1960). The effect of PCP and NaPCP on increasing the resistance of waterproof plywood to decay and on its physical and mechanical properties. *Drevarsky Vyskum* 5(1):43–60.
Lagergren, S., Rydholm, S., and Stockman, L. (1957). Studies on the interfibre bonds of wood. Part 1. Tensile strength of wood after heating, swelling and delignification. *Sven. Papperstidn.* 60:632–644.
Lam, F., and Morris, P.I., (1991). Effect of double-density incising on bending strength. *Forest Products Journal,* 41(6):43–47.
Larson, P.G., and Cox, W.T. (1938). *Mechanics of Material.* John Wiley and Sons, New York.
Larsen, M.J., Winandy, J.E., and Green, F. III. (1995). A proposed model of the tracheid cell wall having an inherent radial ultrastructure in the S_2 layer. *Materials and Organism* 29(3):197–210.
Lebow, S.T., and Winandy, J.E. (1998). The role of grade and thickness in the degradation of fire-retardant-treated plywood. *Forest Products Journal* 48(6):88–94.
Lebow, P.K., and Winandy, J.E. (1999a). Verification of the kinetics-based model for long-term effects of fire retardants on bending strength at elevated temperatures. *Wood and Fiber Science* 31(1): 49–61.
Lebow, S.T., and Winandy, J.E. (1999b). Effect of fire-retardant treatment on plywood pH and the relationship of pH to strength properties. *Wood Science and Technology* 33:285–298.
LeVan, S.M., Ross, R.J., and Winandy, J.E. (1990). Effects of fire retardant chemicals on the bending properties of wood at elevated temperatures. USDA Research Paper FPL-RP-498, Madison, WI.
LeVan, S.M., and Evans, J.W. (1996). Mechanical properties of fire-retardant treated plywood after cyclic temperature exposure. *Forest Products Journal* 46(5):64–71.
Mark, R.E. (1967). *Cell Wall Mechanics of Tracheids.* Yale Press, New Haven, CT.
McCarthy, W.G., Seaman, E.W., DaCosta, B., and Bezemer, L.D. (1972). Development and evaluation of a leach resistance fire retardant preservative for pine fence posts. *Inst. Wood Set.* 6(1):24–31.
MacLean, J.D. (1945). Effect of heat on the properties and serviceability of wood. Experiments on thin wood specimens. USDA, Forest Service, Forest Products Laboratory, Report R1471.
MacLean, J.D. (1953). Effect of steaming on the strength of wood. Proceedings Annual Meeting Am. Wood-Preserv. Assoc. 49:88–112.
MacLean, J.D. (1951). Preservative treatment of wood by pressure processes. USDA Agric. Handbook No. 40. Washington, DC.
Millett, M.A., and Gerhards, C.C. (1972). Accelerated aging: Residual weight and flexural properties of wood heated in air at 115 to 175 C. *Wood Sci.* 4(4):193–201.
Morrell, J.J., Gupta, R., Winandy, J.E., and Riyanto, D.S. (1998). Effects of incising on torsional shear strength of lumber. *Wood and Fiber Science* 30(4):374–381.
Pasek, E., and McIntyre C.R. (1990). Heat effects on fire-retardant treated wood. *J. Fire Sci.* 8:405–420.
Prentice, I.W. (1949). *Annual Report.* East Mailing Res. Station, Kent, UK,122–25, 1949.
Rowell, R.M. (1980). *How the Environment Affects Lumbar Design,* Lyon, D.E. and Galligan, W.L. (Eds.), USDA, Forest Service, Forest Products Laboratory Report.
Siau, J.F. (1969). The swelling of basswood by vinyl monomers. *Wood Sci.*1(4):250–253.
Soltis, L.A., and Winandy, J.E. (1989). Long-term strength of CCA-treated lumber. *Forest Products Journal* 39(5):64–68.

Sweet, M., and Winandy, J.E. (1999). Influence of degree of polymerization of cellulose and hemicellulose on strength loss in fire-retardant-treated southern pine. *Holzforschung* 53(3):311–317.

Stamm, A.J. (1964). *Wood and Cellulose Science.* The Ronald Press Co. New York.

Stamm, A.J., Burr, A.K., Kline, H.A. (1946). Staybwood: Heat-stabilized wood. *Ind. Eng. Chem.* 38:630–637.

Thomas, R.J. (1981). In: *Wood: Its Structure and Properties.* Wangaard, F.F. (Ed.), Penn State Univ. Press, University Park, PA, 101–146.

U.S. Department of Agriculture, Forest Service. (1999). Wood handbook. USDA General Technical Report GTR-113, Madison, WI.

U.S. Department of Defense, Department of the Navy, Civil Engineering Laboratory. (1979). GEL Tech. Data Sheet 79-07. USN, GEL, Port Hueneme, CA.

Wilson, T.R.C. (1932). Strength-moisture relations for wood. USDA Tech. Bull. No. 282, Washington, DC.

Wangaard, F.F. (1950). *The Mechanical Properties of Wood.* John Wiley and Sons, New York.

Winandy, J.E. (1989). The effects of CCA preservative treatment and redrying on the bending properties of 2 by 4 Southern Pine lumber. *Forest Products J.*, 39(9):14–21.

Winandy, J.E. (1994). Effects of long-term elevated temperature on CCA-treated Southern Pine lumber. *Forest Products J.* 44(6):49–55.

Winandy, J.E. (1995a). Effects of fire retardant treatments after 18 months of exposure at 150°F (66°C). USDA, Research Note FPL– RN– 0264. Madison, WI.

Winandy, J.E. (1995b). The Influence of time-to-failure on the strength of CCA-treated lumber. *Forest Products J.* 45(3):82–85.

Winandy, J.E. (1996a). Effects of treatment, incising and drying on mechanical properties of timber. In: *Proceedings: National Conference on Wood in Transportation Structures.* USDA. General Technical Report GTR-94. Ritter, M., Ed. 371–378.

Winandy, J.E. (1996b). Treatment-processing effects model for WBP-treated wood. In: Gopu, V.K.A. (Ed.), *Proceedings: International Wood Engineering Conference.* Louisiana State University, Baton Rouge, LA, 3:125–133.

Winandy, J.E. (2001). Effects of fire retardant retention, borate buffers, and re-drying temperature after treatment on thermal-induced degradation. *Forest Products J.* 47(6):79–86.

Winandy, J.E. (2001). Thermal degradation of fire-retardant-treated wood: Predicting residual service-life. *Forest Products J.* 51(2):47–54.

Winandy, J.E., and Barnes, H.M. (1991). The influence of initial kiln-drying temperature on CCA-treatment effects on strength. *Proc. Am. Wood Pres. Assn.* 87:147–152.

Winandy, J.E., Bendtsen, B.A., and Boone, R.S. (1983). The effect of time-delay between CCA preservative treatment and redrying on the toughness of small clear specimens of Southern Pine. *Forest Products J.* 33(6):53–58.

Winandy, J.E., and Boone, R.S. (1988). The effects of CCA preservative treatment and redrying on the bending properties of 2 by 6 Southern Pine lumber. *Wood Fib. Sci.* 20(3):50–64.

Winandy, J.E., Boone, R.S., and Bendtsen, B.A. (1985). The interaction of CCA preservative treatment and redrying: Effects on the mechanical properties of Southern Pine. *Forest Products J.* 35(10):62–68.

Winandy, J.E., Boone, R.S., Gjovik, L.R., and Plantinga, P.L. (1989). The effects of ACA and CCA preservative treatment and redrying: Effects on the mechanical properties of small clear specimens of Douglas Fir. *Proc. Am. Wood Pres. Assn.* 85:106–118.

Winandy, J.E., and Lebow, P.K. (1996). Kinetic models for thermal degradation of strength of fire-retardant-treated wood. *Wood and Fiber Science* 28(1):39–52.

Winandy, J.E., and Lebow, P.K. (2001). Modeling wood strength as a function of chemical compostion: An individual effects model. *Wood and Fiber Science.* 33(2):239–254.

Winandy, J.E., LeVan. S.M., and Ross, R.J. (1991). Thermal degradation of fire-retardant-treated plywood—Development and evaluation of a test protocol. USDA, Res. Paper FPL-RP-501.

Winandy, J.E., and Morrell, J.J. (1993). Relationship between incipient decay, strength, and chemical composition of Douglas-fir heartwood. *Wood and Fiber Science* 25(3):278–288.

Winandy, J.E., and Morrell, J.J. (1988). Effects of incising on lumber strength and stiffness: Relationship between incision density and depth, species and MSR-grade. *Wood and Fiber Science* 30(2): 185–197.

Winandy, J.E., and Rowell, R.M. (1984). The chemistry of wood strength. In: Rowell, R.M. (Ed.), *The Chemistry of Solid-Wood.* ACS Sym Series #208. Washington DC, 211–255.

Winandy, J.E., and Schmidt, E.L. (1995). Preliminary development of remedial treatments for thermally degraded fire-retardant-treated wood. *Forest Products J.* 45(2):51–52.

Woo, J.K. (1981). Effect of thermal exposure on strength of wood treated with fire retardants, Dept. Wood Science. Univ. of California Ph.D thesis. Berkeley, CA.

Wood, L.W. (1951). Relation of strength of wood to duration of load. USDA, Forest Service, Forest Products Lab. Report No. R-1916.

Wood, M.W., Kelso, K.C., Jr., Barnes, H.M., and Parikh, S. (1980). Effects of the MSU process and high preservative retentions on Southern Pine treated with CCA–Type C. *Proc. Am. Wood Pres. Assn.* 76:22–37.

12 Fiber Webs

Roger M. Rowell[1,2], James S. Han[1], and Von L. Byrd[1]
[1]USDA, Forest Service, Forest Products Laboratory, Madison, WI and
[2]Biological Systems Engineering Department, University of Wisconsin, Madison, WI

CONTENTS

12.1	Webs and Mats	350
	12.1.1 Forming Options	352
	12.1.1.1 Layering	352
	12.1.1.2 Fiber Mixing	352
	12.1.1.3 Use of Additives	353
	12.1.1.4 Scrim Addition	353
	12.1.1.5 Card Combined with Air Forming	353
	12.1.1.6 Melt Blown Polymer Unit (MBP) Combined with Air Forming	353
12.2	Pulp Molding	353
12.3	Geotextiles	355
	12.3.1 Erosion Control	355
12.4	Filters	357
	12.4.1 Types	357
	12.4.1.1 Physical Types	357
	12.4.1.2 Chemical Types	357
	12.4.2 Applications	357
	12.4.3 Testing Protocols for Filters	359
	12.4.3.1 Kinetic Tests	359
	12.4.3.2 Isotherms	359
	12.4.4 Biofilters for Organic Compounds	360
12.5	Sorbents	360
	12.5.1 Density	360
	12.5.2 Porosity and Surface Area	360
	12.5.3 Selectivity	360
	12.5.4 Retention	361
12.6	Mulch Mats	361
References		361

Wood fibers can be used to produce a wide variety of low-density three-dimensional webs, mats, and fiber-molded products. Short wood fibers blended with long fibers can be formed into flexible fiber mats, which can be made by physical entanglement, nonwoven needling, or thermoplastic fiber melt matrix technologies. The most common types of flexible mats are carded, air-laid, needle-punched, and thermobonded mats. In carding, the fibers are combed, mixed, and physically entangled into a felted mat. These are usually of high density but can be made at almost any density. Air-laid webs are made by laying down layers of wood fibers combined with a low-melting thermoplastic fiber, then passing the webs through a heated chamber that melts the thermoplastic. The heated web is then passed through calender rolls that press the melted fibers together with the wood fibers, holding the web together. A needle-punched mat is produced in a machine that passes a randomly formed machine-made web through a needle board that produces a mat in which the fibers are mechanically entangled. The density of air-laid webs and needled mats can be controlled by the amount of fiber going through the processes or by overlapping webs or mats to give the desired density. A thermobonded mat is made by combining natural fibers with a thermoplastic fiber in the needled mat technology that is then melted in a heated press holding the mat together. The webs and mats can be used as filters, geotextiles, sorbents, and mulch mats.

Wood fibers can also be formed into fiber-based products using air or water as a carrier. Fibers can be sprayed in an air stream and used as insulation or ground cover. Fibers can be slurried in water, molded into wide variety of shapes (pulp molding), and dewatered to form the final dry product.

12.1 WEBS AND MATS

Early information indicates that the Russians experimented with air forming during the 1930s. During this period, a patent was issued to two Russians describing a method for the production of a dry web using synthetic fiber and air (Pusyrev and Dimitriev 1960). In the early 1960s, a patent was issued to James Clark for the air forming of fibrous material and consolidating it into a web or sheet (Clark 1960). Later in the decade, a Finnish inventor named H.J. Hieldt developed an air forming method that involved the use of an electrostatic current to help guide the fibers.

In the mid-1960s, the Rando Corporation in the United States developed a different system in order to process long synthetic fibers for use in medium density fiberboard (Curlator Corp. 1967). The Rando system is shown in Figure 12.1. Section A in Figure 12.1 is where the fiber is fed into the system. Section B is a fiber opener where fiber bundles are separated and mixed with other fibers. Between A and B, the fibers are formed into a continuous mat, which is fully formed at C. There are several options at C. A liquid or powdered adhesive can be added if the final product is a web to be thermally formed into a three-dimensional composite. Another option is to place a seed applicator here to incorporate different types of seeds into the mat to be used as seeding geotextiles. At D, the web can go on through a needle board where the web is "needled" together in a nonwoven process. Some of the Rando systems were modified to make lighter weight webs

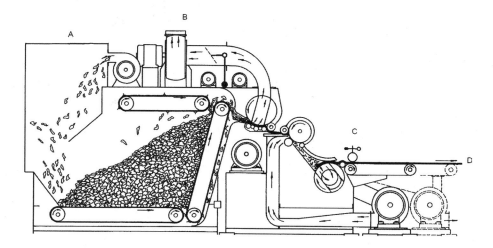

FIGURE 12.1 Schematic of Rando fiber mat forming system.

FIGURE 12.2 Fiber mat made using the Rando system.

that include wood pulp (Rando 1993). Both of these early systems were relatively speed limited. Figure 12.2 shows a web that has been made using the Rando system and a needle board.

In the early 1970s, a Japanese firm, Honshu, developed a process for making a variety of nonwovens using wood pulp and synthetic fibers. Most recently, Danweb Forming International, Ltd., developed a drum former capable of utilizing synthetic or natural fibers of various lengths (Danweb 2003, Figure 12.3). In addition, this firm has made use of a new and simpler horizontal machine layout (Wolff and Byrd 1990, Byrd 1990a,b).

The essential features of the drum system comprise two perforated counter-rotating drums (Figure 12.3B) located transversely above the forming wire within a square-section box. These drums are connected to a fixed pipe, such that the drums and their pipes form a track. A series of brush rolls are located inside the drums, transverse to the wire. The box itself is sealed in the transverse direction by means of seal rolls, and in the longitudinal direction with side plates. Figure 12.4 shows a typical mat made using the Danweb system.

In operation, the fibers, dispersed in air, are fed into the rotating drums via the fixed pipes. As the fiber stream passes into the drums, the brush rolls will force the fibers—partly as a result of a centrifugal effect and partly as a result of turbulence—through the perforated walls of the drum. The air from the suction box then draws the fibers onto the forming wire. The supply of single or mixed fibers is connected to the "horse-track" in such a way that a circular movement is ensured within the system. This guarantees a completely uniform distribution.

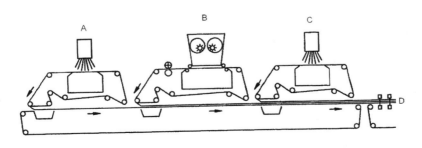

FIGURE 12.3 Schematic of Danweb fiber mat forming system.

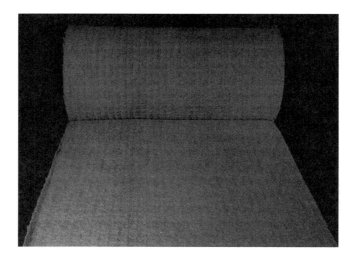

FIGURE 12.4 Fiber mat made using the Danweb system.

The drum system also has the advantage that it allows partially opened fibers to be separated out of the flow, because, as a result of their inertia, they remain on the outer wall of the track. This feature of the system makes it particularly useful in handling recycled fibers.

An additional advantage of the drum-type former is its ability to handle fibers up to 25 mm in length. This permits the addition of regular staple fibers or bonding fibers without modifying the former. Forming capacity remains fairly constant with fiber lengths up to 12 mm and then declines somewhat with fiber length from 12–25 mm.

One of the unique features of modern air forming systems, unlike those of a decade ago, is their flexibility with respect to feedstocks. Much of the early work in air forming centered around recycled and virgin cellulose fibers. This work has advanced to the point where a number of full-scale commercial air forming systems are in operation, producing a wide variety of absorbent and decorative disposables. More recent work with advanced air forming systems has focused on a wide variety of natural and synthetic fibers, which could logically be expected to be used in more advanced composite materials.

12.1.1 Forming Options

The identification of forming options and understanding their flexibility can conserve materials as well as process steps throughout the manufacturing process. Some of the composite web or molded product options that would be available with forming unit layouts are discussed next.

12.1.1.1 Layering

By placing different feedstocks in different forming heads, a manufacturer can produce a composite material with high-performance, high-cost materials on the exterior of the web and low-performance, low-cost material on the interior of the web. Figure 12.3 shows a Danweb system with several different options. Section A and/or Section C can be used to add a top or bottom layer of fiber or other film to the fiber mat that is formed in Section B.

12.1.1.2 Fiber Mixing

The inherent ability of the air former to handle streams of mixed fibers of different fiber types, deniers, and lengths permits in-line forming of a composite. Also, the fiber mixture can be varied for each forming head, which provides additional product flexibility.

Fiber Webs

12.1.1.3 Use of Additives

Work has been done that demonstrates how additives, such as superabsorbent powders and binders, can be added to the web during the forming process (Figure 12.2 Section C, Figure 12.3 Section A or C). In the case of superabsorbents, one advantage of this approach is that the superabsorbent powder, when near the area of maximum void space in the web, can absorb liquids faster and in greater quantity than if added to a finished web as part of a laminate in an off-line process. Also, because of their uniform dispersion, powdered binders can perform in much the same manner to ensure maximum strength with a minimum add-on.

12.1.1.4 Scrim Addition

Composite materials can be further enhanced through the addition of an open net or scrim between two of the air forming heads. This can simplify the process and possibly result in the use of less total raw material (Figure 12.3 Section A or C).

12.1.1.5 Card Combined with Air Forming

The first technique discussed is a means of combining a card with an air forming unit. From the standpoint of size and speed, these two processes are quite compatible. The advantages of a process combination such as this for composites are many: combination of long and short fibers, increased uniformity and bulk of composite materials, the ability to adjust the quantity of different fibers to be used in the process and process simplicity. Both are proven processes and compatible in terms of speed and machine trim.

Economic advantages obtained from a combination of air forming techniques include (a) the ability to make a thermally bondable composite for either single or multistage bonding; (b) the ability to substitute low-cost raw materials such as wood pulp and waste synthetic fiber in place of higher cost fibers; and (c) the ability to limit capital expenditures by upgrading an existing card line versus purchasing an entirely new nonwoven line.

12.1.1.6 Melt Blown Polymer Unit (MBP) Combined with Air Forming

Another technique is a combination of melt blown and air forming processes. This approach permits a composite to be produced in-line as a single process. If required, the process could include a provision for blanking out the center wood pulp core where the top and bottom layers of melt blown are thermally bonded.

This technology represents options to produce such products as (a) oil absorbent pads, (b) backing, laminating composites, and (c) wipes.

Air forming systems offer many product and process advantages in the production of both flexible and rigid composite materials. The units are of manageable size and can be combined with other process equipment to offer significant materials flexibility. From a commercial standpoint, air forming is a relatively young technology. We can expect current and improved systems to play an increasing role in future production of composite materials.

12.2 PULP MOLDING

Composites can be made using wood fibers in a wet process that forms a composite by dewatering the slurry (Laufenberg 1996). This is used today to make such products as egg cartons and nesting packaging where products are kept apart with a thin wall of molded pulp. Pulp molding is done using a forming slurry of pulp fibers with a consistency of 0.5 to 5% (dry fiber weight/water weight). Pulp molding is done in two steps: forming a dense fiber network from a wet fiber slurry onto a

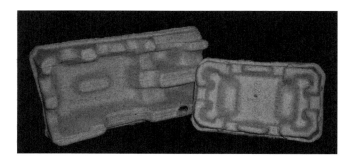

FIGURE 12.5 Products made by pulp molding.

configured surface or mold, and drying. Most of the pulp molding is done using a drainable surface (fine mesh screen), initial dewatering by gravity, vacuum dewatering and finally, heat drying. The pulp slurry is either poured into the mold or the mold is dipped into the pulp slurry and withdrawn or the slurry is pumped into the mold. For highly uniform surfaces, the pulp slurry consistency should be very low (see Figure 12.5).

Drying the wet formed product is done in one of several ways. Densification and drying can be done by applying pressure to the molded product. Minimal density and strength will result without the use of pressure in the drying step. The partially dewatered product can be dried in an oven. The strength of the final product comes from interfiber bonding similar to the bonds formed in the paper-making process. Starch can be added for increased bonding strength; however, since these products were formed in a wet medium, the products have very little wet strength or wet stiffness.

Structural pulp molded products can also be made using a similar process. Setterholm (1985) developed a method of forming a three-dimensional, waffle-like structure using a hard flexible rubber forming head. The product was called "Spaceboard" and could be used for structural applications. Further advancements were made by Hunt and Gunderson (1988) and Scott and Laufenberg (1994), as shown in Figure 12.6.

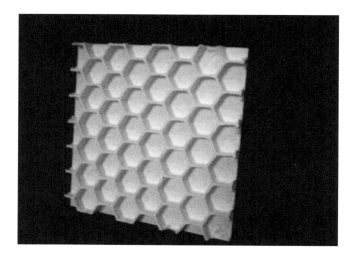

FIGURE 12.6 Spaceboard—A structural pulp molded product.

12.3 GEOTEXTILES

Geotextiles derive their name from the two root words *geo* and *textile*, referring to the use of fabrics in association with the earth. Modern erosion control is mainly done using geotextiles made using synthetic materials such as polypropylene and polyethylene. Wood and other agricultural fibers can be made into geotextiles that can be used to control erosion by using either the Rando or the Danweb process.

12.3.1 Erosion Control

Soil type and vegetation coverage are critical factors in the ability of the land to sorb water. With a healthy forest, where at least 60–75% of the ground is covered with vegetation, only about 2% or less of rainfall becomes surface runoff and erosion is low (less than 0.05 tons loss per acre). When the vegetation cover is between 35 and 50%, the surface runoff increases to 14% with a soil loss of 0.5 tons per acre. Finally, when the vegetation cover is less than 10%, the surface runoff increases to 73% with a soil loss of 5.55 tons per acre (Sedell et al. 2000). From these simple statistics, it is easy to see the effect catastrophic forest fires can have on our fresh water supply.

When there is a severe fire and little ground vegetation remains, surface runoff can increase over 70% and erosion can increase by three orders of magnitude (Bailey and Copeland 1961). There are three components to erosion: detachment, transport, and deposition. The rate of erosion will depend on the geology, topography, vegetation, and climate. In flat terrain, erosion may be minimal after a fire. However, in steep terrain, surface soil loss can be severe following a fire (Figure 12.7).

After the fire, burned land usually sorbs water more slowly than unburned land (Anderson and Brooks 1975). A severe fire in a forest can easily create a ground condition in which surface runoff can lead to flash floods, and erosion can result in not only loss of soil but also in badly contaminated water. Increases in water flow after a fire can result in more solids and dissolved materials in the water (DeBano et al. 1998). Water-soluble and insoluble nutrients can increase aquatic plant growth that may decrease water flow. Inorganic compounds leaked into the water increase the soluble ions that may increase both turbidity and toxicity of the water (Robichaud et al. 2000).

The addition of such chemicals could be based on silvicultural prescriptions to ensure seedling survival and early development on planting sites where severe nutritional deficiencies, animal and fire damage, insect attack, and weed problems are anticipated. Medium-density fiber mats can also

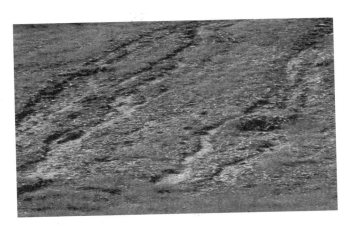

FIGURE 12.7 Example of severe soil erosion.

FIGURE 12.8 Application of fiber-based geotextiles.

be used to replace dirt or sod for grass seeding around new homesites and along highway embankments and stream beds. Grass or other types of seed can be incorporated in the fiber mat. Fiber mats promote seed germination and good moisture retention. Low- and medium-density fiber mats can be used for soil stabilization around new or existing construction sites. Steep slopes, without root stabilization, lead to erosion and loss of top soil.

In one type of geotextile, seeds are added to the geotextiles while the web is being formed. Grass and wildflower seeds can be added so that the geotextiles not only prevent erosion by forming a surface physical barrier but also allow grass to grow, establishing a new layer of plant growth with a root system to stabilize the soil after the geotextiles have degraded. Seeds can also be planted under the geotextile so that the seeds can germinate and grow above the geotextile. This type of geotextile should have enough physical strength to endure strong wind and at the same time about 30% of sunlight should pass through the geotextiles. Wood fiber–based geotextiles can hold moisture to help germinate the seeds. Other chemicals, such as fertilizers, can also be added to the web during its formation. Figure 12.8 shows a geotextiles application on a very steep embankment beside a highway.

Medium- and high-density fiber mats can also be used below ground in road and other types of construction as a natural separator between different materials in the layering of the backfill. It is important to restrain slippage and mixing of the different layers by placing separators between the various layers. Jute and kenaf geotextiles have been shown to work very well in these applications, but the potential exists for any of the long agro-based fibers.

Geotextiles in general are expected to be biodegraded within a given period of time. The timing of the biodegradation is dependent upon the materials used. High lignin content contributes to the biodegradability. Other factors such as density, hydrophobicity, extractive contents, etc. also play a role. Usually, the geotextiles are expected to last until the germination of seeds—between 4 and 6 weeks. The biodegradability can be controlled by addition of preservatives to prolong the decay and addition of fungi to speed up the decay.

It has been estimated that the global market for geotextiles is about 800 million square meters, but this estimate has not been broken down into use categories, so it is impossible to determine the portion that is available for natural geotextiles.

12.4 FILTERS

12.4.1 TYPES

12.4.1.1 Physical Types

Fiber-based filters can be used to remove suspended solids from both air and water. The physical types of wood fiber filters can be in several forms. Fibers can be made into webs or mats, or packed into a column or chamber. Webs or mats increase the surface area of the filter and stabilize hydraulic pressure. The suspended solids are physically captured and held in the webs until the filters are cleaned.

12.4.1.2 Chemical Types

Fiber-based filters can also remove dissolved inorganic ions, organic chemicals, and other soluble contaminates from water. Most of the wood fiber–based mats have limited capacities for removing soluble contaminates from water but their capacity can be greatly improved with chemical or plasma modification.

12.4.2 APPLICATIONS

We live in a water-based world. Water sculpts our landscape, provides navigational opportunities, transports our goods, and is the medium of life. Water is the basis of all life on earth, so it is not surprising that one of humankind's highest priorities is to ensure a long-term supply of clean water.

Seventy percent of the earth's surface is covered with water. Most of this water, 97.5%, is in the oceans and seas and is too salty to drink or to use to irrigate crops. Of the remaining 2.5%, 1.73% is in the form of glaciers and icecaps, leaving only about 0.77% available for our fresh water supply. Said another way, of the total water on earth, only 0.0008% is available and renewable in rivers and lakes for human and agricultural use. This is the water that falls as rain or snow or that has been accumulated and stored as groundwater that we depend on for our "clean" water resource.

For 1.5 to 2.5 billion people in the world, clean water is a critical issue (Lepkowski 1999). It is estimated that by the year 2025, there will be an additional 2.5 billion people on the earth that will live in regions already lacking sufficient clean water. In the United States, it is estimated that 90% of all Americans live within 10 miles of a body of contaminated water (Hogue 2000b). The U.S. Environmental Protection Agency (EPA) is working on guidelines and regulations to establish total maximum daily load (TMDL) for each pollutant that remain a problem (Hogue 2000a). The materials that the EPA has listed as water impairments include sediments, nutrients, pathogens, dissolved oxygen, metals, suspended solids, pesticides, turbidity, fish contamination, and ammonia. Other conditions to be considered for clean water on the list include pH, temperature, habitat, and noxious plants. Of these, sediments, nutrients, pathogens, and dissolved oxygen contribute the greatest to our contaminated water (1998 EPA data, cited in Hogue 2000a).

On one specific issue, that of arsenic in drinking water, the EPA has proposed lowering the maximum allowed level of arsenic from 50 ppb to 5 ppb due to concerns about bladder, lung, and skin cancer (Hileman 2000). Meeting these targets will not be easy. Arsenic in water is a global concern especially in countries like Bangladesh where most of their water wells are contaminated with arsenic (Lepkowski 1999).

About 80% of the fresh water in the United States originates on the 650 million acres of forestlands that cover about one third of the nation's land area. The nearly 192 million acres of national forest and grasslands are the largest single source of fresh water in the United States. In many cases, the headwaters of large river basins are located in national forests. In 1999, the EPA estimated that 3400 public drinking-water systems were located in watersheds contained in national forests and about 60 million people lived in these 3400 communities (Sedell et al. 2000).

The structure of wood and bark is very porous and has a very high free surface volume that should allow accessibility of aqueous solutions to the cell wall components. One cubic inch of a lignocellulosic material, for example, with a specific gravity of 0.4, has a surface area of 15 square feet. However, it has been shown that breaking wood down into finer and finer particles does increase sorption of heavy metal ions.

Lignocellulosics are hygroscopic and have an affinity for water. Water is able to permeate the noncrystalline portion of cellulose and all of the hemicellulose and lignin. Thus, through absorption and adsorption, aqueous solutions come into contact with a very large surface area of different cell wall components.

Laszlo and Dintzis (1994) have shown that wood has ion-exchange capacity and general sorptive characteristics, which are derived from their constituent polymers and structure. The polymers include extractives, cellulose, hemicelluloses, pectin, lignin, and protein. These are adsorbents for a wide range of solutes, particularly divalent metal cations (Laszlo and Dintzis 1994). Wood contains, as a common property, polyphenolic compounds, such as tannin and lignin, which are believed to be the active sites for attachment of heavy metal cations (Waiss et al. 1973, Masri et al. 1974, Randall et al. 1974, Bhattacharyya and Venkobachar 1984, Phalman and Khalafalla 1988, Verma et al. 1990, Shukla and Sakhardande 1991, Maranon and Sastre 1992, Lalvani et al. 1997, Vaughan et al. 2001). Sawdust has been used to remove cadmium and nickel (Basso et al. 2002) and several types of barks have been used to remove heavy metal ions from water (Randall 1977, Randall et al. 1974, Kumar and Dara 1980, Pawan and Dara 1980, Vazquez et al. 1994, Seki et al. 1997, Tiwari et al. 1997, Gaballah and Kibertus 1998, Bailey et al. 1999) from aqueous solution. Cellulose can also sorb heavy metals from solution (Acemioglu and Alma 2001). Isolated kraft lignin has been used to remove copper and cadmium (Verma et al. 1990, Cang et al. 1998) and organosolv lignin has been used to remove copper (Acemioglu et al., unpublished data) from aqueous solutions.

Acemioglu et al. postulate that metal ions compete with hydrogen ions for the active sorption sites on the lignin molecule (Acemioglu et al., unpublished data). They also conclude that metal sorption onto lignin is dependent on both sorption time and metal concentration. Basso et al. (2002) studied the correlation between lignin content of several woods and their ability to remove heavy metals from aqueous solutions. The efficiency of removing Cd(II) and Ni(II) from aqueous solutions was measured and they found a direct correlation between heavy metal sorption and lignin content. Reddad et al. (2002) showed that the anionic phenolic sites in lignin had a high affinity for heavy metals. Mykola et al. (1999) also showed that the galacturonic acid groups in pectins were strong binding sites for cations.

Extracting fibers with different solvents will change both the chemical and physical properties of the fibers. It is known, for example, that during the hot water and 1% sodium hydroxide extraction of fibers, the cell walls delaminate (Kubinsky 1971). A simple base treatment has been shown to greatly increase the sorption capacity of wood fibers (Tiemann et al. 1999, Reddad et al. 2002). At the same time, some of the amorphous matrix and part of the extractives, which have a bulking effect, are removed (Kubinsky and Ifju 1973), so that the individual microfibrils become more closely packed and shrunken (Kubinsky and Ifju 1974). Therefore, delamination and shrinkage may also change the amount of exposed cell wall components that may affect the heavy metal ion sorption capacity of the fibers.

Shen et al. (2004) have shown that phosphorus can be removed from water using a juniper-fiber-based web that is first saturated with a heavy metal. Figure 12.9 shows a plot of phosphorus uptake versus time with webs made of juniper fiber, base-treated juniper fiber, and juniper fiber that has been saturated with iron. The filter made using the heavy metal–loaded fiber removed much more phosphorus than the webs without the heavy metal.

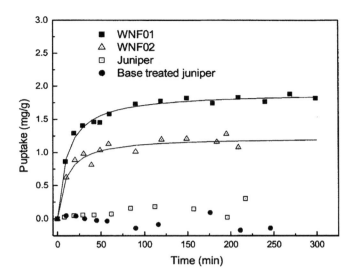

FIGURE 12.9 Plot of phosphorus removal from solution using juniper fiber filters.

12.4.3 Testing Protocols for Filters

12.4.3.1 Kinetic Tests

A kinetic test determines the time it takes to remove a given concentration of a contaminant. Two approaches are used. One uses a constant weight of sorbent that is applied to various concentrations of solution, and the other uses a constant concentration of solution applied to variable weights of the sorbent (Eaton-Dikeman 1960).

12.4.3.2 Isotherms

Adsorption occurs on the surface of the fibers and when the rate of adsorption equals the rate of desorption, equilibrium has been achieved and the capacity of the fiber has been reached. The theoretical adsorption capacity of the fiber for a particular contaminant can be determined by calculating its adsorption isotherm. Wood samples are carefully weighed (M) and placed in a known volume (V) of standard solution of a given metal ion and shaken at 150 rpm. After 24 hours the mixture of the standard solution and fiber sample are filtered and the final concentration of the solution is determined by ICP.

The quantity of adsorbate that can be taken up by an adsorbent is a function of both the characteristics and concentration of adsorbate and the pH (some authors also include temperature, but pH has more impact than temperature in wood fiber filtration). Generally, the amount of material adsorbed is determined as a function of the concentration at a constant pH, and the resulting function is called an adsorption isotherm. One equation that is often used to describe the experimental isotherm data was developed by the isotherm models of Langmuir and Freundlich (Weber 1970).

A characteristic of the Langmuir isotherm is that the loading of sorbate onto the sorbent approaches a limiting value, Q_{max}, as the concentration increases. This corresponds to monolayer coverage of the adsorbent surface with a Langmuir constant, b, related to the free energy of adsorption:

$$q_e = \frac{bQ_{max}C_e}{1+bC_e}, \qquad (12.1)$$

where b is Langmuir constant (L/g) and Q_{max} is maximum adsorbate loading (grams adsorbate/grams adsorbent).

The Freundlich isotherm is empirical and is used for heterogeneous surface energies:

$$q_e = KC_e^{1/n}, \qquad (12.2)$$

where K (L/mg), a measure of sorption capacity, and $1/n$, a measure of sorption intensity, are Freundlich constants. To obtain the best estimate of all constants for the Langmuir and Freundlich isotherms, data were fitted with nonlinear regressions using a least-squares fitting program.

12.4.4 BIOFILTERS FOR ORGANIC COMPOUNDS

A biofilter is a filter that contains a microorganism capable of degrading the organic compound that is trapped by the filter. The filter first sorbs the compound by physical or chemical entrapment and then the microorganism decomposes the chemical into smaller chemicals such as carbon dioxide and water. Langseth and Pflum (1994) reported that biofilters could be used to remove organic compounds with an estimated 95% efficiency. The chemicals removed were alcohols, aldehydes, organic acids, and small amounts of low–molecular weight volatile organics such as benzene and toluene. The retention times for small simple volatile compounds are very short, with a 90% reduction in low–molecular weight compounds in about 10 seconds. Higher molecular weight organics decompose at a slower rate (English 1996).

12.5 SORBENTS

Wood-based sorbents have been used for years as cleaning aids. For example, bark has been used to sorb oil from seawater, and treated sawdust and wood shavings have been used to clean industrial floors. There are several important factors that influence sorptive capacity, including density, porosity, surface area, selectivity, and retention (English 1996).

12.5.1 DENSITY

Bulk density is simply a measure of wood mass per unit volume. Wood bulk density varies from about 0.2 gm/cm³ to greater than 1 gm/cm³. In general, the lower the bulk density, the greater the sorption capacity. The true density of the wood cell wall is between 1.3 and 1.5 and does not take into consideration any internal voids.

12.5.2 POROSITY AND SURFACE AREA

Porosity can be defined in several ways. It can be a measure of the size of the voids in the wood or it can be a measure of how quickly and easily a liquid or gas can penetrate through a piece of wood of a given size. Porosity can be measured as the time it takes for a gas or liquid to travel through a piece of wood or it can be a measurement of the sizes of internal voids in the wood. Woods with a low bulk density have a very large internal void space and therefore large internal surface area. One cubic inch of wood with a specific gravity of 0.4 has a surface area of 15 square feet.

12.5.3 SELECTIVITY

Wood is hydrophilic, so it strongly hydrogen bonds to polar liquids such as water. Selectivity is a measure of the ability of wood to preferentially sorb one gas or liquid, or one chemical in a gas or liquid, over another. Selectivity is also influenced by the wood pore size, ability to wet the wood surface, and capillary pressure.

FIGURE 12.10 Fiber-based mulch mat.

12.5.4 RETENTION

Retention is the ability of a saturated wood sample, after it has been used as a sorbent, to retain the sorbed gas, liquid, or specific chemical. Retention can be a critical factor in choosing a sorbent system. If the wood has the ability to remove a given contaminant but quickly releases it upon, for example, exposure to air, the sorbed contaminant will be released into the environment before it can be disposed of.

12.6 MULCH MATS

Mulching materials have been used around plants for many years (Waggoner et al. 1960) and are generally of two types: particle mulches and sheet mulches. Particle mulches, such as bark, wood shavings, or sawdust, are used to heat and hold moisture close to the plant. Sheet mulches are similar to geotextiles but are usually thicker and last longer in the environment. They are used to enhance seedling survival in wood species where the seedling remains in a "grass-like" state for several years, i.e., the southern pines (Figure 12.10). The mulch mats, like the loose mulches, hold moisture and heat as well as reducing weed growth around the seedling. Mulch mats must remain intact for as long as competition from unwanted plants remains. For this reason, many mulch mats are made from plastics such as polyethylene or polypropylene.

REFERENCES

Acemioglu, B. and Alma, M.H. (2001). Equilibrium studies on the adsorption of Cu(II) from aqueous solution onto cellulose. *J. Colloid Interf. Sci.*, 243:81.

Anderson, W.E. and Brooks, L.E. (1975). Reducing erosion hazard on a burned forest in Oregon by seeding. *J. Range Manage.* 28(5):394.

Bailey, S.E., Olin, T.J., Bricka, R.M., and Adrian, D.D. (1999). A review of potentially low-cost sorbents for heavy metals. *Water Res.* 33:2469.

Bailey, R.W. and Copeland, O.L. (1961). Vegetation and engineering structures in flood and erosion control. In: Proceedings, 13th Congress, International Union of Forest Research Organization. September 1961, Vienna, Austria. Paper 11-1: 23.

Basso, M.C., Cerrella, E.G., and Cukierman, A.L. (2002). Lignocellulosic materials as potential biosorbents of trace toxic metals from wastewater. *Ind. Eng. Chem. Res.* 3580–3585.

Bhattacharyya, A.K. and Venkobachar, C. (1984). Removal of cadmium (II) by low cost adsorbents. *J. Environ. Eng. Division (ASCE)* 110:110.

Browning B.L. (1967). The extraneous components of wood. In: *Methods of Wood Chemistry*, John Wiley & Sons, New York, pp. 75–90.

Byrd, V.L. (1990a). Dan-Web process revolutionizes dry-forming. *Nonwovens Ind.* 21(5):74.

Byrd, V.L. (1990b). New air-laid process revolutionizes industry. *The Business Edition* March, pp. 1,3,5.

Clark, J.A. (1960). Apparatus and method for producing fibrous structures. U.S. Patent 2,931,076.

Curlator Corp. (1967). Method of making a random fiber web. British Patent 1,088,991.

Danweb. (2003). Research and development, Danweb, Aarhus, Denmark.

DeBano, L.F., Neary, D.G., and Folliott, P.F. (1998). *Fire's Effects on Ecosystems.* John Wiley & Sons, New York.

Eaton-Dikeman Company. (1960). *Handbook of Filtration*, The Eaton-Dikeman Company, Mt. Holly Springs, PA.

English, B. (1996). Filters, sorbents and geotextiles. In: Rowell, R.M., Young, R.A., and Rowell, J.K. (Eds.), *Paper and Composites from Agro-based Resources,* CRC Lewis Publishers, Boca Raton, FL, pp. 403–426.

Gaballah, I. and Kibertus, G. (1998). Recovery of heavy metal ions through decontamination of synthetic solutions and industrial effluents using modified barks. *J. Geochem. Expl.* 62:241.

Hogue, C. (2000a). Clearing the water. *Chem. Eng. News* March, 31.

Hogue, C. (2000b). Muddied waters. *Chem. Eng. News* August, 19–20.

Hileman, B. (2000). Rules in the fast lane. *Chem. Eng. News* October, 43–44.

Hunt, J.F. and Gunderson, D.E. (1998). FPL Spaceboard development. In: Tappi Proceedings of the 198th Corrugated Containers Conference, Tappi Press, Atlanta, GA, p. 11.

Kubinsky. E., (1971). Influence of steaming on the properties of *Quercus rubra* L. wood. *Holzforschung* 25(3):78.

Kubinsky E. and Ifju, G. (1973). Influence of steaming on the properties of red oak. Part I. Structural and chemical changes. *Wood Science* 6(1):87.

Kubinsky E. and Ifju, G. (1974). Influence of steaming on the properties of red oak. Part I. Changes of structural and related properties. *Wood Science* 7(2):103.

Kumar, P. and Dara, S.S. (1980). Modified barks for scavenging toxic heavy metal ions. *Indian J. Environ. Health,* 22:196.

Laufenberg, T.L. (1996). Packaging and light weight structural composites. In: Rowell, R.M., Young, R.A., and Rowell, J.K. (Eds.), *Paper and Composites from Agro-based Resources,* CRC Lewis Publishers, Boca Raton, FL, pp. 337–350.

Lalvani, S.B., Wiltowski, T.S., Murphy, D., and Lalvani, L.S. (1997). Metal removal from process water by lignin. *Environ. Technol.* 18(11):1163.

Langseth, S. and Pflum, D. (1994). Weyerhauser tests large pilot biofilters for VOCs removal, *Panel World* 3(2):22.

Laszlo, J.A. and Dintzis, F.R. (1994). Crop residues as ion-exchange materials. Treatment of soybean hull and sugar beet fiber (pulp) with epichlorohydrin to improve cation-exchange capacity and physical stability. *J. Appl. Polymer Sci.,* 52:521.

Lepkowski, W. (1999). Science meets policy in shaping water's future. *Chem. Eng. News* December, 127.

Maranon, E. and Sastre, H. (1992). Behaviour of lignocellulosic apple residues in the sorption of trace metals in packed beds. *Reactive Polymers*, 18:172.

Masri, M.S., Reuter, F.W., and Friedman, M. (1974). Binding of metal cations by natural substances. *J. Appl. Polymer Sci.* 18:675.

Mykola, T.K., Kupchik, L.A., and Veisoc, B.K. (1999). Evaluation of pectin binding of heavy metal ions in aqueous solutions. *Chemosphere* 38(11):2591.

Pawan, K. and Dara, S.S. (1980). Modified barks for scavenging toxic heavy metal ions. *Indian J. Environ. Health* 22:196.

Phalman, J.E. and Khalafalla, J.E. (1988). Use of ligochemicals and humic acids to remove heavy metals from process waste streams. RI 9200, U.S. Department of Interior, Bureau of Mines.

Pusyrev, S.A. and Dimitriev, M. (1960). Production of paper and board by the dry process. *Pulp Paper Mag. Can.* 61(1):T3.

Randall, J.M. (1977). Variations in effectiveness of barks as scavengers for heavy metal ions. *Forest Prod. J.* 27(11):51.

Randall, J.M., Bermann, R.L., Garrett, V., and Waiss, A.C. (1974). Use of bark to remove heavy metal ions from waste solutions. *Forest Prod. J.* 24(9):80.

Rando Machine Corporation. (1993). Product improvement bulletin, PIB-018. Rando Machine Corporation, Macedon, NY.

Rando Machine Corporation. (n.d.) Installation, Operation and Maintenance Manual No.631 for the Rando-Webber, The Commons, Macedon, NY.

Reddad, Z., Gerente, C., Andres, Y., Ralet, M-C., Thibault, J-F., and Cloirec, P.L. (2002). Ni(II) and Cu(II) binding properties of native and modified sugar beet pulp. *Carbohydrate Polymers* 49:23.

Robichaud, P.R., Beyers, J.L., and Neary, D.G. (2000). Evaluating the effectiveness of postfire rehabilitation treatments. USDA, Forest Service, Rocky Mountain Research Station, General Technical Report RMRS-GTR-63.

Rowell, R.M. and Han, J.S. (1999). Changes in kenaf properties and chemistry as a function of growing time. In: *Kenaf Properties, Processing and Products*. Mississippi State University, Ag & Bio Engineering, pp. 33–41, Mississippi State, MS.

Scott, C.T. and Laufenberg, T.L. (1994). Spaceboard II panels: Preliminary evaluation of mechanical properties. In: PTEC 94 Timber Shaping the Future, Proceedings of the Pacific Timber Engineering Conference, July, Gold Coast, Fortitude Valley, Queensland, Australia, TRADA 632, p. 2.

Sedell, J.S. (2000). Water and the Forest Service, ES 660. USDA, Forest Service Washington Office, Washington, DC.

Seki, K., Saito, M., and Aoyama, M. (1997). Removal of heavy metals from solutions by coniferous barks. *Wood Sci. Technol.* 31:441.

Setterholm, V.C. (1985). FPL spaceboard new structural sandwich concept. *Tappi* 68(6):40.

Shin, E.W., Han, J.S., and Min, H. (2004). Removal of phosphorus from water using lignocellulosic material modified with iron species from acid mine drainage. *Environ. Technol.*

Shukla, S.R. and Sakhardande, V.D. (1991). Metal ion removal by dyed cellulosic materials. *J. Appl. Polymer Sci.* 42:825.

Sun, R-C., Lawther, J.M., and Banks, W.B. (1998). Isolation and characterization of organosolv lignins from wheat straw. *Wood and Fiber Sci.* 30(1):56.

Tiemann, K.J., Gardea-Torresdey, J.L., Gamez, G., Dokken, K., Sias, S., Renneer, M.W., and Furenlid, L.R. (1999). Use of x-ray absorption spectroscopy and esterification to investigate Cr (III) and Ni(II) ligands in alfalfa biomass. *Environ. Sci. Technol.* 33:150.

Tiwari, D.P., Saksena, D.N., and Singh, D.K. (1997). Kinetics of adsorption of Pb(II) on used tea leaves and Cr(VI) on *Acacia arabica* bark. *Dev. Chem. Eng. Miner. Process.* 5:79.

Vaughan, T., Seo, C.W., and Marshall, W.E. (2001). Removal of selected metal ions from aqueous solution using modified corncobs. *Bioresource Technol.* 78:133.

Vazquez, G., Antorrena, G., Gonzalez, J., and Doval, M.D. (1994). Adsorption of heavy metal ions by chemically modified *Pinus pinaster* bark. *Bioresource Technol.* 48:251.

Verma, K.V.R., Swaminathan T., and Subrahmnyam, P.V.R. (1990). Heavy metal removal with lignin. *J. Environ. Sci. Health* A25(2):242.

Waggoner, P.E., Miller, P.M., and DeRoo, H.C. (1960). Plastic mulching-principles and benefits, Bull. No. 634. Connecticut Agricultural Experiment Station, New Haven, CT.

Waiss, A.C., Wiley, M.E., Kuhnle, J.A., Potter, A.L., and McCready, R.M. (1973). Adsorption of mercuric cation by tannins in agricultural residues. *J. Environ. Quality* 2:369.

Weber, W.J. (1970). Physicochemical Process: For water quality control. Wiley-Interscience, New York, NY.

Wolff, H. and Byrd, V.L. (1990). Flexible composite and multicomponent air-formed webs. *Tappi J.* 73(9):159.

13 Wood Thermoplastic Composites

Daniel F. Caulfield[1], Craig Clemons[1], Rodney E. Jacobson[2], and Roger M. Rowell[1,3]

[1]USDA Forest Service, Forest Products Laboratory, Madison, WI
[2]A-J Engineering, LLC, Middleton, WI
[3]Department of Biological Systems Engineering, University of Wisconsin, Madison, WI

CONTENTS

13.1 Brief History 366
13.2 Materials 367
    13.2.1 Wood Flour 367
    13.2.2 Wood Fiber 367
    13.2.3 Thermoplastic Matrix Materials 367
    13.2.4 Added Compounds 368
        13.2.4.1 Compatibilizers 368
13.3 Processing 368
13.4 Performance 371
    13.4.1 Mechanical Properties 372
    13.4.2 Moisture Properties 373
    13.4.3 Biological Properties 374
    13.4.4 Weathering and Fire Properties 374
13.5 Higher Performance Materials 374
13.6 Markets 375
References 377

The term "wood-plastic composites" refers to any number of composites that contain wood (of any form) and either thermoset or thermoplastic polymers. Thermosets or thermoset polymers are plastics that, once cured, cannot be remelted by heating. These include cured resins, such as epoxies and phenolics, plastics with which the forest products industry is most familiar (see Chapter 10). Wood-thermoset composites date to the early 1900s. An early commercial composite marketed under the trade name Bakelite was composed of phenol-formaldehyde and wood flour. Its first commercial use was reportedly as a gearshift knob for Rolls Royce in 1916 (Gordon 1988). Thermoplastics are plastics that can be repeatedly melted, such as polyethylene, polypropylene, and polyvinyl chloride (PVC). Thermoplastics are used to make many diverse commercial products, such as milk jugs, grocery bags, and siding for homes. In contrast to the wood-thermoset composites, wood-thermoplastic composites have seen phenomenal growth in the United States in recent years. This chapter deals exclusively with wood-thermoplastic composites, which are now most often simply referred to as wood-plastic composites (WPCs) with the common understanding that the plastic always refers to a thermoplastic.

The birth of the WPC industry involved the interfacing of two industries that historically knew little about each other and had very different knowledge bases, expertise, and perspectives. The forest products industry has greater experience and resources in the building products market and its production methods center around the typical wood processes: sawing, veneering, chipping, flaking, and gluing. The plastics industry has knowledge of plastics processing that centers around extrusion, compression-molding, and injection-molding technologies. Not surprisingly, some of the earliest companies to produce WPCs were window manufacturers that had experience with both wood and plastics.

13.1 BRIEF HISTORY

In the United States, WPCs have been produced for several decades, but they were produced even earlier in Europe. Major growth in the United States, however, did not occur until fairly recently. In 1983, American Woodstock, now part of Lear Corporation in Sheboygan, Wisconsin, began producing WPC panel substrates for automotive interiors using Italian extrusion technology (Schut 1999). Polypropylene with approximately 50% wood flour was extruded into a flat sheet that was then formed into various shapes for interior automotive paneling. This was one of the first major applications of WPC technology in the United States.

In the early 1990s, Advanced Environmental Recycling Technologies (AERT, Junction, TX) and a division of Mobil Chemical Company that later became Trex (Winchester, VA) began producing solid WPCs consisting of approximately 50% wood fiber in polyethylene. These composites were sold as deck boards, landscape timbers, picnic tables, and industrial flooring (Youngquist 1995). Similar composites were milled into window and door component profiles. Today, the decking market is the largest and fastest growing WPC market.

Also in the early 1990s, Strandex Corporation (Madison, WI) patented technology for extruding high wood fiber content composites directly to final shape without the need for milling or further forming. Strandex has continued to license its evolving technology.

Andersen Corporation (Bayport, MN) began producing wood fiber-reinforced PVC subsills for French doors in 1993. Further development led to a wood-PVC composite window line (Schut 1999). These products allowed Andersen to recycle wastes from both wood and plastic processing operations. The market for WPC window and door profiles has continued to grow.

In 1996, several U.S. companies began producing a pelletized feedstock from wood (or other natural fibers) and plastic. These companies provide compound pellets for many processors who do not want to blend their own material. Since the mid-1990s, activity in the WPC industry has increased dramatically. Technology is developing quickly and many manufacturers have begun to produce WPCs. Although the WPC industry is still only a fraction of a percent of the total wood products industry (Smith 2001), it has made significant inroads in certain markets. Current end product manufacturers are an interesting mix of large and small manufacturers from both the plastics and forest products industries. According to a recent market study, the WPC market was 320,000 metric tons (700 million lb) in 2001, and the volume is expected to more than double by 2005 (Mapleston 2001b). The status and developments in the WPC industry are now also closely monitored by Principia Partners, an international business consulting firm for the industry. In 2003, Principia Partners reviewed recent developments and trends (Morton et al. 2003). In North America more than 67 WPC enterprises produce 590,000 tons valued at $700,000. The vast majority of production is extruded materials for decks, with lesser amounts of extruded profiles for windows, railings, transportation, and other infrastructure uses. Demand is projected to increase at an annual rate of 14% over the next few years, reaching a total increase of 290% by 2010. Significant production increases in Europe and Japan are also anticipated.

In 1991, the First International Conference on Woodfiber-Plastic Composites was convened in Madison, Wisconsin (about 50 attendees), with the intent of bringing together researchers and industrial representatives from both the plastics and forest products industries to share ideas and technology on WPCs. A similar, sister conference (Progress in Woodfiber-Plastic Composites) began in Toronto, Ontario, the following year and is being held in alternating years. These conferences have grown steadily through the 1990s, and in 2003 in Madison the 7th International Conference on Wood- and Natural-Fiber-Plastic Composites Conference was attended by over 400 conferees. Additional conferences have been held in North America and elsewhere around the world as both interest and the market have grown.

13.2 MATERIALS

13.2.1 Wood Flour

The wood used in WPCs is most often in particulate form (e.g., wood flour) or very short fibers and bundled fibers rather than long individual wood fibers. Products typically contain approximately 50% wood, although some composites contain very little wood and others as much as 70%. The relatively high bulk density and free-flowing nature of wood flour compared with wood fibers or other longer natural fibers, as well as its low cost, familiarity, and availability, is attractive to WPC manufacturers and users. Common species used include pine, maple, and oak. Typical particle sizes are 10 to 80 mesh. There is a wide range of wood flour suppliers and they cater to a number of different industries. These are both large companies that have broad distribution networks as well as small, single source suppliers catering to single customers. Because of the varied and diverse nature of wood flour suppliers, there are currently few good resources that list wood flour manufacturers. Wood-plastic composite manufacturers obtain wood flour either directly from forest products companies such as lumber mills and furniture, millwork, or window and door manufacturers that produce it as a byproduct or from companies that specialize in wood flour production. With a growing number of wood flour suppliers targeting the wood-plastic composites industry, they are beginning to be listed in plastics industry resources.

As with most materials, wood flour costs are variable and depend on such factors as volume, availability, particle size, and shipping distance. However, wood flour is typically about $0.11–0.22/kg ($0.05–0.10/lb) in the United States. Narrow particle size distributions and fine wood flour sizes tend to increase cost. Because there are many small manufacturers of wood flour and the volume is small relative to other wood products (solid wood, wood composites, and paper), reliable statistical information on wood flour availability is scarce.

13.2.2 Wood Fiber

Because of the potential of improved mechanical properties with fillers of greater l/d ratios, there has been a continuing interest in the use of individual wood fibers rather than wood flour as reinforcement in WPC. Recently, two wood-pulp fiber suppliers, Rayonier Corp. (Savannah, GA) and Creafill Fiber Corp. (Chesterton, MD) have been marketing wood-pulp fibers for use as reinforcements for thermoplastics. The fibers that they supply have been promoted for use in WPC of higher performance potentials. Adding fibers rather than flour increases mechanical properties such as strength, elongation, and unnotched Izod impact energy. However, processing difficulties, such as feeding and metering low-bulk-density fibers, have limited the use of fibers in WPCs. There have been some developments in pelletizing fibers for ease of handling and processing (Jacobson et al. 2002).

13.2.3 Thermoplastic Matrix Materials

The WPC presently consumes over 685,000 tons of thermoplastic resins annually (Morton et al. 2003). There are some differences in processing technologies between Europe and North America and this is reflected in the consumption totals for the various resins. Polyethylene resin (PE) is the dominant resin in North America amounting to 83% of the 685,000 tons consumed. Polypropylene

(PP) is 9% of the total, and 7% of the thermoplastic resin used in the WPC industry is polyvinyl chloride (PVC). Only about 1% of other resins are used.

13.2.4 ADDED COMPOUNDS

Wood and thermoplastic are not the only components in WPCs. These composites also contain additional materials that are added in small amounts to affect processing and performance. Although formulations are highly proprietary, additives such as coupling agents, light stabilizers, pigments, lubricants, fungicides, and foaming agents are all used to some extent. Some additive suppliers are specifically targeting the WPC industry (Mapleston 2001a).

13.2.4.1 Compatibilizers

Since the wood component of a wood-thermoplastic composite is hydrophilic and the plastic is hydrophobic, a compatibilizer is often used to improve interfacial bonding of the two different phases. One of the most common compatibilizers used today is a maleic anhydride grafted polypropylene (MAPP). There are many different types of MAPPs differing in molecular weight and the degree of maleic anhydride substitution. One used most often had a number average molecular weight of 20,000, a weight average molecular weight of 40,000 and had about 6% by weight of maleic anhydride in the polymer.

This compatibilizer it thought to work by two different mechanisms. First, the anhydride reacts with a cell wall polymer hydroxyl group to form an ester bond then the PP polymer attached to the anhydride intertangles into the PP or PE network in the melt (see Figures 13.1 and 13.2). Figure 13.3 shows a composite where a compatibilizer has been used. It can be seen that when the composite was broken in two pieces, the fiber shown broke rather than be pulled out of the thermoplastic matrix. There is a small gap at the base of the fiber showing that there was some slippage but the fiber broke rather than being pulled out of the matrix due to weak interfacial bonding between the fiber and the matrix.

13.3 PROCESSING

The plastics industry has traditionally used inorganic materials like talc, calcium carbonate, mica, and glass fibers to fill and to modify the performance of plastic; about 2.5 billion kg of fillers and reinforcements are used annually (Eckert 2000). The industry was reluctant to use wood or wood fiber as filler because of its low bulk density, low thermal stability, and tendency to absorb moisture. The majority of thermoplastics arrive at a manufacturer as free-flowing pellets or granules with a

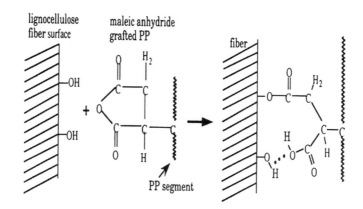

FIGURE 13.1 Reaction of an anhydride end group on MAPP with a cell wall hydroxyl group.

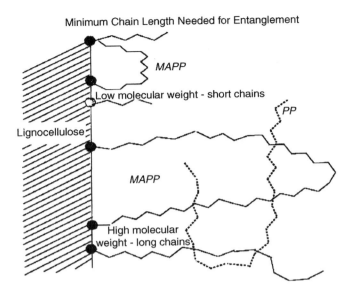

FIGURE 13.2 Entanglement of the PP on MAPP with the matrix PP.

bulk density of about 500 kg/m. The plastics processor is faced with the problem of how to consistently meter and force the low bulk density wood fiber into small feed openings typical of plastics processing equipment. In addition, the processing temperature for even low melting point plastics is often too high for incorporating wood fiber without thermal degradation. The high moisture content of wood and other natural fibers is also problematic to the plastics industry, which considers about 1–2% moisture content high. Even plastics processors with vented equipment capable of removing moisture during processing were averse to removing 5–7% moisture from wood fibers. Resin dryers, which are occasionally needed to dry plastics, are not appropriate for wood particles or fibers, and drying the fine wood particles poses a fire hazard. Plastics processors who tried to use wood or other natural fibers often lacked knowledge about wood, and their failed attempts made the industry generally skeptical of combining wood and plastic.

For the wood products industry, thermoplastics were a foreign world, albeit one that occasionally intruded on traditional markets (e.g., vinyl siding). Competing in different markets, forest products and plastics industries had few material and equipment suppliers in common and they processed materials very differently and on entirely different scales (Youngquist 1995).

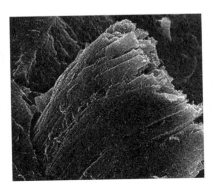

FIGURE 13.3 Fiber broken in a test for compatibilization.

The perspective of some plastics industries has changed dramatically in the last decade. Interest has been fueled by the success of several WPC products, greater awareness and understanding of wood, developments from equipment manufacturers and additive suppliers, and opportunities to enter new markets, particularly in the large-volume building applications sector. Forest products industries are changing their perspective as well. They view WPCs as a way to increase the durability of wood with little maintenance on the consumer's part (one of the greatest selling points). Some forest products companies are beginning to manufacture WPC lumber and others are distributing this product. These ventures into WPCs are being driven by customer demand and opportunities based on the industry's experience in building products (Anonymous 2001).

Because of the limited thermal stability of wood, it was believed initially that only thermoplastics that melt or can be processed at temperatures below 200°C could be used in WPCs and currently that is the practice. Most WPCs are made with polyethylene, both recycled and virgin, for use in exterior building components. However, WPCs made with wood-polypropylene are typically used in automotive applications and consumer products, and these composites have recently been investigated for use in non-structural building profiles. Wood-PVC composites typically used in window manufacture are now being sometimes used in decks and railings as well. Polystyrene and acrylonitrile-butadiene-styrene (ABS) are also being used. The plastic is often selected based on its inherent properties, product need, availability, cost, and the manufacturer's familiarity with the material. Small amounts of thermoset resins such as phenol-formaldehyde or diphenyl methane diisocyanate are also sometimes used in composites with a high wood content (Wolcott and Adcock 2000).

The manufacture of thermoplastic composites is often a two-step process. The raw materials are first mixed together in a process called compounding, and the compounded material is then formed into a product. Compounding is the feeding and dispersing of fillers and additives in the molten polymer (see Figure 13.4). Many options are available for compounding, using either batch or continuous mixers. The compounded material can be immediately pressed or shaped into an end product or formed into pellets for future processing. Some product manufacturing options for WPCs force molten material through a die (sheet or profile extrusion), into a cold mold (injection molding), between calenders (calendering), or between mold halves (thermoforming and compression molding) (Youngquist 1999). Combining the compounding and product manufacturing steps is called in-line processing.

The majority of WPCs are manufactured by profile extrusion, in which molten composite material is forced through a die to make a continuous profile of the desired shape (see Figure 13.5).

FIGURE 13.4 Davis Standard twin-screw extruder with side-stuffer crammer for introducing wood flour or wood fiber into molten thermoplastic matrix.

FIGURE 13.5 Examples of extruded wood fiber plastics.

Extrusion lends itself to processing the high viscosity of the molten WPC blends and to shaping the long, continuous profiles common to building materials. These profiles can be a simple solid shape, or highly engineered and hollow. Outputs up to 3 m/min. (10 ft./min.) are currently possible (Mapleston 2001b).

Although extrusion is by far the most common processing method for WPCs, the processors use a variety of extruder types and processing strategies (Mapleston 2001c). Some processors run compounded pellets through single-screw extruders to form the final shape. Others compound and extrude final shapes in one step using twin-screw extruders. Some processors use several extruders in tandem, one for compounding and the others for profiling (Mapleston 2001c). Moisture can be removed from the wood component before processing, during a separate compounding step (or in the first extruder in a tandem process), or by using the first part of an extruder as a dryer in some in-line process. Equipment has been developed for many aspects of WPC processing, including materials handling, drying and feeding systems, extruder design, die design, and downstream equipment (i.e., equipment needed after extrusion, such as cooling tanks, pullers, and cut-off saws). Equipment manufacturers have partnered to develop complete processing lines specifically for WPCs. Some manufacturers are licensing new extrusion technologies that are very different from conventional extrusion processing (Mapleston 2001c,d).

Compounders specializing in wood and other natural fibers mixed with thermoplastics have fueled growth in several markets. These compounders supply preblended, free-flowing pellets. that can be reheated and formed into products by a variety of processing methods. The pellets are a boon to manufacturers who do not typically do their own compounding or do not wish to compound in-line (for example, most single-screw profilers or injection molding companies).

Other processing technologies such as injection molding and compression molding are also used to produce WPCs, but the total poundage produced is much less (English et al. 1996). These alternative processing methods have advantages when processing of a continuous piece is not desired or a more complicated shape is needed. Composite formulation must be adjusted to meet processing requirements (e.g., the low viscosity needed for injection molding can limit wood content).

13.4 PERFORMANCE

The wide variety of WPCs make it difficult to discuss the performance of these composites. Performance depends on the inherent properties of the constituent materials, interactions between these materials, processing, product design, and service environment. Moreover, new technologies are continuing to improve performance (Mapleston 2001d).

TABLE 13.1
Composition of Aspen-Polypropylene Composite Specimens

	% by Weight	
Aspen Fiber	Polypropylene	MAPP
30	70	0
30	68	2
40	60	0
40	58	2
50	50	0
50	48	2
60	40	0
60	38	2

13.4.1 MECHANICAL PROPERTIES

As an example of mechanical properties of a wood fiber-thermoplastic composite, Table 13.1 shows the composition of aspen fiber, propylene, with and without MAPP compounded using a high intensity thermo-kinetic mixer, where the only source of heat is generated through the kinetic energy of rotating blades. The blending was accomplished at 4600 rpm and then automatically discharged at 190°C. The total residence time of the blending operation averaged about 2 min. The mixed blends were then granulated and dried at 105°C for 4 hours. Test specimens were injection molded at 190°C using pressures varying from 2.75 MPa to 8.3 MPa depending on the constituents of the blend. Test specimen dimensions were according to the respective ASTM standards. The specimens were stored under controlled conditions (20% relative humidity and 32°C) for three days before testing. Tensile tests were conducted according to ASTM 638-90, Izod impact strength tests according to ASTM D 256-90, and flexural testing using the ASTM 790-90 standard. The cross-head speed during the tension and flexural testing was 12.5 mm/min.

Table 13.2 shows the results of mechanical tests done on the composite specimens compounded according to Table 13.1. The addition of MAPP has a great positive effect on flexural strength tensile strength and Izod unnotched toughness. As the percentage of aspen fiber increases, flexural strength and modulus and tensile modulus increase. Unnotched Izod results show a decrease in toughness as the percentage of aspen fiber increases.

TABLE 13.2
Mechanical Properties of Aspen Fiber-Polypropylene Composites

Specimen	Flexural Strength (MPa)	Flexural Modulus (GPa)	Tensile Strength (MPa)	Tensile Modulus (GPa)	Izod Notched (J/m)	Izod Unnotched (J/m)
PP	27.9	1.38	26.2	1.69	22.4	713.5
PP/MAPP	34.6	1.79	29.3	1.82	18.6	563.3
30A/70PP	49.5	4.12	29.3	4.52	24.8	101.7
30A/68PP/MAPP	60.2	3.82	44.9	4.10	21.1	128.3
40A/60PP	54.6	4.60	34.9	5.22	19.6	85.5
40A/58PP/MAPP	66.4	4.66	47.7	5.14	19.8	108.7
50A/50PP	50.2	5.48	28.4	5.81	26.4	67.1
50A/48PP/MAPP	75.7	5.88	53.1	6.68	21.9	98.5
60A/40PP	45.9	6.09	25.6	6.95	23.9	55.2
60A/38PP/MAPP	75.8	6.73	48.1	7.19	21.3	81.1

TABLE 13.3
Weight Gain in Aspen-Polypropylene Composites at 90% Relative Humidity after D Days

Aspen/PP/MAPP	Weight Gain %					
	25 D	50 D	75 D	100 D	150 D	200 D
0/100/0	0	0	0	0	0.2	0.4
30/70/0	0.7	1.4	1.7	2.1	2.4	2.8
30/68/2	0.7	0.7	1.1	1.5	1.5	2.2
40/60/0	0.7	1.4	1.7	2.0	2.4	3.0
40/58/2	0.4	1.2	1.5	1.9	2.7	3.5
50/50/0	1.3	2.0	2.6	3.6	4.3	5.3
50/48/2	1.5	1.8	2.2	2.9	4.0	5.1
60/40/0	3.7	4.5	5.6	6.0	6.3	6.7
60/38/2	1.6	2.2	3.5	4.4	5.1	6.0

13.4.2 Moisture Properties

The specimens listed in Table 13.1 were subjected to 90% relative humidity (RH) for an extended period of time. Table 13.3 shows that even after 200 days, all of the composites continued to gain weight and equilibrium was not reached. The higher the wood fiber content, the more moisture was picked up by the specimen.

Table 13.4 shows data on a cyclic humidity test where the specimens were subjected to 30% RH for 60 days, measured, and then subjected to 90% RH for an additional 60 days. This cycle was repeated four times. As with the static 90% RH tests, the specimens continued to gain weight with each 90% RH cycle.

Soaking the specimens listed in Table 13.1 in liquid water showed that the composites continued to gain small amounts of weight during a 200-day test. As with the RH tests, equilibrium was not reached in the 200 days. The maximum weight gain due to water soaking was approximately 11% in the specimens with the highest percentage of wood fiber.

From the moisture data, it is clear that as the percentage of hydrophilic wood fiber increases in the wood-thermoplastic composites, there is a corresponding increase in moisture gain. The moisture gain is slow, even in liquid water, but continues over a very long period of time. The data

TABLE 13.4
Weight Changes in Repeated Humidity Tests on Aspen-Polypropylene Composites Cycled between 30% and 90% Relative Humidity

Aspen/PP/MAPP	Weight Gain %							
	30%	90%	30%	90%	30%	90%	30%	90%
0/100/0	0	0	0	0	0	0.2	0	0.4
30/70/0	0.6	0.9	0.7	1.4	0.7	1.4	0.9	1.9
30/68/2	0.4	0.9	0.7	1.2	0.7	1.3	0.9	1.8
40/60/0	0.4	0.9	0.7	1.4	0.7	1.6	0.7	2.1
40/58/2	0.2	1.2	0.8	1.6	1.0	2.0	1.2	2.2
50/50/0	0.5	1.5	1.2	2.5	1.2	2.7	1.2	3.2
50/48/2	0.6	1.3	1.1	2.2	1.3	2.2	1.3	2.6
60/40/0	0.7	2.5	1.5	4.1	1.6	4.1	1.3	4.8
60/38/2	0.2	1.5	1.0	2.6	1.3	2.6	1.3	3.3

also suggest that at about 50% wood, the rate and extent of moisture pickup increases. At this point, the fiber content has reached a point where there are a lot of fibers touching each other to wick water faster and further into the composite.

13.4.3 BIOLOGICAL PROPERTIES

Because WPCs absorb less moisture and do so more slowly than solid wood, they have better fungal resistance and dimensional stability when exposed to moisture. For composites with high wood contents, some manufacturers incorporate additives such as zinc borate to improve fungal resistance. Unfilled plastics absorb little, if any, moisture, are very resistant to fungal attack, and have good dimensional stability when exposed to moisture. However, most plastics expand when heated and adding wood decreases thermal expansion.

ASTM standard laboratory tests using isolated brown- and white-rot fungi show a very small weight loss during a 12-week test. These tests, however, are complicated in that the test specimens are not first soaked in water to bring the moisture content of the test specimens up to a moisture content where fungi are able to attack. Outdoor tests show that wood thermoplastics are subject to mold growth.

13.4.4 WEATHERING AND FIRE PROPERTIES

Light stability in outdoor exposures is an area of considerable investigation (Rowell et al. 2000, Lundin 2001). Most WPCs tend to lighten over time (Falk et al. 2002). Some manufacturers add pigments to slow this effect. Others add a gray pigment so that color change is less noticeable. Still others co-extrude a UV-stable plastic layer over the WPCs. In laboratory weatherometer tests, wood-PP composites lightened within 400 to 600 hours of UV exposure. After 2000 hours of UV and water exposure, specimens were white and the white color continued throughout the 10 mm specimen. This in-depth color change may be due to stabilization of the free radicals formed from UV radiation by the lignin in the wood allowing the free radicals to penetrate deeper into the wood.

The fire performance of WPC materials and products is just beginning to be investigated (Malvar et al. 2001, Stark et al. 1997). These composites are different from many building materials in that they can melt as well as burn, making testing for fire resistance difficult.

13.5 HIGHER PERFORMANCE MATERIALS

There has been a continuing interest in achieving improved mechanical properties for WPC. This has involved the use of long individual fibers for their reinforcing potential and the use of thermoplastics with better engineering properties. Thermoplastics with improved strength properties tend to be higher melting thermoplastics, such as polyamides (nylons). The earliest reported attempt to use cellulose flour and cellulosic fibers to reinforce polyamides was that of Klason et al. (1984). That paper followed up an earlier report by these authors that demonstrated the potential of wood flour and cellulosic fibers in commodity thermoplastics, such as polyethylene and polypropylene. The initial results with polyamides were generally discouraging. Although there was some success in reinforcing PA-12 (melting range 176–180°C) with cellulose, when PA-6 was used (melting point 215°C), cellulosic fibers showed poorer reinforcing potential than wood flour and cellulosic flour. In all cases the PA-6 materials exhibited severe discoloration and pronounced pyrolytic degradation. These authors concluded that for the higher melting thermoplastics, such as PA-6, "cellulosic fibers do not produce any significant degree of reinforcement despite their obvious stiffness and strength potential." Since those initial experiments, it was commonly believed that the use of cellulosic fibers as reinforcement in thermoplastics is limited to the low-melting commodity thermoplastics (melting points below 180°C). Furthermore, it has been believed that the higher melting engineering thermoplastics (m.p. > 220°C) cannot be effectively reinforced with

TABLE 13.5
Comparison of Some Plastic/Composite Mechanical Properties

	Tensile Strength MPa	Flexural Strength MPa	Tensile Modulus GPa	Flexural Modulus GPa	Notched Izod J/m	Unnotched Izod J/m
Polypropylene	27.6	28.7	1.39	1.39	16.1	556.4
PP + 33% Wood flour	33.1	49.3	3.38	3.19	18.7	75.4
Nylon 6 (PA-6)	60.2	64.2	2.75	2.38	24	746.3
PA-6 + 33% Wollastonite	62.7	105.7	6.51	6.27	25.8	173.9
PA-6 + 33% Glass fiber	111.2	146.7	8.02	7.55	45.9	406.1
PA-6 + 33% Hardwoodfiber	86.5	121.6	5.71	5.88	25.3	318.3
PA-6 + 33% Softwood fiber	81.9	113.9	5.35	5.45	25.1	247.2
Clear pine	90.4	88.3	7.6	12.34	—	—

cellulosic fibers because of the severe thermal degradation of the cellulose that occurs at temperatures needed to process these high-melting engineering thermoplastics.

In recent patents, it was demonstrated that these common beliefs are erroneous. The patents described materials and methods for achieving composites containing cellulosic fibers in high-melting engineering thermoplastics (Sears et al. 2001, 2004). Other reports demonstrated that not only the high-melting engineering plastics, nylon-6 and nylon-66, but also other high-melting point thermoplastics, ECM (an aliphatic polyketone) and PBT (a polyester), can be reinforced by wood pulp fibers (Caulfield et al. 2001) to produce composites of promising structural potential (see Table 13.5).

Control of viscous shear heating during the extrusion processing of these composites is essential for avoiding the thermal degradation of the cellulosic component. These composites (without compatibilizers and other additives) possess stiffness and bending strength properties that are intermediate between similar composites prepared with wollastonite and glassfiber reinforcements. It is expected that with continuing research on coupling agents/compatibilizers that significant improvements in mechanical properties, especially tensile strength, will be achieved.

13.6 MARKETS

The greatest growth potential for WPCs is in building products that have limited structural requirements. Products include decking (see Figure 13.6), fencing, industrial flooring, landscape timbers, railings, moldings, and roofing (see Figure 13.7). Pressure-treated lumber remains by far the most commonly used decking and railing material (80% of the approximately $3.2 billion market), but the market for WPC decking is growing rapidly (Smith 2002). Market share grew from 2% of the decking market in 1997 to 8% in 2001 (Smith 2002), and it is expected to more than double by 2005 (Eckert 2000, Smith 2002, Mapleston 2001e).

Although WPC decking is more expensive than pressure-treated wood, manufacturers promote its lower maintenance, lack of cracking or splintering, and high durability. The actual lifetime of WPC lumber is currently being debated; most manufacturers offer a 10-year warranty. Compared with unfilled plastic lumber, the advantages of WPC lumber include increased stiffness and reduced thermal expansion. However, mechanical properties such as creep resistance, stiffness, and strength are lower than those of solid wood. Hence, these composites are not currently being used in applications that require considerable structural performance. For example, WPCs are used for deck boards but not the substructure. Solid, rectangular profiles are manufactured as well as more complex hollow and ribbed profiles. Wood fiber, wood flour, and rice hulls are the most common organic fillers used in decking. About 50% wood is typically used, and some products contain as

FIGURE 13.6 Wood fiber plastic deck.

much as 70% wood. A polyethylene matrix is used most often, but manufacturers of decking made with PVC and polypropylene have recently entered the market. At least 20 manufacturers produce decking from WPCs; the market is currently dominated by large manufacturers (Smith 2002).

Window and door profile manufacturers form another large industrial segment that uses WPCs. Fiber contents vary considerably. PVC is most often used as the thermoplastic matrix in window applications, but other plastics and plastic blends are also used. Although more expensive than unfilled PVC, wood-filled PVC is gaining favor because of its balance of thermal stability, moisture resistance, and stiffness (Defosse 1999).

Several industry leaders are offering WPC profiles in their product line. Their approaches vary. One manufacturer co-extrudes a wood-filled PVC with an unfilled PVC outside layer for increased durability. Another manufacturer co-extrudes a PVC core with a wood-filled PVC surface that can be painted or stained (Schut 1999). Yet another manufacturer offers two different composites: a wood-filled PVC and a composite with a foamed interior for easy nailing and screwing (Defosse 1999).

In Europe, decks are not yet common and the WPC decking market is in its infancy. However, other product areas are possible. Anti-PVC sentiment (because PVC is a chlorinated compound) and fears over possible legislation are concerning PVC window manufacturers and creating possiblities

FIGURE 13.7 Wood fiber plastic roofing tiles.

for replacing PVC with WPCs (Mapleston 2001d). The European market for wood profiles, particularly door frames and furniture, is actively being pursued.

In Japan, promising end uses such as decking, walls, flooring, louvers, and indoor furniture have been reported (Leaversuch 2000). At least one Japanese company is licensing WPC extrusion technology in the United States (Mapleston 2001c).

Wood-polypropylene sheets for automobile interior substrates are still made in the United States, but European manufacturers are beginning to use natural fibers other than wood (e.g., kenaf or flax) in air-laid processes. Growth in the use of natural-fiber-reinforced thermoplastics, rather than unfilled plastics, in automotive applications has been slower in the United States than in Europe, where environmental considerations are a stronger driving force. One market analyst cites the lack of delivery channels and high transportation costs as major factors that slow growth in the United States (Eckert 2000). One major U.S. company has used German technology to produce automotive door quarter panels from natural fiber composites with polypropylene and polyester; the doors achieved a 4-star side impact rating (Manolis 1999). A number of other interior automotive components are being made with similar technology. Nonwoven mat technology has been used to make rear shelf trim panels with flax-reinforced polypropylene (Manolis 1999). Other products being tested include instrument panels, package shelves, load floors, and cab back panels (Manolis 1999).

REFERENCES

Anonymous. (2001). Latest LP engineered product is alternative to traditional decking. *PanelWorld,* March, pp. 30–32.
Caulfield, D., Jacobson, R., Sears, K, and Underwood, J. (2001). Fiber reinforced engineering plastics, 2nd International Conference on Advanced Wood Composites, Bethel, ME, August 13–16.
Defosse, M. (1999). Processors focus on differentiation in window profiles. *Modern Plastics*, September, pp. 74–79.
Eckert, C. (2000). Opportunities for natural fibers in plastic composites. Proceedings, Progress in Woodfibre-Plastic Composites, Toronto, Ontario, Canada.
English, B., Clemons, C.M., Stark, N., and Schnieder, J.P. (1996). Waste-wood derived fillers for plastics. Gen. Tech. Rept. FPL-GTR-91. USDA FS, Forest Products Laboratory, Madison, WI, pp. 282–291.
Falk, R.H., Lundin, T., and Felton, C. (2002). Accelerated weathering of natural fiber-thermoplastic composites: Effects of ultraviolet exposure on bending strength and stiffness. Proceedings, Sixth International Conference on Woodfiber-Plastic Composites. Forest Products Society, Madison, WI, pp. 87–93.
Gordon, J.E. (1988). *The New Science of Strong Materials (or Why You Don't Fall Through the Floor)* (2nd ed.). Princeton University Press, Princeton, NJ.
Jacobson, R., Caulfield, D., Sears, K., and J. Underwood, J. (2002). Low temperature processing (LTP) of ultra-pure cellulose fibers into nylon 6 and other thermoplastics, Proceedings, Sixth International Conference on Woodfiber/Plastic Composites, Madison WI.
Klason, C., Kubàt, J., and Strömvall, H.-E. (1984). The efficiency of cellulosic fillers in common thermoplastics. Part 1, Filling without processing aids or coupling agents. *Intern. J. Polymeric Mater.* 10:159–187.
Leaversuch, R.D. (2000). Wood-fiber composites build promising role in extrusion. *Modern Plastics*, December, p. 56.
Lundin, T. (2001). Effect of accelerated weathering on the physical and mechnical properties of natural-fiber thermoplastic composites. M.S. thesis. University of Wisconsin, Madison.
Malvar, L.J., Pendleton, D.E., and Tichy, R. (2001). Fire issues in engineered wood composites for naval waterfront facilities. *Sampe J.* 37(4):70–75.
Manolis, S.L. (1999). Natural fibers: the new fashion in automotive plastics. *Plastics Technol.*, October, pp. 62–68.
Mapleston, P. (2001a). Additive suppliers turn their eyes to wood/plastic composites. *Modern Plastics*, August, p. 52.
Mapleston, P. (2001b). It's one hot market for profile extruders. *Modern Plastics*, June, 49–p. 52.
Mapleston, P. (2001c) Processing technology: A wealth of options exist. *Modern Plastics*, June, pp. 56–60.
Mapleston, P. (2001d). Wood composite suppliers are poised for growth in Europe. *Modern Plastics*, October, p. 41.

Mapleston, P. (2001e). Wood composites in decking, the view from Principia. *Modern Plastics*, June, pp. 62–65.

Morton, J., Quarmley, J., and Rossi, L. (2003). Current and emerging applications for natural and woodfiber plastic composites, 2004 Seventh International Conference on Woodfiber/Plastic Composites, Madison, WI.

Schut, J. (1999). For compounding, sheet & profile: Wood is good. *Plastics Technology*, March, pp. 46–52.

Schut, J. (2001). Foaming expands possibilities for wood-fiber composites. *Modern Plastics*, July, pp. 58–65.

Rowell, R.M., Lange, S.E., and Jacobson, R.E. (2000). Weathering performance of plant-fiber thermoplastic composites. *Mol. Cryst. And Liquid Cryst.* 353:85–94.

Sears, K., Jacobson, R., Caulfield, D., and Underwood, J. (2001). Composites containing cellulosic pulp fibers and methods of making and using same. U.S. Patent 6,270,883.

Sears, K., Jacobson, R., Caulfield, D., and Underwood, J. (2004). Methods of making composites containing cellulosic pulp fibers. U.S. Patent 6,730,249.

Smith, P.M. (2002). U.S. woodfiber-plastic composite decking market. In: Proceedings, Sixth International Conference on Woodfiber-Plastic Composites. Forest Products Society, Madison, WI, pp. 13–17.

Stark, N.M. (1999). Wood fiber derived from scrap pallets used in polypropylene composites. *Forest Products J.* 49(6):39–46.

Stark, N.M., White, R.H., and Clemons, C.M. (1997). Heat release rate of wood-plastic composites. *Sampe J.* 33(5):26–31.

Wolcott, M.P., and Adcock, T. (2000). New advances in wood fiber-polymer formulations: Proceedings Wood-Plastic conference. *Plastics Technology Magazine and Polymer Process Communications,* Baltimore, MD, pp. 107–114.

Youngquist, J.A. (1995). Unlikely partners? The marriage of wood and nonwood materials. *Forest Products J.* 45(10):25–30.

Youngquist, J.A. (1999). Wood-based composites and panel products. In: *Wood Handbook: Wood as an Engineering Material.* Chapter 10. USDA Forest Serv. Forest Prod. Lab. Forest Products Society, Madison, WI, pp. 27–28.

Part IV

Property Improvements

14 Chemical Modification of Wood

Roger M. Rowell
USDA, Forest Service, Forest Products Laboratory,
and Department of Biological Systems Engineering,
University of Wisconsin, Madison, WI

CONTENTS

14.1	Degradation of Wood	383
14.2	Chemical Reaction Systems	383
14.3	Chemical Reactions with Wood	385
	14.3.1 Acetylation	386
	14.3.2 Acid Chlorides	387
	14.3.3 Other Anhydrides	387
	14.3.4 Carboxylic Acids	388
	14.3.5 Isocyanates	388
	14.3.6 Formaldehyde	388
	14.3.7 Other Aldehydes	389
	14.3.8 Methylation	389
	14.3.9 Alkyl Chlorides	389
	14.3.10 β-Propiolactone	390
	14.3.11 Acrylonitrile	390
	14.3.12 Epoxides	390
	14.3.13 Reaction Rates	391
	14.3.14 Effect of Moisture	392
14.4	Properties of Chemically Modified Wood	392
	14.4.1 Changes in Wood Volume Resulting from Reaction	392
	14.4.2 Stability of Bonded Chemical to Chemical Leaching	393
	14.4.3 Accessibility of Reaction Site	395
	14.4.4 Acetyl Balance in Acetylated Wood	395
	14.4.5 Distribution of Bonded Chemical	396
	14.4.6 Moisture Sorption	397
	14.4.7 Dimensional Stability	398
	14.4.8 Loss of Dimensional Stability	401
	14.4.9 Biological Resistance	402
	14.4.9.1 Termite Resistance	402
	14.4.9.2 Decay Resistance	403
	14.4.9.3 Marine Organisms Resistance	405
	14.4.10 Thermal Properties	405
	14.4.11 Ultraviolet Radiation and Weathering Resistance	406
	14.4.12 Mechanical Properties	408
	14.4.13 Adhesion of Chemically Modified Wood	410

14.4.14 Acoustical Properties 410
14.5. Commercialization of Chemically Modified Wood 411
   14.5.1 The Fiber Process 411
   14.5.2 The Solid Wood Microwave Process 412
References 413

For the most part, humans have made the most of wood while putting up with the "natural defects" that accompany it, such as dimensional instability and degradation due to weathering, fire, and decay. Wood is a hygroscopic resource designed to perform, in nature, in a wet environment. Nature is programmed to recycle wood in a timely way through biological, thermal, aqueous, photochemical, chemical, and mechanical degradations. In simple terms, nature builds wood from carbon dioxide and water and has all the tools to recycle it back to the starting chemicals. We harvest a green tree and convert it into dry products, and nature, with its arsenal of degrading reactions, starts to reclaim it at its first opportunity.

The properties of any resource are, in general, a result of the chemistry of the components of that resource. In the case of wood, the cell wall polymers (cellulose, hemicelluloses, and lignin) are the components that, if modified, would change the properties of the resource. If the properties of the wood are modified, the performance of the modified wood would be changed. This is the basis of our chemical modification of wood—to change its properties and improve its performance. This idea is applied to both solid wood and wood composites.

In order to produce wood-based materials with a long service life, it is necessary to interfere with the natural degradation processes for as long as possible (Figure 14.1). This can be done in several ways. For example, traditional methods for decay resistance and fire retardancy treat the product with toxic or corrosive chemicals, which although are effective in providing decay and fire resistance can result in environmental concerns. In order to make property changes, you must first understand the chemistry of the components and the contributions each play in the properties of the resource. Following this understanding, you must then devise a way to modify what needs to be changed to get the desired change in property.

Properties of wood, such as dimensional instability, flammability, biodegradability, and degradation caused by acids, bases, and ultraviolet radiation, are all a result of chemical degradation reactions which can be prevented or, at least, slowed down if the cell wall chemistry is altered (Rowell 1975a, Rowell and Youngs 1981, Rowell 1983, Rowell and Konkol 1987, Rowell et al. 1988a, Hon 1992, Rowell 1992, Kumar 1994, Banks and Lawther 1994).

```
Biological Degradation        - Fungi, Bacteria, Insects, Termites
     Enzymatic Reactions      - Oxidation, Hydrolysis, Reduction
     Chemical Reactions       - Oxidation, Hydrolysis, Reduction
     Mechanical               - Chewing
Fire Degradation              - Lightning, Sun
     Pyrolysis Reactions      - Dehydration, Hydrolysis, Oxidation
Water Degradation             - Rain, Sea, Ice, Acid Rain, Dew
     Water Interactions       - Swelling, Shrinking, Freezing, Cracking
Weather Degradation           - Ultraviolet radiation, Water, Heat, Wind
     Chemical Reactions       - Oxidation, Hydrolysis
     Mechanical               - Erosion
Chemical Degradation          - Acids, Bases, Salts
     Chemical Reactions       - Oxidation, Reduction, Dehydration, Hydrolysis
Mechanical Degradation        - Dust, Wind, Hail, Snow, Sand
     Mechanical               - Stress, Cracks, Fracture, Abrasion
```

FIGURE 14.1 Degradation reactions which occur when wood is exposed to nature.

BIOLOGICAL DEGRADATION
    Hemicelluloses >>>> Accessible Cellulose >> Non-CrystallineCellulose >>>>>>> Crystalline Cellulose >>>>>>>>>>>>>> Lignin
MOISTURE SORPTION
    Hemicelluloses >>> Accessible Cellulose >>>> Non-Crystalline Cellulose >> Lignin >>> Crystalline Cellulose
ULTRAVIOLET DEGRADATION
    Lignin >>>>>>>>>>>>>>> Hemicelluloses >>> Accessible Cellulose >> Non-Crystalline Cellulose >>>>>>>> Crystalline Cellulose
THERMAL DEGRADATION
    Hemicelluloses > Cellulose >>>>>>>>>>>>>>> Lignin
STRENGTH
    Crystalline Cellulose >>> Matrix [Non-Crystalline Cellulose + Hemicelluloses + Lignin] >> Lignin

FIGURE 14.2 Cell wall polymers responsible for the properties of wood.

14.1 DEGRADATION OF WOOD

Figure 14.2 shows the cell wall polymers involved in each property as we understand it today (Rowell 1990). Wood changes dimensions with changing moisture content because the cell wall polymers contain hydroxyl and other oxygen-containing groups that attract moisture through hydrogen bonding (Stamm 1964, Rowell and Banks 1985) (see Chapter 4). The hemicelluloses are mainly responsible for moisture sorption, but the accessible cellulose, non-crystalline cellulose, lignin, and surface of crystalline cellulose also play major roles. Moisture swells the cell wall, and the fiber expands until the cell wall is saturated with water (*fiber saturation point*, or FSP). Beyond this saturation point, moisture exists as free water in the void structure and does not contribute to further expansion. This process is reversible, and the fiber shrinks as it loses moisture below the FSP.

Wood is degraded biologically because organisms recognize the carbohydrate polymers (mainly the hemicelluloses) in the cell wall and have very specific enzyme systems capable of hydrolyzing these polymers into digestible units (see Chapter 5). Biodegradation of the cell wall matrix and the high molecular weight cellulose weakens the fiber cell (Rowell et al. 1988b). Strength is lost as the cell wall polymers and its matrix undergoes degradation through oxidation, hydrolysis, and dehydration reactions.

Wood exposed outdoors undergoes photochemical degradation caused by ultraviolet radiation (see Chapter 7). This degradation takes place primarily in the lignin component, which is responsible for the characteristic color changes (Rowell 1984). The lignin acts as an adhesive in the cell walls, holding the cellulose fibers together. The surface becomes richer in cellulose content as the lignin degrades. In comparison to lignin, cellulose is much less susceptible to ultraviolet light degradation. After the lignin has been degraded, the poorly bonded carbohydrate-rich fibers erode easily from the surface, which exposes new lignin to further degradative reactions. In time, this "weathering" process causes the surface of the composite to become rough and can account for a significant loss in surface fibers.

Wood burns because the cell wall polymers undergo pyrolysis reactions with increasing temperature to give off volatile, flammable gases (see Chapter 6). The gases are ignited by some external source and combust. The hemicelluloses and cellulose polymers are degraded by heat much before the lignin (Rowell 1984). The lignin component contributes to char formation, and the charred layer helps insulate the composite from further thermal degradation.

14.2 CHEMICAL REACTION SYSTEMS

Different authors have used the term "chemical modification" to mean different things over the years. For this chapter, chemical modification will be defined as *a chemical reaction between some reactive part of wood and a simple single chemical reagent, with or without catalyst, to form a covalent bond between the two*. This excludes all simple chemical impregnation treatments, which do not form covalent bonds, monomer impregnations that polymerize *in situ* but do not bond with the cell wall, polymer inclusions, coatings, heat treatments, etc.

There are several approaches to chemically modifying the wood cell wall polymers. The most abundant single site for reactivity in these polymers is the hydroxyl group; most reaction schemes have been based on the reaction of hydroxyl groups. Sites of unsaturation in the lignin structure can also be used as a point of reactivity, as well as free radical additions and grafting. However, the most studied class of chemical reactions are those involving hydroxyl substitutions.

In modifying wood for property improvement, one must consider several basic principles when selecting a reagent and a reaction system. Of the thousands of chemicals available, either commercially or by synthetic means, most can be eliminated because they fail to meet the requirements or properties listed below.

If hydroxyl reactivity is selected as the preferred modification site, the chemical must contain functional groups, which will react with the hydroxyl groups of the wood components. This may seem obvious, but there are several failed reaction systems in the literature using a chemical that could not react with a hydroxyl group.

The overall toxicity of the chemicals must be carefully considered. The chemicals must not be toxic or carcinogenic to humans in the finished product, and should be as nontoxic as possible in the treating stage. The chemical should be as non-corrosive as possible to eliminate the need for special stainless steel or glass-lined treating equipment.

In considering the ease with which excess reagents can be removed after treatment, a liquid treating chemical with a low boiling point is advantageous. Likewise, if the boiling point of a liquid reagent is too high, it will be very difficult to remove the chemical after treatment. It is generally true that the lowest member of a homologous series is the most reactive and will have the lowest boiling point. The boiling point range for liquids to be considered is 90–150°C. In some cases, the lowest member of a homologous series is a gas. It is possible to treat wood with a gas system; however, there may be processing challenges in handling a pressurized gas in a continuous reactor. Gases do not penetrate deeply or quickly into the wood cell wall, so penetration of the reaction system may be limited.

Accessibility of the reagent to the reactive chemical sites is a major consideration. In some cases, this may be the rate-limiting step in a reaction system. To increase accessibility to the reaction site, the chemical should be able to swell the wood structure. If the reagents do not swell the structure, then another chemical or co-solvent can be added to aid the penetration of chemicals. Accessibility to the reactive site is a major consideration in a gas system unless there is a condensation step in the procedure.

If the chemical reaction system requires a catalyst, a strong acid or base catalyst cannot be used as they cause extensive degradation. The most favorable catalyst from the standpoint of wood degradation is a weakly alkaline one. The alkaline medium is also favored because in many cases these chemicals swell the cell wall matrix structure and give better penetration. The properties of the catalyst parallel those of reagents—low boiling point liquid, nontoxic, effective at low temperatures, etc. In most cases, the organic tertiary amines or weak organic acids are best suited.

The experimental reaction condition that must be met in order for a given reaction to go is another important consideration. If the reaction time is long, the temperature required for complete reaction must be low enough so there is little or no fiber degradation. The reaction must also have a relatively fast rate of reaction with the cell wall components. It is important to get as fast a reaction as possible at the lowest temperature without wood degradation. High temperature reactions are possible (up to 170°C) if the reaction time is very short and no strong acid or base catalysts are used. Wood degrades rather quickly at temperatures above 175°C (Stamm 1964) (see Chapter 6).

The moisture present in the wood is another consideration in the reaction conditions. It is costly to dry wood to less than one percent moisture, but it must be remembered that the -OH group in water is more reactive than the -OH group available in the wood components; that is, hydrolysis is faster than substitution. The most favorable condition is a reaction which requires a trace of moisture and the rate of hydrolysis is relatively slow.

Another consideration in this area is to keep the reaction system as simple as possible. Multicomponent systems will require complex separation after reaction for chemical recovery. The optimum choice would be a reactive chemical that swells the wood structure that also acts as both the solvent and catalyst.

If possible, avoid by-products during the reaction that have to be removed. If there is not a 100 percent reagent skeleton add-on, then the chemical cost is higher and will require recovery of the by-product for economic and environmental reasons.

The chemical bond formed between the reagent and the wood components is of major importance. For permanence, this bond should have great stability to withstand nature's recycling system if the product is used outdoors. In order of stability, the types of covalent chemical bonds that may be formed are ethers > acetals > esters. The ether bond is the most desirable covalent carbon-oxygen bond that can be formed. These bonds are more stable than the glycosidic bonds between sugar units in the wood polysaccharides, so the polymers would degrade before the grafted ether. It may be desired, however, to have the bonded chemical released by hydrolysis or enzyme action in the final product so that an unstable bond may be required from the modification.

The hydrophobic nature of the reagent needs to be considered. The chemical added to the wood should not increase the hydrophilic nature of the wood components unless that is a desired property. If the hydrophilicity is increased, the susceptibility to microorganism attack increases. The more hydrophobic the component can be made, the better the moisture exclusion properties of the substituted wood will be.

Single-site substitution versus polymer formation is another consideration. For the most part, a single reagent molecule that reacts with a single hydroxyl group is the most desirable. Crosslinking can occur when the reagent contains more than one reactive group or results in a group which can further react with a hydroxyl group. Crosslinking can cause the wood to become more brittle. Polymer formation within the cell wall after the initial reaction with the hydroxyl groups of the wood components gives, through bulking action, dimensional stabilization. The disadvantage of polymer formation is that a higher level of chemical add-on is required for biological resistance than in the single site reactions.

The treated wood must still possess the desirable properties of wood—its strength should not be reduced, there should be no change in color, its good electrical insulation properties are retained, the final product not dangerous to handle, there are no lingering chemical smells, and it is still gluable and finishable—unless one or more of these properties are the object of change in the product.

A final consideration is, of course, the cost of chemicals and processing. In laboratory-scale experimental reactions, the high cost of chemicals is not a major factor. For the commercialization of a process, however, the chemical and processing costs are very important factors. Laboratory-scale research is generally done using small batch processing. But rapid, continuous processes should always be studied for scale-up. An economy of scale can make an expensive laboratory process economical.

In summary, the chemicals to be laboratory tested must be capable of reacting with wood hydroxyls under neutral, mildly alkaline, or acidic conditions at temperatures below 170°C. The chemical system should be simple and capable of swelling the structure to facilitate penetration. The complete molecule should react quickly with wood components yielding stable chemical bonds, and the treated wood must still possess the desirable properties of untreated wood.

14.3 CHEMICAL REACTIONS WITH WOOD

As stated before, because the properties of wood result from the chemistry of the cell wall components, the basic properties of wood can be changed by modifying the basic chemistry of the cell wall polymers. There are several approaches to cell wall modification, depending on what property is to be modified. For example, if the objective is water repellency, then the approach might be to reduce the hydrophilic nature of the cell wall by bonding on hydrophobic groups. If dimensional stability is to be improved, the cell wall can be bulked with bonded chemicals, or cell wall polymer components crosslinked together to restrict cell wall expansion, or groups can be bonded that reduce hydrogen bonding or increase hydrophobicity. If fire retardancy is desired, chemical groups can be bonded onto cell wall polymers that contain retardants or flame suppressants. If resistance to ultraviolet radiation is desired, UV blockers or absorbers could be bonded to lignin. The chemical modification system selected must perform the desired chemical change to achieve the desired change in performance.

Many chemical reaction systems have been published for the modification of wood, and these systems have been reviewed in the literature several times in the past (Rowell 1975a, 1983, 1991, Kumar 1994, Hon 1996, Rowell 1999). These chemicals include anhydrides such as phthalic, succinic, malaic, propionic, and butyric anhydride; acid chlorides; ketene carboxylic acids; many different types of isocyanates; formaldehyde; acetaldehyde; difunctional aldehydes; chloral; phthaldehydic acid; dimethyl sulfate; alkyl chlorides; β-propiolactone, acrylonitrile; epoxides, such as ethylene, propylene, and butylene oxide; and difunctional epoxides.

14.3.1 Acetylation

Acetylation of wood using acetic anhydride has been done mainly as a liquid phase reaction. The early work was done using acetic anhydride catalyzed with either zinc chloride (Ridgway and Wallington 1946) or pyridine (Stamm and Tarkow 1947). Through the years, many other catalysts have been tried using both liquid and vapor systems. Some of the catalysts used include urea-ammonium sulphate (Clermont and Bender 1957), dimethylformamide (Clermont and Bender 1957; Risi and Arseneau 1957a; Baird 1969), sodium acetate (Tarkow 1959), magnesium persulfate (Arni et al. 1961; Ozolina and Svalbe 1972; Truksne and Svalbe 1977), trifluoroacetic acid (Arni et al. 1961a,b), boron trifluoride (Risi and Arseneau 1957a), and γ-rays (Svalbe and Ozolina 1970). The reaction has also been done without a catalyst (Rowell et al. 1986b), and by using an organic cosolvent (Goldstein et al. 1961). Gas-phase reactions have also been reported using acetic anhydride; however, the diffusion rate is so very slow that this technology has only been applied to thin veneers (Tarkow et al. 1950, Tarkow 1959, Baird 1969, Avora et al. 1979, Rowell et al. 1986a).

The reaction with acetic anhydride results in esterification of the accessible hydroxyl groups in the cell wall with the formation of by-product acetic acid (Rowell et al. 1994).

$$\text{Wood-OH} + \text{CH}_3\text{-C(=O)-O-C(C=O)-CH}_3 \rightarrow \text{Wood-O-C(=O)-CH}_3 + \text{CH}_3\text{-C(=O)-OH}$$

Acetic Anhydride Acetylated Wood Acetic Acid

Acetylation is a single-site reaction, which means that one acetyl group is on one hydroxyl group with no polymerization. All of the weight gain in acetyl can be directly converted into units of hydroxyl groups blocked. This is not true for a reaction where polymer chains are formed (epoxides and isocyanates, for example). In these cases, the weight gain can not be converted into units of blocked hydroxyl groups.

Acetylation has also been done using ketene gas (Tarkow 1945; Karlson and Svalbe 1972, 1977, Rowell et al. 1986c). In this case, esterification of the cell wall hydroxyl groups takes place, but there is no formation of by-product acetic acid:

$$\text{Wood-OH} + \text{CH}_2\text{=C=O} \rightarrow \text{Wood-O-C(=O)-CH}_3$$

Ketene Acetylated Wood

While this is interesting chemistry that eliminates a by-product, it has been shown that reactions with ketene gas result in poor penetration of reactive chemicals, and the properties of the reacted wood are less desirable than those of wood reacted with acetic anhydride (Rowell et al. 1986c).

Chemical Modification of Wood

The preferred method of acetylating wood today is to use a limited amount of liquid acetic anhydride without a catalyst or cosolvent. The rest of this chapter will describe properties of wood that have been acetylated using this technology, unless otherwise noted (Rowell et al. 1986b, Rowell et al. 1986d, Rowell et al. 1991a). Variations of this procedure have been used to modify fibers, particles, flakes, chips, veneers, and wood of various sizes. The fact that only a limited quantity of acetic anhydride is used means that less chemical has to be heated during the reaction, and less chemical has to be cleaned up after the reaction. A small amount of acetic acid seems to be needed in the reaction mixture to swell the cell wall.

Many different types of wood have been acetylated using a variety of procedures, including several types of wood (Narayanamurti and Handa 1953, Rowell 1983, Rowell et al. 1986b, Rowell and Plackett 1988, Imamura et al. 1989, Plackett et al. 1990, Beckers and Militz 1994, Chow et al. 1996) as well as other lignocellulosic resources such as bamboo (Rowell and Norimoto 1987, 1988), bagasse (Rowell and Keany 1991), jute (Callow 1951, Andersson and Tillman 1989, Rowell et al. 1991a), kenaf (Rowell 1993, Rowell and Harrison 1993), wheat straw (Gomez-Bueso et al. 1999, 2000), pennywort, and water hyacinth (Rowell and Rowell 1989).

14.3.2 Acid Chlorides

Acid chlorides can also be used to esterify wood (Suida 1930). The product is the ester of the related acid chloride, with hydrochloric acid as a by-product. Using lead acetate as a catalyst with acetyl chloride,

$$\text{Wood–OH} + \text{R–C(=O)–Cl} \rightarrow \text{Wood–O–C(=O)–R} + \text{HCl}$$

Singh et al. (1981) found a lower acetyl content than with acetic anhydride. Using a 20% lead acetate solution in the reaction reduced the amount of free hydrochloric acid released in the reaction. The very strong acid released as a by-product in this reaction causes extensive degradation of the wood; as such, very little work has been done in this area.

14.3.3 Other Anhydrides

Other anhydrides have been used to react with wood. Risi and Arseneau (1958) reacted wood with phthalic anhydride, resulting in high dimensional stability. Popper and Bariska (1972), however, found that chemical weight gain was lost after soaking in water—showing that the phthaly group was lost to hydrolysis. Phthaly groups have a greater affinity for water than hydroxyl groups in wood, so phthalylated wood is more hydroscopic than untreated wood (Popper and Bariska 1972, 1973). While dimensional stability resulting from acetylation is due to bulking of the cell wall, dimensional stability from phthalylation seems to be due to mechanical bulking of the submicroscopic pores in the wood cell wall (Popper and Bariska 1975). Very high weight gains are achieved by phthalylation (40–130%), which may be a result of polymerization (Risi and Arseneau 1958, Popper 1973).

Goldstein et al. (1961) reacted ponderosa pine with propionic and butyric anhydrides in xylene without a catalyst. After 10 hours at 125°C, the reaction weight gain was 4% for propionylation and 0% for butyrylation. After 30 hours reaction time, propionylation only resulted in a weight gain of 10%. Papadopoulas and Hill (2002) reacted Corsican pine with acetic, propionic, butric, valeric, and hexanoic anhydrides. Butylation has also been done using microwave heating to improve dimensional stability and lightfastness (Chang and Chang 2003).

Succinic and malaic anhydrides have also been reacted with wood fiber (Clemons et al. 1992, Rowell and Clemons 1992). Reaction of these two anhydrides with wood made the wood thermoplastic, and it was possible to thermoform the wood fiber into a high-density composite under pressure.

14.3.4 CARBOXYLIC ACIDS

Carboxylic acids have been esterified to wood catalyzed with trifluoroacetic anhydride (Arni et al. 1961a, 1961b, Nakagami et al. 1974).

$$\text{Wood–OH} + (CH_3)_2\text{–C=CH–COOH} \rightarrow \text{Wood–O–C(=O)–CH–C}(CH_3)_2$$

Several unsaturated carboxylic acids were found to react with wood by the "impelling" method to increase the oven dry volume without a change in color or a decrease in either crystallinity or moisture content (Nakagami and Yokota 1975). Reaction with α-methylcrotonic acid gave a degree of substitution high enough to make the reacted wood soluble in acetone and chloroform to the extent of 30% (Nakagami et al. 1976). Further esterification increased the solubility but was accompanied by considerable degradation of wood components. Solubilization seems to be hindered by both lignin and hemicelluloses (Nakagami 1978, Nakagami and Yokota 1978).

14.3.5 ISOCYANATES

The wood reacts with hydroxyls with isocyanates, forming a nitrogen-containing ester. Clermont and Bender (1957) exposed wood veneer swollen in dimethylformamide to vapors of phenyl isocyanate at 100–125°C. The resulting wood was very dimensionally stable and showed increased mechanical strength with little change in color. Baird (1969) reacted dimethylformamide-soaked cross sections of white pine and Englemann spruce with ethyl, allyl, butyl, *t*-butyl, and phenyl isocyanate.

$$\text{Wood–OH} + \text{R–N=C=O} \rightarrow \text{Wood–O–C(=O)–NH–R}$$
$$\text{Isocyanate}$$

Vapor phase reactions of butyl isocyanate in dimethylformamide gave the best results.

White cedar was reacted with 2,4-tolylene diisocyanate with and without a pyridine catalyst to a maximum nitrogen content of 3.5 and 1.2, respectively (Wakita et al. 1977). Compressive strength and bending modulus increased with increasing nitrogen content. Beech wood was reacted with a diisocyanate that gave the resulting wood very high decay resistance (Lutomski 1975).

Methyl isocyanate reacts very quickly with wood without a catalyst to give high add-on weight gains (Rowell and Ellis 1979). Ethyl, *n*-propyl, and phenyl isocyanates react with wood without the need for a catalyst, but *p*-tolyl-1,6-diisocyanate and tolylene-2,4-diisocyanate require either dimethylformamide or triethylamine as a catalyst (Rowell and Ellis 1981).

14.3.6 FORMALDEHYDE

The reaction between wood hydroxyls and formaldehyde occurs in two steps. Because the bonding is with two hydroxyl groups, the reaction is called cross-linking.

$$\text{Wood–OH} + \text{H–C(=O)–H} \rightarrow \text{Wood–O–C(OH)–H}_2 + \text{Wood–OH} \rightarrow \text{Wood–O–CH}_2\text{–O–Wood}$$
$$\text{Formaldehyde} \quad\quad \text{Hemiacetal} \quad\quad\quad\quad \text{Acetal}$$

The two hydroxyl groups can come from (1) hydroxyls within a single sugar unit; (2) hydroxyls on different sugar residues within a single cellulose chain; (3) hydroxyls between two different cellulose chains; (4) same as 1, 2, and 3, except for reactions occurring on the hemicelluloses; (5) hydroxyl groups on different lignin residues; and (6) interaction between cellulose, hemicelluloses, and lignin hydroxyls. The possible cross-linking combinations are large and theoretically all of them are possible. Since the reaction is a two-step mechanism, some of the added formaldehyde will be in a non–cross-linked hemiacetal form. These chemical bonds are very unstable and would not survive long after reaction.

The reaction is best catalyzed by strong acids such as hydrochloric acid (Tarkow and Stamm 1953, Burmester 1970, Ueyama et al. 1961, Minato and Mizukami 1982), nitric acid (Tarkow and Stamm 1953), sulfur dioxide (Dewispelaere et al. 1977, Stevens et al. 1979), p-toluene sulfonic acid, and zinc chloride (Stamm 1959, Ueyama et al. 1961). Weaker acids such as sulfurous and formic acid do not work (Tarkow and Stamm 1953). Schuerch (1968) speculated that bases such as lime water and tertiary amines can initiate the reaction.

14.3.7 OTHER ALDEHYDES

Acetaldehyde (Tarkow and Stamm 1953) and benzaldehyde (Tarkow and Stamm 1953, Weaver et al. 1960) were reacted with wood using either nitric acid or zinc chloride as the catalyst. Glyoxal, glutaraldehyde, and α-hydroxyadipaldehyde have also been catalyzed with zinc chloride, magnesium chloride, phenyldimethylammonium chloride, and pyridinium chloride (Weaver et al. 1960). Chloral (trichloroacetaldehyde) without catalyst and phthaldehydic acid in acetone catalyzed with p-toluenesulfonic acid have also been used as reaction systems with wood (Weaver et al. 1960, Kenaga 1957). Other aldehydes and related compounds have also been tried either alone or catalyzed with sulfuric acid, zinc chloride, magnesium chloride, ammonium chloride, or diammonium phosphate (Weaver et al. 1960). Compounds such as N,N'-dimethylol-ethylene urea, glycol acetate, acrolein, choroacetaldehyde, heptaldehyde, o- and p-chlorobenzaldehyde, furfural, p-hydroxybenzaldehyde, and m-nitrobenzaldehyde all bulk the cell wall, but no cross-linking seems to occur.

14.3.8 METHYLATION

The simplest ether that can be formed is the methyl ether. Reactions of wood with dimethyl sulfate and sodium hydroxyl (Rudkin 1950, Narayanamurti and Handa 1953) or methyl iodide and silver oxide (Narayanamurti and Handa 1953) are two systems that have been reported.

$$\text{Wood–OH} + \text{CH}_3\text{I} \rightarrow \text{Wood–O–CH}_3$$

Methylation up to 15% weight gain did not affect the adhesive properties of casein glues.

14.3.9 ALKYL CHLORIDES

In the reaction of alkyl chlorides with wood, hydrochloric acid is generated as a by-product. Because of this, the strength properties of the treated wood are poor. Reaction of allyl chloride in pyridine (Kenaga et al. 1950, Kenaga and Sproull 1951) or aluminum chloride resulted in good dimensional stability; but on soaking in water, dimensional stability was lost (Kenaga and Sproull 1951). In the case of allyl chloride-pyridine,

$$\text{Wood–OH} + \text{R–Cl} \rightarrow \text{Wood–O–R} + \text{HCl}$$

Alkyl Chloride

the dimensional stability was not due to the bulking of the cell wall with bonded chemicals but to the formation of allyl pyridinium chloride polymers, which are water soluble and easily leached out (Risi and Arseneau 1957d).

Other alkyl chlorides reported are crotyl chloride (Risi and Arseneau 1957b) and n- and t-butyl chlorides (Risi and Arseneau 1957c) catalyzed with pyridine.

14.3.10 β-Propiolactone

The reaction of β-propiolactone with wood is interesting in that different products are possible depending on the pH of the reaction. Under acid conditions, an ether bond is formed with the hydroxyl group on wood along with the formation of a free acid-end group.

$$\text{Wood-OH} + \begin{array}{c} \text{CH}_2\text{-CH}_2 \\ |\quad\quad| \\ \text{O-C=O} \end{array} \xrightarrow[\text{OH}^-]{\text{H}^+} \begin{array}{l} \text{Wood-O-CH}_2\text{-CH}_2\text{-COOH} \\ \\ \text{Wood O-C(=O)-CH}_2\text{-CH}_2\text{-OH} \end{array}$$

β-Propiolactone

Under basic conditions, an ester bond is formed with the wood hydroxyl giving a primary alcohol end group.

Southern yellow pine was reacted with -propiolactone under acid conditions to give a carboxyethyl derivative (Goldstein et al. 1959, Goldstein 1959). High concentrations of β-propiolactone caused delamination and splitting of the wood due to very high swelling (Rowell unpublished data).

β-Propiolactone has now been labeled as a very active carcinogen. For this reason, this very interesting chemical reaction system will probably not be studied again.

14.3.11 Acrylonitrile

When acrylonitrile reacts with wood in the presence of an alkaline catalyst, cyanoethylation occurs.

$$\text{Wood-OH} + \text{CH}_2 = \text{CH-CN} \rightarrow \text{Wood-O-CH}_2\text{-CH}_2\text{-CN}$$

Acrylonitrile

With sodium hydroxide, weight gains up to 30% have been reported (Goldstein et al. 1959, Baechler 1959a,b, Fuse and Nishimoto 1961, Kenaga 1963). Ammonium hydroxide has also been used giving lower weight gains as compared to sodium hydroxide.

14.3.12 Epoxides

The reaction between epoxides and the hydroxyl groups is an acid- or base-catalyzed reaction; however, all work in the wood field has been alkali catalyzed.

$$\text{Wood-OH} + \text{R-CH(-O-)CH}_2 \rightarrow \text{Wood-O-CH}_2\text{-CH(OH)-R}$$

Epoxide

The simplest epoxide (ethylene oxide) catalyzed with trimethylamine has been used as a vapor phase reaction (McMillin 1963, 1965, Aktiebolag 1965) Liu and McMillin (1965). Barnes et al. (1969) showed that an oscillating pressure was better than a constant pressure in this reaction. Sodium hydroxide has also been used as a catalyst with ethylene oxide (Zimakov and Pokrovski 1954). Other epoxides that have been studied include 1,2-epoxy-3,3,3-trichloropropane, 1,2-epoxy-4,4, 4-trichlorobutane, 1-allyloxy-2,3-epoxypropane, *p*-chlorophenyl-2,3-eposypropyol ether, 1,2:3,4-diepoxybutane, 1,2:7, 8-diepoxyoctane, 1,4-butanediol diglycidal ether, and 3-epoxyethyl-7-oxabicyclo heptane (Rowell and Ellis 1984b).

Propylene and butylenes oxides have also been reacted with wood along with epichlorohydrin and mixtures of epichlorohydrin and propylene oxide (Rowell 1975a, Rowell et al. 1976, Rowell and Ellis 1984a,b). It is theoretically possible for cross-linking to occur with epichlorohydrin, resulting in the splitting out of hydrochloric acid, but this has never been observed (Rowell and Ellis 1981, Rowell and Chen 1994).

TABLE 14.1
Reaction Rates for Pine Using Different Chemicals

Chemical	Temperature (°C)	Time (min)	Weight Percent Gain
Propylene oxide–TEA	120	40	35.5
	120	120	45.5
	120	240	52.2
Butylene oxide–TEA	120	40	24.6
	120	180	36.9
	120	360	42.2
Methyl isocyanate	120	10	25.7
	120	20	40.4
	120	60	51.8
Butyl isocyanate	120	120	5.0
	120	180	16.0
	120	360	24.3
Acetic anhydride (liquid)	100	60	7.8
	100	120	11.2
	100	360	19.9
	120	60	17.2
	120	180	21.4
	140	60	17.9
	140	180	22.1
	160	60	21.4
Acetic anhydride (vapor)	120	480	7.2
	120	1440	22.1
Ketene	50–60	60	6.8
	50–60	120	21.7

In the case of the epoxy system, after the initial reaction with a cell wall hydroxyl group, a new hydroxyl group originating from the epoxide is formed. From this new hydroxyl, a polymer begins to form. Because of the ionic nature of the reaction and the availability of alkoxyl ions in the wood components, the chain length of the new polymer is probably short owing to chain transfer.

14.3.13 REACTION RATES

Table 14.1 shows the rate of reaction of southern pine reacted with propylene and butylene oxide, methyl and butyl isocyanate, and acetylated under different reaction conditions using liquid acetic anhydride, acetic anhydride vapor, and ketene gas. The propylene and butylenes oxide reactions were done using 5% triethyl amine as a catalyst. The methyl and butyl isocyanates were uncatalyzed. In the case of acetic anhydride vapor reactions, the pine veneer was suspended above a supply of acetic acid anhydride (Rowell et al. 1986a). In the case of ketene gas, the pine chips were subjected to ketene gas in batches (Rowell et al. 1986c).

In the reactions of the two epoxides and two isocyanates, the longer the reaction is run, the higher the weight percent gain (WPG). Acetylation, however, reaches a maximum WPG of about 22 no matter how it is done; further reaction time does not increase this value. The acetylation of aspen follows a similar pattern, except a maximum WPG of about 17–18 is reached. Other softwood and hardwood that have been acetylated using these procedures have followed a similar pattern.

It has been shown that the acetic anhydride reaction solution could contain up to 30% acetic acid without any detrimental effect on the reaction rate (Rowell et al. 1990). There is actually an increase in reaction rate at 10–20% acetic acid in the impregnation solution due to the swelling effect of the acetic acid and, possibly, as an effect of the increased acidity.

Since the reaction of wood with acetic anhydride is an exothermic reaction, once the acetylation starts to take place, the temperature of the system goes up. This heating can be taken advantage of to reduce the external heat supplied to the system.

14.3.14 EFFECT OF MOISTURE

Since all of the chemicals that react with wood cell wall hydroxyl groups can also react with water, the amount of water in the wood before reaction is very important (Rowell and Ellis 1984a). Hydrolysis with water is much faster than the reaction with a cell wall hydroxyl group. There is a trade off between the cost and energy to remove water below about 5% moisture content and the cost of chemical lost to hydrolysis. In the case of epoxides, reaction with water leads to the formation of nonbonded polymers in the cell wall. In the reaction with anhydrides, free acid is generated, which in the case of acetylation can be an advantage so long as the amount of acetic acid generated is not too large.

14.4 PROPERTIES OF CHEMICALLY MODIFIED WOOD

It is not possible to describe all of the properties of chemically modified wood in a short chapter. An entire book would be needed to describe all of the changes in properties from all of the reaction systems in the published literature. The next section will select some of the major property changes with a few examples from several of the chemical reactions systems. Since most of the historical and recent research in chemical modification of wood has been in the area of acetylation, much of the property and performance data will focus on this technology.

14.4.1 CHANGES IN WOOD VOLUME RESULTING FROM REACTION

Table 14.2 shows the increase in volume of pine wood after reaction with several liquid chemicals and the calculated volume of chemical added to the cell wall after re-drying the wood. Since the

TABLE 14.2
Changes in Pine Volume and Volume of Chemical Added as a Result of Chemical Reactions

Chemical	WPG	Increase in Wood Volume[1] (cm^3)	Calculated Volume of Added Chemical[2] (cm^3)
Propylene oxide	26.5	7.1	7.5
	36.2	8.9	9.0
Butylene oxide	25.3	6.9	6.9
Methyl isocyanate	12.4	0.16	0.14
	25.7	0.21	0.27
	47.7	0.46	0.54
Acetic anhydride	17.5	3.0	2.9
	22.8	3.9	4.0
Acrylonitrile	25.7	0.46	0.77
	36.0	0.74	1.2

[1] Difference in oven-dry volume between reacted and nonreacted wood.
[2] Density used in volume calculations: Propylene and butylenes oxides:1.01, methyl isocycnate:0.967, acetic anhydride:1.049, and acrylonitrile:0.806

Chemical Modification of Wood

volume increase due to reactions with propylene and butylenes oxide, methyl and butyl isocyanates, and acetic anhydride is equal to the calculated volume of chemical added, this shows that the reaction has taken place in the cell wall and not in the void spaces of the wood (Rowell 1983). This correlation is not true for reactions of wood with, for example, acrylonitrile (Rowell 1983). In this case, the chemical is located in the voids in the wood structure.

14.4.2 Stability of Bonded Chemical to Chemical Leaching

One of the indirect ways to tell if the reaction has resulted in bonding with the cell wall is to leach the reacted wood with several solvents and determine loss of weight in the reacted sample. At least two different solvents are used. One is a solvent that the starting chemicals are soluble in, and a second is one in which any polymers that might be formed in the reaction are soluble.

Table 14.3 shows the weight loss due to leaching in benzene/ethanol, water in a soxhlet extractor for 24 hours, and water soaking for 7 days. The data shows that reactions with methyl isocyanate, butylene oxide, and acetic anhydride form chemical bonds that are stable to solvent extraction, while propylene oxide reacted wood loses some chemical in water leaching. The reaction with acrylonitrile catalyzed with either ammonium or sodium hydroxyl forms bonds that are not stable to solvent extraction.

Table 14.4 shows the stability of acetyl groups on acetylated pine and aspen to various pHs and temperatures, and Table 14.5 shows the activation energy for the deacetylation of pine and aspen (Rowell et al. 1992a, 1992b). These results show that acetylated wood is much more stable under slightly acidic conditions as opposed to slightly alkaline conditions. The data can be used to predict the stability of acetylated wood (or other wood) at any pH and temperature combinations to estimate the life expectancy of the product in its service environment. However, it must be considered that the data from this study were collected using either finely ground powder or small fiber. As a result, a very large surface area came into contact with the various pH and temperature environments. Therefore, this data represents the fastest possible degradation rates. It is well documented that in archaeological wood, acetyl groups are stable over thousands of years and little loss of acetyl has been observed.

TABLE 14.3
Oven-Dry Weight Loss of Chemically Reacted Wood Extracted with Several Solvents

Chemical	WPG	Benzene/Ethanol 4 hours Soxhlet 20 mesh (% wt loss)	Water 24 hrs Soxhlet 40 mesh (% wt loss)	Water 7 days Soxhlet 20 mesh (% wt loss)
None	0	2.3	11.2	0.6
Methyl isocyanate	23.5	6.5	11.6	1.0
Propylene oxide	29.2	5.2	10.7	4.0
Butylene oxide	27.0	3.8	11.7	1.6
Acetic anhydride	22.5	2.8	12.2	1.2
Acrylonitrile (NH_4OH)	26.1	22.3	20.4	21.7
Acrylonitrile (NaOH)	25.7	17.6	18.7	13.5

TABLE 14.4
Effects of Various pH and Temperature Conditions on the Stability of Acetyl Groups in Acetylated Pine and Aspen (Acetyl Content: Pine 20.2 WPG, Aspen 20.6 WPG)

Wood	Temperature (°C)	pH	Rate Constant (k × 10³)	Half Life (Days)
Pine	24	2	0.26	2,640
		4	0.15	4,630
		6	0.06	10,900
		8	1.40	500
Aspen	24	2	0.23	3,083
		4	0.15	4,697
		6	0.06	10,756
		8	1.11	623
Pine	50	2	2.07	340
		4	1.06	650
		6	1.71	410
		8	7.31	95
Aspen	50	2	1.18	590
		4	0.66	1,050
		6	1.29	540
		8	8.51	80
Pine	75	2	11.3	61
		4	8.41	82
		6	15.5	45
		8	32.2	22
Aspen	75	2	7.33	95
		4	4.78	145
		6	12.6	55
		8	32.5	21

Table 14.6 shows the stability of acetyl groups in pine and aspen flakes to cyclic exposure to 90% and 30% relative humidity (RH) (Rowell et al. 1992a,b). Each cycle represents exposing the flakes for three months at 30% RH and then three months at 90% RH. Within experimental error, there is no loss of acetyl over 41 cycles of humidity changes. This data was collected in 1992 and this experiment has continued. After almost 10 years of cycling these chips from 30% to 90% RH, recent acetyl analysis shows that there is still no loss of acetyl resulting from humidity cycling.

TABLE 14.5
Activation Energy for Deacetylation of Pine and Aspen over a Temperature Range of 24–75°C

Wood	pH	Activation Energy (kJ/mole)
Pine	2	58
	4	58
	6	89
	8	57
Aspen	2	63
	4	68
	6	93
	8	53

TABLE 14.6
Stability of Acetyl Groups in Pine and Aspen Flakes after Cyclic Exposure between 90% Relative Humidity (RH) and 30% RH

Wood	Acetyl Content (%) after Cycle (Number)				
	0	13	21	33	41
Pine	18.6	18.2	16.2	18.0	16.5
Aspen	17.9	18.1	17.1	17.8	17.1

14.4.3 ACCESSIBILITY OF REACTION SITE

The rate controlling step in the chemical modification of wood is the rate of penetration of the reagent into the cell wall (Rowell et al. 1994). In the reaction of liquid acetic anhydride with wood, at an acetyl weight percent gain of about 4, there is more bonded acetyl in the S_2 layer than in the middle lamella. At a weight gain of about 10, there is an equal distribution of acetyl throughout the S_2 layer and the middle lamella. At a WPG over 20, there is a slightly higher concentration of acetyl in the middle lamella than in the rest of the cell wall. This was found using chloroacetic anhydride and following the fate of the chlorine by energy-dispersive X-ray analysis (Rowell et al. 1994).

When larger pieces of wood are used, the time for the anhydride to penetrate increased in proportion to the size of wood used. Of course, the critical dimension is the size in the longitudinal dimension. Table 14.7 shows that if spruce chips are reacted for 30 minutes at 120°C, the WPG is 14.2 with an acetyl content of 15.6. If the chips are first reduced to fiber and then acetylated under the same reaction conditions, the WPG is 22.5 and the acetyl content is 19.2. If the acetylated chips were reduced to fiber and then reacetylated under the same conditions, the WPG is 20.4 with an acetyl content of 20.5. Since no acetyl groups were lost in the refining step, this shows that new -OH sites become available when the acetylated chips are reduced to fiber (Rowell and Rowell 1989). This also shows that some -OH groups are not accessible in a chip that becomes accessible in a fiber.

14.4.4 ACETYL BALANCE IN ACETYLATED WOOD

The mass balance in the acetylation reaction shows that all of the acetic anhydride going into the acetylation of hardwood and softwood could be accounted for as increased acetyl content in the wood, acetic acid resulting from hydrolysis by moisture in the wood, or as unreacted acetic anhydride (Rowell et al. 1990). The consumption of acetic anhydride can be calculated stoichiometrically based

TABLE 14.7
Effects of Size of Spruce Specimen on Acetyl Content

Specimen	WPG	Acetyl Content
Chips Acetylated	14.2	15.6
Acetylated Chips to Fiber	14.2	15.4
Chips to Fiber and then Acetylated	22.5	19.2
Acetylated Chips to Fiber	22.5	19.4
Acetylated Chips to Fiber and again Acetylated	20.4	20.5

TABLE 14.8
Weight Percent Gain and Acetyl Analysis of Acetylated Pine and Aspen

Pine		Aspen	
WPG	Acetyl Content	WPG	Acetyl Content
0	1.4	0	3.9
6.0	7.0	7.3	10.1
14.8	15.1	14.2	16.9
21.1	20.1	17.9	19.1

on the degree of acetylation and the moisture content of the wood. This is true for all wood acetylated to date.

Table 14.8 shows the comparison of the weight gain due to acetylation, with the acetyl content found by chemical analysis. At the lower weight-gain levels, there is always a higher acetyl content as compared to the WPG. This may be due to the removal of extractives and some cell wall polymers into the acetic anhydride solution resulting in an initial specimen weight loss. At WPGs greater than about 15, the acetyl content and the WPG values are almost the same.

14.4.5 DISTRIBUTION OF BONDED CHEMICAL

Table 14.9 shows the distribution of bonded chemicals and the degree of substitution (DS) of hydroxyl groups in methyl isocyanate reacted pine (Rowell 1980, 1982, Rowell and Ellis 1981, Rowell et al. 1994). This analysis is based on many assumptions: (1) that pine has a holocellulose content of 67% and a lignin content of 27%; (2) that the holocellulose content is 87% hexosans, 13% pentosans; (3) that the cellulose content of the holocellulose is 71.8%, 14.8% hemicellulose hexosans, and 13.4% hemicellulose pentosans; (4) that whole pine wood is 67% holocellulose, of which 48.1% is cellulose, 9.0% pentosans, and 9.9% hexosans; (5) that the calculated theoretical acetyl content of the holocellulose is 77.7% and 25.7% for the lignin; and (6) that there are 1.1 hydroxyl groups for each nine-carbon unit of lignin.

Based on this analysis, the lignin is completely substituted at a WPG of about 47 and at that point, only about 20% of the hydroxyls in the holocellulose are substituted. These calculations are also based on the added assumption that all of the theoretical hydroxyl groups are accessible during the acetylation reaction. Assuming 100% accessibility of the hemicellulose fraction but only 35% accessibility of the cellulose (based on the degree of crystallinity), the degree of substitution of the

TABLE 14.9
Distribution of Methyl Isocyanate Nitrogen and Degree of Substitution (DS) of Hydroxyl Groups in Modified Pine

	MeIsoN in Lignin		MeIsoN in Holo	
WPG	%	DS	%	DS
5.5	1.42	0.17	0.59	0.03
10.0	2.36	0.28	1.19	0.05
17.7	3.44	0.41	2.11	0.08
23.5	4.90	0.59	2.94	0.12
47.2	7.46	0.89	5.24	0.21

TABLE 14.10
Fiber Saturation Point for Acetylated Pine and Aspen

WPG	Pine (%)	Aspen (%)
0	45	46
6	24	—
8.7	—	29
10.4	16	—
13.0	—	20
17.6	—	15
18.4	14	—
21.1	10	—

accessible holocellulose would be 0.48. In experiments acetylating isolated cell wall components, lignin reacts faster than the hemicelluloses and whole wood and cellulose did not react at all (Rowell et al. 1994). It is possible that the cellulose is modified during the isolation procedure, so this result does not necessarily mean that no cellulose modification takes place during the acetylation of whole wood.

14.4.6 MOISTURE SORPTION

By replacing some of the hydroxyl groups on the cell wall polymers with bonded acetyl groups, the hygroscopicity of the wood material is reduced. Table 14.10 shows the fiber saturation point for acetylated pine and aspen (Rowell 1991). As the level of acetylation increases, the fiber saturation point decreases in both the soft and hard wood.

Table 14.11 shows the equilibrium moisture content (EMC) of control and several types of chemically-modified pine at three levels of relative humidities. In all cases, as the level of chemical weight gain increases, the EMC of the resulting wood goes down (Rowell et al. 1986b). Reaction of formaldehyde with wood is the most effective in reducing EMC as a function of chemical weight gain. Of the other chemicals, acetylation is the most effective, followed by butylene oxide reaction. Propylene-oxide–reacted wood has only a slight decrease in EMC as compared to non-reacted wood. Table 14.12 shows the EMC of acetylated pine and aspen at several levels of acetylation.

TABLE 14.11
Equilibrium Moisture Content of Control and Chemically Modified Pine

Chemical	WPG	Equilibrium Moisture Content at 27°C		
		35% RH	60% RH	85% RH
Control	0	5.0	8.5	16.4
Propylene oxide	21.9	3.9	6.1	13.1
Butylene oxide	18.7	3.5	5.7	10.7
Formaldehyde	3.9	3.0	4.2	6.2
Acetic anhydride	20.4	2.4	4.3	8.4

TABLE 14.12
Equilibrium Moisture Content of Acetylated Pine and Aspen

Specimen	WPG	Equilibrium Moisture Content at 27°C		
		30% RH	65% RH	90% RH
Pine	0	5.8	12.0	21.7
	6.0	4.1	9.2	17.5
	10.4	3.3	7.5	14.4
	14.8	2.8	6.0	11.6
	18.4	2.3	5.0	9.2
	20.4	2.4	4.3	8.4
Aspen	0	4.9	11.1	21.5
	7.3	3.2	7.8	15.0
	11.5	2.7	6.9	12.9
	14.2	2.3	5.9	11.4
	17.9	1.6	4.8	9.4

In the case of acetylated wood, if the reductions in EMC at 65% RH of many different types of acetylated woods referenced to unacetylated fiber is plotted as a function of the bonded acetyl content, a straight line plot results (Rowell and Rowell 1989). Even though the points represent many different types of wood, they all fit a common curve. A maximum reduction in EMC is achieved at about 20% bonded acetyl. Extrapolation of the plot to 100% reduction in EMC would occur at about 30% bonded acetyl. This represents a value not too different from the water fiber saturation point in these fibers. Because the acetate group is larger than the water molecule, not all hygroscopic hydrogen-bonding sites are covered; thus it would be expected that the acetyl saturation point would be lower than that of water.

The fact that EMC reduction as a function of acetyl content is the same for many different types of wood indicates that reducing moisture sorption may be controlled by a common factor. The lignin, hemicellulose, and cellulose contents of all the woods are different.

Figure 14.3 shows the sorption-desorption isotherms for acetylated spruce fibers (Stromdahl 2000). The 10-minute acetylation curve represents a WPG of 13.2, and the 4-hour curve represents a WPG of 19.2. The untreated spruce reaches an adsorption/desorption maximum at about 35% moisture content, the 13.2 WPG a maximum of about 30%, and the 19.2 WPG a maximum of about 10%. There is a very large difference between the adsorption and desorption curves for both the control and the 13.2 WPG fibers, but much less difference in the 19.2 WPG fibers. The sorption of moisture is presumed to be sorbed either as primary water or secondary water. *Primary water* is the water sorbed to primary sites with high-binding energies, such as the hydroxyl groups. *Secondary water* is that water sorbed to less-binding energy sites—water molecules sorbed on top of the primary layer. Since some of the hydroxyl sites are esterified with acetyl groups, there are fewer primary sites to which water sorbs. And since the fiber is more hydrophilic due to acetylation, there may also be less secondary binding sites.

14.4.7 DIMENSIONAL STABILITY

Changes in dimensions in tangential and radial direction in solid wood, and in thickness and linear expansion for composites, are a great problem for wood composites (see Chapter 4). Composites not only undergo normal swelling (reversible swelling), but also swelling caused by the release of residual compressive stresses imparted to the board during the composite pressing process (irreversible swelling). Water sorption causes both reversible and irreversible swelling, with some of the reversible shrinkage occurring when the board dries.

Chemical Modification of Wood

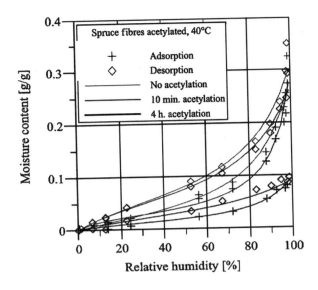

FIGURE 14.3 Sorption/desorption isotherms for control and acetylated spruce fiber.

Thickness swelling at three levels of relative humidity is greatly reduced as a result of acetylation (Table 14.13). Linear expansion is also greatly reduced as a result of acetylation (Krzysik et al. 1992, 1993). Increasing the adhesive content can reduce the thickness swelling, but not to the extent that acetylation does.

The rate and extent of thickness swelling in liquid water of fiberboards made from control and acetylated fiber is shown in Table 14.14. Both the rate and extent of swelling are greatly reduced as a result of acetylation. At the end of 5 days of water soaking, control boards swelled 36%, whereas boards made from acetylated fiber swelled less than 5%. Drying the boards at the end of the test showed that control boards exhibit a greater degree of irreversible swelling as compared to boards made from acetylated fiber (Rowell et al. 1991b).

Table 14.15 shows the thickness swelling of pine, beech, and wheat straw fiberboards after 24 hours of water swelling and the thickness of the boards after drying. In all cases, the control fiberboards swelled on water soaking and remained almost as thick after drying, while the acetylated fiberboards showed very little swelling in water and little residual swelling after drying.

TABLE 14.13
Thickness Swelling (TS) of Aspen Fiberboards Made from Control and Acetylated Fiber

WPG	Phenolic Resin Content (%)	TS at 27°C		
		30% RH (%)	65% RH (%)	90% RH (%)
0	5	0.7	3.0	12.6
	8	1.0	3.1	11.2
	12	0.8	2.5	9.7
17.9	5	0.2	1.8	3.2
	8	0.2	1.7	3.1
	12	0.1	1.7	2.9

TABLE 14.14
Rate and Extent of Thickness Swelling in Liquid Water of Pine Fiberboards Made from Control and Acetylated Fiber (8% Phenolic Resin)

	Percent Thickness Swelling at						
	Minutes			Hours			Days
Fiberboard	15	30	60	3	6	24	5
Control	25.7	29.8	33.5	33.8	34.0	34.0	36.2
Acetylated (21.6 WPG)	0.6	0.9	1.2	1.9	2.5	3.7	4.5

TABLE 14.15
Thickness Swelling of Fiberboards in Water and Thickness after Drying (10% Phenolic Resin)

Fiber	WPG	Thickness Swelling (24 hrs in Water) (%)	Residual Thickness Swelling (Oven Dry) (%)
Pine	0	21.3	19.7
	21.5	2.1	1.0
Beech	0	17.0	13.0
	19.7	2.2	0.9

Table 14.16 shows the antishrink efficiencies (ASE—see Chapter 4 for definition and calculations) of several different types of chemically modified solid pine wood. All chemically reacted pine wood shows an ASE of approximately 70 at weight gains between 22–26 WPG.

If the water-swelling test is continued through several cycles of water soaking and oven drying, the ASE may change if chemicals are lost in the leaching process. Table 14.17 shows the ASE for several types of chemically-modified solid pine. ASE_1 is calculated from the wood going from oven-dry to water-soaked (Rowell and Ellis 1978). ASE_2 is then calculated from the wood going from water-soaked back to the oven-dry state. ASE_3 and ASE_4 are the values on the second complete cycle.

TABLE 14.16
Antishrink Efficiencies of Chemically Modified Solid Pine Wood

Chemical	WPG	ASE
None	0	—
Propylene oxide	29.2	62
Butylene oxide	27.0	74.3
Acetic anhydride	22.5	70.3
Methyl isocyanate	26.0	69.7
Acrylonitrile		
NH_4OH	26.1	80.9
NaOH	25.7	48.3

Chemical Modification of Wood

TABLE 14.17
Repeated Antishrink Efficiency (ASE) of Chemically Modified Solid Pine

Chemical	WPG	ASE$_1$	ASE$_2$	ASE$_3$	ASE$_4$	Weight Loss after Test
Propylene oxide	29.2	62.0	43.8	50.9	50.3	5.7
Butylene oxide	26.7	74.3	55.6	59.7	48.1	4.6
Acetic anhydride	22.5	70.3	71.4	70.6	69.2	<.2
Methyl isocyanate	26.0	69.7	62.8	65.0	60.7	4.3
Acrylonitrile						
NH$_4$OH	26.1	80.9	0	0	0	22.6
NaOH	25.7	48.3	0	0	0	14.7

This data in Table 14.17 shows that the highest ASE is for acrylonitrile catalyzed with ammonium in the first wetting cycle. But redrying results in a loss of much of the chemical, so ASE$_2$ is zero with a final weight loss of 22.6%—almost a complete loss of reacted chemical. Propylene and butylene oxides, methyl isocyanate, and acetic-anhydride–reacted wood are more stable to swelling and shrinking cycles. Acetylation is the most stable treatment, with less than 0.2% weight loss after the two-cycle test.

14.4.8 Loss of Dimensional Stability

At very high WPG with isocyanates and epoxide-modified wood, the ASE values started to drop (Rowell and Ellis 1979, 1981, 1984a,b). Figure 14.4 shows a plot of ASE versus WPG for propylene- and butylene-oxide–reacted wood. As the WPG increases up to about 25 to 30, the ASE increases. At WPGs higher than 25–30, the ASE starts to decrease. The same effect is also observed in methyl-isocyanate–reacted wood.

Explanation of this phenomenon is observed in an electron microscopic examination of the epoxide and isocyanate reacted wood at different WPG levels. Figure 14.5 shows the electron micrographs for propylene-oxide–reacted wood. Figure 14.5a shows the unreacted wood. Figure 14.5b shows wood reacted with propylene oxide to 25 WPG. The cell wall is swollen in

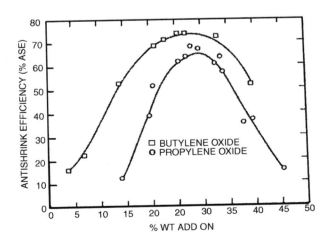

FIGURE 14.4 Loss of antishrink efficiency at high chemical weight gains.

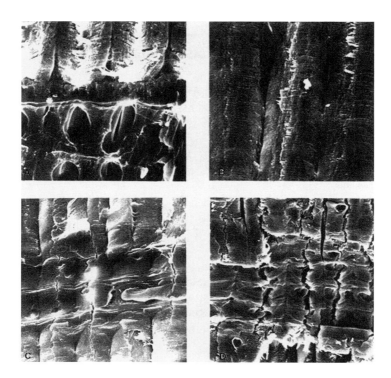

FIGURE 14.5 Scanning electron micrographs of radial-split southern pine showing swelling of wood when reacted with propylene oxide-triethylamine. A = control (1100×), B = WPG 25 (1100×), C = WPG 32.6 (600×), D = WPG 45.3 (550×).

Figure 14.5b, but no cracks are observed. Figure 14.5c shows small cracks starting to form in the tracheid walls at a WPG of 33, and major cracks are observed in Figure 14.5d at a WPG of 45. The largest cracks in Figure 14.5d are between cells. Similar results are observed in methyl-isocyanate–reacted wood.

14.4.9 Biological Resistance

Various types of woods, particleboards, and flakeboards made from chemically-modified wood have been tested for resistance to several different types of organisms (Imamura et al. 1987, Rowell et al. 1989, Chow et al. 1994, Beckers et al. 1994, Militz 1991, Wang et al. 2002).

14.4.9.1 Termite Resistance

Table 14.18 shows the results of a two-week termite test using *Reticulitermes flavipes* (subterranean termites) on several types of chemically-modified pine (Rowell et al. 1979, 1988a). Propylene and butylene oxide and acetic anhydride–modified wood becomes somewhat resistant to termite attack at about 30% for propylene oxide and about 20–25% for butylene oxide and acetic-anhydride–reacted wood. There was not a complete resistance to attack, and this may be attributed to the severity of the test. However, since termites can live on acetic acid and decompose cellulose to mainly acetate, perhaps it is not surprising that acetylated wood is not completely resistant to termite attack. Termite survival was quite high at the end of the tests, showing that the modified wood was not toxic to them.

TABLE 14.18
Wood Weight Loss in Chemically Modified Pine after a Two-Week Exposure Test to *Reticulitermes flavipes*

Chemical	WPG	Wood Weight Loss (%)
Control	0	31
Propylene oxide	9	21
	17	14
	34	6
Butylene oxide	27	4
	34	3
Acetic anhydride	10.4	9
	17.8	6
	21.6	5

14.4.9.2 Decay Resistance

Chemically-modified wood has been tested with several types of decay fungi in an ASTM standard 12-week soil-block test using either the brown rot fungus *Gloeophyllum trabeum* or the white rot fungus *Trametes versicolor* (Nilsson et al. 1988, Rowell et al. 1987, 1988a). Table 14.19 shows that all of the chemical modifications demonstrate good resistance to white-rot fungi, and all but propylene oxide shows good resistance to brown-rot fungi.

Weight loss resulting from fungal attack is the method most used to determine the effectiveness of a preservative treatment to protect wood composites from decaying. In some cases, especially for brown-rot fungal attack, strength loss may be a more important measure of attack, since large strength losses are known to occur in solid wood at very low wood-weight loss (Cowling 1961). A dynamic bending-creep test has been developed to determine strength losses when wood composites are exposed to a brown- or white-rot fungus (Imamura and Nishimoto 1985, Norimoto et al. 1987, Norimoto et al. 1992).

Using this bending-creep test on aspen flakeboards, control boards made with phenol-formaldehyde adhesive failed in an average of 71 days using the brown-rot fungus *Tyromyces palustris* and 212 days using the white-rot fungus *Traetes versicolor* (Imamura et al. 1988, Rowell et al. 1988b).

TABLE 14.19
Resistance of Chemically Modified Pine against Brown- and White-Rot Fungi

Chemical	WPG	Weight Loss After 12 Weeks	
		Brown-rot Fungus (%)	White-rot Fungus (%)
Control	0	61.3	7.8
Propylene oxide	25.3	14.2	1.7
Butylene oxide	22.1	2.7	0.8
Methyl isocyanate	20.4	2.8	0.7
Acetic anhydride	17.8	1.7	1.1
Formaldehyde	5.2	2.9	0.9
β-Propiolacetone	25.7	1.7	1.5
Acrylonitrile	25.2	1.9	1.9

At failure, weight losses averaged 7.8% for *T. palustris* and 31.6% for *T. versicolor*. Isocyanate-bonded control flakeboards failed in an average of 20 days with *T. palustris* and 118 days with *T. versicolor*, with an average weight loss at failure of 5.5% and 34.4%, respectively (Rowell et al. 1988b). Very little or no weight loss occurred with both fungi in flakeboards made using either phenol-formaldehyde or isocyanate adhesive with acetylated flakes. None of these specimens failed during the 300-day test period.

Mycelium fully covered the surfaces of isocyanate-bonded control flakeboards within one week, but mycelial development was significantly slower in phenol-formaldehyde-bonded control flakeboards. Both isocyanate- and phenol-formaldehyde-bonded acetylated flakeboards showed surface mycelium colonization during the test time, but little strength was lost since the fungus did not attack the acetylated flakes.

In similar bending-creep tests, both control and acetylated pine particleboards made using melamine-urea-formaldehyde adhesive failed because *T. palustris* attacked the adhesive in the glue line (Imamura et al. 1988). Mycelium invaded the inner part of all boards, colonizing in both the wood and glue lines in control boards, but only in the glue line in acetylated boards. These results show that the glue line is also important in protecting composites from biological attack.

After a 16-week exposure to *T. palustris*, the internal bond strength of control aspen flakeboards made with phenol-formaldehyde adhesive was reduced over 90%, and that of flakeboards made with isocyanate adhesive was reduced 85% (Imamura et al. 1987). After six months of exposure in moist unsterile soil, the same control flakeboards made with phenol-formaldehyde adhesive lost 65% of their internal bond strength, and those made with isocyanate adhesive lost 64% internal bond strength. Failure was due mainly to great strength reductions in the wood caused by fungal attack. Acetylated aspen flakeboards lost much less internal bond strength during either the 16-week exposure to *T. palustris* or a 6-month soil burial. The isocyanate adhesive was somewhat more resistant to fungal attack than the phenol-formaldehyde adhesive. In the case of acetylated composites, the loss in internal bond strength was mainly due to fungal attack in the adhesive and moisture, which caused a small amount of swelling in the boards.

The mechanism of resistance to fungal attack by chemical modification has been suggested to be due to the specific enzymatic reactions that can take place due to a change in configuration and conformation of the modified wood and/or when the reduction in equilibrium moisture content becomes too low to support fungal growth. In the specific case of brown-rot fungal attack, a mechanism has been suggested that explains the loss of strength before there is very much weight loss (Nilsson 1986, Winandy and Rowell 1984). This mechanism is consistent with the data from the soil block weight-loss tests and the strength-loss tests.

Enzymes → Hemicelluloses
(Energy source for
  generation of
  chemical oxidation
  system) → Hemicellulose Matrix
    (Strength losses)
    (Energy source for generation
     of β-glucosidases) → Weight Loss

Another test for fugal and bacterial resistance that has been done on acetylated composites is with brown-, white-, and soft-rot fungi and tunneling bacteria in a fungal cellar (Table 14.20). Control blocks were destroyed in less than 6 months, while flakeboards made from acetylated furnish above 16 WPG showed no attack after 1 year. (Nilsson et al. 1988, Rowell et al. 1988a). This data shows that no attack occurs until the wood begins to swell (Rowell and Ellis 1984, Rowell et al. 1988a). This fungal cellar test was continued for an additional 5 years with no attack at

TABLE 14.20
Fungal Cellar Tests of Aspen Flakeboards Made from Control and Acetylated Flakes[1,2]

WPG	Rating at Intervals (Months)[3]							
	2	3	4	5	6	12	24	36
0	S/2	S/3	S/3	S/3	S/4	—	—	—
7.3	S/0	S/1	S/1	S/2	S/3	S/4	—	—
11.5	0	0	S/0	S/1	S/2	S/3	S/4	—
13.6	0	0	0	0	S/0	S/1	S/2	S/3
16.3	0	0	0	0	0	0	0	0
17.9	0	0	0	0	0	0	0	0

[1] Nonsterile soil containing brown-, white-, and soft-rot fungi and tunneling bacteria.
[2] Flakeboards bonded with 5% phenol-formaldehyde adhesive.
[3] Rating system: 0 = no attack; 1 = slight attack; 2 = moderate attack; 3 = heavy attack; 4 = destroyed; S = swollen.

17.9 WPG. Further evidence suggests that the moisture content of the cell wall is critical before attack can take place (Ibach and Rowell 2000, Ibach et al. 2000).

In-ground tests have also been done on acetylated solid wood and flakeboards (Hadi et al. 1995, Rowell et al. 1997, Larsson Brelid et al. 2000). Specimens have been tested in the United States, Sweden, New Zealand, and Indonesia. The specimens in the United States, Sweden, and New Zealand are showing little or no attack after ten years, while the specimens in Indonesia failed in less than three years (Hadi et al. 1996). The failure in Indonesia was mainly due to termite attack.

Recent results show that acetylated pine at a WPG of 21.2 is outperforming CCA (copper-chromium-arsenic) at 10.3 kg/m^3 after 8 years in test in Sweden (Larsson Brelid et al. 2000).

14.4.9.3 Marine Organisms Resistance

Table 14.21 shows the data for chemically-modified pine in a marine environment (Johnson and Rowell 1988). As with the termite test, all types of chemical modifications of wood help resist attack by marine organisms. Control specimens were destroyed in six months to one year, mainly because of attack by *Limnoria tripunctata*, while propylene and butylene oxide, butyl isocyanate, and acetic-anhydride–reacted wood showed good resistance. Similar tests were run in Sweden on acetylated wood, and the modified wood failed in marine tests after two years (Larsson Brelid et al. 2000). The failure was due to attack by crustaceans and mollusks.

14.4.10 THERMAL PROPERTIES

Table 14.22, Figure 14.6, and Figure 14.7 shows the results of thermogravimetric and evolved gas analysis of chemically-modified pine (see Chapter 6). The unreacted, acetylated, and methyl isocyanate samples show two peaks in the thermogravimetric runs while propylene and butylene oxide only show one peak. The lower temperature peak represents the hemicellulose fraction, and the higher peak represents the cellulose in the fiber. Acetylated pine fibers pyrolyze at about the same temperature and rate (Rowell et al. 1984). The heat of combustion and rate of oxygen consumption are approximately the same for control and acetylated fibers, which means that the acetyl groups added have approximately the same carbon, hydrogen, and oxygen content as the cell wall polymers. The two peaks in control, acetylated, and methyl isocyanate samples are combined into only one large peak in the epoxide-reacted pine (Rowell et al. 1984). The hemicelluloses seem to become more thermally stable when epoxidized. The heat of combustion and rate of oxygen consumption

TABLE 14.21
Ratings of Chemically Modified Southern Pine Exposed to a Marine Environment[1]

			Mean Rating Due to Attack by	
Chemical	WPG	Years of Exposure	Limnoriid and Teredinid Borers[2]	Shaeroma terebrans[3]
Control	0	1	2–4	3.4
Propylene oxide	26	11.5	10	—
		3	—	3.8
Butylene oxide	28	8.5	9.9	—
		3	—	8.0
Butyl isocyanate	29	6.5	10	—
Acetic anhydride	22	3	8	8.8

[1] Rating system: 10 = no attack; 9 = slight attack; 7 = some attack; 4 = heavy attack; 0 = destroyed
[2] Installed in Key West, FL. 1975 to 1987
[3] Installed in Tarpon Springs, FL 1984 to 1987

is almost double for the epoxide-reacted samples as compared to control, acetylated, and methyl-isocyanate–modified pine.

Reactive fire retardants could be bonded to the cell wall hydroxyl groups in reactions similar to this technology. The effect would be an improvement in dimensional stability, biological resistance, and fire retardancy (Rowell et al. 1984, Ellis and Rowell 1989, Ellis et al. 1987, Lee et al. 2000).

14.4.11 ULTRAVIOLET RADIATION AND WEATHERING RESISTANCE

Reaction of wood with epoxides and acetic anhydride has also been shown to improve ultraviolet resistance of wood (Feist et al. 1991a, 1991b). Table 14.23 shows the weight loss, erosion rate,

TABLE 14.22
Thermal Properties of Control and Acetylated Pine Fiber

Chemical	WPG	Temperature of Maximum Weight Loss (°C)	Heat of Combustion (KCal/g)	Rate of Oxygen Consumption (MM/g sec)
Control	0	335/375	2.9	0.06/0.13
Acetic anhydride	21.1	338/375	3.1	0.08/0.14
Methyl isocyanate	24.0	315/375	2.6	0.07/0.12
Propylene oxide	32.0	380	4.3	0.23
Butylene oxide	22.0	385	4.1	0.24

Chemical Modification of Wood

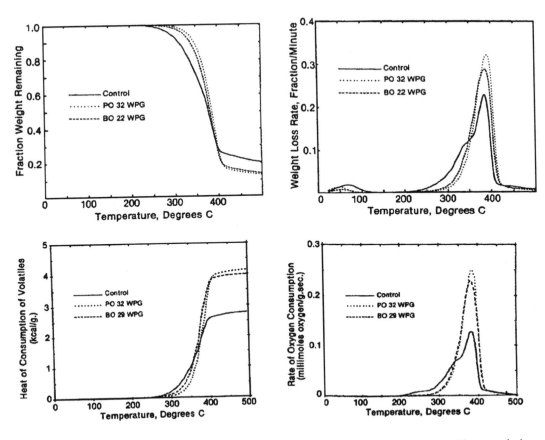

FIGURE 14.6 Thermogravimetric and evolved gas analysis of propylene and butylene oxide reacted pine.

and depth of penetration resulting from 700 hours of accelerated weathering for acetylated aspen. Control specimens erode at about 0.12 μm/hr, or about 0.02 %/hr. Acetylation reduces surface erosion by 50 percent. The depth of the effects of weathering is about 200 μm into the fiber surface for the unmodified boards and about half that of the acetylated boards.

Table 14.24 shows the acetyl and lignin content of the outer 0.5 mm surface and of the remaining specimen after the surface had been removed before and after accelerated weathering. The acetyl content is reduced in the surface after weathering, showing that the acetyl blocking group is removed during weathering (Hon 1995). UV radiation does not remove all of the blocking acetyl group, so some stabilizing effect to photochemical degradation still is in effect. The loss of acetate is confined

TABLE 14.23
Weight Loss and Erosion of Acetylated Aspen after 700 Hours of Accelerated Weathering

WPG	Weight loss in Erosion (%/hr)	Erosion Rate (μm/hr)	Reduction in Erosion (%)	Depth of Weathering (μm)
0	0.019	0.121	—	199–210
21.2	0.010	0.059	51	85–105

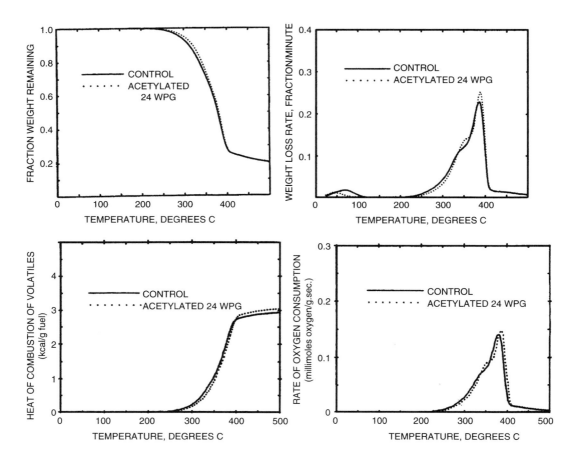

FIGURE 14.7 Thermogravimetric and evolved gas analysis of acetylated pine.

to the outer 0.5 mm since the remaining wood has the same acetyl content before and after accelerated weathering. The lignin content is also greatly reduced in the surface as a result of weathering; the main cell wall polymer has been degraded by UV radiation. Cellulose and the hemicelluloses are much more stable to photochemical degradation.

In outdoor tests, flakeboards made from acetylated pine wood maintain a light yellow color after one year while control boards turn dark orange to light gray during this time (Feist et al. 1991b). Within two years the acetylated pine had started to turn gray. Acetylated pine in indoor test settings retains its bright color after ten years, while control pine turns light orange after a few months.

14.4.12 Mechanical Properties

Strength properties are modified as a result of many of the chemical reactions performed on wood. For example, shear strength parallel to the grain is decreased in acetylated wood (Dreher et al. 1964), a slight decrease in the modulus of elasticity (Narayanamurti and Handa 1953) but no change in impact strength (Koppers 1961) or stiffness (Dreher et al. 1964). Wet and dry compressive strength (Koppers 1961, Dreher et al. 1964), hardness, fiber stress at proportional limit, and work to proportional limit are increased (Dreher et al. 1964). The modulus of rupture is increased in softwoods but decreased for hardwoods (Dreher et al. 1964, Minato et al. 2003). For isocyanate-reacted wood, compressive strength and bending modulus are increased (Wakita et al. 1977). The mechanical properties for formaldehyde-reacted wood are all reduced as compared to unreacted

TABLE 14.24
Acetyl and Lignin Analysis before and after 700 Hours of Accelerated Weathering of Aspen Fiberboards Made from Control and Acetylated Fiber

	Before Weathering		After Weathering	
WPG	Surface (%)	Remainder (%)	Surface (%)	Remainder (%)
Acetyl				
0	4.5	4.5	1.9	3.9
19.7	17.5	18.5	12.8	18.3
Lignin				
0	19.8	20.5	1.9	17.9
19.7	18.5	19.2	5.5	18.1

wood. Toughness and abrasion resistance are greatly reduced (Tarkow and Stamm 1953, Stamm 1959), crushing and bending strength are reduced 20%, and impact bending strength is reduced up to 50% (Burmester 1967). A definite embrittlement is observed in formaldehyde-reacted wood. This may be due to the short, inflexible crosslinking unit of –O-C-O– type. Some of the strength loss may also be polymer hydrolysis due to the strong acid catalyst. Strength properties of methylated wood are also reduced as a result of the strong acid used as a catalyst (Narayanamurti and Handa 1953). Cyanoethylated wood, using sodium hydroxide as a catalyst, has a lower impact strength (Goldstein et al. 1959, Kenaga 1963). Many of the mechanical properties of propylene oxide–reacted wood are reduced. Modulus of elasticity is reduced 14%, modulus of rupture is reduced 17%, fiber stress at proportional limit is reduced 9% and maximum crushing strength is reduced 10% (Rowell et al. 1982).

Wood composites made from acetylated furnish, however, do not lose mechanical properties as compared to nonreacted wood. The modulus of rupture (MOR), modulus of elasticity (MOE) in bending, and tensile strength (TS) parallel to the board surface are shown in Table 14.25 for fiberboards made from control and acetylated pine fiber. Acetylation results in an increase in MOR and MOE, but equal values in IBS (Youngquist et al. 1986a, 1986b). The MOR value is above the minimum standard as given by the American Hardboard Association (ANSI 1982). It has been shown that there is very little effect on the strength properties of thin flakes as a result of acetylation (Rowell and Banks 1987). The small decrease in some strength properties resulting from acetylation may be

TABLE 14.25
Equilibrium Moisture Content (EMC), Thickness Swelling (TS), Biological Resistance Modulus of Rupture (MOR), Modulus of Elasticity (MOE), and Internal Bond Strength (IBS) of Fiberboards Made from Control and Acetylated Pine Fiber (10% Phenolic Resin)

WPG	MOR (MPa)	MOE (GPa)	IBS (MPa)
0	53	3.7	2.3
19.6	61	4.1	2.3
ANSI Standard	31	—	

attributed to the hydrophobic nature of the acetylated furnish, which may not allow the water-soluble phenolic or isocyanate resins to penetrate into the flake.

It should also be pointed out that strength properties of wood are very dependent on the moisture content of the cell wall. Fiber stress at proportional limit, work to proportional limit, fiber stress at proportional limit, and maximum crushing strength are the mechanical properties most affected by changing moisture content by only +/− one percent below FSP (Rowell 1984, USDA 1999). Since the EMC and FSP are much lower for acetylated fiber than for control fiber, strength properties will be different due to this fact alone.

14.4.13 Adhesion of Chemically Modified Wood

Acetylated wood is more hydrophobic than natural wood, so studies have been done to determine which adhesives might work best to make composites (Vick and Rowell 1990, Larsson et al. 1992, Vick et al. 1993, Rowell et al. 1987, Youngquist et al. 1988, Youngquist and Rowell 1990, Gomez-Bueso et al. 1999, Rowell et al. 1996). Shear strength and wood failure was measured using acetylated yellow poplar at 0, 8, 14, and 20 WPG and adhesives including emulsion polymer isocyanate cold set, polyurethane cold set, polyurethane hot-melt, polyvinyl acetate emulsion, polyvinyl acetate cold set, polyvinyl acetate cross-link cold set, rubber-based contact-bond, neoprene contact-bond cold set, waterborne contact-bond cold set, casein, epoxy-polyamide cold set, amino resin, melamine-formaldehyde hot set, urea-formaldehyde hot set, urea-formaldehyde cold set, resorcinol-formaldehyde cold set, phenol-resorcinol-formaldehyde cold set, phenol-resorcinol-formaldehyde hot set, phenol-formaldehyde hot set, and acid-catalyzed phenol-formaldehyde. In all cases, adhesive strength was reduced by the level of acetylation—some adhesives to a minor degree and others to a severe degree.

Many adhesives were capable of strong and durable bonds at the 8 WPG level of acetylation, but not at 14 and 20 WPG levels. Most of the adhesives tested contained polar polymers, and all but four were aqueous systems, so their adhesion was diminished in proportion to the presence of the non-polar and hydrophobic acetate groups in acetylated wood (Rudkin 1950). Thermosetting adhesives produced the strongest bonds in both dry and wet conditions, but thermoplastic adhesives were capable of high shear strengths in the dry condition. With the exception of the acid-catalyzed phenol-formaldehyde adhesive, thermosetting adhesives that were hot pressed became highly mobile and tended to over-penetrate the wood because of the limited capacity of the acetylated wood to sorb water from the curing bond-line. The abundance of hydroxyl groups in the highly reactive resorcinol adhesive permitted excellent adhesion at room temperature, despite the limited availability of hydroxyl groups in the acetylated wood.

An emulsion polymer-isocyanate, a cross-linking polyvinyl acetate, a resorcinol-formaldehyde, a phenol-resorcinol-formaldehyde, and an acid-catalyzed phenolic-formaldehyde adhesive developed bonds of high shear strength and wood failure at all levels of acetylation in the dry condition. A neoprene contact-bond adhesive and a moisture-curing polyurethane hot-melt adhesive performed as well on acetylated wood as untreated wood in tests of dry strength. Only the cold-setting resorcinol-formaldehyde, the phenol-resorcinol-formaldehyde adhesive, and the hot-setting acid-catalyzed phenolic adhesive developed bonds of high strength and wood failure at all levels of acetylation when tested in a water-saturated condition.

14.4.14 Acoustical Properties

The acetylation of wood slightly reduced both sound velocity and sound absorption as compared to unreacted wood (Zhao et al. 1987, Norimoto et al. 1988, Yano et al. 1993). Acetylation greatly reduces variability in the moisture content of the cell wall polymers, thereby stabilizing the physical dimensions of wood and its acoustic properties. It was not possible to determine if acetylation effected sound quality. A violin and guitar in Japan, a violin in Sweden, and a recorder in the United States have been made from acetylated wood, and actual changes in sound quality are presently being investigated.

Chemical Modification of Wood

14.5 COMMERCIALIZATION OF CHEMICALLY MODIFIED WOOD

In spite of the vast amount of research on chemical modification of wood, only acetylation of wood has been tested for commercialization. Two attempts, one in the United States (Koppers 1961) and one in Russia (Otlesnov and Nikitina 1977, Nikitina 1977), came close to commercialization but were discontinued presumably because they were not cost-effective. There is a commercial acetylation plant for solid wood flooring in Japan and a pilot plant for solid wood in Holland.

Two new processes are presently under way in Sweden to commercialize the acetylation of wood. One is a fiber process and the second a process to acetylate wood of large dimensions using microwave technology. One of the concerns about the acetylation of wood, using acetic anhydride as the reagent, has been the by-product acetic acid. Many attempts have been made for the "complete removal" of the acid to eliminate the smell, make the process more cost effective, and to remove a potential ester-hydrolysis–causing chemical. Complete removal of by-product acetic acid has now been achieved in both the fiber process and the solid wood microwave process.

14.5.1 THE FIBER PROCESS

There is a pilot plant in Sweden with a capacity of approximately 4000 tons of acetylated fiber a year. Figure 14.8 shows the schematic of the new continuous fiber acetylation process (Nelson et al. 1994, 1995a, 1995b, 1999, Simonson and Rowell 2000). The fiber is first dried in an optional dryer section to reduce the moisture content to as low a level as is economically feasible, realizing that the anhydride will react with water to form acetic acid and that a certain amount of acetic acid is needed to swell the fiber wall for chemical access.

The dried fiber is then introduced by a screw feeder into the reactor section and the acetylating agent is added. The temperature in this section is within the range of 110–140°C, so the acetylating agent is in the form of a vapor/liquid mixture. Back flow of the acetylating agent is prevented by a fiber plug formed in the screw feeder. A screw-conveyor or similar devise is used to move the material through the reactor and to mix the fiber-reagent mixture. During the acetylation reaction, which is exothermic, the reaction temperature can be maintained substantially constant by several conventional methods. The contact time in the reactor section is from 6 to 30 minutes. The bulk of the acetylation reaction takes place in this first reactor.

The resultant acetylated fiber from the first reactor contains excess acetylating agent and formed acetic acid as it is fed by a star feeder into the second reactor, designed as a long tube and working as an anhydride stripper. The fiber is transported through the stripper by a stream of superheated vapor of anhydride and acetic acid. The temperature in the stripper is preferably in the range of 185–195°C. The primary function of this second step is to reduce the content of the unreacted

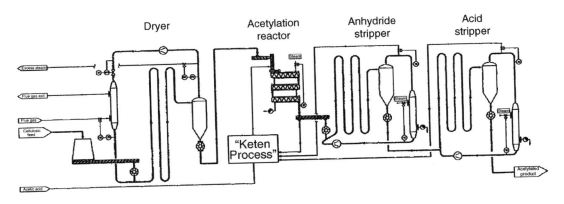

FIGURE 14.8 Schematic of the new fiber acetylation process.

acetylating medium remaining in the fiber emerging from the first reactor. An additional acetylation of the fiber is, however, also achieved in this step. The residence in this step is relatively short and normally less than one minute. After the second reactor (stripper), superheated vapor and fiber are separated in a cyclone. Part of the superheated vapor is recirculated after heating to the stripper fiber inlet, and the rest is transferred to the system for chemical recovery.

The acetylated fiber from the second reactor may still contain some anhydride and acetic acid that is sorbed or occluded in the fiber. In order to remove remaining chemicals and the odor from them, the acetylated fiber is introduced into a second stripper step, which also acts as a hydrolysis step. The transporting medium in this step is superheated steam, and any remaining anhydride is rapidly hydrolyzed to acetic acid, which is evaporated. The acetylated fiber emerging from the second stripper is essentially odor-free and completely dry. The acetylated fiber, as a final treatment, can be resinated for fiberboard production or conditioned and baled for other uses as desired. The steam and acetic acid removed overhead from this step is processed in the chemical recovery step.

The preferred recovery of chemicals includes separation of acetic anhydride from acetic acid by distillation, and conversion of acetic acid, recovered as well as purchased, by the ketene process into anhydride. The raw materials entering the production site are thus fiber and acetic acid to cover the acetyl groups introduced in the fiber. This minimizes the transportation and chemical costs and makes the process much more cost-effective.

Figure 14.9 shows the pilot plant being assembled. The plant was built during the spring of 2000, taken apart, and reassembled in Kvarntorp, Sweden in the summer. The designated production rate is 500 Kg per hour, which equals 12 tons per day or about 4000 tons per year of acetylated wood fiber. The process can be applied to any wood fiber, and fiber other than wood will be used.

14.5.2 THE SOLID WOOD MICROWAVE PROCESS

Microwave energy has been shown to heat acetic anhydride and acetic anhydride–impregnated wood (Larsson-Brelid and Simonson 1997, 1999, Larsson-Brelid et al. 1999, Risman et al. 2001).

FIGURE 14.9 Acetylation pilot plant being assembled in Kvarntorp, Sweden.

Chemical Modification of Wood

FIGURE 14.10 Microwave solid wood acetylation reactor.

Figure 14.10 shows the microwave reactor. The absorption of microwave energy in acetic anhydride–impregnated wood is preferred over other methods of heating since it heats only specific parts of the wood, provides some self-regulation of the overall temperature rise, and promotes a more uniform heating pattern. Acetic anhydride is supplied to the reactor, under a vacuum, then pressure is applied for a short time, followed by another vacuum step to remove excess anhydride. Microwave energy is then applied to heat the anhydride-soaked wood.

The penetration depth of the microwaves at 2450 MHz is approximately 10 cm, which means this technology can be used to acetylate large wood members. The variation in acetyl content, both within and between samples, is less than 2%. Microwave energy can also be used to remove the excess acetic anhydride and byproduct acetic acid after acetylation.

REFERENCES

Andersson, M. and Tillman, A.-M. (1989). Acetylation of Jute. Effects on strength, rot resistance and hydrofobicity. *J. Appl. Polymer Sci.* 37:3437–3447.

American National Standards Institute. (1982). Basic Hardboard. ANSI/AHA 135.4 (reaffirmed Jan. 1988), American Hardboard Association, Palatine, IL.

Arni, P.C., Gray, J.D., and Scougall, R.K. (1961a). Chemical modification of wood. I. Use of trifluoroacetic acid in the esterification of wood by carboxylic acids. *Journal of Appl. Chem.* 11:157–163.

Arni, P.C., Gray, J.D., and Scougall, R.K. (1961b). Chemical modification of wood. II. Use of trifluoroacetic acid as catalyst for the acetylation of wood. *J. Appl. Chem.* 11:163–170.

Avora, M., Rajawat, M.S., and Gupta, R.C. (1979). Studies on the acetylation of wood. *Holzforschung und Holzverwertung* 31(6):138–141.

Baechler, R.H. (1959a). Improving wood's durability through chemical modification. *Forest Products J.* 9:166–171.

Baechler, R.H. (1959b). Fungus-resistant wood prepared by cyanoethylation. U.S. Patent 2,959, 496.

Baird, B.R. (1969). Dimensional stabilization of wood by vapor phase chemical treatments. *Wood and Fiber* 1(1):54–63.

Banks, W. B. and Lawther, J.M. (1994). Derivation of wood in composites, In Gilbert, R.G. (Ed.), *Cellulosic polymers, blends and composites*. Hanser Publishers, New York, p. 131.

Barnes, H.M., Choong, E.T., and McIlhenny, R.L. (1999). An evaluation of several vapor phase chemical treatments for dimensional stabilization of wood. *Forest Products J.* 19(3):35–39.

Beckers, E.P.J., Militz, H., and Stevens, M. (1994). Resistance of acetylated wood to basidiomycetes, soft rot and blue stain. International Research Group on Wood Preservation, 25th Annual Meeting, Bali, Indonesia, Document No: WP 94-40021.

Beckers, E.P.J. and Militz, H. (1994). Acetylation of solid wood–Initial trials on lab and semi-industrial scale. Proceedings: Second Pacific Rim Bio-Based Composite Symposium, Vancouver, Canada, 125–134.

Burmeister, A. (1967). Tests for wood treatment with monomeric gas of formaldehyde using gamma rays. *Holzforschung* 21(1):13–20.

Burmeister, A. (1970). Wood of dimensional stability to moisture. German Patent 1,812,409.

Callow, H. J. (1951). Acetylation of cellulose and lignin in jute fiber. *J. Indian Chem. Soc.* 43:605–610.

Chang, H-T and Chang, S-T.(2003). Improvements in dimensional stability and lightfastness of wood by butylation using microwave heating. *J. Wood Sci.* 49:455–469.

Chow, P., Bao, Z., Youngquist, J.A., Rowell, R.M., Muehl, J.H., and Krzysik, A.M. (1996). Properties of hardboards made from acetylated aspen and southern pine. *Wood and Fiber Sci.* 28(2):252–258.

Chow, P., Harp, T., Meimban, R., Youngquist, J.A., and Rowell, R.M. (1994). Biodegradation of acetylated southern pine and aspen composition board. Proceedings: International Research Group on Wood Preservation 25th Annual Meeting, IRG/WP94-40020, May–June 1994, Bali, Indonesia.

Clemons, C., Young, R.A., and Rowell, R.M. (1992). Moisture sorption properties of composite boards from esterified aspen fiber. *Wood and Fiber Sci.* 24(3):353–363.

Clermont, L.P. and Bender, F. (1957). Effect of swelling agents and catalysts on acetylation of wood. *Forest Products J.* 7(5):167–170.

Cowling, E. B. (1961). Comparative biochemistry of the decay of Sweetgum sapwood by white-rot and brown-rot fungus. U.S. Department of Agriculture, Forest Serv. Technol. Bull., No. 1258, p. 50.

Dewispelaere, W., Raemedondk, J. van, and Stevens, M. (1977). Decay resistance of wood treated for dimensional stabilization with monomers and formaldehyde. *Material und Organismen* 12(3): 211–222.

Dreher, W.A., Goldstein, I.S., and Cramer, G.R. (1964). Mechanical properties of acetylated wood. *Forest Products J.* 14(2):66–68.

Ellis, W.D. and Rowell, R.M. (1989). Flame-retardant treatment of wood with a diisocyanate and an oligomer phosphonate. *Wood and Fiber Sci.* 21(14):367–375.

Ellis, W.D., Rowell, R.M., LeVan, S.L., and Susott, R.A. (1987). Thermal degradation properties of wood reacted with diethylchlorophosphate or phenylphosphonic dichloride as potential flame retardants. *Wood and Fiber Sci.* 19(4):439–445.

Feist, W.C., Rowell, R.M., and Ellis, W.D. (1991a). Moisture sorption and accelerated weathering of acetylated and/or methyl methacrylate treated aspen. *Wood and Fiber Sci.* 23(1):128–136.

Feist, W. C., Rowell, R.M., and Youngquist, J.A. (1991b). Weathering and finish performance of acetylated aspen fiberboard. *Wood and Fiber Sci.* 23(2):260–272.

Fuse, G. and Nishimoto, K. (1961). Preservation of wood by chemical modification. I. Preservative properties of cyanoethylated wood. *J. Japan Wood Res. Soc.* 7(4):157–161.

Goldstein, I.S. (1960). Improving fungus resistance and dimensional stability of wood by treatment with β-propiolactone. U.S. Patent 2,931,741.

Goldstein, I.S., Dreher, W.A., and Jeroski, E.B. (1959). Wood processing inhibition against swelling and decay. *Ind. Eng. Chem.* 5(10):1313–1317.

Goldstein, I.S., Jeroski, E.B., Lund, A.E., Nielson, J.F., and Weater, J.M. (1961). Acetylation of wood in lumber thickness. *Forest Products J.* 11(8):363–370.

Gomez-Bueso, J., Westin, M., Torgilsson, R., Olesen, P.O., and Simonson, R. (1999). Composites made from acetylated wood fibers of different origin, Part II. The effect of nonwoven fiber mat composition upon molding ability. *Holz als Roh- und Werkstoff* 57:178–184.

Gomez-Bueso, J., Westin, M., Torgilsson, R., Olesen, P.O., and Simonson, R. (2000). Composites made from acetylated wood fibers of different origin, Part I. Properties of dry-formed fiberboards. *Holz als Roh- und Werkstoff*, 58:9–14.

Hadi, Y.S., Darma, I.G.K.T., Febrianto, F., and Herliyana, E.N. (1995). Acetylated rubberwood flakeboard resistance to bio-deterioration. *Forest Products J.* 45(10):64–66.

Hadi, Y.S., Rowell, R.M., Nelsson, T., Plackett, D.V., Simonson, R., Dawson, B., and Qi, Z.-J. (1996). In-ground testing of three acetylated wood composites in Indonesia. Toward the new generation of bio-based composite products, In: Proceedings of the Third Pacific Rim Bio-Based Composites Symposium, Kyoto, Japan, Dec. 1996.

Hon, D. N.-S. (1992). Chemical modification of wood materials: old chemistry, new approaches. *Polymer News* (17):102.

Hon, D.N.-S. (1995). Stabilization of wood color: Is acetylation blocking effective? *Wood and Fiber Sci.* 27(4):360–367.

Hon, D.N.-S. (1996). *Chemical Modification of Wood Materials.* Marcel Dekker, New York.

Ibach, R.E. and Rowell, R.M. (2000). Improvements in decay resistance based on moisture exclusion. *Mol. Cryst. Liq. Cryst.* 353:23–33.

Ibach, R.E., Rowell, R.M., and Lee, B.-G. (2000). Decay protection based on moisture exclusion resulting from chemical modification of wood. Proceedings of the 5th Pacific Rim Bio-Based Composites Symposium, Canberra, Australia, 197–204.

Imamura, Y. and Nishimoto, K. (1985). Bending creep test of wood-based materials under fungal attack. *J. Soc. Materials Sci.* 34(38):985–989.

Imamura, Y., Nishimoto, K., and Rowell, R.M. (1987). Internal bond strength of acetylated flakeboard exposed to decay hazard. *Mokuzai Gakkaishi* 33(12):986–991.

Imamura, Y., Rowell, R.M., Simonson, R., and Tillman, A.-M. (1988). Bending-creep tests on acetylated pine and birch particleboards during white- and brown-rot fungal attack. *Paperi ja Puu* 9:816–820.

Imamura, Y., Subiyanto, B., Rowell, R.M., and Nelsson, T. (1989). Dimensional stability and biological resistance of particleboard made from acetylated albizzia wood particles. *Japan J. Wood Res.* 76:49–58.

Johnson, B. R. and Rowell, R. M. (1988). Resistance of chemically-modified wood to marine borers, *Material und Organismen* 23(2):147–156.

Karlson, I. and Svalbe, K. (1972). Method of acetylating wood with gaseous ketene. *Uchen. Zap. Latv. Univ.* 166:98–104.

Karlson, I. and Svalbe, K. (1977). Method of acetylating wood with gaseous ketene. *Latv. Lauksiamn. Akad. Raksti.* 130:10–21.

Kenaga, D.L. (1957). Stabilized wood. U.S. Patent 2,811,470.

Kenaga, D.L. (1963). Dimenstioal stabilization of wood and wood products. U.S. Patent 3,077,417.

Kenaga, D.L. and Sproull, R.C. (1951). Further experiments on dimensional stabilization of wood by allylation. *J. Forest Products Res. Soc.* 1:28–32.

Kenaga, D.L., Sproull, R.C., and Esslinger, J. (1950). Preliminary experiments on dimensional stabilization of wood by allylation. *Southern Lumberman* 180(2252):45.

Koppers' Acetylated Wood. (1961). Dimensionally stabilized wood. New Materials Technical Information No. (RDW-400) E-106.

Krzysik, A. M., Youngquist, J.A., Muehl, J.M., Rowell, R.M., Chow, P., and Shook, S.R. (1992). Dry-process hardboards from recycled newsprint paper fibers. In: Rowell, R.M., Laufenberg, T.L., and Rowell, J.K. (Eds.), *Materials Interactions Relevant to Recycling of Wood-Based Materials.* Materials Research Society, Pittsburgh, PA, 266, 73–79.

Krzysik, A.M., Youngquist, J.A., Muehl, J.M., Rowell, R.M., Chow, P., and Shook, S.R. (1993). Feasibility of using recycled newspaper as a fiber source for dry-process hardboards. *Forest Products J.* 43(7/8):53–58.

Kumar, S. (1994). Chemical modification of wood. *Wood and Fiber Sci.* 26(2):270–280.

Larsson, P., Mahlberg, R., Vick, C., Simonson, R., and Rowell, R. (1992). Adhesive bonding of acetylated pine and spruce. In: Plackett, D.V. and Dunningham, E.A. (Eds.), *Forest Research Institute Bulletin No. 176,* Rotorua, New Zealand, pp. 16–24.

Larsson Brelid, P. and Simonson, R. (1997). Chemical modification of wood using microwave technology. Proceedings of the Fourth International Conference of Frontiers of Polymers and Advanced Materials, Cairo, Egypt.

Larsson Brelid, P. and Simonson, R. (1999). Acetylation of solid wood using microwave heating. Part II: Experiments in laboratory scale. *Holz als Roh- und Werkoff* 57:383–389.

Larsson Brelid, P., Simonson, R., Bergman, O., and Nilsson, T. (2000). Resistance of acetylated wood to biological degradation. *Holz als Roh- und Werkstoff* 58:331–337.

Larsson Brelid, P., Simonson, R, and Risman, P.O. (1999). Acetylation of solid wood using microwave heating. Part I: Studies of dielectric properties. *Holz als Roh- und Werkstoff* 57:259–263.

Lee, H-L., Chen, G.C., and Rowell, R.M. (2000). Chemical modification of wood to improve decay and thermal resistance. Proceedings: Fifth Pacific Rim Bio-Based Composite Symposium, Evans, P.D., (Ed.), Canberra, Australia, pp. 179–189.

Liu, C. and McMillin, C.W. (1965). Treatment of wood with ethylene oxide gas and propylene oxide gas. U.S. Patent 3,183,114.

Lutomski, K. (1975). Resistance of beechwood modified with styrene, methyl methacrylate and diisocyanates against the action of fungi. *Material und Organismen* 10(4):255–262.

McMillin, C.W. (1963). Dimensional stabilization with polymerizable vapor of ethylene oxide. *Forest Products J.* 13(2):56–61.

Militz, H. (1991). Improvements of stability and durability of beechwood (*Fagus sylvatica*) by means of treatment with acetic anhydride. Proceedings of the International Research Group on Wood Preservation, 22nd Annual Meeting, Kyoto, Japan, Document No: WP 3645.

Minato, K. and Mizukami, F. (1982). A kinetic study of the reaction between wood and vaporous formaldehyde. *Journal of the Japan Wood Research Society* 28(6):346–354.

Mo and Domsjo Aktiebolag. (1965). Wood impregnated with gaseous ethylene oxide. French Patent 1,408,170.

Minato, K, Takazawa, R., and Ogura, K. (2003). Dependence of reaction kinetics and physical and mechanical properties on the reaction systems of acetylation. II: Physical and mechanical properties. *J. Wood Sci.* 49:519–524.

Nakagami, T. (1978). Esterification of wood with unsaturated carboxylic acids. V. Effect of delignification treatments on dissolution of wood esterified by the TFAA method. *J. Japan Wood Res. Soc.* 24(5):318–323.

Nakagami, T., Amimoto, H., and Yokata, T. (1974). Esterification of wood with unsaturated carboxylic acids. I. Preparation of several wood esters by the TFAA method. Bulletin, Kyoto University Forests No. 46, 217–224.

Nakagami, T., Ohta, M., and Yokata, T. (1976). Esterification of wood using unsaturated carboxylic acids. III. Dissolution of wood by the TFAA esterification method. Bulletin, Kyoto University Forests No. 48, 198–205.

Nakagami, T. and Yokata, T. (1975). Esterification of wood using unsaturated carboxylic acids. II. Reaction conditions of esterification and properties of the prepared esters of wood. Bulletin, Kyoto University Forests No. 47, 178–183.

Nakagami, T. and Yokata, T. (1978). Esterification of wood using unsaturated carboxylic acids. IV. Chemical composition and molecular weight of the acetone soluble fraction of β-methylcrotonylated woods. *J. Japan Wood Rese. Soc.* 24(5):311–317.

Narayanamurti, V.D. and Handa, B.K. (1953). Acetylated wood. *Das Papier* 7:87–92.

Nelson, H.L., Richards, D.I., and Simonson, R. (1994). Acetylation of lignocellulosic materials. European Patent 650,998.

Nelson, H.L., Richards, D.I., and Simonson, R. (1995a). Acetylation of lignocellulosic materials. European Patent 746,570.

Nelson, H.L., Richards, D.I., and Simonson, R. (1995b). Acetylation of lignocellulosic fibers. European Patent 799,272.

Nelson, H.L., Richards, D.I. and Simonson, R. (1999). Acetylation of wood materials. European Patent EP 0 650 998 B1.

Nikitina, N. (1977). Quality control of acetylated wood. *Latvijas Lauksaimniecibas Akademijas Raksi* 130:54–55.

Nilsson, T. (1986). Personal communication, Upsalla, Sweden.

Nilsson, T., Rowell, R.M., Simonson, R., and Tillman, A.-M. (1988). Fungal resistance of pine particle boards made from various types of acetylated chips. *Holzforschung* 42(2):123–126.

Norimoto, M., Grill, J., Minato, K., Okamura, K., Mukudai, J., and Rowell, R.M. (1987). Suppression of creep of wood under humidity change through chemical modification. *Wood Industry* (Japan) 42(18):14–18.

Norimoto, M., Grill, J., and Rowell, R.M. (1992). Rheological properties of chemically modified wood: Relationship between dimensional stability and creep stability. *Wood and Fiber Sci.* 24(1):25–35.

Norimoto. M., Grill, J., Sasaki, T., and Rowell, R.M. (1988). Improvement of acoustical properties of wood through chemical modification. In: Proceedings of the European Scientific Colloquium, Mechanical Behavior of Wood, Bordeaux, France, June 8–9, 1988, pp. 37–44.

Otlesnov, Y. and Nikitina, N. (1977). Trial operation of a commercial installation for modification of wood by acetylation. *Latvijas Lauksaimniecibas Akademijas Raksi* 130:50–53.

Ozolina, I, and Svalbe, K. (1972). Acetylation of wood by an anhydrone catalyst. *Latvijas Lauksaimniecibas Akademijas Raksi* 65:47–50.

Papadopoulos, A.N. and Hill, C.A.S. (2002). The biological effectiveness of wood modified with linear chain carboxylic acid anhydrides against *Coniophora puteana*. *Holz als Roh- und Werkstoff* 60:329–332.

Plackett, D.V., Rowell, R.M., and Close, E.A. (1990). Acetylation and the development of new products from radiata pine. Proceedings of the Composite Products Symposium, Rotorua, New Zealand, November, 1988. *Forest Research Institute Bulletin* No. 153, pp. 68–72.

Popper, R. (1973). Treatments to increase dimensional stability of wood. SAH Bulletin. *Schweizerische Arbeitsgemeinschaft fur Holzforschung* 1(1):3–12.

Popper, R. and Bariska, M. (1972). Acylation of wood. I. Water vapour sorption properties. *Holz als Roh- and Werkstoff* 30(8):289–294.

Popper, R. and Bariska, M. (1973). Acylation of wood. II. Thermodynamics of water vapour sorption properties. *Holz als Roh- and Werkstoff* 31(2):65–70.

Popper, R. and Bariska, M. (1975). Acylation of wood. III. Swelling and shrinkage behaviour. *Holz als Roh- and Werkstoff* 33(11):415–419.

Ridgway, W.B. and Wallington, H.T. (1946). Esterification of wood. British Patent 579,255.

Risi, J. and Arseneau, D.F. (1957a). Dimensional stabilization of wood. I. Acetylation. *Forest Products J.* 7(6):210–213.

Risi, J. and Arseneau, D.F. (1957b). Dimensional stabilization of wood. II. Crotonylation and crotylation. *Forest Products J.* 7(7):245–246.

Risi, J. and Arseneau, D.F. (1957c). Dimensional stabilization of wood. III. Butylation. *Forest Products J.* 7(8):261–265.

Risi, J. and Arseneau, D.F. (1957d). Dimensional stabilization of wood. IV. Allylation. *Forest Products J.* 7(9):293–295.

Risi, J. and Arseneau, D.F. (1958). Dimensional stabilization of wood. V. Phthaloylation. *Forest Products J.* 8(9):252–255.

Risman, P.O., Simonson, R., and Laarsson-Bredil, P. (2001). Microwave system for heating voluminous elongated loads. Swedish Patent 521,315.

Rowell, R. M. (1975a). Chemical modification of wood: advantages and disadvantages, *Proc. Am. Wood Preservers' Assoc.* 71:1–10.

Rowell, R.M. (1975b). Chemical modification of wood: Reactions of alkylene oxides with southern yellow pine. *Wood Sci.* 7(3):240–246.

Rowell, R.M. (1980). Distribution of reacted chemicals in southern pine modified with methyl isocyanate. *Wood Sci.* 13(2):102–110.

Rowell, R. M. (1982). Distribution of reacted chemicals in southern pine modified with acetic anhydride. *Wood Sci.* 15(2):172–182.

Rowell, R. M. (1983). Chemical modification of wood: A review, Commonwealth Forestry Bureau, Oxford, England, 6(12):363–382.

Rowell, R. M. (1984). The Chemistry of Solid Wood, Advances in Chemistry Series No. 207. American Chemical Society, Washington, DC.

Rowell, R. M. (1990). Chemical modification of wood: Its application to composite wood products, Proceedings of the Composite Products Symposium, Rotorua, New Zealand, November, 1988. *Forest Research Institute Bulletin* No. 153, pp. 57–67.

Rowell, R. M. (1991). Chemical modification of wood. In: Hon, D. N.-S. and Shiraishi, N. (Eds.), *Handbook on Wood and Cellulosic Materials*. Marcel Dekker, New York, pp. 703–756.

Rowell, R. M. (1992). Property enhancement of wood composites. In Vigo, T. L. and Kinzig, B.J. (Eds.), *Composites Applications: The Role of Matrix, Fiber, and Interface*. VCH Publishers, New York, pp. 365–382.

Rowell, R. M. (1993). Opportunities for composite materials from jute and kenaf. International consultation of jute and the environment, Food and Agricultural Organization of the United Nations, ESC:JU/IC 93/15, pp. 1–12.

Rowell, R.M. (1999). Chemical modification of wood. In: Gatenholm, P., and Chihani, T., (Eds.), Proceedings of the International Workshop on Frontiers of Surface Modification and Characterization of Wood Fibers, Fiskebackskil, Sweden, May 30–31, 1996. pp. 31–47.

Rowell, R. M. and Banks, W.B. (1985). Water repellency and dimensional stability of wood. USDA Forest Service General Technical Report FPL 50, Forest Products Laboratory, Madison, WI.

Rowell, R. M. and Banks, W.B. (1987). Tensile strength and work to failure of acetylated pine and lime flakes, *British Polymer J.* 19:479–482.

Rowell, R.M. and Chen, G.C. (1994). Epichlorohydrin coupling reactions with wood. I. Reaction with biologically active alcohols. *Wood Sci. Technol.* 28:371–376.

Rowell, R.M. and Clemons, C.M. (1992). Chemical modification of wood fiber for thermoplasticity, compatibilization with plastics and dimensional stability, In: Maloney, T.M. (Ed.), Proceedings of the International Particleboard/Composite Materials Symposium, Pullman, WA, pp. 251–259.

Rowell, R.M., Dawson, B.S., Hadi, Y.S., Nicholas, D.D., Nilsson, T., Plackett, D.V., Simonson, R., and Westin, M. (1997). Worldwide in-ground stake test of acetylated composite boards. International Research Group on Wood Preservation, Section 4, Processes, Document no. IRG/WP 97-40088, Stockhold, Sweden, pp. 1–7.

Rowell, R.M. and Ellis, W.D. (1978). Determination of dimensional stabilization of wood using the water-soak method. *Wood and Fiber* 10(2):104–111.

Rowell, R.M. and Ellis, W.D. (1979). Chemical modification of wood: Reaction of methyl isocyanate with southern yellow pine. *Wood and Fiber* 12(1):52–58.

Rowell, R.M. and Ellis, W.D. (1981). Bonding of isocyanates to wood. American Chemical Society Symposium Series 172:263–284.

Rowell, R.M. and Ellis, W.D. (1984a). Effects of moisture on the chemical modification of wood with epoxides and isocyanates. *Wood and Fiber Sci.* 16(2):257–276.

Rowell, R. M. and Ellis, W.D. (1984b). Reaction of epoxides with wood. USDA Forest Service Research Paper, FPL 451, Forest Products Laboratory, Madison, WI.

Rowell, R.M., Esenther, G.R., Nicholas. D.D., and Nilsson, T. (1987). Biological resistance of southern pine and aspen flakeboards made from acetylated flakes, *J. Wood Chem. Tech.* 7(3):427–440.

Rowell, R. M., Esenther, G.R., Youngquist, J.A., Nicholas, D.D., Nilsson, T., Imamura, Y., Kerner-Gang,W., Trong, L., and Deon, G. (1988a). Wood modification in the protection of wood composites, Proceedings of the IUFRO wood protection subject group, Honey Harbor, Ontario, Canada. Canadian Forestry Service, pp. 238–266.

Rowell, R. M. and Harrison, S.E. (1993). Property enhanced kenaf fiber composites. In: Bhangoo, M. S. (Ed.), Proceedings of the Fifth Annual International Kenaf Conference, California State University Press, Fresno, CA, pp. 129–136.

Rowell, R. M., Hart, S.V., and Esenther, G.R. (1979). Resistance of alkylene-oxide modified southern pine to attack by subterranean termites, *Wood Sci.* 11(4):271–274.

Rowell, R.M., Imamura, Y., Kawai, S., and Norimoto, M. (1989). Dimensional stability, decay resistance and mechanical properties of veneer-faced low-density particleboards made from acetylated wood, *Wood and Fiber Sci.* 21(1):67–79.

Rowell, R. M., and Keany, F. (1991). Fiberboards made from acetylated bagasse fiber. *Wood and Fiber Sci.* 23(1):15–22.

Rowell, R. M. and Konkol, P. (1987). Treatments that enhance physical properties of wood. USDA, Forest Service, Forest Products Laboratory Gen. Technical Report FPL-GTR-55, Madison, WI.

Rowell, R.M., Lichtenberg, R.S., and Larsson, P. (1992a). Stability of acetyl groups in acetylated wood to changes in pH, temperature, and moisture. In: Plackett, D.V. and Dunningham, E.A. (Eds.), Forest Research Institute Bulletin No. 176, Rotorua, New Zealand, pp. 33–40.

Rowell, R.M., Lichtenberg, R.S, and Larsson, P. (1992b). Stability of acetylated wood to environmental changes. *Wood and Fiber Sci.* 25(4):359–364.

Rowell, R.M., Moisuk, R., and Meyer, J.A. (1982). Wood polymer composites: Cell wall grafting of alkylene oxides and lumen treatments with methyl methacrylate. *Wood Sci.* 15(2):90–96.

Rowell, R. M. and Norimoto, M. Acetylation of bamboo fiber. (1987). *J. Jap. Wood Res. Soc.* 33(11):907–910.

Rowell, R. M. and Norimoto, M. (1988). Dimensional stability of bamboo particleboards made from acetylated particles. *Mokuzai Gakkaishi* 34(7):627–629.

Rowell, R.M. and Plackett, D. (1988). Dimensional stability of flakeboards made from acetylated *Pinus radiata* heartwood and sapwood flakes. *New Zealand J. For. Sci.* 18(1):124–131.

Rowell, R. M and Rowell, J.S. (1989). Moisture sorption of various types of acetylated wood fibers. In Schuerch, C. (Ed.), *Cellulose and Wood*. John Wiley & Sons, New York, pp. 343–356.

Rowell, R.M., Simonson, R., and Tillman, A.-M. (1986d). A simplified procedure for the acetylation of chips for dimensionally stable particle board products. *Paperi ja Puu.* 68(10):740–744.

Rowell, R.M., Simonson, R., and Tillman, A.-M. (1990). Acetyl balance for the acetylation of wood particles by a simplified procedure. *Holzforschung* 44(4):263–269.

Rowell, R. M., Simonson, R., and Tillman, A.-M. (1991a). A process for improving dimensional stability and biological resistance of wood materials, European Patent 0213252.

Rowell, R. M., Simonson, R., Hess, S., Plackett, D.V., Cronshaw, D., and Dunningham, E. (1994). Acetyl distribution in acetylated whole wood and reactivity of isolated wood cell wall components to acetic anhydride. *Wood and Fiber Sci.* 26(1):11–18.

Rowell, R. M., Susott, R.A., De Groot, W.G., and Shafizadeh, F. (1984). Bonding fire retardants to wood. Part I. *Wood and Fiber Sci.* 16(2):214–223.

Rowell, R.M., Tillman, A.-M., and Simonson, R. (1986a). Vapor phase acetylation of southern pine, Douglas-fir, and aspen wood flakes. *J. Wood Chem. Tech.* 6(2):293–309.

Rowell, R.M., Tillman, A.-M., and Simonson. R. (1986b). A simplified procedure for the acetylation of hardwood and softwood flakes for flakeboard production. *J. Wood Chem. Tech.* 6(3):427–448.

Rowell, R.M., Wang, R.H.S., and Hyatt, J.A. (1986c). Flakeboards made from aspen and southern pine wood flakes reacted with gaseous ketene. *J. Wood Chem. Tech.* 6(3):449–471.

Rowell, R.M., Young, R.A., and Rowell, J.K. (1996). *Paper and Composites from Agro-based Resources.* CRC Lewis Publishers, Boca Raton, F.

Rowell, R. M., Youngquist, J.A., and Imamura, Y. (1988b). Strength tests on acetylated flakeboards exposed to a brown rot fungus. *Wood and Fiber Sci.* 20(2):266–271.

Rowell, R.M., Youngquist, J.A., Rowell, J.S., and Hyatt, J.A. (1991b). Dimensional stability of aspen fiberboards made from acetylated fiber. *Wood and Fiber Sci.* 23(4):558–566.

Rowell, R.M., Youngquist, J.A., and Sachs, I.B. (1987). Adhesive bonding of acetylated aspen flakes. Part I. Surface changes, hydrophobicity, adhesive penetration, and strength. *Int. J. Adhesion Adhesives* 7(4):183–188.

Rowell, R. M. and Youngs, R.L. (1981). Dimensional stabilization of wood in use. USDA, Forest Serv. Res. Note. FPL-0243. Forest Products Laboratory, Madison, WI.

Rudkin, A.W. (1950). The role of hydroxyl group in the gluing of wood. *Austral. J. Appl. Sci.* 1:270–283.

Schuerch, C. (1968). Treatment of wood with gaseous reagents. *Forest Products J.* 18(3):47–53.

Singh, S.P., Dev, I., And Kumar, S. (1981). Chemical modification of wood. II. Vapour phase aceylation with acetyl chloride. *Int. J. Wood Preservation* 1(4):169–171.

Simonson, R. and Rowell, R.M. (2000). A new process for the continuous acetylation of wood fiber. In: Evans, P.D. (Ed.), Proceedings of Fifth Pacific Rim Bio-Based Composite Symposium, Canberra, Australia, pp. 190–196.

Stamm, A.J. (1959). Dimensional stabilization of wood by thermal reactions and formaldehyde crosslinking. *Tappi* 42:39–44.

Stamm, A. J. (1964). *Wood and Cellulose Science.* The Ronald Press Co., New York.

Stamm, A.J. and Baecher, R.H. (1960). Decay resistance and dimensional stability of five modified woods. *Forest Products J.* 10(1):22–26.

Stamm, A.J. and Tarkow, H. (1947). Acetylation of wood and boards. U.S. Patent 2,417,995.

Stevens, M., Schalck, J., and Raemdonck, J.V. (1979). Chemical modification of wood by vapour-phase treatment with formaldehyde and sulfur dioxide. *Int. J. Wood Preservation* 1(2):57–68.

Stromdahl, K. (1930). Water sorption in wood and plant fibers. PhD thesis, The Technical University of Denmark, Department of Structural Engineering and Materials, Copenhagen, Denmark.

Suida, H. (1930). Acetylating wood. Austrian Patent 122,499.

Svalbe, K. and Ozolina, I. (1970). Modification of wood by acetylation. *Plast. Modif. Drev.* 145–146.

Tarkow, H. (1945). Acetylation of wood with ketene. Office Report, Forest Products Laboratory, USDA, Forest Service.

Tarkow, H. (1959). A new approach to the acetylation of wood. Office Report, Forest Products Laboratory, USDA, Forest Service.

Tarkow, H., Stamm, A.J., and Erickson, E.C.O. (1950). Acetylated wood. USDA, Forest Service, Forest Products Laboratory Report 1593.

Tarkow, H. and Stamm, A.J. (1953). Effect of formaldehyde treatments upon the dimensional stabilization of wood. *J. Forest Products Res. Soc.* 3:33–37.

Truksne, D. and Svalbe, K. (1977). Water-repellent properties and dimensional stability of acetylated pine wood in relation to the degree and method of acetylation. *Latvijas Lauksaimniecibas Akademijas Raksi* 130:26–31.

Ueyama, A., Araki, M., and Goto, T. (1961). Dimensional stability of woods. X. Decay resistance of formaldehyde treated wood. *Wood Res.* No. 26:67–73.

United States Department of Agriculture, Forest Service. Wood Handbook. (1999) USDA Agri. Handbook 72, General Technical Report FPL-GTR-113, Washington, DC.

Vick, C. B. and Rowell, R.M. (1990). Adhesive bonding of acetylated wood. *Int. J. Adhesion and Adhesives* 10(4):263–272.

Vick, C.B., Larsson, P.Ch., Mahlberg, R.L., Simonson, R., and Rowell, R.M. (1993). Structural bonding of acetylated Scandinavian softwood for exterior lumber laminates. *Int. J. Adhesion and Adhesives* 13(3):139–149.

Wakita, H., Onishi, H., Jodai, S., and Goto, T. (1977). Studies on the improvement of wood materials. XVIII. Vapor phase reactions of 2,4-tolyene diisocyanate in wood. *Zairyo* 26(284):460–464.

Wang, C.-L., Lin, T.-S, and Li, M.-H. (2002). Decay and termite resistance of planted tree sapwood modified by acetylation. *Taiwan J. Forest Sci.* 17(4):483–490.

Weaver, J.W., Nielson, J.F., and Goldstein, I.S. (1960). Dimensional stabilization of wood with aldehydes and related compounds. *Forest Products J.* 10(6):306–310.

Westin, M. (1998). High-performance composites from modified wood fiber. Ph.D. Thesis, Department of Forest Products and Chemical Engineering, Chalmers University of Technology, Goteborg, Sweden.

Winandy, J.E. and Rowell, R.M. (1984). The chemistry of wood strength. In: Rowell, R.M. (Ed.), American Chemical Society Advances in Chemistry Series No. 207. American Chemical Society, Washington, DC, pp. 211–255.

Yano, H., Norimoto, M., and Rowell, R.M. (1993). Stabilization of acoustical properties of wooden musical instruments by acetylation. *Wood and Fiber Sci.* 25(4):395–403.

Youngquist, J.A. and Rowell, R.M. (1990). Adhesive bonding of acetylated aspen flakes. Part III. Adhesion with isocyanates. *Int. J. Adhesion and Adhesives* 10(4):273–276.

Youngquist, J.A., Rowell, R.M., and Krzysik, A. (1986a). Mechanical properties and dimensional stability of acetylated aspen flakeboards. *Holz als Roh- und Werkstoff* 44:453–457.

Youngquist, J.A., Rowell, R.M., and Krzysik, A. (1986b). Dimensional stability of acetylated aspen flakeboard. *Wood and Fiber Sci.* 18:90–98.

Youngquist, J.A., Sachs, I.B., and Rowell, R.M. (1988). Adhesive bonding of acetylated aspen flakes, Part II. Effects of emulsifiers on phenolic resin bonding. *Int. J. Adhesion and Adhesives* 8(4):197–200.

Zhao, J., Norimoto, M., Tanaka, F., Yamada, T., and Rowell, R.M. (1987). Structure and properties of acetylated wood. I. Changes in degree of crystallinity and dielectric properties by acetylation, *J. Japan Wood Res. Soc.* 33(2):136–142.

Zimakov,, P.V. and Pokrovski, V.A. (1954). A peculiarity in the reaction of ethylene oxide with wood. *Zhurnal Prikladnoi Khimii* 27:346–348.

15 Lumen Modifications

Rebecca E. Ibach and W. Dale Ellis
USDA, Forest Service, Forest Products Laboratory, Madison, WI

CONTENTS

15.1 *In Situ* Polymerization of Liquid Monomers in the Lumens	422
15.2 Polymerization Methods	423
15.2.1 Chemical Initiators	424
15.2.1.1 Peroxides	424
15.2.2.2 Vazo Catalysts	424
15.2.2.3 Radiation	425
15.2.2.3.1 Gamma Radiation	425
15.2.2.3.2 Electron Beam	426
15.3 Monomers	426
15.3.1 Acrylic Monomers	426
15.3.2 Styrene	429
15.3.3 Polyesters	431
15.3.4 Melamine Resins	432
15.3.5 Acrylonitrile	432
15.4 Crosslinking Agents	433
15.4.1 Isocyanates	433
15.4.2 Anhydrides	433
15.5 Properties of Wood-Polymer Composites	434
15.5.1 Hardness	434
15.5.2 Toughness	434
15.5.3 Abrasion Resistance	437
15.5.4 Dimensional Stability	437
15.5.5 Moisture Exclusion	438
15.5.6 Fire Resistance	438
15.5.7 Decay Resistance	438
15.5.8 Weathering Resistance	439
15.5.9 Mechanical Properties	439
15.6 Applications	439
15.7 Polymer Impregnations	439
15.7.1 Epoxy Resins	440
15.7.2 Compression of Wood during Heating and Curing with Resin	440
15.7.2.1 Staypak	440
15.7.2.2 Compreg	440
15.7.2.3 Staybwood	440
15.8 Water soluble polymers and synthetic resins	441
15.8.1 Polyethylene Glycol (peg)	441
15.8.2 Impreg	441
References	441

Wood is used to produce many products (structural and nonstructural) for applications in which its natural properties are adequate. With the decrease of wood availability and the increase of less durable, younger and faster-growing trees, it is possible to modify wood in various ways to improve the properties, depending on the ultimate application. Wood-polymer composites (WPCs) can be any combination of wood and polymer, from polymer filled with wood fiber to solid pieces of wood filled with polymer. This chapter refers to WPCs made from solid wood such as small pieces of wood or veneer. When wood and polymer are combined, the physical properties, surface hardness, water repellency, dimensional stability, abrasion resistance, and fire resistance can be improved over those of the original wood.

15.1 *IN SITU* POLYMERIZATION OF LIQUID MONOMERS IN THE LUMENS

When wood is vacuum-impregnated with liquid vinyl monomers that do not swell wood, and then *in situ* polymerized either by chemical catalyst, heat, or gamma radiation, the polymer is located almost solely in the lumen of the wood. Figure 15.1 is a scanning electron microscopy (SEM) micrograph of unmodified wood showing open cells that are susceptible to indentation and wear. In contrast, Figure 15.2 is a micrograph of wood after impregnation and polymerization, showing the voids filled with polymer that will resist indentation and wear.

The process for impregnating wood with acrylics involves drying the wood (usually at 105°C) overnight to remove moisture, and then weighing. The wood is placed in a container (large enough for the wood and an equal volume of solution), and a weight is placed on top of the wood to hold it under the solution. A vacuum (0.7–1.3 kPa) is applied to the wood for 30 minutes (or longer, depending on the size of the wood to be treated). The acrylic monomer solution (containing the acrylic monomer, a catalyst such as an azo compound, and perhaps a cross-linking agent) is introduced into the container. The vacuum is maintained for 5–10 minutes to remove air from the monomer. The vacuum is then released, and the chamber returned to atmospheric conditions. The wood and solution are allowed to stand, usually for 30 minutes. If the

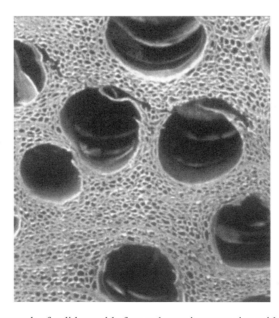

FIGURE 15.1 SEM micrograph of solid wood before polymer impregnation with open lumens.

Lumen Modifications

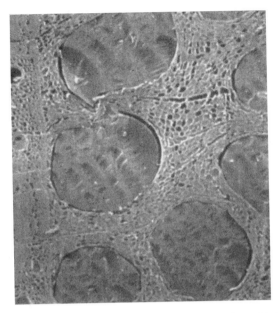

FIGURE 15.2 SEM micrograph of solid wood after polymer impregnation with filled lumens.

wood specimens have large dimensions, or are hard-to-penetrate species, then pressure is applied. The amount of pressure and time under pressure depends again on the size of the wood to be treated. Pressure is applied for 30 minutes, released, and again the wood is allowed to soak in the solution for 30 minutes. The treated wood is removed from the solution, drained, and wiped to remove excess chemical from the outside of the specimens. The monomer in the wood can be cured by either heat or irradiation. For heat, the chamber itself can be heated, or the wood can be removed from the chamber and heat cured in an oven or heated press. If polymerizing by heat, the temperature is prescribed by the catalyst, i.e., Vazo 67 is heated at around 67°C. Those monomers that do not polymerize in the presence of air require a curing environment without air present. Heat is applied usually overnight or until the monomer has polymerized in the wood. Samples are weighed again, and percentage weight gain is calculated. Some polymer will be on the surface of the wood; this is sanded off.

There are many sources of acrylics and many different types of acrylics. Some monomers will be discussed in this chapter. The general one is methyl methacrylate (MMA). The thickness of the piece of wood being treated will determine the amount of pressure and/or vacuum needed. Small thin pieces of wood may not require any vacuum.

15.2 POLYMERIZATION METHODS

Free radicals used to initiate polymerization can be generated in two ways: by temperature-sensitive catalysts or radiation curing. Chemical curing is a cheaper method with small-scale productions, whereas gamma radiation is more economical on a larger scale (Lee 1969).

A free radical catalyst or gamma-irradiated monomer generates the free radicals (R• + R•).

$$\text{Initiation step:} \quad R\bullet + M \text{ (monomer)} \rightarrow R\text{-}M\bullet$$
$$\text{Propagation step:} \quad R\text{-}(M)_n\text{-}M\bullet + M \rightarrow R\text{-}(M)_{n-1}\text{-}M\bullet$$
$$\text{Termination step:} \quad R\text{-}(M)_n\text{-}M\bullet + R\text{-}(M)_n\text{-}M\bullet \rightarrow R\text{-}(M)_n\text{-}M\text{-}M\text{-}(M)_n\text{-}R$$

15.2.1 CHEMICAL INITIATORS

15.2.1.1 Peroxides

Peroxides form free radicals when thermally decomposed. These radicals initiate polymerization of vinyl monomers. Some peroxides used to initiate polymerization of monomers in wood include *t*-butyl hydroperoxide, methyl ethyl ketone peroxide, lauroyl peroxide, isopropyl hydroperoxide, cyclohexanone peroxide, hydrogen peroxide, and benzoyl peroxide. Each of the radicals generated from these peroxides has a different reactivity. The phenyl radical is more reactive than the benzyl radical, and the allyl radical is unreactive. Benzoyl peroxide is the most commonly used initiator. Usually the amount of peroxide added ranges from 0.2–3% by weight of monomer.

15.2.2.2 Vazo Catalysts

Dupont manufactures a series of catalysts with the trade name Vazo® that are substituted azonitrile compounds. The catalysts are white crystalline solids that are soluble in most vinyl monomers. Upon thermal decomposition, the catalysts decompose to generate two free radicals per molecule. Nitrogen gas is also generated. The grade number is the Celsius temperature at which the half-life in solution is 10 hours. The series consists of the following compounds:

Vazo® 52, the low-temperature polymerization initiator (2,2'-azobis-2,4-dimethylvaleronitrile),

Vazo® 64 (2,2'-azobisisobutyronitrile), also known as AIBN (toxic tetramethylsuccinonitrile (TMSN) is produced, therefore better to substitute Vazo 67),

Vazo® 67 (2,2'-azobis-(2-methylbutyronitrile)), best solubility in organic solvents and monomers,

and Vazo® 88 (1,1'-azobis-cyclohexanecarbonitrile),

Vazo® free radical initiators are solvent soluble and have a number of advantages over organic peroxides. They are more stable than most peroxides, so they can be stored under milder conditions, and are not shock-sensitive. They decompose with first-order kinetics; are not sensitive to metals, acids, and bases; and are not susceptible to radical-induced decompositions. The Vazo catalysts produce less energetic radicals than peroxides, so there is less branching and cross-linking. They are weak oxidizing agents, which allows them to be used to polymerize unsaturated amines, mercaptans, and aldehydes without affecting pigments and dyes.

Catalysts are most frequently used in concentrations of 1% or less by weight of the monomers. The rate of free radical formation is dependent on the catalyst used and is controlled by regulating the temperature. For Vazo 52, the temperature range is 35–80°C; for Vazo 64 and 67, 45–90°C; and for Vazo 88, 80–120°C.

AIBN and cyclohexanone peroxide were used to initiate the polymerization of styrene in birch wood (Okonov and Grinberg 1983). Benzoyl peroxide or AIBN were used as initiators for beech wood impregnated with trimethylolpropane trimethacrylate and polyethylene glycol dimethacrylate (Nobashi et al. 1986). Buna sapwood was impregnated with tetraethylene glycol dimethacrylate containing AIBN as initiator (Nobashi et al. 1986). WPCs were prepared from wood by impregnating with a mixture of unsaturated polyester, MMA, styrene, and AIBN or benzoyl peroxide followed by heat-curing (Pesek 1984). Wood materials such as birch wood, basswood, and oak wood were impregnated with MMA or unsaturated polyester-styrene containing AIBN or benzoyl peroxide and polymerized. The polymerization was faster in the presence of AIBN than with benzoyl peroxide (Kawakami and Taneda 1973). Free radical copolymerization of glycidyl methacrylate (GMA) and N-vinyl-2- pyrrolidone was carried out using AIBN, in chloroform at 60°C (Soundararaian and Reddy 1991).

15.2.2.3 Radiation

There are two main radiation-initiated polymerization methods used to cure monomers in wood: gamma radiation and electron beam.

15.2.2.3.1 Gamma Radiation

Wood is a mixture of high-molecular-weight polymers; therefore, exposure to high-energy radiation will depolymerize the polymers, creating free radicals to initiate polymerization. With gamma radiation, polymerization rate and extent of polymerization are dependent on the type of monomer, other chemical additives, wood species, and radiation dose rate (Aagaard 1967). An example of radiation polymerization of the vinyl monomer MMA using cobalt 60 gamma ray dose rates of 56, 30, and 9 rad/s produced exotherms at 120°C, 90°C, and 70°C, respectively, with reaction times of 5, 7, and 12 h, respectively, produced 70–80% wood weight gain (Glukhov and Shiryaeva 1973). A 1.5–2.5 Mrad dose of gamma irradiation from a cobalt-60 source of isotope activity 20,000 Ci can be used to polymerize MMA in wood.

Addition of a solid organic halogen-compound with a high content of Cl or Br, accelerates the polymerization (Pesek et al. 1969). Addition of tributyl phosphate accelerates the polymerization rate of MMA 2.5 times and decreases the required radiation dosage. Addition of alkenyl phosphonates or alkenyl esters of phosphorus acids increases the polymerization rate and imparts fire resistance and bioresistance to the resultant WPC (Schneider et al. 1990). Pietrzyk reports the optimum irradiation conditions for MMA in wood are: irradiation dose 1.5 Mrad and dose strength approximately 0.06 Mrad/h (Pietrzyk 1983). It is best if the irradiation is done in a closed container without turning the samples in order to minimize the escape of the monomer from the wood. Beech wood impregnated with MMA alone or in carbon tetrachloride or methanol solutions, can be cured with cobalt-60 gamma radiation giving polymer loadings of up to 70% by weight. Radiation doses of 2–4 Mrad are necessary for complete conversion (Proksch 1969).

Moisture in wood accelerates polymerization (Pesek et al. 1969). A small amount of water in the wood or monomer improves the properties of the WPC (Pietrzyk 1983). The polymerization rate of MMA in beech wood is increased by using aqueous emulsions containing 30% MMA and 0.2% oxyethylated fatty alcohol mixture instead of 100% MMA. The complete conversion of MMA required ~5 kJ/kg radiation dose when the aqueous emulsions were used, in comparison to >16 kJ/kg when 100% MMA was used. The radiation polymerization of MMA in wood is inhibited by lignin (Pullmann et al. 1978).

Polymerization rate of vinyl compounds in wood, by gamma-ray irradiation, decreases in the presence of oxygen giving 50–90% conversion for styrene, methyl-, ethyl-, propyl-, and butyl methacrylates, and 4–8% conversion of vinyl acetate. Toluene diisocyanate addition increases monomer conversion, and decreases benzene extractives from the composite (Kawase and Hayakawa 1974).

The U.S. Atomic Energy Commission sponsored research that used gamma radiation to make WPCs in the early 1960s, but drawbacks include safety concerns and regulations needed when using radiation. Some advantages are that the monomer can be stored at ambient conditions, as long as inhibitor is included, and the rate of free radical generation is constant for cobalt-60 and does not increase with temperature (Meyer 1984).

15.2.2.3.2 Electron Beam

High-energy electrons are another way of generating free radicals to initiate polymerization, and have been used with some success. Electron-beam irradiation was used to make WPCs of beech sapwood veneers with styrene, MMA, acrylonitrile, butyl acrylate, acrylic acid, and unsaturated polyesters (Handa et al. 1973; Handa et al. 1983). Gotoda used electron beam to polymerize several different monomers and monomer combinations in wood (Gotoda et al. 1970a, 1970b, 1971, 1974, 1975; Gotoda and Takeshita 1971; Gotoda and Kitada 1975). Increasing the wood moisture content has a positive effect on electron curing. For example, the monomer conversion in the electron beam-induced polymerization of MMA pre-impregnated in beech veneer increases with increases of moisture content in the wood up to 20–30% moisture, and is proportional to the square root of the electron dosage. The polymerization of styrene and acrylonitrile in veneer is also affected similarly by moisture content (Handa et al. 1973).

Some studies have indicated that curing of monomer systems in wood causes some interaction of the polymer with the wood. WPCs made with MMA, MMA–5% dioxane, and vinyl acetate impregnation into the wood cellular structure, followed by electron-beam irradiation show an increase in the compressive and bending strength, indicating some interaction at the wood-polymer interface (Boey et al. 1985). The dynamic modulus of WPC made from beech veneer impregnated with acrylic acid and acrylonitrile containing unsaturated polyester or polyethylene glycol methacrylate by electron beam irradiation increased logarithmically as the weight polymer fraction increased, suggesting an interaction between the polymer and cell wall surface. The temperature dispersion of the dynamic viscoelasticity of composites also indicates an interaction between polymer and wood cell walls (Handa et al. 1981).

15.3 MONOMERS

15.3.1 ACRYLIC MONOMERS

Methyl methacrylate (MMA), shown in Figure 15.3, is the most commonly used monomer for WPCs (Meyer 1965). It is one of the least expensive and most readily available monomers and is used alone or in combination with other monomers to crosslink the polymer system. MMA has a low boiling point (101°C) that can result in significant loss of monomer during curing and it must be cured in an inert atmosphere, or at least in the absence of oxygen. MMA shrinks about 21% by volume after polymerization, which results in some void space at the interface between the cell wall of the wood and the polymer. Adding crosslinking monomers such as di- and tri-methacrylates

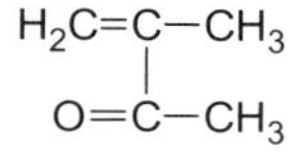

$$H_2C=C-CH_3$$
$$|$$
$$O=C-CH_3$$

FIGURE 15.3. Structure of methyl methacrylate (MMA).

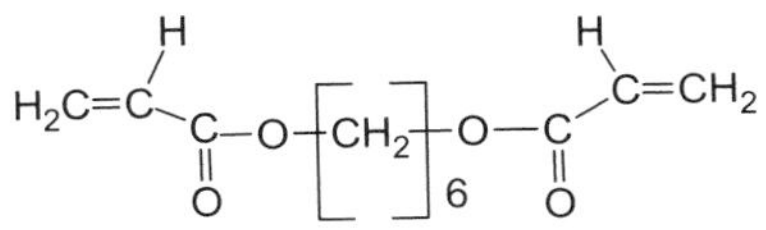

FIGURE 15.4 Structure of 1,6-hexanediol diacrylate (HDDA).

increases the shrinkage of the polymer, which results in larger void spaces between the polymer and cell walls (Kawakami et al. 1981).

MMA can be polymerized in wood using catalysts (Vazo or peroxides) and heat, or radiation. Curing of MMA using cobalt-60 gamma radiation requires a longer period of time (8–10 h depending upon the radiation flux); catalyst-heat initiated reactions are much faster (30 min or less at 60°C) (Meyer 1981).

Hardness modulus values determined for untreated and polymethyl methacrylate-treated red oak, aspen, and sugar maple, on both flat and edge-grained faces show untreated wood hardness values are related to sample density. There are significant relationships between treated wood hardness modulus, wood density, and loading. Large variations in hardness modulus of treated aspen and maple are related to their diffuse-porous structure. In contrast, the hardness modulus of treated red oak is predictable on the basis of density or polymer loading (Beall et al. 1973).

The compressive and bending strengths of a tropical wood (*Kapur-Dryobalanops* sp.) are improved significantly by impregnation of MMA (Boey et al. 1985). Using a gamma irradiation method, some tropical wood–poly(methyl methacrylate) and –poly(vinyl acetate) composites are produced that exhibit a significant improvement in uniaxial compressive strength (Boey et al. 1987). Samples with an average polymer content of 63% (based on dry wood) show increases in compressive strength, toughness, radial hardness, compressive strength parallel to the grain, and tangential sphere strength (Bull et al. 1985). Hardness and mechanical properties of poplar wood are improved by impregnation with MMA and polymerization of the monomer by exposure to gamma irradiation, the hardness of the product increases with impregnation pressure and weight of polymer (Bull et al. 1985; Ellis 1994).

Various other acrylic monomers have been investigated (Ellis and O'Dell 1999). WPCs were made with different chemical combinations and evaluated for dimensional stability, ability to exclude water vapor and liquid water, and hardness. Different combinations of hexanediol diacrylate (HDDA, Figure 15.4), hydroxyethyl methacrylate (HEMA, Figure 15.5), hexamethylene diisocyanate (Desmodur N75, DesN75), and maleic anhydride (MA) were *in situ* polymerized in solid pine, maple, and oak wood. The rate of water vapor and liquid water absorption was slowed, and the rate of swelling was less than that of unmodified wood specimens, but the dimensional stability was not permanent (see Figure 15.6). The WPCs were much harder than unmodified wood (see Table 15.1). Wetting and penetration of water into the wood was greatly decreased, and hardness and dimensional stability increased with the chemical combination of hexanediol diacrylate, hydroxyethyl methacrylate, and hexamethylene diisocyanate. Treatments containing hydroxyethyl methacrylate were harder and excluded water and moisture more effectively. This is probably due to the increased interfacial adhesion between the polymer and wood, due to the polarity of HEMA monomer.

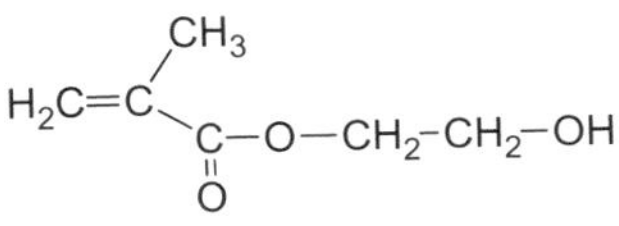

FIGURE 15.5 Structure of 2-hydroxyethyl methacrylate (HEMA).

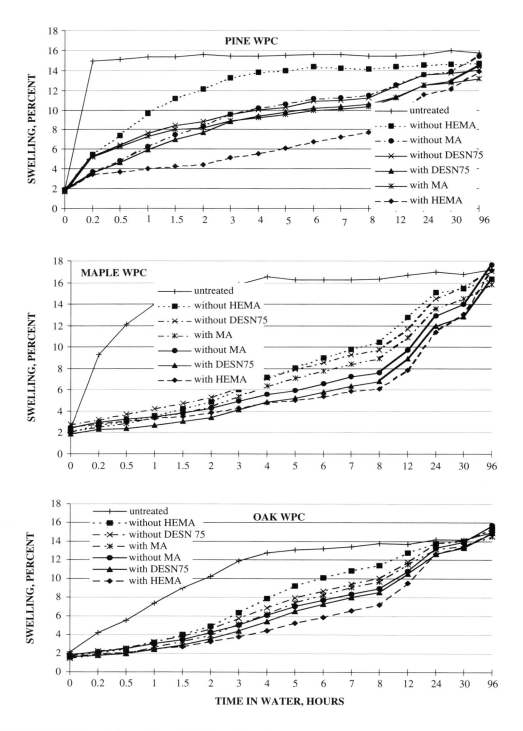

FIGURE 15.6 Volumetric swelling of WPC specimens in water.

TABLE 15.1
Rockwell Hardness of WPC Specimens

Treatment	Earlywood	Latewood
	Pine	
HDDA / DesN75 (3:1)	37.3	45.0
HDDA	32.2	41.9
HDDA / MA (3:1)	31.7	26.0
HDDA / HEMA / MA (1:2:1)	61.7	67.8
HEMA / DesN75 / MA (2:1:1)	61.0	70.1
HDDA / DesN75 / MA (2:1:1)	31.7	49.3
HDDA / HEMA (1:1)	47.0	55.3
HDDA / HEMA / DesN75 (1:2:1)	63.2	74.2
Control (untreated)	−15.5	−10.6
	Maple*	
HDDA / DesN75 (3:1)	44.8	
HDDA	46.8	
HDDA / MA (3:1)	49.2	
HDDA / HEMA / MA (1:2:1)	60.0	
HEMA / DesN75 / MA (2:1:1)	56.4	
HDDA / DesN75 / MA (2:1:1)	46.6	
HDDA / HEMA (1:1)	49.5	
HDDA / HEMA / DesN75 (1:2:1)	65.6	
Control (untreated)	−9.4	
	Red Oak	
HDDA / DesN75 (3:1)	23.3	22.7
HDDA	27.9	25.1
HDDA / MA (3:1)	23.1	20.5
HDDA / HEMA / MA (1:2:1)	38.6	46.3
HEMA / DesN75 / MA (2:1:1)	26.8	35.4
HDDA / DesN75 / MA (2:1:1)	23.0	13.6
HDDA / HEMA (1:1)	29.6	25.5
HDDA / HEMA / DesN75 (1:2:1)	39.3	40.8
Control (untreated)	−17.1	−25.1

Rockwell hardness * of the longitudinal face of 25 mm by 25 mm by 0.6 mm specimens.

* 1/4-inch ball indenter and 60 Kgf (Rockwell scale L)
* Maple measured without regard to earlywood or latewood

15.3.2 STYRENE

Styrene (Figure 15.7) is another monomer that is commonly used for WPCs. It can be polymerized in wood using catalysts (Vazo or peroxides) and heat, or radiation. Other monomers are commonly added to control the polymerization rate, extent of polymerization, and to crosslink the styrene for improved physical properties of the WPCs.

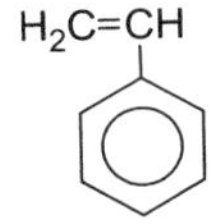

FIGURE 15.7 Structure of styrene.

Hardness, impact strength, compression and shear strength, and bending and cleavage strengths of styrene-treated wood are better than for untreated samples and the same as, or better than, those for samples impregnated with MMA. The treated wood is sometimes unevenly colored and more yellow than the original samples (Autio and Miettinen 1970).

Modification of several types of hard- and softwoods with polystyrene improves their resistance to wear. Wood-polystyrene composites made from the softwood species birch, gray and black alder, and spruce exhibit abrasion resistance comparable to that of natural oak wood (Dolacis 1983). The flexural strength, hardness, and density of alder wood are increased by impregnating it with styrene and heating to obtain the polystyrene-saturated wood (Lawniczak 1979). Poplar wood modified with polystyrene has increased hardness, static bending strength, and toughness; the increase in toughness depends on the polymer content to a certain limit (Lawniczak 1973).

WPC can be prepared from a mixture of acrylonitrile-styrene-unsaturated polyester in wood. This mixture gives a tough crosslinked polymer, and is more favorable for radiation polymerization than the systems of MMA, MMA-unsaturated polyester, and acrylonitrile-styrene (Czvikovszky 1977, 1981). Composite materials obtained by evacuation of wood (beech, spruce, ash, and tropical wood *Pterocarpus vermalis*) followed by its impregnation with an unsaturated polyester-MMA-styrene mixture or unsaturated polyester-acrylonitrile-styrene mixture and gamma irradiation-induced curing exhibit decreased water vapor absorption and improved dimensional stability, hardness, compression strength, and wear resistance, compared to untreated wood (Czvikovszky 1982).

Curing of unsaturated polyester-styrene mixture can be affected by the initiator-heat technique by using either 0.1–0.2% benzoyl peroxide or 1% methyl ethyl ketone peroxide (Doss et al. 1991). Polymer-reinforced alder wood can be prepared by impregnating it with styrene and peroxide catalyst, followed by thermal polymerization for 3–7 h. The addition of 1.0% divinylbenzene, triallyl phosphate, or trimethylolpropane trimethacrylate crosslinking agent to styrene results in an increased polymerization rate, with divinylbenzene having the most pronounced effect on the polymerization rate (Lawniczak and Szwarc 1987). Gamma ray-induced polymerization of styrene in impregnated samples of beech wood in the presence of carbon tetrachloride requires a minimum dose of 159 kGy for full monomer hardening. The polymer content in the resulting samples is 53% at a monomer conversion of >90%. A modified sample has ~50% increase in density, ~90% increase in hardness, and ~125% decrease in absorptivity, compared to unmodified wood (Raj and Kokta 1991).

The impregnation of beech wood with ternary mixtures of styrene, dioxane, acetone, or ethanol, and water followed by curing by ionizing radiation gives a product with some dimensional stability due to chemical fixation of the polymer on the lignocellulosic material. This change is accompanied by a marked change in the structure of the cell wall. Pure styrene or styrene in aqueous solution gives a composite with low dimensional stability (Guillemain et al. 1969). Wood polymers based on aqueous emulsion polyester-styrene mixtures are dimensionally more stable than those produced with a pure polyester-styrene mixture (Jokel 1972). Impregnation of poplar wood with styrene-ethanol-water followed by polymerization at 70°C gives a 50% increase in dimensional stability with 30–40% polystyrene content in the wood. Use of styrene alone increased wood dimensional stability by only 10% even with >100% styrene retention. Dimensional stability of poplar wood is also significantly increased (~40%) by a 90:5 styrene-ethanol system (Katuscak et al. 1972). The use of a mixture of polar solvents with styrene to make WPCs seems to improve the dimensional stability of the composites. The use of styrene alone for the modification of wood was not as favorable as using a styrene-methanol-water system which gives greater bending strength and better dimensional stability. Hardness increases with increasing polystyrene in the wood (Varga and Piatrik 1974).

Untreated woods of ash, birch, elm, and maple absorb about four times more water than woods containing acrylonitrile-styrene copolymer (Spindler et al. 1973). Addition of acrylonitrile and butyl methacrylate to styrene does not affect significantly the maximum amount of water sorbed by the composites but decreases their swelling rate and increases their dimensional stability and bending strength (Lawniczak and Pawlak 1983). WPCs prepared using styrene-acrylonitrile have increased hardness, substantially improved dimensional stability, and give no difficulties in machining and gluing (Singer et al. 1969).

Monomer- and polyester prepolymer-impregnated beech wood veneer irradiated with 3–6 Mrad and cured at 80°C has improved shrinkage resistance and water repellency and provides laminates suitable for flooring and siding. Various mixtures of unsaturated polyester and styrene, as well as the individual monomers MMA, ethyl acrylate, butyl acrylate, acrylonitrile, and vinyl acetate, have been used to treat veneers. A variety of tests of physical properties show the styrene-polyester system to be superior (Handa et al. 1972).

WPCs can be prepared by gamma irradiation of hardwood impregnated with a styrene-unsaturated polyester mixture, MMA, or acrylonitrile-styrene mixture. The addition of chlorinated paraffin oil to any of these monomer systems imparts fire resistance to the composites and reduces the gamma ray dosage needed for total polymerization of the monomers. Styrene-unsaturated polyester mixtures containing about 30% chlorinated paraffin oil are suitable systems for large-scale preparation of composites (Iya and Majali 1978).

Modification of wood samples with polystyrene increases the resistance of the composites to degradation in contact with rusting steel (Helinska-Raczkowska and Molinski 1983). Conversion of unsaturated polyester-styrene mixture and dimensional stability of the wood-styrene-unsaturated polyester composites decreases with an increase in moisture content of the wood to be treated (Yamashina et al. 1978).

Polymerization of styrene in wood can result in the grafting of styrene to cellulose, lignin, and pentosans (Lawniczak et al. 1987). The treatment of wood with diluted hydrogen peroxide solution leads to an increase in the viscosity-average molecular weight of the polystyrene, and to the graft polymerization of the monomer, which, in turn, enhances the stress properties of wood-polystyrene composites (Manrich and Marcondes 1989).

Wood impregnated with a styrene-ethylene glycol dimethacrylate mixture under full vacuum (0.64 kPa) has higher densities and hardness in the earlywood than that of latewood, and early wood shows hardness increases roughly double those in latewood at the same density, indicating styrene uptake is predominantly in earlywood (Brebner et al. 1985).

Kenaga (1970) researched high-boiling styrene-type monomers including vinyltoluene, t-butylstyrene, and o-chlorostyrene. In the preparation of WPCs the cure rate, monomer loss, and composite physical properties can be varied by appropriate selection and concentration of catalyst, comonomers, and crosslinking agents. The composite can be bonded to untreated crossbanded veneers simultaneously with polymerization in a press because these three styrene-type monomers have boiling points from 27 to 74°C higher than styrene's boiling point. The monomer t-butylstyrene has the highest boiling point at 219°C and the least shrinkage, 7%, on polymerization. Crosslinking agents increase reaction rate and improve the WPC physical properties. Effects of the crosslinking agents trimethylolpropane triacrylate, trivinyl isocyanurate, trimethylolpropane trimethacrylate, ethylene glycol dimethacrylate, trimethylene glycol dimethacrylate, tetraethylene glycol dimethacrylate, polyethylene glycol dimethacrylate, and divinylbenzene were studied. Generally 10% or more crosslinking agent is needed to give the best improvement in abrasion resistance. Copolymers of t-butylstyrene with diethyl maleate, diethyl fumarate, and acrylonitrile were studied in basswood and birch wood blocks. All the copolymers except acrylonitrile improved the abrasion resistance of the composite. Polyesters lowered cure time and styrene monomer loss during cure but increased the exotherm temperature to a level that could be unacceptable for larger pieces of wood.

15.3.3 Polyesters

Unsaturated polyester resins are most often used in combination with other monomers, making them less expensive and improving their properties. Many polyester resins are available as commercial products. Polyester, MMA, and styrene were polymerized individually and in combinations by gamma radiation or benzoyl peroxide (Miettinen et al. 1968; Miettinen 1969). MMA composites had higher tensile strength and abrasion resistance, but lower bending strength and impact strength, compared to the polyester composites. Styrene is frequently mixed with polyester resins to reduce viscosity, thus enabling better penetration into the wood. Polyesters decrease the loss of styrene monomer, and the time to heat-cure (Kenaga 1970).

15.3.4 MELAMINE RESINS

Melamine resins have many uses with paper and wood products. Paper can be impregnated with melamine resins, and then laminated to the surface of wood veneers, fiber boards, or other panel products resulting in a hard, smooth, and water-resistant surface. Wood veneers can also be impregnated with melamine resins to improve dimensional stability, water resistance, and hardness (Inoue et al. 1993; Takasu and Matsuda 1993). The melamine resin-modified wood is 1.5–4 times harder than untreated wood (Inoue et al. 1993). Yet, the maximum hardness achieved by melamine-modified wood is less than that of wood modified with acrylate, methacrylate, and other vinyl monomers which increase hardness 7–10 times that of unmodified wood (Mizumachi 1975). But, the hardness of melamine resin impregnated wood can be increased by compressing the wood (Inoue et al. 1993). The melamine resins decrease the abrasion resistance of wood (Inoue et al. 1993; Takasu and Matsuda 1993).

15.3.5 ACRYLONITRILE

Acrylonitrile (Figure 15.8) is used in the production of WPCs mostly in combination with other monomers because the polymer does not improve properties by itself. It is most frequently used with styrene, and less frequently with MMA, methyl acrylate, unsaturated polyester, diallyl phthalate, and vinylidene chloride. WPCs made with MMA-acrylonitrile or styrene-acrylonitrile mixtures were cured using either gamma radiation or catalyst, and the resultant composites were found to be very similar (Yap et al. 1990; Yap et al. 1991).

Styrene-acrylonitrile WPCs show high dimensional stability, probably due to swelling of the wood by the acrylonitrile during treatment creating a bulking action (Loos 1968). Addition of acrylonitrile to styrene gives substantial improvement in hardness and compressibility of the wood (Rao et al. 1968). Ratios of acrylonitrile to styrene between 7:3 and 4:1 give the most substantial improvements in dimensional stability, compressibility, and hardness. The anti-swell efficiencies of wood-styrene-acrylonitrile combinations are 60–70% (no swelling is 100%) (Ellwood et al. 1969). Moisture absorption increases with increase of acrylonitrile in WPCs made with various ratios of acrylonitrile and methyl acrylate (Gotoda et al. 1970).

Copolymerization of bis(2-chloroethyl) vinylphosphonate with vinyl acetate or acrylonitrile in beech wood improves the dimensional stability of the WPCs (Ahmed et al. 1971). A ternary resin mixture of styrene, acrylonitrile, and unsaturated polyester that can be cured in the wood with a low dose of gamma radiation also has favorable properties (Czvikovszky 1977, 1981, 1982).

The addition of acrylonitrile to a diallyl phthalate prepolymer improves the glueability of the composite against a substrate, such as plywood or particleboard. The weatherability of a wood composite laminate containing diallyl phthalate prepolymer and acrylonitrile is improved by incorporating polyethylene glycol dimethacrylate (Gotoda et al. 1971).

Acrylonitrile is highly toxic and is a carcinogen; therefore attempts have been made to find chemicals that can be substituted for acrylonitrile in the treating solutions. These attempts have

$$H_2C = CH$$
$$|$$
$$CN$$

FIGURE 15.8 Structure of acrylonitrile.

been only partially successful. *N*-vinyl carbazol can be used as a partial replacement of acrylonitrile. Several other compounds including acryloamide, *N*-hydroxy acryloamide, and 1-vinyl-2-pyrrolidone were tried unsuccessfully (Schaudy and Proksch 1982).

15.4 CROSSLINKING AGENTS

Some of the crosslinking agents frequently used with MMA, styrene, or other vinyl monomers are trimethylolpropane triacrylate, trivinyl isocyanurate, trimethylolpropane trimethacrylate, ethylene glycol dimethacrylate, trimethylene glycol dimethacrylate, tetraethylene glycol dimethacrylate, polyethylene glycol dimethacrylate, and divinylbenzene. Crosslinking agents generally increase reaction rate and improve the physical properties of WPCs (Kenaga 1970).

Several crosslinking monomers, including 1,3-butylene dimethacrylate ethylene dimethacrylate and trimethylolpropane trimethacrylate and the polar monomers 2-hydroxyethyl methacrylate and glycidyl methacrylate, were added at 5–20% concentration to MMA, and their effects upon the polymerization and properties of the composites were examined (Kawakami et al. 1977). WPCs with only MMA show a void space at the interface between cell wall and polymer. With addition of crosslinking esters such as di- and trimethacrylate, the shrinkage (and hence void spaces) of the polymer during polymerization increases. On the other hand, in the WPCs containing polar esters having hydroxyethyl and glycidyl groups, the voids due to the shrinkage of polymer were found to form inside the polymer itself, suggesting better adhesion of the polymer to the inner surface of cell wall (Kawakami et al. 1981).

Impregnation with ethyl α-hydroxymethylacrylate (EHMA) plus another multifunctional monomer, 2-vinyl-4,4-dimethyl 2-oxazolin-5-one (vinyl azlactone), results in improved mechanical properties of wood samples. Improvements of 38–54% in impact strength and 27–44% in compression modulus are achieved depending on the relative amount of vinyl azlactone incorporated (Mathias et al. 1991).

15.4.1 ISOCYANATES

The addition of isocyanate compounds with acrylic monomers reduces the brittleness of WPCs consisting only of acrylic compounds (Schaudy and Proksch 1981). WPC properties improve by adding a blocked isocyanate to a mixture of MMA and 2-hydroxyethyl methacrylate (Fujimura et al. 1990). The isocyanate compound crosslinks the copolymer.

The mechanical properties of a wood-polystyrene composite are improved by the addition of an isocyanate compound to the styrene treating mixture. Polymethylene (polyphenyl isocyanate) forms a bridge between wood and polymer on the interfaces. The isocyanate compound then becomes instrumental in efficient stress transfer between the wood and polymer (Maldas et al. 1989).

15.4.2 ANHYDRIDES

A maleic anhydride and styrene mixture has been used to make WPCs (Ge et al. 1983). Also, a mixture of tetraethylene glycol dimethacrylate and chlorendic anhydride increased the fire, chemical, and abrasion resistance as well as the hardness (Paszner et al. 1975). A process has been developed that is designed to prepare crosslinked oligoesterified wood with improved dimensional stability and surface properties. Maleic, phthalic, and succinic anhydrides are used. The wood is reacted with the anhydride, then impregnated with glycidyl methacrylate, and then heated to cause crosslinking. In a one-step process the anhydride and glycidyl methacrylate are impregnated into the wood together, then polymerized and reacted with the wood simultaneously. The resulting wood is hard and has smooth surfaces. As the anhydride in the anhydride:glycidyl methacrylate ratio increases, the dimensional stability increases (Ueda et al. 1992).

15.5 PROPERTIES OF WOOD-POLYMER COMPOSITES

WPCs can improve many properties of solid wood, and therefore be tailored for a specific application. Some of these properties are surface hardness, toughness, abrasion resistance, dimensional stability, moisture exclusion, and fire, decay, and weather resistance. Table 15.2 is a summary of some of the properties of woods modified by five different treatments.

15.5.1 Hardness

Hardness is the property that resists crushing of wood and the formation of permanent dents. The hardness or indent resistance of WPC is measured by any of several methods. The test method used depends on the WPC and the expected final product. Measurement can be made using a hand-held Shore Durometer tester, ball indenters such as Brinell and Rockwell hardness, the Janka ball indenter, or the Gardner Impact tester that uses a falling dart to make dents that can be measured (Miettinen et al. 1968; Beall et al. 1973; Schneider 1994).

Hardness of a WPC depends on the polymer loading and the hardness of the polymer. Polymer loading is affected by wood porosity and density. For example, a more porous and lower density wood will require a higher polymer loading. Generally, a higher polymer loading will give a greater WPC hardness. Figure 15.9 shows an SEM micrograph of a WPC with no polymer attachment and the lumens incompletely filled. Figure 15.10 shows an SEM micrograph of a WPC with filled lumens and some interaction of the polymer with the wood. The hardness of a WPC is improved when the cells are completely filled and there is attachment of the polymer to the wood.

The type of polymer, crosslinking chemicals, and method and extent of polymerization affect polymer hardness. A 7–10-fold increase in hardness can be expected by most treatments, for example, MMA-impregnated alder wood has more than a 10-fold increase in hardness of the sides and more than a 7-fold increase in hardness of the cross-cut areas (Miettinen et al. 1968).

15.5.2 Toughness

Increasing the toughness of wood with polymer increases the crack resistance and brittleness at room temperature. Impact strength and toughness are closely related; both refer to the ability of WPCs to resist fracturing. Measurements of impact strength are made using the Izod and the Charpy impact

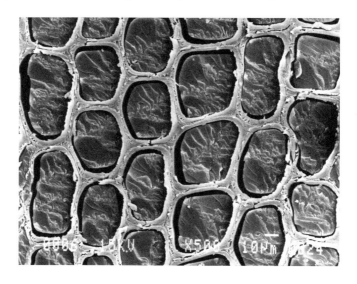

FIGURE 15.9 SEM micrograph of a WPC having the wood cells incompletely filled with polymer and having no attachment of polymer to the wood.

TABLE 15.2
Properties of Wood after Five Different Modifications

Property	Water-Soluble Polymers and Synthetic Resins		Compression		Heat	Organic Chemicals or Crosslinking Agents		Liquid Monomers	
	Polyethylene Glycol (PEG)	Impreg	Staypak	Compreg	Staywood	Bulking	Crosslinking	Methyl Methacrylate	Epoxy Resin
Specific gravity	Slightly increased	15 to 20 pct greater than normal wood	1.2 to 1.4	1.0 to 1.4	Unchanged	Slightly increased	Unchanged	Increased	Increased
Permeability to water vapor	Hygroscopic	Better than normal	Better than normal	Greatly improved	Better than normal	Unchanged	Unchanged	Greatly improved	Greatly improved
Liquid water repellency	Hygroscopic	Better than normal	Better than normal	Greatly improved	Better than normal	Better than normal	Better than normal	Greatly improved	Greatly improved
Dimensional stability	80 pct	60 to 70 pct	Slightly improved	80 to 85 pct	40 pct	66 to 75 pct	80 to 90 pct	10 pct	Slightly improved
Decay resistance	Better than normal	Better than normal	Unchanged	Much batter than normal	Better than normal	Much better than normal	Better than normal	Somewhat increased	Somewhat increased
Heat resistance	No data	Greatly increased	No data	Greatly increased	No data	Not data	No data	Increased	No data
Fire resistance	No data	Unchanged	Unchanged	Unchanged	Unchanged	Unchanged	Unchanged	Unchanged	No data
Chemical resistance	No data	Better than normal	Slightly better than normal	Much better than normal	Better than normal	No data	No data	Much better than normal	Much better than normal
Compression strength	Slightly increased	Increased	Increased	Greatly increased	Reduced	Slightly reduced	Slightly reduced	Greatly increased	Greatly increased
Hardness	Unchanged	Increased	Increased	10 to 20 times greater	Reduced	Slightly reduced	Slightly reduced	Greatly increased	Greatly increased

(*Continued*)

TABLE 15.2
Properties of Wood after Five Different Modifications (Continued)

Property	Water-Soluble Polymers and Synthetic Resins		Compression		Heat	Organic Chemicals or Crosslinking Agents		Liquid Monomers	
	Polyethylene Glycol (PEG)	Impreg	Staypak	Compreg	Staybwood	Bulking	Crosslinking	Methyl Mathacrylate	Epoxy Resin
Abrasion resistance	Slightly reduced	Reduced	Increased	Increased	Greatly reduced	Slightly reduced	Greatly reduced	Greatly increased	Greatly increased
Machinability	Unchanged	Better than normal but dulls tools	Metalworking tools required	Metalworking tools required	Unchanged	Unchanged	Unchanged	Metalworking tools preferred	Metalworking tools preferred
Glueability	Special glues required	Unchanged	Unchanged	Same as normal after sanding	Unchanged	Unchanged	Unchanged	Special glues required	Epoxy used as adhesive
Finishability	Requires polyurethane, oil, or 2 parts polymer	Unchanged	Unchanged	Plastic-like surface (can be polished without finish)	Unchanged	Unchanged	Unchanged	Plastic-like surface (no finish required)	Plastic-like surface (no finish required)
Color change	Little change	Reddish brown	Little change	Reddish brown	Darkened	Little change	Little change	Little change	Little change

Lumen Modifications

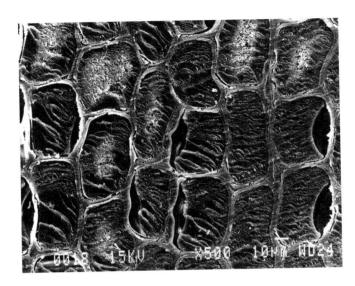

FIGURE 15.10 SEM micrograph of a WPC with the wood cells filled with polymer and having some interaction of the polymer with the wood.

test instruments. The test involves striking the specimen with a pendulum and measuring the impact energy necessary to initiate fracture. Treating sugar maple wood with a vinyl polymer increases toughness in both radial and tangential impact directions compared to untreated wood. Increased polymer load results in increased toughness. Microscopy indicates brittle polymer fracture extends across lumens but stops at the polymer and cell wall interface (Schneider et al. 1989). Brittleness of a composite can be severely increased by increasing the amount of a crosslinker such as ethylene glycol dimethacrylate even to as little as 1.5% in MMA (Schaudy and Proksch 1982). WPCs with high toughness (low brittleness) have been prepared by using a treating mixture consisting of MMA and an isocyanate that has an acrylic functionality. This treating mixture increased the impact bending strength of the WPC by about 100% (Schaudy et al. 1982).

15.5.3 Abrasion Resistance

Abrasion resistance is determined by the Taber wear index, which is the weight loss (mg/1000 cycles) caused by an abrasive wheel turning on a specimen. The lower the weight loss value, the better the resistance to wear. In general, abrasion resistance increases with increasing polymer content in the wood (Kawakami and Taneda 1973). Softwood species such as birch, gray and black alder, and spruce when made into a composite with polystyrene have abrasion resistance comparable to that of natural oak wood (Dolacis 1983). Alder wood and birch wood impregnated with MMA had up to 85% less weight loss than untreated wood (Miettinen et al. 1968).

15.5.4 Dimensional Stability

Dimensional stability is the property of wood that allows it to resist changes in dimensions when exposed to various moisture conditions. Dimensional stability is reported as percent volumetric swelling or as antishrink efficiency (ASE). ASE is the percent reduction in volumetric swelling of treated wood compared to untreated wood at equilibrium water- or moisture-saturated conditions (see Chapter 4). Many WPCs are not dimensionally stable so that with time in water or high humidity, most WPCs will swell to the same amount as untreated wood.

There are two approaches to improve the dimensional stability of WPCs. One approach is to direct the penetration into the wood cell walls to bulk the wood at or near its wet or green dimensions.

Aqueous and non-aqueous solvents have been used to swell the wood and carry the monomers into the cell walls, and polar monomers have been used to increase the swelling of the wood and penetration of the monomers into the wood. The second approach is to react the chemicals with the cell wall hydroxyl groups, therefore decreasing its affinity for moisture (Loos 1968; Rowell et al. 1982). WPCs with polymer located just within the lumen do not make a significant contribution to dimensional stability compared with chemical modifications of the cell wall (Fujimura and Inoue 1991).

15.5.5 Moisture Exclusion

Moisture exclusion efficiency (MEE) is the property of a WPC to exclude moisture and is related to the rate at which the composite absorbs moisture and swells and not to the maximum extent of swelling or moisture uptake (see Chapter 4). If the WPCs are not allowed to reach equilibrium with respect to moisture or water, then MEE can be mistaken for dimensional stability or ASE values (Loos 1968). Many WPCs will absorb water and swell at a slower rate than untreated wood, but in most cases the maximum swelling is nearly the same as that of untreated wood.

15.5.6 Fire Resistance

There are several methods of measuring different aspects of the property of fire retardancy (see Chapter 6). Thermogravimetry measures char formation and decomposition temperatures by heating small specimens in an inert atmosphere. More char generally indicates greater fire retardancy. The oxygen index test measures the minimum concentration of oxygen, in an oxygen and nitrogen atmosphere, that will just support flaming combustion. Highly flammable materials are likely to have a low oxygen index. Flame spread tests are those in which the duration of flaming and extent of flame spread are measured. The results of any of the test methods that use small specimens often do not correlate with the actual performance of materials in a real fire situation. The surface burning characteristics of WPCs used as building materials are best measured by flame spread tests that use large specimens, such as the ASTM E84 test that requires specimens approximately 514 mm wide by 7.3 m long. The test chamber in this test also has a photometer system built in to measure smoke and particulate density. Smoke evolution is very important because many fire deaths are due to smoke inhalation.

Polymethyl methacrylate enhances the flammability of wood (Calleton et al. 1970; Lubke and Jokel 1983) but not styrene and acrylonitrile (Schaudy et al. 1982). Bis(2-chloroethyl)vinyl phosphonate with vinyl acetate or acrylonitrile improves the fire retardancy, but is less effective than poly(dichlorovinyl phosphate) or poly(diethyl vinyl phosphonate). Wood impregnated with dimethylaminoethyl methacrylate phosphate salt and then polymerized in the presence of crosslinking agents has high fire retardancy (Ahmed et al. 1971) as does trichloroethyl phosphate (Autio and Miettinen 1970). The addition of chlorinated paraffin oil to monomer systems imparts fire retardancy to composites (Iya and Majali 1978). The limiting oxygen index values of the MMA-bis(2-chloroethyl)vinyl phosphonate copolymer and MMA-bis(chloropropyl)-2-propene phosphonate copolymer wood composites are much higher than that of untreated wood and other composites, indicating the effectiveness of the phosphonates as fire retardants (Yap et al. 1991). WPC specimens made with MMA are smoke-free, but styrene-type monomers create dense smoke (Siau et al. 1972). The presence of aromatic polymers, such as poly(chlorostyrene), and fire retardants having aromatic benzene rings in wood increase the smoke evolution, flame spread, and fuel contribution in a modified tunnel furnace test (Siau et al. 1975). In all specimens tested, the smoke evolution increased markedly after the flame is extinguished.

15.5.7 Decay Resistance

Most WPCs are not decay-resistant because the polymer merely fills the lumens and does not enter the cell walls, which makes the cell walls accessible to moisture and decay organisms (see Chapter 5). WPCs prepared using MMA and several kinds of crosslinking monomers (1,3-butylene dimethacrylate, ethylene dimethacrylate, and trimethylolpropane trimethacrylate) and polar monomers (2-hydroxyethyl methacrylate and glycidyl methacrylate) added at 5–20% concentration have little resistance to brown rot decay (Kawakami et al. 1977). Using methanol with MMA or styrene allows the polymer to penetrate the cell walls. The amount of polymer in the cell wall is important for decay resistance. Some protection against biological degradation is possible at cell wall polymer contents of 10% or more (Rowell 1983). Acrylate monomers with various bioactive moieties were synthesized (Ibach and Rowell 2001a). Pentachlorophenol acrylate and Fyrol 6 acrylate polymers provided no protection against decay, whereas tributyltin acrylate, 8-hydroxyquinolyl acrylate, and 5,7-dibromo-8-hydroxyquinolyl acrylate were found to be resistant to the brown-rot fungus *Gloeophyllum trabeum* at low polymer loading of 2–5% retention (Ibach and Rowell 2001b).

15.5.8 Weathering Resistance

WPCs made with birch and pine impregnated with MMA or styrene-acrylonitrile were exposed in a weatherometer for 1000 hours (Desai and Juneja 1972). The specimens were more resistant to surface checking than untreated wood and the styrene-acrylonitrile treatment performed better than MMA. A combination of cell wall-modifying treatments (butylene oxide or methyl isocyanate) with MMA lumen-filled treatments results in a dual treatment that resists the degradative effects of accelerated weathering in a weatherometer (see Chapter 7). The use of MMA in addition to the cell wall-modifying chemical treatments provides added dimensional stability and lignin stabilization and has a significant effect on weatherability (Rowell et al. 1981).

15.5.9 Mechanical Properties

The strength properties of WPCs are enhanced compared to untreated wood. The hardness, compression, and impact strength of wood composites increase with increasing monomer loading (Mohan and Iyer 1991). Crosslinking monomers increases static bending properties, compressive strength, and torsional modulus, but reduces dimensional stability, while polar monomers improve dimensional stability and static bending properties but have no significant effect on compressive strength and torsional modulus (Kawakami et al. 1977).

15.6 APPLICATIONS

The major uses of WPCs are for flooring, sports equipment, musical instruments, and furniture (Fuller et al. 1997). Flooring has the largest volume that includes solid plank flooring, top veneers of laminated flooring, and fillets for parquet flooring. As for sports equipment, patents have been issued for golf club heads (Katsurada and Kurahashi 1985), baseball bats, hockey sticks (Yamaguchi 1982), and parts of laminated skis. WPCs are used for wind instruments, mouthpieces of flutes and trumpets, and finger boards of stringed instruments (Knotik and Proksch 1971; Knotik et al. 1971). One area with potential is the use of veneer laminates for furniture, such as desk writing surfaces and tabletops (Maine 1971; Kakehi et al. 1985). A history of the commercialization of WPCs and future opportunities is covered by Schneider and Witt (Schneider and Witt 2004).

15.7 POLYMER IMPREGNATIONS

Wood polymer composites are usually formed by impregnating the wood with a monomer that is polymerized *in situ* mainly in the lumen. Because the monomers are small, almost complete penetration of chemical is achieved. However, it is possible to impregnate wood with oligomers or polymers; however, penetration depends on the size of the oligomer or polymer. Chemical retention is often limited by the inability of the large polymers to penetrate into the wood structure. Several resin systems have been used to treat wood to improve performance properties.

15.7.1 Epoxy Resins

Epoxy resin is a partially polymerized, clear solution, with a consistency slightly thicker than varnish at room temperature (21°C). Just before treating wood, the epoxy resin is mixed with hardener. It is cured or hardened within the lumen structure from a few minutes to a few hours depending upon the hardener and temperature. Treatment with epoxy resin is usually performed on veneers because of its large molecular size and high viscosity which does not allow for deep penetration into larger specimens. Veneers are either vacuum-treated or soaked in the epoxy resin-hardener solution and then cured. Mechanical properties are greatly increased with epoxy resin treatments, especially hardness (Rowell and Konkol 1987). Epoxy resins are used for wooden boat hulls, the outer ply of plywood, and strengthening softened or decayed wood.

15.7.2 Compression of Wood during Heating and Curing with Resin

Wood can be compressed using heat either with or without resin that improves strength, stiffness, and stability.

15.7.2.1 Staypak

Compressed wood containing no resin is called Staypak (Seborg et al. 1945). During compression with 1400–1600 lb/in² pressure, with temperatures of 170–177°C, lignin will flow relieving internal stresses, but also causing a darkening of the wood. The compression time varies with the thickness of the wood, but usually to a specific gravity of at least 1.3 (Rowell and Konkol 1987). The resultant wood product has a slower moisture absorption and hence a reduction in swelling. It is more dimensionally stable, but not necessarily more biologically resistant. Tensile strength (both parallel and perpendicular to the grain), modulus of rupture, elasticity in bending, and impact bending strength of the wood are increased. Staypak is used for tool handles, mallet heads, and various tooling jigs and dies.

15.7.2.2 Compreg

Compreg is resin-treated, compressed wood, and made with layers of treated veneers (Stamm and Seborg 1951). The most common resin is phenol-formaldehyde. The veneers are treated to 25–30% weight gain based on oven-dried weight. The veneers are dried at 30°C or less to prevent the resin from curing. The resin is cured during the heating (140–150°C) and compression process (pressures of 1000–1200 lb/in²). Water absorption is greatly reduced; biological resistance to decay, termite, and marine borer is increased; electrical, acid, and fire resistance are increased. Strength properties of Compreg are increased, except for impact bending strength. The abrasion resistance and hardness are also increased compared to untreated wood. Compreg has many uses from knife handles and tools to musical instruments (Rowell and Konkol 1987).

15.7.2.3 Staybwood

Staybwood is one product that is made by heating wood in a vacuum at high temperatures (93–160°C) in a bath of molten metal (Stamm et al. 1960). The high temperature causes the lignin to flow and the hemicelluloses to decompose, producing water-insoluble polymers. The process increases dimensional stability, but decreases strength and therefore has not been used commercially (Rowell and Konkol 1987).

15.8 WATER SOLUBLE POLYMERS AND SYNTHETIC RESINS

Wood can be impregnated with water-soluble treatments such as PEG or resins (Impreg) that become insoluble after curing. Keeping the wood in a partially or completely swollen state increases dimensional stability, as well as strength and water repellency.

15.8.1 POLYETHYLENE GLYCOL (PEG)

PEG is a chemical with the following structure:

$$HO\text{-}CH_2CH_2\text{-}O\text{-}CH_2CH_2\text{-}(O\text{-}CH_2CH_2)_n\text{-}O\text{-}CH_2CH_2\text{-}OH$$

PEG 1000 is most commonly used when treating wood, usually undried, green wood (Mitchell 1972). It is a waxy, white solid that has an n value (average molecular weight) of 1,000 and it can penetrate the cell wall because of its small size. It melts at 40°C, readily dissolves in warm water, is noncorrosive, odorless, and colorless, and has a very high fire point (305°C) (Rowell and Konkol 1987). Molecular weights up to 6000 are soluble in water.

To treat wood with PEG 1000, the wood is placed in a container and covered with a 30–50 weight percent solution dissolved in water. The treatment is based on diffusion and therefore soak time will vary depending upon the thickness of the specimen. PEG remains in the cell walls when the wood is dried because of its low vapor pressure. The rate of diffusion into the cell wall increases as water evaporates from the solution (Stamm 1964).

Treatment temperature is usually from 21–60°C, but diffusion can be accelerated with increasing the temperature and/or the concentration of the solution. After treatment, the wood is air-dried in a ventilated room. The drying time also depends upon specimen size.

PEG is not chemically attached to the wood, and because it is water soluble, it will leach out if it gets wet (see Chapter 4). Glycol attracts moisture, so if the relative humidity reaches above 70%, the wood becomes sticky.

PEG has many uses especially in prevention or reduction of cracking due to drying sound wood for tabletops, to partially decomposed wooden artifacts, or archeological waterlogged wood.

15.8.2 IMPREG

Impreg is wood that has been treated with a thermosetting, fiber-penetrating resin and is cured without compression (Stamm and Seborg 1962). Phenol-formaldehyde resin-forming systems with low molecular weights are the most successful thermosetting agents. The resins penetrate the cell wall (25–25% weight gain) and keep the wood in a swollen state, dried at 80–93°C for 30 minutes, and then are polymerized by heat (155°C for 30 min) to form a water-insoluble resin in the cell wall (Rowell and Konkol 1987). Treatments are usually done on thin veneers (less than 9 mm thick) due to time.

The cured product is usually reddish brown with reduced swelling, shrinkage, grain raising, and surface checking. It improves the compression strength, but reduces the impact bending strength. Impreg shows resistance to decay, termite, and marine-borer attack, and it has a high resistance to acid. It is suited for pattern and die models as well as electrical control equipment.

REFERENCES

Aagaard, P. (1967). Swedish studies on wood-polymer composites. *Svensk Kem. Tidskr.* 79(9):501–510.
Ahmed, A.U., Takeshita, N., and Gotoda, M. (1971). Fire-retardant wood-polymer composite based on radiation-induced polymerization of phosphorous-containing vinyl monomers particularly bis(2-chloroethyl) vinylphosphonate. *Nippon Genshiryoku Kenkyusho Nempo* 5027:82–90.

Autio, T. and Miettinen, J.K. (1970). Experiments in Finland on properties of wood-polymer composites. *Forest Products J.* 20(3):36–42.

Beall, F.C., Witt, A.E., and Bosco, L.R. (1973). Hardness and hardness modulus of wood-polymer composites. *Forest Products J.* 23(1):56–60.

Boey, F.Y.C., Chia, L.H.L., and Teoh, S.H. (1985). Compression, bend, and impact testing of some tropical wood-polymer composites. *Radiat. Phys. Chem.* 26(4):415–421.

Boey, F.Y.C., Chia, L.H.L., and Teoh, S.H. (1987). Model for the compression failure of an irradiated tropical wood-polymer composite. *Radiat. Phys. Chem.* 29(5):337–348.

Brebner, K.I., Schneider, M.H., and St-Pierre, L.E. (1985). Flexural strength of polymer-impregnated eastern white pine. *Forest Products J.* 2:22–27.

Bull, C., Espinoza, B.J., Figueroa, C.C., and Rosende, R. (1985). Production of wood-plastic composites with gamma radiation polymerization. *Nucleotecnica* 4:61–70.

Calleton, R.L., Choong, E.T., and McIlhenny, R.C. (1970). Treatments of southern pine with vinyl chloride and methyl methacrylate for radiation-produced wood-plastic combinations. *Wood Sci. Technol.* 4(3):216–225.

Czvikovszky, T. (1977). Pilot-scale experiments on radiation processing of wood-plastic composites. *Proc. Tihany Symp. Radiat. Chem.* 4:551–560.

Czvikovszky, T. (1981). Wood-polyester composite materials. II. Dependence of the processing parameters on the initiation rate. *Angew. Makromol. Chem.* 96:179–191.

Czvikovszky, T. (1982). Composite materials made of wood and a polyester resin. *Plast. Massy,* 7:37. [Also in *Plast. Manufacture Process* 50:1–43.]

Desai, R.L. and Juneja, S.C. (1972). Weatherometer studies on wood-plastic composites. *Forest Products J.* 22(9):100–103.

Dolacis, J. (1983). Comparative abrasive wear of wood of various species modified radiochemically with polystyrene. In: Rotsens, K.A. (Ed.), *Modif. Svoistv Drev. Mater 107-11.* Zinatne, Riga, USSR 37.

Doss, N.L., El-Awady, M.M., El-Awady, N.I., and Mansour, S.H. (1991). Impregnation of white pine wood with unsaturated polyesters to produce wood-plastic combinations. *J. Appl. Polymer Sci.* 42:2589–2594.

Ellis, W.D. (1994). Moisture sorption and swelling of wood polymer composites. *Wood Fiber Sci.* 26(3):333–341.

Ellis, W.D. and O'Dell, J.L. (1999). Wood-polymer composites made with acrylic monomers, isocyanate, and maleic anhydride. *J. Appl. Polymer Sci.* 73(12):2493–2505.

Ellwood, E., Gilmore, R., Merrill, J.A., and Poole, W.K. (1969). An investigation of certain physical and mechanical properties of wood-plastic combinations. U.S. Atomic Energy Commission ORO-638. [Available from Dept.; CFSTI: *Nucl. Sci. Abstr.* 23(23):48583.]

Fujimura, T. and Inoue, M. (1991). Improvement of the durability of wood with acryl-high-polymer. III. Dimensional stability of wood with crosslinked epoxy-copolymer. *Mokuzai Gakkaishi* 37:719–726.

Fujimura, T., Inoue, M., and Uemura, I. (1990). Durability of wood with acrylic-high-polymer. II. Dimensional stability with cross linked acrylic copolymer in wood. *Mokuzai Gakkaishi* 36(10):851–859.

Fuller, B.S., Ellis, W.D., and Rowell, R.M. (1997). Hardened and fire retardant treatment of wood for flooring. U.S. Patent 5,605,767, February 25, 1997 and U.S. Patent 5,609,915, March 11, 1997.

Ge, M., Peng, H., Dai, C., and Li, J. (1983). Heating wood-plastic composites. *Linye Kexue* 19(1):64–72.

Glukhov, V.I. and Shiryaeva, G.V. (1973). Parameters of the radiation polymerization of vinyl monomers in wood. *Plast. Massy* 6:35–36.

Gotoda, M., Harada, O., Yagi, T., and Yoshizawa, I. (1975a). Radiation curing of a mixture of diallyl phthalate prepolymer and vinyl monomer. X. Application of electron beam curing of low molecular weight diallyl phthalate prepolymer-vinyl monomer mixtures to the preparation of American hemlock-polymer composite laminated board. I. Thermal curing. *Nippon Genshiryoku Kenkyusho Nempo* 5030: 92–102.

Gotoda, M., Horiuchi, Y., and Urasugi, H. (1974). Radiation curing of mixtures of diallyl phthalate prepolymer and vinyl monomers. VIII. Application to the manufacture of resin composite-veneered plywood of electron beam curing within wood. 2. *Nippon Genshiryoku Kenkyusho Nempo* 5029:79–91.

Gotoda, M. and Kitada, Y. (1975). Radiation curing of mixtures of diallyl phthalate prepolymer and vinyl monomer. IX. Fundamental examination of the application of low molecular weight diallyl phthalate prepolymer. *Nippon Genshiryoku Kenkyusho Nempo* 5030:85–91.

Gotoda, M., Okugawa, H., Yagi, T., and Yoshizawa, I. (1975b). Radiation curing of a mixture of diallyl phthalate prepolymer and vinyl monomer. XI. Application of electron beam curing of low molecular weight diallyl phthalate prepolymer-vinyl monomer mixtures to the preparation of American hemlock-polymer composite laminated board. 2. Electron beam curing. *Nippon Genshiryoku Kenkyusho Nempo* 5030:103–113.

Gotoda, M. and Takeshita, N. (1971a). Preparation of a wood-polymer composite by ionizing radiation. VII. Improvement of thermal stability of vinylidene chloride copolymers and fire retardancy of the wood-vinylidene chloride copolymer composites. *Nippon Genshiryoku Kenkyusho Nempo* 5027:63–71.

Gotoda, M. and Takeshita, N. (1971b). Preparation of a wood-polymer composite by ionizing radiation. VIII. Radiation curing of monomer solution of chlorinated allyl chloride oligomer and its application to the preparation of a flame-retardant wood-polymer composite. *Nippon Genshiryoku Kenkyusho Nempo* 5027:72–81.

Gotoda, M., Tsuji, T., and Toyonishi, S. (1970a). Preparation of wood-polymer-composite by ionizing radiation. V. Effect of solvent extraction of wood as a pretreatment on the gamma-induced polymerization of vinylidene chloride in wood. *Nippon Genshiryoku Kenkyusho Nempo* 5026:86–93.

Gotoda, M., Yokoyama, K., Takeshita, N., and Senzaki, Y. (1971). Radiation curing of mixture of diallyl phthalate prepolymer and vinyl monomer. VI. Electron-beam curing of diallyl phthalate prepolymer/monomer mixture in the preparation of wood-polymer composite piled board. *Nippon Genshiryoku Kenkyusho Nempo* 5027:91–99.

Gotoda, M., Yokoyama, K., and Toyonishi, S. (1970b). Radiation curing of a mixture of diallyl phthalate prepolymer and vinyl monomer. IV. Radiation (especially electron-beam) curing of the mixture, impregnated in wood, for preparing wood-polymer composites. *Nippon Genshiryoku Kenkyusho Nempo* 5026:108–120.

Guillemain, J., Laizier, J., Marchand, J., and Roland, J.C. (1969). Polymer-substrate bonding obtained by radiation in resin-treated woods. *Large Radiat. Sources Ind. Processes, Proc. Symp. Util. Large Radiat. Sources Accel. Ind. Process.*, pp. 417–433.

Handa, T., Otsuka, N., Akimoto, H., Ikeda, Y., and Saito, M. (1973). Characterization on the electron beam-induced polymerization of monomers in wood. I. Effect of moisture in the polymerization of vinyl monomer in presoaked beech veneer. *Mokuzai Gakkaishi.* 19(10):493–498.

Handa, T., Seo, I., Akimoto, H., Saito, M., and Ikeda, Y. (1972). Physical properties of wood-polymer composite materials prepared by I.C.T-type electron-accelerator. *Proc. Jpn. Congr. Mater. Res.* 15:158–163.

Handa, T., Seo, I., Ishii, T., and Hashizume, Y. (1983). Polymer-performance on the dimensional stability and the mechanical properties of wood-polymer composites prepared by an electron beam accelerator. *Polym. Sci. Technol.* 20(3):167–190.

Handa, T., Yoshizawa, S., Seo, I., and Hashizume, Y. (1981). Polymer performance on the dimensional stability and the mechanical properties of wood-polymer composites evaluated by polymer-wood interaction mode. *Org. Coat. Plast. Chem.* 45:375–381.

Helinska-Raczkowska, L. and Molinski, W. (1983). Effect of atmospheric corrosion in contact with rusting iron on the impact strength of lignomer. *Zesz. Probl. Postepow Nauk Roln.* 260:199–210.

Ibach, R.E. and Rowell, R.M. (2001a). Wood preservation based on in situ polymerization of bioactive monomers—Part 1. Synthesis of bioactive monomers, wood treatments and microscopic analysis. *Holzforschung* 55(4):358–364.

Ibach, R.E. and Rowell, R.M. (2001b). Wood preservation based on in situ polymerization of bioactive monomers—Part 2. Fungal resistance and thermal properties of treated wood. *Holzforschung* 55(4):365–372.

Inoue, M., Ogata, S., Kawai, S., Rowell, R.M., and Norimoto, M. (1993a). Fixation of compressed wood using melamine-formaldehyde resin. *Wood Fiber Sci.* 25(4):404–410.

Inoue, M., Ogata, S., Nishikawa, M., Otsuka, Y., Kawai, S., and Norimoto, M. (1993b). Dimensional stability, mechanical properties, and color changes of a low molecular weight melamine-formaldehyde resin impregnated wood. *Mokuzai Gakkaishi* 39(2):181–189.

Iya, V.K. and Majali, A.B. (1978). Development of radiation processed wood-polymer composites based on tropical hardwoods. *Radiat. Phys. Chem.* 12(3–4):107–110.

Jokel, J. (1972). Wood plastics made from aqueous emulsions of a polyester-styrene mixture emulsion. *Drevarsky Vyskum* 17:247–260.

Kakehi, M., Yoshida, Y., and Minami, K. (1985). *Modified Wood Manufacture.* Daiken Kogyo Co., Ltd.

Katsurada, S. and Kurahashi, K. (1985). Process and Apparatus for Producing Wood Heads of Golf Clubs. Sumitomo Rubber Industries Ltd. (Patent Number 6B2141753) June 4, 1984.

Katuscak, S., Horsky, K., and Mahdalik, M. (1972). Phase diagrams of ternary monomer-solvent-water systems used for the preparation of wood-plastic combinations (WPC). *Drevarsky Vyskum* 17(3):175–186.

Kawakami, H. and Taneda, K. (1973). Impregnation of sawn veneers with methyl methacrylate and unsaturated polyester-styrene mixture, and the polymerization by a catalyst-heat technique. Polymerization in several wood species and properties of treated veneers. *J. Hokkaido Forest Prod. Res. Inst.* 10:22–27.

Kawakami, H., Taneda, K., Ishida, S., and Ohtani, J. (1981). Observation of the polymer in wood-polymer composite II. On the polymer location in WPC prepared with methacrylic esters and the relationship between the polymer location and the properties of the composites. *Mokuzai Gakkaishi.* 27(3): 197–204.

Kawakami, H., Yamashina, H., and Taneda, K. (1977). Production of wood-plastic composites by functional resins. I. Effects of adding crosslinking and polar monomers to methyl methacrylate. *J. Hokkaido Forest Prod. Res. Inst.* 306:10–17.

Kawase, K. and Hayakawa, K. (1974). Manufacturing of a wood-plastic combination by irradiation of microwave. Polymerization of impregnated monomer in wood with microwave irradiation. *Mokuzai Kogyo.* 29(32):12–17.

Kenaga, D.L. (1970). Heat cure of high boiling styrene-type monomers in wood. *Wood Fiber Sci.* 2(1):40–51.

Knotik, K. and Proksch, E. (1971). Polymer-impregnated wood for musical instruments, Oesterreichische Studiengesellschaft fuer Atomenergie G.m.b.H. (Patent Number DE2054730) Nov. 14, 1969.

Knotik, K., Proksch, E., and Bresimair, K. (1971). Polymer-impregnated wood mouthpieces for wind instruments, Oesterreichische Studiengesellschaft fuer Atomenergie G.m.b.H. (Patent Number 2069273) Nov. 14, 1969.

Lawniczak, M. (1973). Poplar wood modification. *Prace Komisji Technologii Drewna* 1:3–42.

Lawniczak, M. (1979). Effect of temperature changes during styrene polymerization in wood on the quality of the lignomer. *Prace Komisji Technologii Drewna* 9:69–82.

Lawniczak, M., Melcerova, A., and Melcer, I. (1987). Analytical characterization of pine and alder wood polymer composites. *Holzforschung und Holzverwertung* 39(5):119–121.

Lawniczak, M. and Pawlak, H. (1983). Effect of wood saturation with styrene and addition of butyl methacrylate and acrylonitrile on the quality of the produced lignomer. *Zesz. Probl. Postepow Nauk Roln.* 260:81–93.

Lawniczak, M. and Szwarc, S. (1987). Crosslinking of polystyrene in wood-polystyrene composite preparation. *Zesz. Probl. Postepow Nauk Roln.* 299(37):115–125.

Lee, T.J. (1969). Wood plastic composites. *Hwahak Kwa Kongop Ui Chinbo.* 9(3):220–224.

Loos, W.E. (1968). Dimensional stability of wood-plastic combinations to moisture changes. *Wood Sci. Technol.* 2(4):308–312.

Lubke, H. and Jokel, J. (1983). Combustibility of lignoplastic materials. *Zesz. Probl. Postepow Nauk Roln.* 260:281–292.

Maine, J. (1971). *Composite Wood-Polymer Product.* Maine, C. W., and Sons. (US Patent Number 3560255) March 18, 1968.

Maldas, D., Kokta, B.V., and Daneault, C. (1989). Thermoplastic composites of polystyrene: Effect of different wood species on mechanical properties. *J. Appl. Polymer Sci.* 38:413–439.

Manrich, S. and Marcondes, J.A. (1989). The effect of chemical treatment of wood and polymer characteristics on the properties of wood polymer composites. *J. Appl. Polymer Sci.* 37(7):1777–1790.

Mathias, L.J., Kusefoglu, S.H., Kress, A.O., Lee, S., Wright, J.R., Culberson, D.A., Warren, S.C., Warren, R.M., Huang, S., Lopez, D.R., Ingram, J.E., Dickerson, C.W., Jeno, M., Halley, R.J., Colletti, R.F., Cei, G., and Geiger, C.C. (1991). Multifunctional acrylate monomers, dimers and oligomers—Applications from contact-lenses to wood-polymer composites. *Makromolekulare Chemie-Macromolecular Symposia* 51:153–167.

Meyer, J.A. (1965). Treatment of wood-polymer systems using catalyst-heat techniques. *Forest Products J.* 15(9):362–364.

Meyer, J.A. (1981). Wood-polymer materials: State of the art. *Wood Sci.* 14(2):49–54.

Meyer, J.A. (1984). Wood polymer materials. *Adv. Chem. Ser. 207 (Chem. Solid Wood)*:257–289.

Miettinen, J.K. (1969). Production and properties of wood-plastic combinations based on polyester/styrene mixtures. *Amer. Chem. Soc., Div. Org. Coatings Plast. Chem., Pa.* 29:182–188.

Miettinen, J.K., Autio, T., Siimes, F.E., and Ollila, T. (1968). Mechanical properties of plastic-impregnated wood made from four Finnish wood species and methyl methacrylate or polyester by using irradiation. *Valtion Tek. Tutkimuslaitos, Julk.* 137:(1–58).

Mitchell, H.L. (1972). How PEG Helps the Hobbyist Who Works with Wood. USDA Forest Service, Forest Products Laboratory.

Mizumachi, H. (1975). Interaction between the components in wood-polymer composite systems. *Nippon Setchaku Kyokai Shi.* 11(1):17–23.

Mohan, H. and Iyer, R.M. (1991). Study of wood-polymer composites. *Conf. Proc., Rad. Tech. Int. North Am.: Northbrook, Ill.:* 93–97.

Nobashi, K., Koshiishi, H., Ikegami, M., Ooishi, K., and Kaminaga, K. (1986a). WPC treatments of several woods with functional monomer. III. Hygroscopicities and dimensional stabilities of Buna WPC. *Shizuoka-ken Kogyo Gijutsu Senta Kenkyu Hokoku* 31:15–26.

Nobashi, K., Koshiishi, H., Ikegami, M., Ooishi, K., Oosawa, T., and Kaminaga, K. (1986b). WPC treatments of wood with functional monomer. II. *Shizuoka-ken Kogyo Gijutsu Senta Kenkyu Hokoku* 30:21–26.

Okonov, Z.V. and Grinberg, M.V. (1983). Modification of the properties of wood by polystyrene as applied to machine parts of the textile industry. In: Rotsens, K.A. (Ed.), *Modif. Svoistv Drev. Mater 102-6.* Zinatne: Riga, USSR 38, 40.

Paszner, L., Szymani, R., and Micko, M.M. (1975). Accelerated curing and testing of some radiation curable methacrylate and polyester copolymer finishes on wood panelings. *Holzforschung und Holzverwertung* 27:67–73.

Pesek, M. (1984). The possibilities of preparation of wood-plastic combinations based on unsaturated polyester resins and methyl methacrylate and styrene by combined radiation and chemical methods. *ZfI-Mitt.* 97:359–363.

Pesek, M., Jarkovsky, J., and Pultar, F. (1969). Radiation polymerization of methyl methacrylate in wood. Effect of halogenated organic compounds on the polymerization rate. *Chem. Prum.* 19(11):503–506.

Pietrzyk, C. (1983). Effect of selected factors on polymerization of methyl methacrylate by the radiation method in alder and birch wood. *Zesz. Probl. Postepow Nauk Roln* 260:71–80.

Proksch, E. (1969). Wood-plastic combinations prepared from beechwood. *Holzforschung* 23:93–98.

Pullmann, M., Jokei, J., and Manasek, Z. (1978). Dimensional stabilization of wood with a synthetic polymer in wood-plastic composites. II. Effect of components of the system wood + impregnant on the radiation polymerization of methyl methacrylate. *Drevarsky Vyskum* 22(4):261–276.

Raj, R.G. and Kokta, B.V. (1991). Reinforcing high-density polyethylene with cellulosic fibers. I. The effect of additives on fiber dispersion and mechanical properties. *Polymer Eng. Sci.* 31:1358–1362.

Rao, K.N., Moorthy, P.N., Rao, M.H., Vijaykumar, and Gopinathan, C. (1968). Wood-plastic combinations. II. Acrylic esters and their copolymers. *India, At. Energy Comm., Bhabha Atomic Research Cent.,* BARC-369.

Rowell, R.M. (1983). Bioactive polymer-wood composites. In: *Controlled Release Delivery Systems.* Marcel Dekker, New York, pp. 347–357.

Rowell, R.M., Feist, W.C., and Ellis, W.D. (1981). Weathering of chemically modified southern pine. *Wood Sci.* 13(4):202–208.

Rowell, R.M. and Konkol, P. (1987). Treatments that enhance physical properties of wood. USDA Forest Service, Forest Products Laboratory, Madison, WI.

Rowell, R.M., Moisuk, R., and Meyer, J.A. (1982). Wood-polymer composites: Cell wall grafting with alkylene oxides and lumen treatments with methyl methacrylate. *Wood Sci.* 15:290–296.

Schaudy, R. and Proksch, E. (1981). Wood-plastic combinations with high dimensional stability. *Oesterr. Forschungszent. Seibersdorf* Report No. 4113.

Schaudy, R. and Proksch, E. (1982). Wood-plastic combinations with high dimensional stability. *Ind. Eng. Chem. Prod. Res. Dev.* 21(3):369–375.

Schaudy, R., Wendrinsky, J., and Proksch, E. (1982). Wood-plastic composites with high toughness and dimensional stability. *Holzforschung* 36(4):197–206.

Schneider, M.H. (1994). Wood polymer composites. *Wood Fiber Sci.* 26(1):142–151.

Schneider, M.H., Phillips, J.G., Brebner, K.I., and Tingley, D.A. (1989). Toughness of polymer impregnated sugar maple at two moisture contents. *Forest Products J.* 39(6):11–14.

Schneider, M.H., Phillips, J.G., Tingley, D.A., and Brebner, K.I. (1990). Mechanical properties of polymer-impregnated sugar maple. *Forest Products J.* 40(1):37–41.

Schneider, M.H. and Witt, A.E. (2004). History of wood polymer composite commercialization. *Forest Products J.* 54(4):19–24.

Seborg, R.M., Millett, M.A., and Stamm, A.J. (1945). Heat-stabilized compressed wood—Staypak. *Mech. Eng.* 67(1):25–31.

Siau, J.F., Campos, G.S., and Meyer, J.A. (1975). Fire behavior of treated wood and wood-polymer composites. *Wood Sci.* 8(1):375–383.

Siau, J.F., Meyer, J.A., and Kulik, R.S. (1972). Fire-tube tests of wood-polymer composites. *Forest Products J.* 22(7):31–36.

Singer, K., Vinther, A., and Thomassen, T. (1969). Some technological properties of wood-plastic materials. Danish Atomic Energy Commission Riso Report No. 211.

Soundararaian, S. and Reddy, B.S.R. (1991). Glycidyl methacrylate and N-vinyl-2-pyrrolidone copolymers: Synthesis, characterization, and reactivity ratios. *J. Appl. Polymer Sci.* 43:251–258.

Spindler, M.W., Pateman, R., and Hills, P.R. (1973). Polymer impregnated fibrous materials. Resistance of polymer-wood composites to chemical corrosion. *Composites* 4(6):246–253.

Stamm, A.J. (1964). *Wood and Cellulose Science.* The Ronald Press Co., New York.

Stamm. A.J., Burr, H.K., and Kline, A.A. (1960). Heat stabilized wood–Staybwood. USDA Forest Service, Forest Products Laboratory, Report No. 1621.

Stamm, A.J. and Seborg, R.M. (1951). Resin-treated laminated, compressed wood—Compreg. USDA Forest Service, Forest Products Laboratory, Report No. 1381.

Stamm. A.J. and Seborg, R.M. (1962). Resin treated wood - Impreg. USDA Forest Service, Forest Products Laboratory, Report No. 1380.

Takasu, Y. and Matsuda, K. (1993). Dimensional stability and mechanical properties of resin impregnated woods. *Aichi-ken Kogyo Gijutsu Senta Hokoku* 29:57–60.

Ueda, M., Matsuda, H., and Matsumoto, Y. (1992). Dimensional stabilization of wood by simultaneous oligoesterification and vinyl polymerization. *Mokuzai Gakkaishi* 38(5):458–465.

Varga, S. and Piatrik, M. (1974). Contribution of the radiation preparation of wood-plastic materials [WPC]. VIII. Testing of the physical and mechanical properties of WPC. *Radiochem. Radio Anal. Lett* 19:255–261.

Yamaguchi, K. (1982). Wood hitting parts for sporting goods. Japan Patent JP 57203460.

Yamashina, H., Kawakami, H., Nakano, T., and Taneda, K. (1978). Effect of moisture content of wood on impregnation, polymerization and dimensional stability of wood-plastic composites. *J. Hokkaido Forest Prod. Res. Inst.* 316:11–14.

Yap, M.G.S., Chia, L.H.L., and Teoh, S.H. (1990). Wood-polymer composites from tropical hardwoods I WPC properties. *J. Wood Chem. Technol.* 10(1):1–19.

Yap, M.G.S., Que, Y.T., and Chia, L.H.L. (1991a). Dynamic mechanical analysis of tropical wood polymer composites. *J. Appl. Polymer Sci.* 43(11):1999–2004.

Yap, M.G.S., Que, Y.T., and Chia, L.H.L. (1991b). FTIR characterization of tropical wood-polymer composites. *J. Appl. Polymer Sci.* 43:2083–2090.

Yap, M.G.S., Que, Y.T., Chia, L.H.L., and Chan, H.S.O. (1991c). Thermal-properties of tropical wood polymer composites. *J. Appl. Polymer Sci.* 43(11):2057–2065.

16 Plasma Treatment of Wood

Ferencz S. Denes, L. Emilio Cruz-Barba, and Sorin Manolache
Biological Systems Engineering and Center for Plasma-Aided Manufacturing, University of Wisconsin, Madison, WI

CONTENTS

16.1 Wood Surface Properties 447
16.2 Surface Treatment of Wood 448
    16.2.1 Wood Protection 448
    16.2.2 Wood Coating 448
16.3 The Plasma State 449
    16.3.1 Classification of Plasmas 450
        16.3.1.1 Natural and Man-Made Plasma 450
    16.3.2 Plasma Parameters 451
    16.3.3 Atmospheric Pressure Cold Plasma 454
16.4 Cold Plasma Modification of Wood 457
References 470

This chapter presents a review of the various plasma-mediated treatments of wood reported in the literature. Some of these works presented extensive detail, and others only described the process with the plasma-parameters. This is because of the high diversity of the plasma tools involved in the modification of wood surfaces, the complexity of the environment in the plasma discharges, and the main focus of many researchers in the characteristics of the plasma-generated surfaces, which result in a black-box-like representation of the plasma processes.

16.1 WOOD SURFACE PROPERTIES

The properties of wood surfaces depend on the species, environmental conditions, age, and mechanical processing (see Chapters 2, 3, and 8). Depending on its final use, wood needs to undergo surface improvement treatments to overcome issues resulting from surface inactivation, formation of weak boundary layers, and material processing such as machining, drying, and aging.

Surface inactivation can be defined as the detriment of wood surface properties as a result of different factors. These factors can be summarized as the migration of hydrophobic extractives to the surface during drying, which results in poor wettability, acidity or reactivity of extractives affecting adhesion, oxidation of the wood surface as a result of aging, molecular reorientation of the surface functional groups, and closure of wood cell micro-pores, which reduces adhesive absorption into the wood (Christiansen 1990, 1991).

Weak boundary layers are the result of chemical and mechanical variations on the wood surfaces. Chemical weak boundary layers are the result of extractive materials migrating to the surface. Resin acids, fats/fatty acids, sterols, and waxes will render the surface hydrophobic, while sugars, phenols, tannins, and proteins will render it hydrophilic. Mechanical weak boundary layers are the result of degradation by light, damaged surfaces from machining processes, and trapped air at the interface.

Morphology of the wood surface reveals important properties such as surface roughness and mechanical weak boundary layers. Surface roughness affects wettability due to capillary force, and changes on it can lead to improved mechanical interlocking. Adhesion is affected by the presence of damaged fibers resulting from machining, as well as cracks and splits.

16.2 SURFACE TREATMENT OF WOOD

Depending on the intended application of wood products, it is generally necessary to treat the wood to provide specific surface properties. A typical problem in softwoods is yellowing, a degradation resulting from UV-radiation damage caused by sunlight exposure. Softwood products that have to be used outdoors need to be treated with a UV-protecting layer.

Hardwood products can undergo surface treatments where the coating is used to cover only the surface or to fill up the pores on the structure. There is an increased use of wood-based board materials in a wide variety of applications. These engineered composite materials have improved properties and structural characteristics. The most common board materials are chipboards, fiberboards, and veneer.

16.2.1 Wood Protection

In order to protect wood from weather conditions and pest attack, it is necessary to treat it with preservatives. The preservatives used can be divided into water-based (e.g., sodium phenylphenoxide, benzalconium chloride, copper chrome arsenate); organic solvent-based (e.g., triazoles, permethrin, copper and zinc naphthenates); borates; and tar oils.

16.2.2 Wood Coating

The main purpose of a surface treatment is to protect the surface of wood and give it a good appearance. Different materials can be used for finishing, such as fillers, stains, primers and top coatings, oils, and waxes. Fillers are used to seal cracks and small cavities on the wood surface. Stains are usually applied under transparent coatings and have the purpose of giving color to the wood to emphasize the natural beauty and structure of it. Lacquers and paints are the most used coating materials and are usually the outermost layer of the treated wood. Coating materials consist of binders, fillers, pigments, flatting agents, solvents, and additives.

The main properties of the materials to be used for surface treatment depend greatly on the binder. Binder materials commonly used for the treatment of wood include: amino resins, polyurethane, acrylate, polyester, and nitrocellulose. Special properties can be achieved by the use of a combination of these materials.

Amino resins, such as urea and melamine, and alkyd resins, which are often modified with nitrocellulose, are used as binders in acid-curing treatments. These materials harden as a result of a polycondensation process that is initiated when a catalyst is added. Such a catalyst (acid hardener) should be added to the coating before application. The hardening process can be significantly accelerated by increasing the drying temperature and boosting air flow. Acid-curing coatings have good durability against chemical attack and mechanical impact.

Emulsions, colloidal dispersions, and totally water-soluble binders can be used for water-borne treatments. Use of these products is mainly because emission of toxic organic solvents is greatly reduced or totally avoided during the treatment. In addition, these materials have good light and fire resistance.

There are a wide variety of water-borne products with different properties that can be used, depending on the application for which the wood is needed. Water-borne coatings can be alkyd-, polyurethane-, acrylate-, or polyester-based. The main disadvantages of these materials are that storage and transportation can take place only at temperatures above 0°C, and the possibility of swelling of the wood.

Polyurethane materials dry as a result of chemical reaction between the isocyanate and hydroxyl groups. Most polyurethane materials are organic solvent-based, although there are some water-based products. Compared to acid-curing materials, polyurethanes dry more slowly. Polyurethanes are known for good resistance to chemicals and mechanical impact, as well as excellent moisture resistance. Their flexibility enables them to provide good resistance against swelling and shrinking.

Acrylate combined with a photo initiator is generally used as a UV-curing coating. Under UV radiation, the photo initiator starts a fast curing process. These materials, which can be acrylates or polyesters, need to be dried before the curing in order to remove all of the solvent.

Moisture content causes wood to swell or shrink, and the results can differ depending on the kind of wood. The moisture content of wood should be kept stable during the treatment processes, as well as during storage, transport, and usage. Cycled or even single swelling and shrinking of the wood surface can cause severe damage to the treated surface, resulting in cracking and splitting which appears on the surface of the coating.

The coatings are usually applied to the wood surfaces by pressurized impregnation; by soaking, brushing; or spraying; by dipping or immersion; and by thermal diffusion (immersion in a hot bath of preservative).

16.3 THE PLASMA STATE

Broadly the plasma state can be considered to be a gaseous mixture of oppositely charged particles with a roughly zero net electrical charge (Chen 1965; Venugopalan 1971; Nasser 1971; Vossen and Kern 1978; Chapman 1980; Chen and Pfender 1983; Herman 1988; Hershkowitz 1989; MacRae 1989; Smith et al. 1989; Cecchi 1990; Chang and Pfender 1990; Leveroni and Pfender 1990; Raizer 1991; Danikas 1993; Grill 1994; Hitchon 1999; Becker 2003). Sir William Crooks suggested the concept of the "fourth state of matter" (1879) for electrically discharged matter, and Irving Langmuir first used the term *"plasma"* to denote the state of gases in discharge tubes.

Ionization processes can occur when, for instance, molecules of a gas are subjected to high-energy radiation, electric fields, or high caloric energy. During these processes the energy levels of particles composing the gas increase significantly and, as a result, electrons are released, and charged heavy particles are produced.

Increasing the energy content of solid phase matter sufficiently leads to phase transformation processes, and, depending on the specific molecular structure, molecular weight and temperature; a system that has been heated will frequently be in solid, liquid, or gaseous phase or a mixture of these. Not all materials can undergo all of these transformation processes. Some of them will undergo structural changes at specific absorbed-energy levels, through molecular fragmentation processes. At atmospheric pressures and temperatures around 5000 K materials exist only in the gaseous phase. Above 10,000 K few atoms, but mainly ions, are the constituent particles of matter. Under these conditions and at even higher temperatures matter is considered to be in the plasma state. More elevated temperatures induce very high degrees of ionization, and temperatures of 10^8 K and higher lead to mixtures of bare nuclei and electrons.

Gas-phase plasmas are the most common plasmas, although the plasma state can be present in liquid and solid phases. Man-made plasmas can be produced in the laboratory by raising the energy content of matter regardless of the nature of the energy source. Thus plasmas can be generated by mechanical (close to adiabatic compression), thermal (electrically heated furnaces), chemical (exothermic reactions, e.g., flames), radiant (high-energy electromagnetic and particle radiations, e.g., electron beams), nuclear (controlled nuclear reactions), and electrical (arcs, coronas, DC and RF discharges) energies, and by the combination of them, as in the combination of mechanical and thermal energies (e.g., explosions).

Most electrical discharge processes are initiated and sustained by electron impact ionization phenomena. The plasma state is created when "omni-present" free electrons (e.g., free electrons generated by cosmic radiation) in a gas environment are accelerated by an electric or an electromagnetic field to energy levels where ionization and molecular fragmentation processes occur through non-elastic collision mechanisms. The electrons involved in non-elastic collision reactions lose part of their energy by generating more free electrons, ions of either polarity, excited species, and free radical species through the fragmentation of molecular structures. Electrons traveling along the free path (λ) with a specific distribution must be accelerated to energy levels by an applied electric field (E) to gain sufficient kinetic energy (ε_{kin}) for the initiation of ionization, excitation, and fragmentation (ε_{ex}) of atoms or molecules of a specific gas environment:

$$\varepsilon_{kin} = e \cdot E \cdot \lambda > \varepsilon_{ex}$$

Accordingly, to initiate the discharge, low-pressure gas environments (large λ values) or very intense electric fields (E) are required. Most of the ionization energies of atoms involved in organic structures are in the range of 10–13 eV, so if an electron must gain a kinetic energy of 10 eV at atmospheric pressure ($\lambda = 10^{-5}$ cm) E must exceed 10^6 V/cm. Under low-pressure conditions like 1 mTorr (at 1 mTorr and room temperature λ_{argon} is about 8 cm) E must exceed only 1.25 V/cm (Gericke et al. 2002; Baars-Hibbe et al. 2003; Scheffler et al. 2004).

16.3.1 Classification of Plasmas

Plasma states can be divided into two main categories: *hot plasmas* (near-equilibrium plasmas) and *cold plasmas* (non-equilibrium plasmas). Hot plasmas are characterized by comparable and very high-temperature electrons and heavy particles, both charged and neutral, and they are close to maximal degrees of ionization (100%). Cold plasmas are composed of low-temperature particles (charged and neutral molecular and atomic species) and relatively high-temperature electrons, and they are associated with low degrees of ionization (10^{-4}–10%). Hot plasmas include electrical arcs, plasma jets of rocket engines, thermonuclear reaction-generated plasmas, etc., while low-pressure DC and RF discharges (silent discharges), discharges from fluorescent (neon) illuminating tubes, and corona discharges can be identified as cold plasmas.

16.3.1.1 Natural and Man-Made Plasma

It is estimated that more than 99% of the known universe is in the plasma state, the exception being cold celestial bodies and planetary systems. The stars of the universe including our sun have extremely high temperatures and they consist entirely of plasma. It is also suggested that the space between the stars and galaxies are star-origin, radiation-induced, rarefied plasmas. Our planet is not an exception to the presence of the plasma state. The ionosphere, "northern lights" (aurora borealis), and lightning also represent plasma states.

Meteoritic impacts on the atmosphere of the earth also create plasma states. An incoming meteorite traveling at extremely high speed will compress the atmospheric gas components to very high densities and convert its tremendous kinetic energy into caloric energy. Molecular fragmentation and ionization processes are initiated as a consequence and the plasma state is generated.

The plasma state can be produced in the laboratory by raising the energy content of matter regardless of the nature of the energy source. Plasmas lose energy to their surroundings through collision and radiation processes; consequently, energy must be supplied continuously to the system in order to sustain the discharge. The easiest way to inject energy into a system in a continuous manner is with electrical energy, and that is the reason why electrical discharges are the most common man-made plasmas. Electrical discharges usually generate non-homogeneous plasma media owing to various shapes and geometrical locations of the electrodes (antennas) and the geometry of the plasma reactor. As a result of a more or less intense interaction (reaction) of plasma species with the materials of the electrodes and reactor walls, which confine the discharge, an even greater degree of chemical non-homogeneity characterizes such plasma.

DC discharges require electrically conductive electrodes. If dielectric layers are deposited on the surfaces (including the electrode surfaces) that confine the DC plasma, the discharge will be quickly extinguished as the electrons accumulate on the insulator films and recombine with the available ions. This problem can be overcome by alternating the polarity of the discharge, using low- or high-frequency driving fields. When the frequency of the driving field increases above values where the time taken by the positive ions to move between the electrodes becomes larger than half the period of the electric field (critical ion frequency), the ions created near a momentary anode cannot reach the cathode before the field is reversed. Consequently, both the positive and negative space charge is partly retained between the two half-cycles of the field and facilitates the re-initiation of the discharge (lower voltages are required to initiate and sustain the discharge). Due to the small mass of the electrons in comparison to those of ions and to their higher mobility, critical electron frequencies are much higher, relative to the critical ion frequencies.

The high-frequency (RF) ranges and, accordingly, the plasmas are recognized as RF plasmas. Specifically, selected frequencies such as 13.56 MHz and its superior harmonics are usually employed to avoid interference with the communication bands.

16.3.2 Plasma Parameters

Plasmas are characterized by *external* and *internal* parameters. External parameters include the material, geometrical shape and dimension of plasma reactor and electrodes, plasma-gas composition, gas flow rates, total and partial pressure, and electrical power (specific amplitude, driving-field frequency, duty cycle, etc.) dissipated to the electrodes. These are the "knobs" of plasma installations. Internal plasma parameters include densities of charged and neutral particles, including electrons and ions of either-polarity, free radical species, atoms, and molecules, and average energy and energy distribution of all species. These parameters are also directly related to the basic plasma properties, such as plasma frequency, and Debye length and degree of ionization.

The energy distribution (or velocity distribution) of electrons is one of the most important plasma characteristics. Because of the low mass and high mobility of electrons they are the "*ab-initio*" species that are responsible for kinetic energy-induced ionization and molecular-fragmentation processes. Accordingly, in the absence of additional energy input (e.g., biased discharges or magnetically confined discharges) all plasma species will have lower energies than the electrons, which are responsible for the initiation and sustaining of the plasma state. The energy distribution of electrons depends on a variety of factors, and it is related to the velocity distribution function defined as the density of particles in the velocity space that satisfy the equation,

$$n(cm^{-3}) = 4\pi \int_0^\infty f(v)v^2 dv$$

where v = velocity, $f(v)$ = velocity distribution function (density in velocity space), and n = the density of the particles in the geometrical space. For a Maxwellian distribution it is assumed that the temperature of the electrons is equal to the temperature of the gas, the distribution of electrons in plasma is isotropic, the effects of inelastic collisions act only as a perturbation of the isotropy, and the effects of the electric fields are negligible. In this case the electron energy distribution function $f(w)$ is related to the velocity distribution function $f(v)$ through the relation,

$$f(v) = n_e \left(\frac{m_e}{2\pi kT_e}\right)^{\frac{3}{2}} \exp\left(-\frac{m_e v^2}{2kT_e}\right)$$

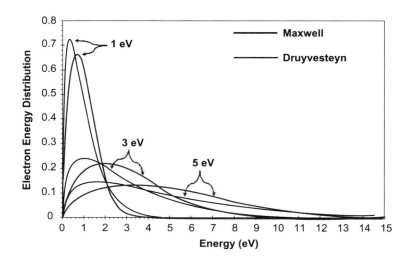

FIGURE 16.1 Electron energy distributions, according to Maxwell and Druyvesteyn, for average electron energies of 1, 3, and 5 eV.

Due to the assumptions in its formulation, the Maxwellian distribution gives only an approximation of the electron energy distribution. In low-pressure plasma conditions this distribution can be replaced by the Druyvesteyn distribution, using the following "correcting" assumptions: the electric field in the plasma is so low that the influence of inelastic collisions can be neglected, the temperature of the electrons is much higher than that of the ions, the driving field is of sufficiently low frequency ω and is much lower than the collision frequency, and the collision frequency is independent of the electron energy. Figure 16.1 presents Maxwellian and Druyvesteyn distributions for situations with 1, 3, and 5 average electron energies.

Average plasma-particle energies $\bar{\varepsilon}$ are often expressed in terms of electron (T_e), ion (T_i), and gas (T_g) temperatures:

$$\bar{\varepsilon}_e = \frac{3}{2}kT_e, \quad \bar{\varepsilon}_i = \frac{3}{2}kT_i, \quad \bar{\varepsilon}_g = \frac{3}{2}kT_g$$

where k is the Boltzmann constant.

It can be observed from Figure 16.1 that a small number of electrons have relatively high energies (5–15 eV) while the bulk of the electrons belong to the low-energy electron range (0.5–5 eV). Since the ionization potentials of the atoms of common organic structures (e.g., $C^+ = 11.26$ eV; $H^+ = 13.6$ eV; $O^+ = 13.6$ eV; $N^+ = 14.53$ eV; etc.) belong to the tail region of the electron energy distribution, the low degrees of ionization of cold plasmas appear obvious. However, this argument is somewhat circular since the inelastic process of ionization is, in large part, what determines the electron temperature needed to sustain the discharge.

It is important to note that the energy range of most of the electrons (2–5 eV) is intense enough to dissociate almost all chemical bonds involved in organic structures (Table 16.1) and organic structures containing main group elements, and to create free radical species capable of reorganizing into macromolecular structures. As a consequence, the structures of all volatile compounds can be altered and/or converted into high-molecular-weight compounds, even if they do not have the functionalities that are present in common monomer structures. Higher energies are usually required for the dissociation of unsaturated linkages and the formation of multiple free radicals. Accordingly, initial or plasma-generated unsaturated bonds will have a better "survival rate" under plasma conditions, in comparison to the σ linkages. Thus, it can be understood why plasma-generated macromolecular structures are usually characterized by unsaturation and branched and cross-linked architecture.

TABLE 16.1
Bond Energies and Enthalpies of Formation of Free Radicals

Bond Energies		Enthalpies of Formation of Free Radicals		
Species	Energy (eV)	Species	Energy (kJ/mol)	Energy (eV)
Diatomic Molecules				
C–H	3.3	·CH·	596.3	6.1
C–N	7.8	CH$_2$:	430.1	4.4
C–Cl	4.0	CH$_3$·	146.0	1.5
C–F	5.7	HC=C·	566.1	5.8
C=O	11.2	HC=CH$_2$·	300.0	3.1
C–C	6.3	NH:	350.0	3.6
		NH$_2$·	185.4	1.9
Polyatomic Molecules				
C=C	7.6	:Si:	456.6	4.7
C≡C	10.0	·SiCl·	195.0	2.0
CH$_3$–H	4.5	SiCl$_2$:	−163.0	−1.7
C$_2$H$_5$–H	4.3	SiCl$_3$·	−318.0	−3.3
CH$_2$CH–H	4.8	C$_6$H$_5$·	328.9	3.4
CHC–H	5.7	C$_6$F$_5$·	−547.7	−5.0

Source: Handbook of Chemistry and Physics, 82nd ed., CRC Press, Boca Raton, Florida.

Due to their low mass and high mobility, the electrons will respond faster than the ions to the electric forces of the driving field (perturbation from neutrality). The response to the perturbation (restoring forces) will be through oscillation. The frequency of these oscillations is called the "plasma frequency" or Langmuir frequency ω_p. This temporal plasma parameter depends on the plasma density (n_e), mass of the electron (m_e), the permittivity of vacuum, and the unit charge (e):

$$\omega_p = \sqrt{\frac{e^2 n_e}{m_e \varepsilon_o}}$$

That means that for plasma densities in the range of 10^9–10^{12} cm^{-3} the plasma frequency is in the range of 90 MHz to 90 GHz.

The response of charged particles to reduce the effect of local electric fields is called the Debye shielding, and this shielding controls the quasi-neutrality of the plasma. Deviations from the global quasi-neutrality are possible only locally in a small confined environment recognized as the "Debye sphere." The "Debye length" (λ_D) is the radius of the Debye sphere and it can be defined as the characteristic dimensions of regions in plasma in which the neutrality rule can be violated.

$$\lambda_D = \sqrt{\frac{kT_e \varepsilon_o}{n_e e}} = 743 \left[\frac{T_e(eV)}{n_e(cm^{-3})} \right]^{\frac{1}{2}}$$

It can be shown for instance that, for average electron temperatures of 1 eV and plasma densities of 10^{10} cm^{-3}, λ_D = 74 μm.

There are three conditions that should be satisfied for the existence of the plasma state: (1) The linear dimensions of a plasma (L) should be much larger in comparison to the Debye length, a requirement for the existence of quasi-neutrality; (2) in order to create a sheath around an extra charge immersed in the plasma or near the walls (a collective phenomenon), the number of electrons in the Debye sphere (N_D) should be large ($N_D \gg 1$); and (3) the product between the perturbation frequency (ω) and the mean time between the collision of charged particles with neutral species (τ) should be larger than 1 ($-\omega\tau > 1$), because if the charged particles collide too frequently with neutral species, their motion is controlled by hydrodynamic forces rather than electromagnetic forces.

Most of the plasma applications are directed for deposition and surface-modification processes, and, accordingly, what happens at a surface immersed in plasma (e.g., plasma reactor walls, substrates and substrate-holder assemblies, diagnostic tool-"sensors", etc.) are the most significant processes that help us to understand the mechanisms of plasma-enhanced reactions. Charged particles (electrons and ions of either polarity) reaching solid surfaces that confine the discharge recombine and are lost from the bulk of plasma. Electrons with higher thermal velocities than those of ions reach the surfaces faster and leave the plasma with a positive charge in the close vicinity of the surfaces. The resulting electric field developed, as a result of the nascent negative surface and the positive plasma at the surface boundary layers, will render the net current zero. The surface will be at a negative self bias relative to plasma. As a result of the Debye shielding effect, the potential developed between the surface and the plasma is confined to layers of thickness of several Debye lengths. This layer of positive space charge "following" the surface is called the plasma sheath. The potential developed across the plasma sheath is recognized as the sheath potential V_s, and it adjusts itself in such a way that the flux of electrons is always equal to the flux of ions:

$$V_s = \frac{kT_e}{2e} \ln\left(\frac{m_e}{2.3m_i}\right)$$

for planar surfaces. The thickness of the plasma sheath d_s is defined as the thickness of the layer where the density of the electrons is negligible and where the V_s potential develops.

It is noteworthy that the degrees of ionization, α, of non-equilibrium plasmas (cold plasmas) are very low. α is defined as the fraction of the gaseous phase plasma particles that are ionized.

$$\alpha = \frac{n_i}{n}$$

where n_i denotes the density of ions or electrons, n stands for $n_e(n_i) + n_g$, and n_g is the density of neutral particles. In the case of low pressure, non-equilibrium discharges, typical degrees of ionizations are between the limits of 10^{-6} to 10^{-3}. That means that only one in the limits of every one million or one thousand neutral particles is ionized.

This is an extremely important conclusion often ignored in plasma chemistry research. It indicates that the free radical species and excited atomic and molecular species, generated as a result of plasma-enhanced molecular fragmentation processes, involving electrons with energies lower than the ionization energies, are the main "players" in the reaction mechanisms of low-pressure, non-equilibrium plasmas, where additional energy (such as electric bias, magnetic field) input is not used.

16.3.3 Atmospheric Pressure Cold Plasma

Most of the previous research related to the plasma-enhanced synthesis and surface modification of materials (deposition, surface functionalization, and etching) has been performed under low-pressure RF-plasma environments due to the high efficiency of RF discharges. However, in these cases the initiation and sustaining of these discharges require complex and expensive vacuum systems, which usually are associated with low-productivity batch-type processes. The use of these technologies is justified only when high-value-added materials are created, when the plasma is the only method suitable, or when benign processing environments are required. To take advantage of the plasma-generated active species for the development of efficient and often novel chemical processes, and still avoid vacuum conditions, the use of atmospheric-pressure discharges was considered as an alternative.

Non-equilibrium atmospheric pressure discharges are often recognized as partial discharges (PD) even if they can operate in a wide range of temperature and pressure (Hollahan and Bell 1974; Goldman et al. 1985; Eliasson et al. 1987; Kreuger et al. 1993; Eliasson et al. 1994; Van Brunt 1994; Boeuf and Pitchford 1996; Gentile and Kushner 1996; Kogelschatz et al. 1997). PDs are localized or confined electrical discharges, and often exhibit a non-stationary character (an unpredictable transition between different plasma modes). PDs are gas discharges, which may or may not occur in the presence of a solid or liquid dielectric. Due to the high diversity of electrode configurations, reactor geometries, nature and dimensions of the dielectric materials separating the electrodes, and the nature of the materials of the electrodes and reactor walls, these discharges are very complex phenomena, and manifest different modes (patterns). Accordingly, dielectric barrier discharges (DBD), corona discharges, constricted glows, electron avalanches, localized Townsend discharges, and streamers are considered as distinctive PDs.

PDs are cold plasmas (non-equilibrium discharges) which have mean electron energies considerably higher than the energies of charged and neutral molecular species (ions, radicals, and atoms). Just like low-pressure cold plasmas, these discharges will not generate extensive heating in their surroundings, and as a result they are also suitable for the processing of organic compounds and for surface modification of organic materials (e.g., polymers). The most representative PDs are dielectric barrier discharges and coronas.

The DBD was invented by W. Siemens in 1857, and it was used for the generation of ozone. Later on, the phenomenon was extensively studied, and it was demonstrated that, in a plane-parallel gap with insulated electrodes, the discharge occurs in a number of individual filamentary breakdown channels. It was shown that the plasma parameters of the breakdown channels (micro-discharges) can be controlled and modeled, and, consequently, the DBD process can be optimized for applications. Barrier discharge installations have various electrode configurations and are characterized by the presence of one or more solid dielectric layers (e.g., glass, quartz, ceramic) located between the metal electrodes in addition to the gap. The gap between the electrodes (including the dielectric) can range from 100 µm to several centimeters. In atmospheric pressure environments a distance of a few mm between the electrodes is common under 10 kV AC conditions. Multiple arrangements of the electrode systems are also common. Joint and non-joint electrode configurations are possible. When toxic gas media represent the discharge environment, the electrode systems are configured in closed chambers or adequate exhaust systems are used.

A large amount of individual filamentary discharges spread at the dielectric surface into surface discharges, which are significantly larger than the original channel diameter, creating charge centers on the surface of the insulator. The dielectric barrier controls the amount of charge and energy content of the micro-discharges and distributes the micro-discharges over the entire electrode surface. During the voltage increase period, new micro-discharges will be generated, which will be "struck" on the dielectric surface at locations different from the location of the former arclets, owing to the fact that the electric field is diminished by the presence of residual charges at the geometrical positions where the micro-discharges already occurred. During the reverse voltage period, new micro-discharges will form at the old micro-discharge locations. As a result, high-voltage, low-frequency conditions tend to spread the arclets, while low-voltage, high-frequency environments regenerate the micro-discharges at "old" locations. Comparative characteristics of BDs and low-pressure RF discharges are presented in Table 16.2.

DBD discharges are initiated when the breakdown field is reached and are extinguished when electron attachment and recombination processes reduce the plasma conductivity. Interruption of the micro-discharges occurs at field values slightly lower than the breakdown field, which is generated by the charge build-up at the plasma filament locations. The filamentary micro-discharges

TABLE 16.2
Comparative Characteristics of DBDs and Low-Pressure RF Discharges

Characteristics	Dielectric-Barrier Discharge	Low-Pressure RF Discharge
Duration	1–10 ns	permanent
Filament radius	0.1 mm	—
Peak current	0.1 A	—
Current density	100–1000 A/cm^2	—
Pressure range	usually 1 atm	10 mT–1000 mT
Total charge	0.1–1 nC	—
Electron density	10^{14}–10^{15} cm^{-3}	10^9–10^{10} cm^{-3}
Electron energy	1–10 eV	0.5–20 eV
Gas temperature	close to ambient	close to ambient
Driving-field frequency	10 kHz–10 MHz	40 kHz–100 MHz
Operational mode	open or closed	closed
Plasma sheath potential	—	around 20 V

can be considered as weakly ionized plasma channels. The transported charge is proportional to the gap spacing and permittivity of the dielectric but does not depend on the pressure, and over a wide range of frequencies and voltage characteristics, micro-discharge properties do not depend on the external driving circuitry. They are controlled by the gas properties, pressure, and the electrode configuration instead. However, at very high frequencies (very fast rising voltages) the dielectric surface cannot absorb the energy of all the micro-discharges, and, as a consequence, a weaker micro-discharge system results. Synchronization of micro-discharge events by adapting a pulsed-field mode can minimize the over-discharge phenomenon.

The most significant difference between the electron-impact processes of conventional RF plasmas and DBDs is that, at atmospheric pressure, the driving field-accelerated electrons undergo an extremely high number of inelastic collisions with neutral and charged gas particles, and, as a result, they will reach equilibrium values within about 10 ps, much before any significant ns-range, electric-field changes could appear. In atmospheric pressure conditions, excited species and free radicals, generated by electron collision processes, will undergo de-excitation and recombination processes much faster (1–100-μs range) than the time required for the removal of active species from the discharge zone by diffusion and convection processes (millisecond time intervals). As a result, free radical-mediated gas-phase and surface reactions will be initiated and sustained.

Coronas are considered gas discharges where electrode geometries control and confine the ionization processes of gases in a high-field ionization environment, in the absence of insulating surfaces or when the dielectric surfaces are far removed from the discharge zone. Corona discharges are often called negative, positive, bipolar, AC, DC, or high-frequency (HF), according to the polarity of the stressed electrodes, to whether one or both positive and negative ions are involved in the current conduction, and to the nature of the driving field. What makes corona discharges unique in comparison to other plasmas is the presence of a large low-field drift region located between the ionization zone and the passive (low-field) electrode. Ions and electrons penetrating the drift space will undergo neutralization, excitation, and recombination reactions involving both electrons and neutral and charged molecular and atomic species. However, owing to the multiple inelastic collision processes in atmospheric pressure environments, the charged active species escaping from the ionization zone (electrons, ions) will have energies lower than the ionization energies, and, as a consequence, neutral chemistry (free radical chemistry) will characterize the drift region. According to various electrode configurations point-to-plane, wire-cylinder, and wire-to-plane corona discharges can be identified.

16.4 COLD PLASMA MODIFICATION OF WOOD

Wood is an "ideal" natural composite material due to its high physical strength, great machinability, aesthetic appeal, and low price. At the same time, wood exhibits characteristics that limit considerably its use in several application areas. Wood-based objects and construction items, including some of the composite materials, are biodegradable, have a low dimensional stability, demonstrate a physical behavior that is moisture content-dependent, and undergo photochemical or chemical degradation in the presence of UV radiation and acid and base environments. To ensure a lasting value, wood materials are usually coated with various protective and decorative/protective coatings. Due to the inherent water content of wood and to the presence of hydrophilic functionalities, such as OH groups, in its predominantly cellulose composite structure, water- and organic solvent-based surface treatments applied for controlling its moisture content are not always adequate. Due to the water-solubility of some of the coating components and to the dissimilarity (incompatibility) existing between the wood-constituting macromolecules and the synthetic polymers considered for the surface treatment, conventional wet-chemical surface-modification approaches are often not effective.

Non-equilibrium plasma conditions open up a novel way for surface modification of wood-based materials, including surface functionalization and coating. Cold-plasma technology being a very efficient, dry chemistry approach, practically avoids disadvantages related to use of large amounts of usually toxic solvents and solutions.

A 2.450-GHz microwave plasma-enhanced surface treatment of scots pine wood samples (155 × 74 × 18 mm), with an initial 12% moisture content, was carried out in gaseous- and liquid-phase environments (Podgorski et al. 2001). The plasma installation consists of a cylindrical-shape vessel connected to a quartz tube, where the microwave plasma is generated, a microwave power supply, a rotary and diffusion pumps-based vacuum installation, and a gas/vapor supply system (Figure 16.2). The gas/vapor supply assembly consists of five gas lines and a line that allows the supply of liquid-phase materials. A wide range of plasma-gas environments were used for various purposes, including cleaning (Ar, He, H_2), oxidation (air, O_2, CO_2), nitriding (N_2, NH_3), fluorination (CF_4, fluorine-containing monomers, such as C_3F_6 and acrylate with a C_8F_{17} chain), and deposition (acetylene, propylene, hexamethyldisiloxane). Liquid-phase materials were also *ex situ* sprayed

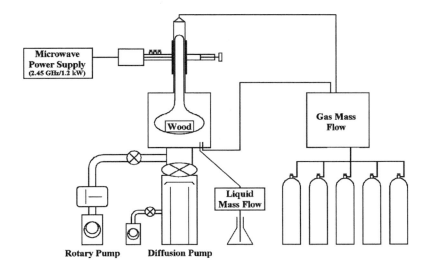

FIGURE 16.2 Schematic of the equipment used to modify wood surfaces by plasma polymerization (Podgorski et al. 2001).

onto selected substrate surfaces and plasma-treated consecutively. An experimental design was used to evaluate the influence of experimental parameters, such as the nature of the plasma gases, geometrical location of the gas/vapor supply connections, gas and liquid flow rates (0–100 cm^3/min), treatment time (1–15 min), distance between the samples and the microwave antenna (37–57 cm), and microwave power dissipated to the antenna (700–1100 W) plasma-enhanced surface modification processes. The authors indicate that both fluorine-containing compounds and hexamethyldisiloxane (HMDSO) generate, under plasma conditions, wood surfaces with high contact angle values (76 to 120 degrees), and that in most of the cases plasma treatments using fluorinated structures are more effective in comparison to those performed under HMDSO-plasma environments. Improvement of coating adhesion using plasma treatment is also suggested (Podgorski and Roux 1999).

Unfortunately, the authors do not present any analytical information related to the chemical nature of the plasma-generated surface layers. The only characteristic they evaluated was the contact angle. Nothing is mentioned of the thickness of the deposited layers, surface topography changes as a result of the plasma exposure, and the stability in time of contact angle values. There were no data presented indicating whether the plasma-generated structures were covalently bound or if they were just deposited onto the substrate surfaces. In the absence of this information it is very difficult to evaluate the significance of these findings.

Pine wood samples (50 × 10 × 2 mm, and 70 × 30 × 20 mm) were exposed to 125–375 kHz CF_4, C_2F_4, and HMDSO plasma reactor environments, using a tubular plasma reactor (Cho and Sjöblom 1990). It was demonstrated that the efficiency of the hydrophobic modification is dependent on the pressure and is controlled by the evacuation time and the physical dimension of the substrates. Water vapor and/or low-molecular-weight compounds present in wood migrate out of the lignocellulosic matrix under low-pressure conditions and limit significantly the penetration of plasma species that are the precursor building blocks of the hydrophobic structures. The plasma treatments significantly diminished the water uptake of wood substrates. At base pressures lower than 18 mTorr the water uptake after two hours was only half of the saturation value, while unmodified samples saturated within 30 minutes. Samples plasma-treated after a 30-minute evacuation exhibited a significantly reduced capillary rise of water (transport rate of water in the direction of wood growth). C_2F_4 plasma reduced the water uptake more than CF_4 and HMDSO discharges when the treatments were performed after a long evacuation period. Extensive (less practical) evacuation times required by HMDSO plasma treatments could be avoided by increasing the HMDSO pressure in the reactor. Evaluation of the adhesion characteristics of a water-borne paint to unmodified and HMDSO-modified wood samples indicated the presence of a poor adhesion. However, the authors' findings show that an additional, consecutive acrylic acid- or oxygen-plasma treatment of HMDSO-modified substrates significantly improve the paint adhesion. Surface morphology, porosity data, information related to the plasma-induced chemical changes on the surface of the substrates, and the chemical nature of the deposited layers, which might be responsible for the water uptake behavior, are not discussed by the authors. In the absence of this information the conclusions drawn by the authors have only a speculative character.

Surface characteristics of wood and cellulose were modified in radio frequency (RF) and atmospheric-pressure plasma conditions (Setoyama 1996). Wood and cellulose substrates (50 × 50 mm) were exposed to plasma environments in a 250-mm-diameter and 650-mm-height bell-jar-type RF reactor provided with parallel-plate electrodes (100 mm gap) and with a substrate holder located at 50 mm between the electrodes. Atmospheric-pressure discharge treatments were performed in a 220-mm-diameter and 500-mm-height bell-jar chamber, equipped with parallel plate, aluminum electrodes with a gap between them of 10 to 20 mm. The upper electrode was covered with a thin glass plate and the lower electrode was also used as the silicon wafer substrate holder. The plasma was initiated and sustained using a high-voltage generator with a frequency range of 3 to 20 kHz. It was shown that stable atmospheric-pressure discharge environments can be achieved in the following experimental conditions: presence of an insulating plate between the electrodes, helium as diluting component in the processing-gas mixture, and driving frequencies higher than 1 kHz. The materials used as substrates were samples of 0.03-mm-thick Hinoki (Japanese cypress) lamina, 0.1-mm-thick Sugi (Japanese cedar) sheets, 34 × 100 × 6–7-mm bamboo specimens, and 0.3-mm-thick cellulose (cellophane). Oxygen, nitrogen, air, argon, and CF_4 were considered as plasma-gas-mixture components.

Oxygen plasma was found to more intensely ablate wood and cellulose substrates, while argon, nitrogen, and CF_4 plasma species resulted in significantly lower weight losses. Based on electron spin resonance (ESR) analysis, the nature of plasma-generated free radicals was suggested. It was indicated that in plasma-treated wood most of the free radicals were produced in the lignin structure. However, the quality of the ESR spectra presented by the author and the absence of data resulting from deconvolution leave more space for investigations in this area. Oxygen-, nitrogen-, and air-plasma exposure of wood samples generated hydrophilic surfaces and enhanced adhesion of varnish of plasma-modified substrates relative to unmodified samples. CF_4 discharges resulted in good water repellency; however, the water-absorption characteristics were not changed notably even if ESCA data indicated the formation of a fluorine-rich surface, as a result of CF_4-plasma treatment. The high porosity of the substrates and an incomplete plasma coating might be responsible for this phenomenon. Differences between the low- and atmospheric-pressure, plasma-induced reaction mechanisms were not discussed.

Solid pine wood was coated with macromolecular layers resulting from ethylene, acetylene, vinyl acetate, and butene-1 plasma treatments (Esteves Magalhães and Ferreira de Souza 2000, 2002). Defect-free softwood samples (2.0 × 2.0 × 1.0 cm) that were cut in radial, tangential, and longitudinal directions and oven-dried at 100°C for 12 hours were plasma-treated in a cylindrical stainless steel, parallel-plate reactor under the following experimental conditions: base pressure in the reactor: 40 mTorr; number of argon purging cycles: 3; working gas pressure: 400 mTorr; power dissipated to the electrodes: 10 W; driving field frequency: 60 Hz; and treatment time 30 minutes. Softwood surface treated by "plasma-polymer" deposition exhibited a different behavior depending on the chemical nature of the plasma gases involved. Contact angle values collected from cross-sectional surfaces (plane transversal to the fibers) of virgin and plasma-treated samples indicated that ethylene-, acetylene-, and butene-1-plasma modifications generate highly hydrophobic surfaces and that contact angle values as high as 140 degrees can be achieved in the case of butene-1 plasma exposures. However, water absorption values resulting from absorption from humid air indicated little change of the hygroscopic character of wood substrates. Results from water vapor uptake by plasma-treated wood suggested that even the most hydrophobic films did not coat the pores and capillary wells of wood. It was concluded that continuous thick-film depositions are required for enhanced protective coatings.

To improve the dimensional stability of wood products, two different low- and corona-plasma-treatments were considered (Podgorski et al. 2000; Podgorki and Roux 2000). First, the wettability of wood (fir, curupixa, pine, lauan-meranti) surfaces was increased in order to enhance the adhesion of the coating; however, no improvement in the coating adhesion was observed. Second, the wettability was decreased by the deposition of coatings from ethylene- and fluorine-based plasma environments. The authors do not present sufficient details related to the plasma treatment. Except for the power range (400–1299 W), treatment time (1–30 minutes), and gas pressure (0.08 mbar), no information is supplied on the CW or pulsed nature of the discharge, frequency of the driving field, geometry and geometrical location of the electrodes, etc. As a consequence it is difficult to judge the relevance of the experiments. Increased wettability of chemically and thermally treated wood samples was achieved under oxygen, air, nitrogen, argon, and carbon dioxide plasma environment or in mixtures of these gases with oxygen. Surprisingly, ammonia plasma resulted in high contact angle values; however, ammonia/oxygen plasma also increased the wettability of wood substrates. It was shown that the power dissipated to the electrodes, treatment time, and the geometrical location of the samples relative to the plasma source have a significant influence on the plasma-induced wettability. Higher powers and longer treatment times resulted in lower contact angle values, while shorter distances between the substrates and the plasma source generated the lowest contact angles. Torrefied (230°C; 60 minutes) wood samples were also plasma-treated. The effects of plasma treatment were more significant on the torrefied substrates in comparison to the non-torrefied samples. Contact angle measurements performed before the plasma treatment, and 7 and 15 days afterward, clearly indicated the efficacy of the plasma process for decreasing surface contact angle values. Results from low-pressure and corona-plasma treatment of wood were compared using air- and nitrogen-discharge environments. The treatment efficiency of corona exposure was lower, even at the highest voltage values (15 kV), relative to the efficiency of the low-pressure plasma modifications. Adhesion of two different solvent- and water-based coatings was evaluated on untreated, as well as oxygen-, carbon dioxide- and air-plasma-treated *Pinus sylvestris* and lauan-meranti (some samples of lauan-meranti were torrefied) samples. The authors did not observe any improvement in the coating characteristics of the plasma-treated specimens. A significant decrease in wettability was noted in the case of CF_4/C_3F_6 low-pressure plasma-exposed pine samples. The potential application of plasma-enhanced surface modification of wood objects was emphasized.

Southern yellow pine wood surfaces ($5 \times 1.5 \times 0.1$ cm) were coated with hydrophobic layers using hexamethyldisiloxane (HMDSO)-RF plasma treatments (Denes et al. 1999). A toluene/ethanol, 8-hour Soxhlet-extraction process was used to remove extractives from wood surfaces. All depositions were performed in a stainless steel, parallel-plate RF plasma reactor (Denes and Young 1999) using the following external plasma-parameter space: base pressure in the reactor chamber: 50 mTorr; pressure in the presence of plasma: 300 mTorr; RF-power dissipated to the electrodes: 150 to 250 W; flow rate of HMDSO vapor: 5.5 sccm; and plasma-exposure time: 1.5 to 10 minutes. ESCA and ATR-FTIR data revealed the presence of macromolecular layers based on Si-C and Si-O-Si bonds on plasma-treated wood surfaces. Molecular fragmentation analysis of the coating of wood using pyrolysis-assisted GC-MS indicated highly cross-linked plasma layers. It was emphasized that short treatment times were enough for the development of efficient surface-coating processes, and that the RF power has a significant effect both on the plasma-induced surface chemistry and on the characteristics of the resulting macromolecular layers. Plasma-modified substrates exhibited very high contact angle values (e.g., 130 degrees), and AFM and DTA/TG evaluation showed a progressive growth of a smooth topography, thermally stable "plasma polymer" on wood substrate surfaces.

In order to avoid the negative influence of surface porosity of wood on the achievement of a complete surface coating, an original plasma-enhanced process for the coating of wood was developed, which resulted in diminished degradation of wood under simulated harsh environmental conditions (Denes and Young 1999). Reflective zinc oxide and electromagnetic radiation (EMR)-absorbent (benzotriazole, 2-hydroxybenzophenone, phtalocianine, and graphite) substances were incorporated into liquid-phase, high-molecular-weight polydimethylsiloxane (PDMSO) and deposited as thin layers onto wood substrate surfaces. The polymer layers containing the dispersed ingredients were converted in the next step into solid-state networks, in a 30-kHz capacitively coupled RF plasma reactor (Figure 16.3), using the following experimental parameters: base pressure: 40–50 mTorr; plasma gas: oxygen; oxygen flow rate: 4.5–5 sscm; pressure in the reaction chamber: 250 mTorr; RF-power dissipated to the electrodes: 250 W; plasma treatment time: 10 minutes; concentration of dispersed materials in the 200,000 mCP PDMSO: 8%; wood substrates: $3 \times 1.5 \times 0.6$ cm southern yellow pine; and amount of PDMSO containing the ingredient materials, deposited onto wood samples: 0.05 g. Weathering behavior of untreated and plasma-coated wood substrates was evaluated according to the ASTM G23, and D1499 standards. The coated surfaces of samples were exposed for 48–336 hours to a 6500 W Xenon arc light source at 45–50°C. Some of the samples were also exposed to 102 minutes of irradiation, followed by 18 minutes of light and water spray. Specimens for weight loss evaluations were vacuum-oven-dried for 24 hours at 105°C. Oxygen-plasma-induced exposure of coated surfaces generally resulted in surface layers that diminished the weathering degradation of wood. ESCA, ATR-FTIR, and SEM results indicated

Plasma Treatment of Wood

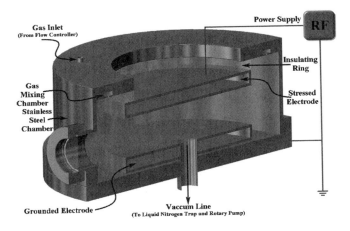

FIGURE 16.3 Schematic of the RF-cold plasma reactor used to reduce weathering degradation of wood (Denes and Young 1999).

less oxidation of the coated substrates and the presence of a smoother surface as a result of the 2-week weathering. It was shown that the dominant chemical bonds (Si-C, Si-O, and C-O) of the plasma-generated matrix-polymer were not degraded by the action of UV radiation in the presence of water. Less weight loss and less degradation of the plasma-treated films were recorded in comparison to unmodified wood, with the exception of the PDMSO- and 2-benzotriazole/PDMSO-coated and plasma-exposed samples (Figure 16.4). It was suggested that the photocatalytic behavior of entrapped 2-benzotriazol molecules might be responsible for this phenomenon. High-water-contact-angle values recorded for both plasma-treated and plasma-treated-and-weathered substrates demonstrated the hydrophobic character of the modified substrates. Generation of a smooth, more compact surface layer during the weathering process could explain the existence of the highest

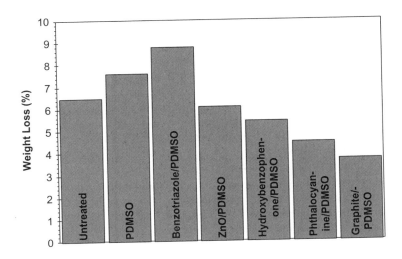

FIGURE 16.4 Percent weight loss of plasma-treated coated wood for the reduction of weathering degradation (Denes and Young 1999).

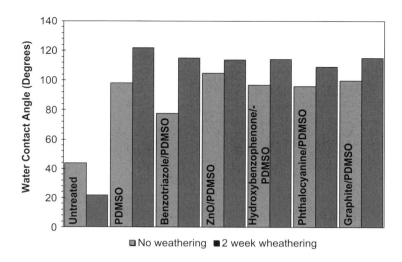

FIGURE 16.5 Contact angle of water on plasma-treated reflective and EMR-containing coatings on wood (Denes and Young 1999).

contact angle values (Figure 16.5). It was emphasized that plasma treatment of wood could become a viable alternative for weathering protection of wood objects.

The effect of oxygen plasma treatment on the surface characteristics of polypropylene (PP) was investigated, and the influence on the adhesion of PP to HMDSO-plasma-treated wood was evaluated (Mahlberg et al. 1998). All plasma treatments were carried out in a cylindrical, stainless steel 40-kHz RF parallel-plate reaction chamber, equipped with disc-shaped stainless steel electrodes. The lower electrode served also as the substrate holder, and it was provided with electrical heating capabilities. Prior to the plasma exposure, decontamination of the reactor was performed using oxygen plasma (250 mTorr, 200 W, 10 minutes at a substrate-holder temperature of 200°C). The oxygen-plasma treatment was followed by argon plasma under similar conditions, then the reactor was cooled down and evacuated to base-pressure level. For oxygen-plasma treatments, PP samples were positioned on the substrate-holder electrode, then the chamber was evacuated to base pressure. The oxygen-plasma environments were created under the following experimental conditions: pressure of oxygen: varied between 250–300 mTorr; RF power dissipated to the electrodes: 60 and 100 W; and treatment time: varied in the range of 15 seconds to 90 seconds. Deposition onto birch veneer surfaces of HMDSO-based macromolecular layers was carried out at a 6.3-sccm HMDSO flow rate, 200 mTorr pressure, 2.5- and 8-minute treatment time, and 200 W RF-power.

Contact angle measurements performed on PP surfaces using deionized water indicated that treatment times as short as 15 seconds decrease the contact angle values from approximately 90° for untreated PP to approximately 65°. The contact angle decreased gradually with the plasma exposure and reached its lowest value of 55° after 60-second surface modification. A similar declining trend of contact angle with plasma-treatment time has also been noted for birch surfaces. Thirty second oxygen-plasma exposure reduced the contact angle of wood substrates from 60° to 45° (Figure 16.6). It was shown that the oxygen-plasma treatment of birch surfaces increases the polar component of the surface energy. The increase of the total surface energy of wood due to the plasma-enhanced surface modification was solely related to the increase of the polar component (Table 16.3).

HMDSO-plasma environments generated at relatively long treatment-time values resulted in a significant hydrophobization of wood sample surfaces. Two-minute plasma exposure was not enough to make the birch surface hydrophobic. However, the treatment for 5 minutes increased the contact angle values from 60° for unmodified substrates to 95–135°.

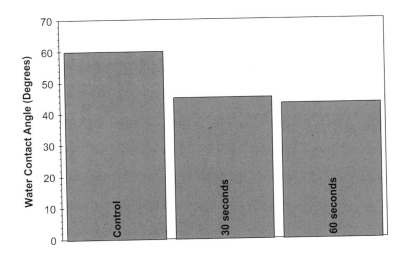

FIGURE 16.6 Contact angle of deionized water on birch surface treated for different times in an oxygen plasma at 100 W (Mahlberg et al. 1998).

Results from evaluation of bonding strength between PP film and birch veneer pre-treated with HMDSO-plasma indicated that the HMDSO plasma-layer present on wood surfaces does not notably improve the adhesion characteristics. AFM studies of HMDSO-plasma-treated wood surfaces indicated that the HMDSO-based macromolecular layer follows the features of the substrate without forming an actual film on the surface. The pinhole-containing topographies of the HMDSO-plasma-coated birch surfaces were emphasized (Figure 16.7). Atomic force microscopy was also used to study surfaces of PP, kraft pulp, filter paper, and wood (Mahlberg et al. 1999). The adhesion properties between PP film and wood surfaces were investigated using the peel test method. The highest adhesion to wood resulted from the shortest plasma-treatment times, and the best adhesion characteristics were recorded when both the PP and wood surfaces were plasma-modified (Table 16.4). AFM images showed that all plasma-treated substrates exposed a nodular surface morphology (Figure 16.8). It was suggested that these nodules have a poor interaction between surfaces that could result in the generation of a weak interface between PP and wood.

TABLE 16.3
Total Surface Energy, Dispersion, and Polar Components of the Surface Energy for Untreated and Treated PP and Birch

Surface	γ_s^p (mJ/m²)	γ_s^d (mJ/m²)	γ_s (mJ/m²)
Untreated PP	1.2	27.6	28.8
PP treated for 30 s	19.4	32.3	51.7
PP treated for 60 s	25.4	26.3	51.7
Untreated birch	12.2	39.8	52.0
Birch treated for 30 s	39.9	24.9	64.8
Birch treated for 60 s	36.9	28.2	65.1

Source: Mahlberg et al. 1998.

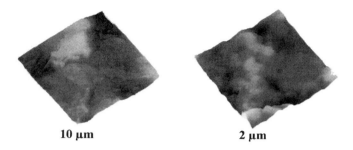

10 µm 2 µm

FIGURE 16.7 AFM image of wood surface coated by plasma polymerization with HMDSO (Mahlberg et al. 1998).

Oak samples treated in oxygen-, helium-, nitrogen-, CO_2-, and tetrafluoromethane-plasma environments (power: 50–70 W; pressure: 1–2 Torr; treatment time: 5, 15, and 30 minutes) led to decreased distilled water contact-angle values (Legeay et al. 1986). Sticking tests carried out using polyvinylacetate emulsion indicated that, except for CF_4-plasma exposure, most of the plasma treatments resulted in an increase of the rupture stress. The adhesive behavior of two different varnishes (glycerophtalic and polyurethane-monoconstituents) was also studied for four varieties of wood samples. Oxygen-plasma treatment increased the adhesion for the glycerophtalic varnish, while tests in the presence of polyurethane varnish exhibited decreased adhesion characteristics.

Extensive structural damage was noticed on cellulose and wood material surfaces during oxygen-plasma treatment in comparison to nitrogen-, argon-, and CF_4-plasma modifications (Harada et al. 1994). It was shown that oxygen-, nitrogen-, and air-plasma conditions render wood and cellulose surfaces with hydrophilic properties. Water repellence of wood surfaces was achieved by modifying the substrate surfaces using tetrafluoroethylene- and tetrafuoromethane-plasma environments. However, the moisture absorption of treated wood samples was not found to be different

TABLE 16.4
Effect of Oxygen Plasma Treatment on the Peel Strength between Wood and PP[a]

Specimen	Treatment Time (s)	Bonding Strength (N/mm) Max	Min	Wood Failure (%)
Spruce				
Untreated spruce + untreated PP		0.20 ± 0.02	0.12 ± 0.02	≤10
Untreated spruce + treated PP	15	0.22 ± 0.05	0.16 ± 0.03	30–40
Treated spruce + treated PP	15	0.27 ± 0.03	0.20 ± 0.04	40–100
Untreated spruce + treated PP	30	0.20 ± 0.11	0.13 ± 0.12	10–40
Treated spruce + treated PP	30	0.22 ± 0.03	0.14 ± 0.03	30–100
Untreated spruce + treated PP	60	0.22 ± 0.06	0.18 ± 0.07	20–30
Treated spruce + treated PP	60	0.24 ± 0.03	0.19 ± 0.03	20–40
Birch				
Untreated birch + untreated PP		0.21 ± 0.02	0.13 ± 0.05	≤10
Treated birch + treated PP	15	0.41 ± 0.02	0.29 ± 0.02	30–70
Treated birch + treated PP	30	0.23 ± 0.03	0.19 ± 0.06	20–40
Treated birch + treated PP	60	0.29 ± 0.02	0.22 ± 0.04	20–50

[a] Treatments were carried out at 100 W and 200–250 mT for spruce and birch, and 200–225 mT for PP.

Source: Mahlberg et al. 1999.

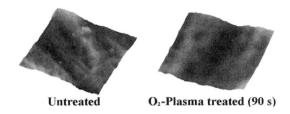

Untreated **O₂-Plasma treated (90 s)**

FIGURE 16.8 AFM image of untreated and oxygen plasma-treated birch wood (Mahlberg et al. 1999).

from that of virgin substrates. ESCA data indicated the presence on plasma-modified wood sample surfaces of CF_2-CF_2- and -CF_3 functionalities. Unfortunately, the authors do not present in their publication the data and experimental parameters that were used during the plasma treatments.

White fir and Douglas fir heartwood and sapwood samples (1.5 cm in diameter and 1.2 cm long) were exposed to oxygen-, nitrogen-, and helium-RF plasmas, and the altered wood permeability was evaluated. The experiments were carried out using an LFE Model LTA-604 low-temperature asher operating at 13.56 MHz and in the range of 0 to 600 W, and provided with a 10-cm-diameter and 21.5-cm-long cylindrical reaction chamber (Chen and Zavarin 1990). Oxygen plasma generated a higher permeability increase, followed by nitrogen and helium plasma treatments (Table 16.5). In all cases permeability values increased with plasma exposure time without a leveling-off behavior. It was emphasized that Douglas fir did not behave like white fir, showing negligible effect of the treatment time on the increase of permeability. Model substrates (discs composed of compressed cellulose, xylan, and lignin powder) were also used to investigate the nitrogen- and oxygen-plasma-mediated rates of ablation of individual wood constituents. Weight-loss results indicated that cellulose and xylan ablate greater than two times more than lignin does. Increased RF energy dissipated to the electrodes increased the permeability values, while an increase in the plasma-gas flow rates had an opposite effect. Extractives present in the wood samples suppressed the plasma-induced permeability increase, and, therefore, removal of these compounds by water and ethanol

TABLE 16.5
Effect of Cold RF Plasma on Permeability of White Fir and Douglas Fir Wood

Wood	Plasma	Gas Flow (mL/min)	Permeability Increase[a]	Number of Samples
White Fir, Sap	N_2	10	1.48	5
White Fir, Heart	N_2	10	1.63	5
White Fir, Sap	O_2	10	1.67	4
White Fir, Heart	O_2	10	2.22	4
White Fir, Heart	He	10	1.44	4
White Fir, Sap	N_2	50	1.41	4
White Fir, Heart	N_2	50	1.41	4
White Fir, Heart	O_2	50	1.81	4
White Fir, Heart	He	50	1.50	4
Douglas-Fir, Sap	N_2	10	2.05	8
Douglas-Fir, Heart	N_2	10	2.04	4
Douglas-Fir, Sap	O_2	10	3.13	4
Douglas-Fir, Heart	O_2	10	8.00	16

[a] K/K_o, where K = permeability after 30 min with plasma at 300 W and K_o = original permeability.

Source: Chen and Zavarin 1990.

extractions, followed by oxygen-plasma exposure, significantly increased the permeability values. The permeability of extracted and plasma-treated Douglas fir increased 32 times, while the permeability of the not-extracted and plasma-treated sample increased only eight times. The authors suggest that since the RF radiation penetrates wood, and because of the porous nature of the cell lumina, it would be expected that the plasma would form both inside and outside of the wood structure. This is probably not the case, given the very low cavity dimensions present in the wood structure and the difficult evacuation of the pores, which would inhibit the initiation and sustainability of the plasma state. The significance of plasma-increased permeability for an improved impregnation of wood (e.g., impregnation with commercial fungicides) was emphasized.

Nineteen species of tropical wood were treated in CF_4-RF plasma in a rotating, capacitively coupled, 1-m-long and 10-cm-diameter, cylindrical glass reactor provided with semicylindrical outside electrodes, and the resistance of modified wood samples against white rot (*Trametes versicolor*) was tested (Wistara et al. 2002). All wood samples were exposed to plasma under the following experimental conditions: base pressure: 30 mTorr; frequency of the driving field: 13.56 MHz; pressure in the absence of plasma: 150 mTorr; pressure in the presence of plasma: 230 mTorr; RF power dissipated to the electrodes: 100 W; and plasma gas: CF_4. The weight loss and after-feeding moisture content of control and treated woods were measured to determine the influence of CF_4 plasma on the resistance of modified wood samples. It was concluded that CF_4 plasma generated hydrophobic surfaces, can significantly increase the resistance of rengas, teak, kapur, kempas, duabanga, pasang, nangka, keruing, gmelina, mahogany, rubber wood, ulai, and African wood against decaying influence of white rot.

Wood surfaces were modified under microwave (MW) plasma environments. Nitrogen, oxygen, and air MW plasma treatments were carried out on dried wood substrates (30- to 100-μm thickness and 19- × 22-mm area) (Guanben et al. 2001). The plasma treatments were performed using a MW installation composed of a MW cavity, tubular quartz reactor, vacuum system, MW generator, MW wave-guide and MW power measuring system, and gas supply assembly. Unfortunately, the authors do not present experimental details related to the MW power, pressure in the system, etc., that was used for the surface-modification reaction. All plasma exposure times were 1.5 minutes. Plasma treatments decreased the contact angle values to zero even when the plasma modifications were carried out in mild conditions. SEM images indicated that a coarse surface structure is generated as a result of plasma exposure, and that oxygen-plasma environments induce the most significant ablation processes. An increased O/C relative atomic ratio of modified surfaces was noted by ESCA measurements in comparison to virgin substrates, and incorporation of nitrogen atoms into the plasma-generated surface layers was demonstrated. The authors suggest that MW plasma treatments increase both the hydroxyl and carbonyl functionalities of cellulose and/or lignin components.

Mica (60/100 mesh) and Aspen ground-wood fibers were MW-plasma-treated and considered as additives for extrusion-grade polypropylene (melt flow index: 1.5) composite preparations (Bialski et al. 1975). The particle and fiber surface-modification reactions were performed in a MW installation equipped with a rotating, tubular, quartz reaction chamber using the following experimental conditions: pressure in the reactor: 0.5 Torr; MW power: 1.5 kW; treatment time: 90 seconds; and plasma gases: argon, nitrogen, ethylene, sulfur-dioxide, ammonia, and a 1/1 mixture of ethylene and ammonia. Polypropylene-based composite materials were prepared using a Brabender Plasticorder operating at 190°C and 60 rpm. Composites of 10% and 20% solids content were prepared in the presence of 0.1% thermal stabilizer. Based on calorimetric measurements it was found that the surface properties of mica were significantly changed as a result of ammonia-, ethylene-, and ammonia/ethylene mixture-plasma exposure. Calorimetric evaluation of plasma-treated wood samples is more difficult due to the significant increase of immersional heat (H_i) as a result of loss of water. Tensile strength properties of composites indicated that there is a close correlation between the plasma-induced change of immersional heat and ultimate tensile strength. Ethylene-plasma treatments in particular seemed effective in increasing the mechanical strength of mica-based composites. A less

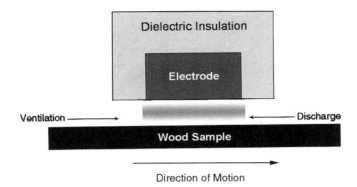

FIGURE 16.9 Schematic of the plasma set up used for the modification of wood under dielectric barrier discharges at atmospheric pressure (Rehn et al. 2003).

pronounced effect of plasma was noted in the case of wood fibers in comparison to mica. The authors suggest that a systematic study would be justified for optimizing the MW treatment for various fillers for the preparation of high-performance composites.

Atmospheric-pressure plasma treatments are very attractive approaches for the large-scale surface modification of various substrates, and for the design and development of continuous-flow-system, plasma-enhanced technologies that can be conveniently scaled up to pilot and industrial levels. Dielectric barrier discharges (DBD) and corona discharges were also recently tested for the generation of modified wood surfaces. A DBD electrode system was tested for surface treatment of wood surfaces (Rehn et al. 2003).

The geometrical configuration of the electrically insulated electrode and the wood sample surface is shown in Figure 16.9. A pulsed high-voltage (30 kV, pulse duration: 2µs; pulse repetition rate: 15 kHz) is applied to the electrode in order to generate the discharge in air between the stressed electrode and wood surface (counter electrode; 1.2- to 3-mm gap between the electrodes) at atmospheric pressure and 35°C gas temperature. Air was blown through the discharge gap with a velocity of approximately 3 m/min. Wood samples (5-mm-thick) with a moisture content between 10 to 15% were moved through the plasma zone at an adjustable velocity between 1.5 to 50 m/min. It was found that as a result of the plasma treatment, the fracture strength of glued robinia wood was increased 28%, and that the moisture content of wood does not have a significant influence on the final characteristics (Figures 16.10–16.11).

The wettability of plasma-treated wood samples was improved, and results from delaminating experiments indicated a 18% improvement when plasma-treated wood samples were involved in the test. According to the opinion of the authors, the short treatment time (1 second) and low energy consumption (0.1 kWh/m^2) of this approach could make this process economical.

Hydrophilic and hydrophobic wood sample surfaces were generated under DBD environments using a parallel-plate electrode configuration, where both electrodes were covered (insulated) with quartz plates (Rehn and Viöl 2003). Pine wood species with an average density of 470 Kg/m^3 were used in all studies. The wood samples were cut and planed into boards with the convex sides of the annual rings parallel to the test surface. Growth-ring width was between 2 and 3 mm, and the substrates were conditioned to an initial moisture content of 6 to 8%. The following experimental parameters were employed for the surface-modification reactions: pulsed high voltage applied to the electrodes; pulse repetition rate: 17 kHz; pulse duration: 2µs; high voltage amplitude: 30 kV with an alternating polarity; a pulse train of 1s followed by a pause of 1s, with this cycle repeated 1 to 60 times; pressure in the discharge zone: atmospheric pressure; plasma gases for hydrophilic surfaces: air, helium, nitrogen, and argon; and plasma gases for hydrophobic surfaces: methane, acetylene, argon/methane = 80/20. Best hydrophilic properties were achieved in air plasma.

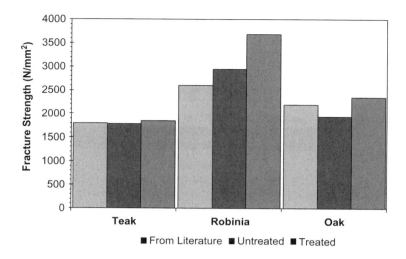

FIGURE 16.10 Fracture strength graph for untreated and DBD-treated woods (Rehn et al. 2003).

One- to 20-second plasma exposures significantly increased the wettability of wood surfaces; the water absorption also was improved 22 times, and the fracture strength of glued wood was increased 68%. Methane and acetylene plasmas generated hydrophobic wood surfaces; the water absorption of plasma-treated wood substrates decreased 32 times in comparison to unmodified substrates. The importance of these findings for potential applications of atmospheric plasma technologies for surface modification of wood was emphasized.

Enhanced performances for outdoor coatings of pine sapwood samples exposed to atmospheric-pressure plasma, corona, and fluorination environments were compared. All treatments increased the wet adhesion of coatings, which was related to the increasing polar components of the surface free energy (Lukowsky and Hora 2002). Unfortunately, the authors do not present sufficient details related to the surface-modification processes, so the comparative relevance of different surface-treatment techniques cannot be evaluated.

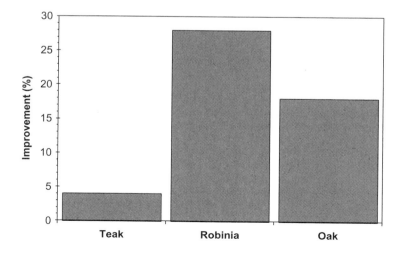

FIGURE 16.11 Fracture strength improvement in wood by DBD treatment (Rehn et al. 2003).

The influence of MW and corona treatments on the surface characteristics of synthetic- and natural-polymer surfaces and related composites, including wood, was reviewed (Goring 1976). The efficiency and importance of plasma approaches for paper and composite preparations were emphasized. The cost-effectiveness of plasma treatments was discussed.

Wood materials considered for the manufacture of stringed and pizzicato musical instruments (upper and lower sounding boards of the case of violin, viola, guitar, etc.) were plasma-treated (4–20 Pa; 0.4-3 mA/cm^2; 5–40 minutes) in acetylene, propane, and benzene glow-discharge environments (Gavrilenko et al. 2002). Hydrophobic properties and improved resonance characteristic of plasma-treated wood substrates were noticed.

Reduced formaldehyde content was generated in various wood substrates including plywood, particle boards, and decorative boards prepared using formaldehyde-based resins and folmaldehyde-urea copolymers as adhesives, by exposing the substrates to electron-beam radiation, followed by cold-plasma treatment (Origasa 1998).

Improved formalination of wood has been achieved by exposure of wood samples to plasma (Ishikawa et al. 1994). It was emphasized that a deep and uniform formalination can be generated.

Close-to-atmospheric-pressure, plasma-generated species from air/moisture or oxygen, nitrogen, and hydrocarbon mixtures were used for surface modification of plastic, metal, paper, wood, non-woven fabric, glass, ceramic, and building material sample surfaces. Low-cost surface hydrophilization of various materials is suggested.

Plasma-treated carbohydrate-based plant fibers were reacted with liquid CO_2 at 20°C, and the polymerized product was mixed with inorganic materials, such as Mg or diatomite, and molded, resulting in high-strength, heat-resistant, non-combustible building materials that also absorb sound and vibration (Togawa 1994).

Lignocellulosic materials, including wood, were wet-chemistry-, ozone-, corona-, plasma-, and UV-treated to generate carboxyl functionalities. Kneaded composites from modified lignocellulosics and UV-absorbers, anti-oxidants, synthetic resins, cement, wood, untreated lignocellulosics, concrete and ceramics were prepared and tested. Preparation of weather-resistant, lignocellulosic-based composite materials is claimed by the authors (Kikuchi et al. 2002).

Water- and moisture-resistant lignocellulosic boards were produced in reduced- and inert-gas-plasma treatment (glow discharge) of lignocellulosic-board substrates that were fabricated from wood powder, wood fiber, wood chips, and wood veneer. Water and moisture resistance of the modified boards are claimed by the authors (Sugawara et al. 1998).

Bamboo culm (*Phyllostachys bambusoides*) surfaces were wet–chemistry-(chromic acid; chromic acid/nitric acid) and plasma-treated (oxygen, nitrogen, and ammonia plasmas) for enhancing the adhesion characteristics to varnish (Kawamura and Kotani 1992). It was shown that both wet-chemistry and plasma treatments implanted surface carbonyl functionalities and increased the O/C relative atomic ratio of the modified surfaces. All treatments were found effective for the improvement of adhesion of varnish to bamboo surfaces.

Wood surfaces were ethylene-plasma-treated. Wettability, color changes, and contact angle modifications were evaluated (Uehara 2002). Wood surfaces were plasma-modified using various gas environments, and the wetting properties, penetration of adhesive material into wood pores, and adhesion characteristics were analyzed (Hatano and Hirabayashi 2001). Wood surfaces were exposed to an atmospheric-pressure-plasma-in-ethylene-gas environment for the generation of hydrophobic surface properties. Deposition thicknesses of 10.5 nm were evaluated from macromolecular layers deposited onto glass surfaces (12.75 kV, 0.1/min flow of ethylene for five minutes, atmospheric pressure). Wood veneer substrates were plasma-exposed under similar experimental conditions. Wettability of wood sample surfaces was decreased with the increase of treatment time, and the hydrophobic character of modified wood samples persisted after soaking them in water (Katakami et al. 2001).

Morphology and properties of carbonaceous macromolecular layers deposited under acetylene-DC discharge conditions onto spruce wood substrate surfaces was investigated (Eruzin et al. 1999).

Hydrophobic character and the formation of globular morphologies were noticed as a result of plasma treatment. Hardness measurement indicated that the hardness of the plasma-generated surface layer is 40% higher, relative to the surface of virgin wood substrates.

The adhesive bond strength between hollow, laminated veneer lumber (LVL) from sugi wood (*Cryptomeria japonica* or japanese cedar) and core filler made of injection-molded waste-paper-PE mixture as a result of various treatments, including mechanical processing, plasma irradiation, and flame treatment, was investigated using resorcinol-formaldehyde resin as the adhesive (Yang et al. 1999). A significant contact angle decrease was noted as a result of flame and plasma treatments, while mechanical processing generated less intense contact angle decreases. Plasma irradiation effectively improved the dry bonding strength. However, the bonding strength under wet conditions was much lower.

Oil- and water-repellent wood surfaces were generated without damaging their sense of touch and beauty by exposing them to CF_4-plasma species. ESCA data indicated the presence of plasma-modified wood surfaces of CF, CF_2, and CF_3 functionalities, in addition to oxygen atoms in atomic vicinities that are not characteristic of wood structures. Surface oxidation during the plasma treatments was avoided by oven-drying the substrates. The modified surface layers were considerably stable and retained their water repellency even after soaking them in water (Matsui et al. 1992).

Electron-beam-induced (EBI) plasma was also applied for modification of surface characteristics of wood. Low enthalpy, EBI plasma (0.5–2.4 MJ/Kg) was used to modify individual wood components and wood samples (Sokolov et al. 1999). Formation of water-soluble lignin, hemicellulose, and cellulose in ratios proportional to these components in virgin wood was observed. Cellulose separated from wood exhibited a higher degree of degradation as compared to cellulose present in the wood matrix. Lignin showed good resistance to the plasma treatment. It was suggested that the EBI plasma technique could be employed in hydrolysis and saccharification of wood.

Treatment of waste lignocellulosic materials, such as sawdust or wood chips, under EBI cold-plasma conditions, followed by extraction with water and 6% NaOH, resulted in extractives that could be used as additives in paper preparation technologies for improvement of paper stock dehydration, mechanical strength, etc. Optimal plasma and extraction conditions were discussed (Sokolov et al. 2001).

To finish this chapter, we can conclude that even if many of the investigations related to plasma treatment of wood do not directly relate the plasma parameters to the chemical mechanisms and the resulting surface characteristics, it is clear that the surface modification of wood by these means is efficient and offers a great potential for the future, where, without a doubt, we will see plasma technologies being used as one important way of attaining the desired characteristics of wood for a specific application.

REFERENCES

Baars-Hibbe, L., Sichler, P., Schrader, C., Gessner, C., Gericke, K.-H., and Büttgenbach, S. (2003). Microstructured electrode arrays: Atmospheric pressure plasma processes and applications. *Surf. Coat. Technol.*, 174–175:519–523.

Becker, K.H. and Belkind, A. (2003). Introduction to plasmas. *Vac. Technol. Coat.* September, 28–37.

Bialski, A., Manley, R.St.J., Wertheimer, M.R., and Schreiber, H.P. (1975). Composite materials with plasma-treated components: Treatment-property correlation. *Polym. Prepr.* 16(1):70–72.

Boeuf, J.P. and Pitchford, L.C. (1996). Calculated characteristics of an ac plasma display panel cell. *IEEE Trans. Plasma Sci.* 24(1):95–96.

Cecchi, J.L. (1990). Introduction to plasma concepts and discharge configurations. In: Rossnagel S.M., Cuomo J.J., and Westwood W.D. (Eds.), *Handbook of Plasma Processing Technology*. Noyes Publications, Park Ridge, NJ, chap. 2.

Chang, C.H. and Pfender, E. (1990). Nonequilibrium modeling in low-pressure argon plasma jet, Part II: Turbulent flow. *Plasma Chem. Plasma Proc.* 10(3):493–500.

Chapman, B.N. (1980). Glow Discharge Processes: Sputtering and Plasma Etching. John Wiley & Sons, New York.

Chen, F.F. (1965.). Electric probes. In: Huddlestone R.H. and Leonard S.L. (Eds.), *Plasma Diagnostic Techniques*. Academic Press, New York.

Chen, H.Y. and Zavarin, E. (1990). Interactions of cold radiofrequency plasma with solid wood, I: Nitrogen permeability along the grain. *J. Wood Chem. Technol.* 10(3):387–400.

Chen, X. and Pfender E. (1983). Behavior of small particles in a thermal plasma flow. *Plasma Chem. Plasma Proc.* 3(3):351–366.

Cho, D. L. and Sjöblom, E. (1990a). Plasma treatment of wood. *Polym. Mater. Sci. Eng.* 62:48–49.

Cho, D.L. and Sjöblom, E. (1990b). Plasma treatment of wood. *J. Appl. Polym. Sci.: Applied Polymer Symposium.* 46:461–472.

Christiansen, A.W. (1990). How overdrying wood reduces its bonding to phenol-formaldehyde adhesives: A critical review of the literature, Part I: Physical responses. *Wood Fiber Sci.* 22(4):441–459.

Christiansen, A.W. (1991). How overdrying wood reduces its bonding to phenol-formaldehyde adhesives: A critical review of the literature, Part II: Chemical reactions. *Wood Fiber Sci.* 23(1):69–84.

Danikas, M.G. (1993). The definitions used for partial discharge phenomena. *IEEE Trans. Elec. Insul.* 28(6):1075–1081.

Denes, A.R. and Young, R.A. (1999). Reduction of weathering degradation of wood through plasma-polymer coating. *Holzforschung* 53(6):632–640.

Denes, A.R., Tshabalala, M.A., Rowell, R.M., Denes, F.S., and Young, R.A. (1999). Hexamethyldisiloxane-plasma coating of wood surfaces for creating water repellent characteristics. *Holzforschung* 53(3):318–326.

Eliasson, B., Egli, W., and Kogelschatz, U. (1994). Modeling of dielectric barrier discharge chemistry. *Pure Appl. Chem.* 66(6):1279–1286.

Eliasson, B., Hirth, M., and Kogelschatz, U. (1987). Ozone synthesis from oxygen in dielectric barrier discharges. *J. Phys. D: Appl. Phys.* 20(11):1421–1437.

Eruzin, A.A., Gavrilenko, I.B, and Udalov, Y.P. (1999). Surface polymerization of acetylene in plasma of a D.C. glow discharge on wood. *Fizika I Khimiya Obrabotki Materialov* 6:36–41.

Esteves Magalhães, W.L. and Ferreira de Souza, M. (2000). Solid softwood coated with plasma-polymer for water repellence. *Proceedings from the 3rd International Symposium on Natural Polymers and Composites.* (Meeting date 2000), 260–264.

Esteves Magalhães, W.L. and Ferreira de Souza, M. (2002). Solid wood coated with plasma-polymer for water repellence. *Surf. Coat. Technol.* 155:11–15.

Gavrilenko, I.B., Eruzin, A.A., Tsoi, Y., Lukin, V.G., Onegin, V.I., and Udalov, Y.P. (2002). Method for working of wood for musical instruments. Russian Patent, RU 2185283.

Gentile, A.C. and Kushner, M.J. (1996). Microstreamer dynamics during plasma remediation of NO using atmospheric pressure dielectric barrier discharges. *J. Appl. Phys.* 79(8):3877–3885.

Gericke, K.-H., Gessner, C., Scheffler, P. (2002). Microstructure electrodes as a means of creating uniform discharges at atmospheric pressure. *Vacuum* 65(3-4):291–297.

Goldman, M., Goldman, A., and Sigmond, R.S. (1985). The corona discharge, its properties and specific uses. *Pure App. Chem.* 57(9):1353–1362.

Goring, D.A.I. (1976). Plasma-induced adhesion in cellulose and synthetic polymers. *Transactions of the Symposium on The Fundamental Properties of Paper Related to its Uses.* (Meeting date 1973), 172–189.

Grill, A. (1994). *Cold Plasma in Materials Fabrication—From Fundamentals to Applications.* IEEE Press, New York.

Guanben, D., Yukun, H., and Zheng, W. (2001). Wood surface treatment with microwave plasma. *Proceedings from the 7th International Symposium on Wood Adhesives.* (Meeting date 2000), 361–366.

Harada, T., Hirata, T., Setoyama, K., Ohkoshi, M. (1994). Utilization of wood and its feasibility as ecomaterials. *Trans. Mater. Res. Soc. Jpn.* 18A(Ecomaterials):677–680.

Hatano, Y. and Hirabayashi, Y. (2001). Wettability on the wood surface treated with plasma and adhesive strength. *Koen Yoshisu-Nippon Setchaku Gakkai Nenji Taikai* 39:5–6.

Herman, H. (1988). Plasma-sprayed coatings. *Sci. Am.* 259(3):112–117.

Hershkowitz, N. (1989). How Langmuir probes work. In: Auciello O. and Flamm D.L. (Eds.), *Plasma Diagnostics.* Vol. 1. Academic Press, New York.

Hitchon, W.N.G. (1999). *Plasma Processes for Semiconductor Fabrication.* Cambridge University Press, Cambridge.

Hollahan, J.R. and Bell, A.T. (Eds.). (1974). *Techniques and Applications of Plasma Chemistry.* John Wiley & Sons, Inc., New York.

Ishikawa, H., Usui, H., Adachi. A., Konishi, S., Oonishi, K., and Pponda, K. (1994). Improved process for formalination of wood. *Jpn. Kokai Tokkyo Koho,* JP 06099409.

Katakami, E., Uehara, T., and Katayama, H. (2001). Plasma polymerization of ethylene on wood at atmospheric pressure. *Nippon Setchaku Gakkaishi* 37(10):380–384.

Kawamura, J. and Kotani, K. (1992). Improvement treatments of bamboo culm surfaces. *Mokuzai Gakkaishi* 38(4):417–423.

Kikuchi, M., Kataoka, A., Akita, K., and Kaneiwa, H. (2002). Weather-resistant lignocellulose, manufacture thereof, and composites therefrom for interiors/exteriors. *Jpn. Kokai Tokkyo Koho,* JP 20022283306.

Kogelschatz, U., Eliasson, B., and Egli, W. (1997). Dielectric-barrier discharges: Principle and applications. XXIII International Conference on Phenomena in Ionized Gases. *J. Phys. IV.* 7(C4):C4/47–C4/66.

Kondo, Y. (2002). Method for surface treatment. *Jpn. Kokai Tokkyo Koho,* JP 2002035574.

Kreuger, F.H., Gulski, E., and Krivda, A. (1993). Classification of partial discharges. *IEEE Trans. Elec. Insul.* 28(6):917–931.

Legeay, G., Epaillard, F., and Brosse, J.C. Surface modification of natural or synthetic polymers by cold plasmas. *Proceedings from the 2nd Annual Int. Conf. Plasma Chem. Technol.* (Meeting date 1984), 29–39.

Leveroni, E. and Pfender, E. (1990). A unified approach to plasma-particle heat transfer under non-continuum and non-equilibrium conditions. *Int. J. Heat Transfer,* 33(7):1497–1509.

Lukowsky, D. and Hora, G. (2002). Pretreatments of Wood to enhance performance of outdoor coatings. *Macromolecular Symposia.* 187:77–85.

MacRae, D.R. (1989). Plasma arc process systems, reactors, and applications. *Plasma Chem. Plasma Proc.* 9(1, Suppl.):85S–118S.

Mahlberg, R., Niemi, H.E.-M., Denes, F.S., and Rowell, R.M. (1999). Application of AFM on the adhesion studies of oxygen-plasma-treated polypropylene and lignocellulosics. *Langmuir* 15(8):2985–2992.

Mahlberg, R., Niemi, H.E.-M., Denes, F.S., and Rowell, R.M. (1998). Effect of oxygen and hexamethyldisiloxane plasma on morphology, wettability and adhesion properties of polypropylene and lignocellulosics. *Int. J. Adhesion Adhesives* 18:283–297.

Matsui, H., Setoyama, K., and Kurosu, H. (1992). Surface modification of wood in fluorine-containing gas plasma, I: Tetrafluoromethane plasma treatment. *Mokuzai Gakkaishi* 38(1):73–80.

Nasser, E. (1971). *Fundamentals of Gaseous Ionization and Plasma Electronics.* Wiley-Interscience, New York.

Origasa, T. (1998). Treatment of wooden substrates containing formaldehyde-based resins. *Jpn. Kokai Tokkyo Koho,* JP 1026410.

Podgorski, L., Bousta, G., Schambourg, F., Maguin, J., and Chevet, B. (2001). Surface modification of wood by plasma polymerization. *Pigment Resin Technol.* 31(1):33–40.

Podgorski, L., Chevet, B., Onic, L., and Merlin, A. (2000). Modification of wood wettability by plasma and corona treatments. *Int. J. Adhesion Adhesives* 20:103–111.

Podgorski, L. and Roux, M. (2000). Wood modification to improve the durability of coatings. *Färg och Lack Scandinavia* 46(1):4–12.

Podgorski, L. and Roux, M. (1999). Wood modification to improve the durability of coatings. *Surf. Coat. Int.* 82(12):590–596.

Raizer, Y.P. (1991). *Gas Discharge Physics.* Springer-Verlag, Berlin.

Rehn, P. and Viöl, W. (2003). Dielectric barrier discharge treatments at atmospheric pressure for wood surface modification. *Holz als Roh- und Werkstoff* 61(2):145–150.

Rehn, P., Wolkenhauer, A., Bente, M., Förster, S., and Viöl, W. (2003). Wood surface modification in dielectric barrier discharges at atmospheric pressure. *Surf. Coat. Technol.* 174–175:515–518.

Scheffler, P., Bräuning-Demian, A., Gericke, K.-H., Gessner, C., Penache, C., Schmidt-Böcking, H., and Spielberger, L. (2004). Micro-structured electrode 2D-arrays as a new device for direct current glow discharges. Available at www.pci.tu-bs.de/aggericke/Forschung/MSE2D_basic.htm.

Setoyama, K. (1996). Surface modification of wood by plasma treatment and plasma polymerization. *J. Photopolymer Sci. Technol.* 9(2):243–250.

Smith, R.W., Wei, D., and Apelian, D. (1989). Thermal plasma materials processing-applications and opportunities. *Plasma Chem. Plasma Proc.* 9(1, Suppl.):135S–165S.

Sokolov, O.M., Komarov, V.I., Kazakov, Y.V., and Visil'ev, M.N. (2001). Use of alkali-soluble extracts of plasma-modified wood for strengthening of paper. *Tsellyuloza, Bumaga, Karton* 9–10:38–31.

Sokolov, O.M., Vasilijev, M.N., and Chukhchin, D.G. (1999). Investigation of changes in wood composition during low-enthalpy treatment by electron-beam-induced plasma. *Izvestiya Vysshikh Uchebnykh Zavedenii, Lesnoi Zhurnal* 2–3:167–175.

Sugawara, A., Sawata, Y., Onishi, K., and Okudaira, Y. (1998). Manufacturing of modified lignocellulosic board with good water and moisture resistance. *Jpn. Kokai Tokkyo Koho*, JP 10305410.

Togawa, S. (1994). Method and apparatus for manufacture of solid building materials and the materials. *Jpn. Kokai Tokkyo Koho*, JP 06100349.

Uehara, T. (2002). Surface treatment of timber by atmospheric plasma polymerization. *Setchaku* 46(2):60–64.

Van Brunt, R.J. (1994). Physics and chemistry of partial discharge and corona—Recent advances and future challenges. *IEEE Trans. on Dielectrics and Electrical Insulators* 1(5):761–784.

Venugopalan, M. (1971). *Reactions Under Plasma Conditions*. Vol. 1. Wiley-Interscience, New York.

Vossen, J.L. and Kern, W. (Eds.). (1978). *Thin Film Processes*. Academic Press, New York.

Wistara, N., Denes, F.S., and Young, R.A. (2002). The resistance of CF_4 plasma treated tropical woods against white-rot (*Trametes versicolor* L. Fr. Pilat) attack. *Proceedings of the International Wood Science Symposium*. (Meeting date 2002), pp. 157–163.

Yang, P., Jinno, K., Nishimoto, A., Ohsako, Y., Yamauchi, H., and Sasaki, H. (1999). Reinforcement of cylindrical LVL, I: Improving the bondability of waste paper-polyethylene composite core fillers to the cylindrical LVL. *Mokuzai Kogyo* 54(9):416–419.

Index

A

Abies spp., 16, 24
    chemical composition of wood of, 58, 71
    pitting in, 24
    polysaccharide content of wood of, 60
Acacia sp., 26
Accelerators, in adhesives, 271
Acer spp., 16, 17, 26
    chemical composition of wood of, 56, 69, 71
    polysaccharide content of wood of, 60
Acetyl content, of common hardwoods and softwoods, 55
Acetyl group determination, by gas liquid chromatography, 67–68
Acetylation
    of wood, 386–387
        acetyl balance in, 395–396
    of wood fiber, 411–412
Acid chlorides, reaction with wood, 387
Acidity, and wood degradation, 163, 328
Acoustical properties, of chemically modified wood, 410
Acrylic monomers, 426–429
Acrylonitrile, reaction with wood, 390, 432–433
Activation energy, 84–87
Adhesion, of wood, 225
    bond strength in, 221, 238–241, 242–245
    bonding surface in, 226–229
    chemical modification and, 410
    durability testing in, 245–246
    and environmental concern, 272
    models of, 224–225
    performance testing in, 242–245
    polymerization of adhesive in, 236–237, 239–240
    self-adherence in, 248–249
    setting of adhesive in, 234–236
    solidification of adhesive in, 237–238
    spatial scales of wood for, 229–230
    stress–strain behavior in, 238–241
        variables and effects on, 241–242
    surface preparation for, 225–226
    theories of, 221–225
    wetting and penetration in, 230–234
Adhesive(s)
    construction, 268–269
    contact, 270
    defined, 220
    formulation of, 270–272
    hot melt, 269
    pressure sensitive, 269–270
    wood, 215–217, 215–268, 246 (*see also* Adhesion)
        application of, 220–221
        biobased, 265–267
        epoxy, 261–263
        fillers in, 240–241, 271
        formaldehyde based, 249–257
        formulation of, 270–272
        isocyanates in, 257–261
        mechanical properties of, 238
        poly(vinyl acetate) and poly(ethylene vinyl acetate), 263–265
        polymerization of, 236–237, 239–240, 246–248
        polyurethane, 260–261
        in products with low polarity plastics, 267–268
        setting of, 234–236
        solidification of, by cooling, 237–238
        stress–strain data for, 238–241
        surface preparation for, 225–226, 231–234
        terminology used with, 219–220
        two-part, 237
        viscoelastic properties of, 238–241
        wetting and penetration by, 230–234
        and wood composites, 217–219
Adhesive joint failure, defined, 219
Adsorptive characteristics, of wood, 207–208, 358–360
    and wood strength, 328–329
Aesculus sp., chemical composition of wood of, 69
Afzelia spp., 27
Air forming, 350–353
Aldehydes, reaction with wood, 388–389
Alkyl chlorides, reaction with wood, 389
Alnus sp., chemical composition of wood of, 56
Alpha-cellulose preparation, 63–64
American Society for Testing and Materials (ASTM), 306
Amino-resins, as leach resistant fire retardants, 136–137
Anhydrides
    as cross-linking agents, 433
    reaction with wood, 387
Anobiid powder post beetles, 108
Antishrink efficiency (ASE), 93–94, *see also* Dimensional stability
Aquilaria sp., 12
Araucaria sp., 17
    pitting in, 24
Arbutus sp., chemical composition of wood of, 56
Arsenicals, inorganic, EPA approved consumer information on wood treated with, 116–120
Ash content, 50
    analysis of, 61–62
    of bark, 50
Aspen, chemically modified
    biological decay resistance of, 405
    dimensional stability of, 399
    equilibrium moisture content of, 398
    fiber saturation point of, 397
    mechanical properties of, 409

Aspen, chemically modified (*Continued*)
  stability of acetyl groups in, 394, 395
  thermal properties of, 407
  weight percent gain in, 396
Assembly time, defined, 219
Atmospheric pressure plasma treatment, 454–456, 467–469
Atomic force microscopy (AFM), in surface characterization, 192–195
Axial loading, and wood strength, 314–315
Axial parenchyma, 13
  of hardwood, 26
  patterns of, 27
  of softwood, 21–23

B

Bacteria, biological degradation by, 100
Bark, 46
  celluloses in, 49
  chemical composition of, 46
  extractives in, 46–48
  hemicelluloses in, 48
  inorganics and pH of, 50
  lignins in, 49
  of softwoods and hardwoods, 46, 47
Base, adhesive, 270
Beetles, wood damage by, 108–109
Below proportional limit, 322, 324
Beta-propiolactone, reaction with wood, 390
Betula spp., 11, 17
  chemical composition of wood of, 56, 71
  polysaccharide content of wood of, 60
Beyond proportional limit, 325
Biobased adhesives, 265–267
Biofilters, 360
Biological degradation
  by bacteria, 100
  by beetles, 108–109
  by carpenter ants, 107–108
  by carpenter bees, 108
  of chemically modified wood, 402–405
  by decay fungi, 101–103
  by insects, 103–105
  by marine borers, 110–111
  by mold and stain fungi, 100–101
  preservation of wood from, 111–120
  resistance of chemically modified wood to, 402–405
  by termites, 105–107
  and weathering, 173
  and wood strength, 332–336
Biological properties, 100–101
  moisture and, 90
Bond dissociation energy(ies), 151–152
  and chemical changes, 152–158
  and radiation wavelength, 153
Bond strengths, 222–224
Bonded chemical distribution, in chemically modified wood, 396–397
Bonded products, 225, *see also* Composites

Bonding performance testing, 243–245
Bonding surface, in wood adhesion, 226–229
Bonding variables, wood, 218
Borax, as fire retardant, 134, 136
Bordered pits, 20, 146–147
  ultraviolet degradation of, 165
Borers, wood damage by
  flathead, 108–109
  marine, 110–111
  old-house, 109
Boron, as fire retardant, 134, 136
Brown rot fungi, 102

C

Cambium, vascular, 14–15
Carbohydrate polymers, 37–43
Carbon dioxide plasma treatment, 464
Carboxylic acids, reaction with wood, 388
Carpenter ants, wood damage by, 107–108
Carpenter bees, wood damage by, 108
Carriers, in adhesives, 271–272
Carya sp., 17
  chemical composition of wood of, 56, 69
Catalysts, in adhesives, 271
Cedrela sp., 17
Ceiba sp., 12, 26
Cell wall
  chemical composition of, 27–29, 35–71, 149, 316–322
    cellulose and hemicelluloses, 37–43
    component distribution, 50–52
    extractives, 45–46
    holocellulose, 37
    inorganics, 50
    lignin, 43–45
    other polysaccharides, 43
  chemical modification of, 176–177
  chemistry of, 35–46
    analytical procedures in, 53–55, 61–71
  elasticity limit of, 87
  in juvenile wood and reaction wood, 52–53
  in North American hardwoods, 56–57, 60, 69–70
  in North American softwoods, 58–59, 60–61
  structure of, 18–19, 316–322 (*see also* Strength)
  ultraviolet degradation of, 165
Cellobiose, chemical structure of, 37
Cellular structure, of wood, 17–21
Cellulose(s), 27–28, 37–39, 320
  amorphous and crystalline regions of, 149
  in bark, 49
  depolymerization of, rate constants for, in air and nitrogen, 125
  in hardwoods, 150
  heat of combustion for, 127
  pyrolysis and combustion of, 125
    tar and char yields from, 126–127
  pyrolytic degradation of, 123–128
  in softwoods, 150
  structure of, 38–39
  in wood cells, 148

Index

Celtis spp., 17
  chemical composition of wood of, 56
Cerambycids, 109
Chamaecyparis spp., 24
  chemical composition of wood of, 58
Char, 122, 123
  formation of, 125–127
    chemicals promoting, 131–133
  heat of combustion for, 127
Checks, and raised grain, in weathering, 170–171
Chemical adsorption, and wood strength, 328–329
Chemical analysis, or preparation, 53
  of acetyl groups, 67–68
  of alpha-cellulose, 63–64
  of ash content, 61–62
  extraction procedure in, 54–55, 61
  of hollocellulose, 62–63
  of Klason lignin, 65
  of methoxyl groups, 65–67
  sampling procedure in, 53–54
Chemical composition, of cell wall, 27–29, 35–71, 149, 316–322
  cellulose and hemicelluloses in, 37–43
  component distribution in, 50–52
  extractives in, 45–46
  holocellulose in, 37
  inorganics in, 50
  of juvenile wood and reaction wood, 52–53
  lignin in, 43–45
  in North American hardwoods, 56–57, 60, 69–70
  in North American softwoods, 58–59, 60–61
  other polysaccharides in, 43
  and wood strength, 322–325
Chemical initiators, 424–425
Chemical leaching, in chemically modified wood, 393–395
Chemical modification, of wood, 176–177, 383, 392
  acetyl balance in, 395–396
  acoustical properties of, 410
  adhesion and, 410
  adhesion of, 410
  biological decay resistance of, 402–405
  chemical leaching of, 393–395
  commercial use of, 411
  dimensional stability of, 398–402
  distribution of bonded chemicals in, 396–397
  mechanical properties of, 408–410
  moisture sorption in, 397–398
  by monomer polymerization in lumen, 421–434
    (*see also* Monomer polymerization; Wood–polymer composites)
  production of, 411–413
  reagent penetration in, 395
  stability of bonded chemicals in, 393–395
  thermal degradation resistance of, 405–406
  ultraviolet degradation resistance of, 406–408
  volume changes in, 392–393
  weathering resistance of, 406–408
Chemical reaction systems, 383–386
  accessibility of reaction sites in, 395
  acetylation, 386–387
  acid chlorides, 387

acrylonitrile, 390
aldehydes, 388–389
alkyl chlorides, 389
anhydrides, 387
beta-propiolactone, 390
carboxylic acids, 388
epoxides, 390–391
formaldehyde, 388–389
isocyanates, 388
methylation, 389
moisture effects on, 392
reaction rates of, 391–392
Chemically bonded stabilizers, 175
Chemistry, of wood, 35–46, 50, 149
  analytical procedures in, 53–55, 61–71
  and wood strength, 316–347 (*see also* Chemical modification)
Chlorite hollocellulose preparation, 62–63
Chromated copper arsenate preservatives, 175
Chromic acid, as weathering retardant, 174–175
Coatings, wood, 448–449
Cohesive failure, defined, 219
Cold plasma, atmospheric pressure, 454–456
Cold plasma surface treatment, 457–470
Combustion
  flaming, 124
  intensity of, 126
Commercial stabilizers, 175–176
Compatibilizers, 368
Composite beams, 283
Composites, wood, 225, 279–281, 300
  adhesives used in, 217–219, 288–289
  applications of, 289–300
  bond strength in, 221, 238–241, 242–245
  bonding surface in, 226–229
  classification of, by particle size, density, and processing, 289–290
  durability testing of, 245–246
  and environmental concern, 272
  fiber web, 349–363 (*see also* Fiber webs)
  minor wood composition in, 267–268
  monomer polymerization in, 421–433
    (*see also* Monomer polymerization)
  performance of, 238–242, 289–300
  performance testing of, 242–245
  polymer-, 421–444 (*see also* Monomer polymerization; Wood–polymer composites)
  polymerization of adhesive in, 236–237, 239–240
  production and properties of, 289–300
    fiberboard, 295–298
    flakeboard, 293–295
    glued laminated timber, 290
    nanocomposites, 298–300
    particleboard, 293
    plywood, 291–293
    structural composite lumber, 290–291
  setting of adhesive in, 234–236
  solidification of adhesive in, 237–238
  spatial scales of, 229–230
  stress–strain behavior in, 238–241
    variables and effects on, 241–242

Composites, wood, 225, 279–281, 300 (*Continued*)
　surface preparation for, 225–226
　theories of adhesion of, 221–225
　thermoplastic, 365–378 (*see also* Thermoplastic composites)
　types of, 281–288
　　composite beams, 283
　　fiberboard, 285–287
　　flakeboard, 283
　　glued laminated timber, 281–282
　　nanocomposites, 287–288
　　other, 287
　　particleboard, 284
　　plywood, 282
　　structural composite lumber, 282–283
　　waferboard, 283
　weathering of, 171–173
　wetting and penetration in, 230–234
Compreg, 440
Compression
　in heating and curing with resins, 440
　parallel to grain, 314
　perpendicular to grain, 314–315
Compression (reaction) wood, 30–31, 146
　cell wall chemistry in, 52–53
Confocal laser scanning microscopy (CLSM), in surface characterization, 188–190
Conidendrin, 45
Construction adhesives, 268–269
Contact angle analysis (CAA), in surface characterization, 204–206
Copper based preservatives, 175
Cornus sp., chemical composition of wood of, 69
Coronas, 456
Creosote, EPA approved consumer information on wood treated with, 116–120
Critical oxygen index test, 129
Cross-field pitting, 24
Crustaceans, wood damage by, 110–111
Cupping, of board, 144
Cure, defined, 219
Curing agents, in adhesives, 271
Cyclic delamination test, 243

D

Dalbergia sp., 12, 27
Dampwood termites, 107
Debye shielding, 453
Decay fungi, biological degradation by, 101–103, 141
　resistance to
　　in chemically modified wood, 403–405
　　in plasma treated wood, 466
　　in wood–polymer composites, 438–439
Degradation, of wood, 383, *see also* Strength
　biological, 100–111 (*see also* Biological degradation)
　environmental, 140–178 (*see also* Weathering)
　thermal, 121–130 (*see also* Thermal degradation)
　ultraviolet, 149–152 (*see also* Ultraviolet degradation)
Degree of polymerization (DP), 37

Density, of wood, 29
Dicyandiamide, as fire retardant, 134
Dielectric barrier discharge (DBD), 455–456
Differential scanning calorimetry (DSC), 129
Differential thermal analysis (DTA), 129
Diisocyanate, polymeric diphenylmethane, 258–260
Diluent, in adhesive, 271
Dimensional stability, 90–94
　of chemically modified wood, 398–402
　loss of, in chemically modified wood, 401–402
　of plasma treated wood, 459–460
　of wood–polymer composites, 437–438
Diospyros spp., 12, 16
Dipole–dipole interactions, 222–223
Dispersive components, 207–208
DRIFT spectroscopy, 157–158
Dry wood, shrinkage of, 79
Dry wood termites, 107
Durability testing, 245–246

E

Earlywood, 16, 144–145
　cell wall chemistry of, 50–51
　ultraviolet degradation of, 167–170
Elastic strength, 322, 324
Elastic theory, 306
Electrical properties, of wood
　moisture and, 90
Electron beam radiation, 426
Electron spectroscopy, 199–202
　for chemical analysis (ESCA), 199
Electron spin resonance (ESR), detection of free radicals, 154–157
Electronic spectroscopy, 196
Elemental composition, of selected woods, 71
Emulsion polymer isocyanates, 260
Energy dispersive x-ray spectroscopy (EDXA, EDX, or EDS), 196
Environmental degradation, of wood, 140–142, 326–337
　acids in, effects of, 163, 328
　and biological degradation, 173
　chemical changes in, 152–154, 328–329, 336–337
　compared to decay, 140–141
　depth of degradation in, 162–163
　free radical formation in, 154–157
　hydroperoxides in, 157–158
　normal and accelerated, studies of, 165–170
　and paint adhesion, 143, 173
　physical aspects of, 164–171
　reaction products of, and chemical analysis, 158–162
　of tropical woods, 173
　ultraviolet radiation in, 149–152, 329
　of wood composites, 171–173
Environmental scanning microscopy (ESEM), 191
Epoxides, reaction with wood, 390–391
Epoxy adhesives, 261–263
Epoxy resins, wood impregnation with, 440
Equilibrium moisture content (EMC), 80–81, 308–309
　of chemically modified wood, 397–398

Index

Erosion control, with geotextiles, 355–356
Erosion rates, for various species, outdoor exposure, 167–171
Eucalyptus sp., chemical composition of wood of, 56
EVA adhesives, 263–265
Extenders, in adhesives, 271
Extraction procedure, 54–55, 61
Extractive content, 149
    in bark, 46–48
    in wood, 45–46
    and wood strength, 336–337

F

Fagus sp.
    chemical composition of wood of, 56, 69
    polysaccharide content of wood of, 60
Fatigue, and wood strength, 311
Fatty acids, in bark, 45, 47
Fiber(s), wood, 18–19, 25–26
    loss of, in ultraviolet degradation, 165–170
    orientation of, 319
    in thermoplastic composites, 367
Fiber acetylation process, 411–412
Fiber mat systems, 350–352
Fiber reinforced plastic, 231
Fiber saturation point (FSP), 28, 79–80, 383
Fiber stress, at proportional limit, 313
Fiber webs, 349–350
    additive use in producing, 353
    card and airforming in producing, 353
    fiber mixing in producing, 352
    in filters, 357–360
    forming, 352–353
    in geotextiles, 355–356
    layering in producing, 352
    melt blown polymer unit and air forming in producing, 353
    in mulch mats, 361
    in pulp molding, 353–354
    scrim addition in producing, 353
    as sorbents, 360–361
    systems for producing, 350–353
Fiberboard, 285–287, 295–298
Fillers, in adhesives, 240–241, 271
Filters, fiber
    applications of, 357–359
    chemical types of, 357
    isotherm protocol for testing, 359–360
    kinetic protocol for testing, 359
    for organic compounds, 360
    physical types of, 357
Fire, resistance to, 121–130
    in chemically modified wood, 405–406
    fire retardants and, 128–137
    in wood–polymer composites, 438
    and wood strength, 330–332
Fire retardants, 131
    barrier promoting chemicals as, 134

combustible gas dilution by, 134
    formulations for, 135–137
    free radical inhibition by, 133–134
    increased char formation by, 131–133
    leach-resistant, 136–137
    noncombustible gas release by, 134
    phosphorus and nitrogen as, 135
    testing, 128–130
    that promote thermal conductivity of wood, 134
    volatile gas heat reduction by, 134–135
    and wood strength, 330–332, 339–343
Flakeboard, 283, 293–295
Flaming combustion, 124
Flathead borers, 108–109
Flavonoids, in bark, 48
Flexural loading, and wood strength, 312–314
Forested areas
    regeneration of, 3
    in United States, 1–5
Formaldehyde, reaction with wood, 388–389
Formaldehyde copolymer adhesives, 249–257
    melamine, 255–257
    phenol-resorcinol in, 252–254
    phenols in, 250–252
    resorcinol in, 252–254
    urea, 254–255
Fortifiers, in adhesives, 271
Fourier transform infrared (FTIR) spectroscopy, 196–199
Fourier transform Raman (FT–Raman) spectroscopy, 196–199
Fraxinus spp., 16, 17, 28
    chemical composition of wood of, 56, 69, 71
Free radical catalyst, 423
Free radical formation
    bond energies and enthalpies of, 453
    in ultraviolet degradation of wood, 154–157
Free radical inhibition, chemicals causing, 133–134
Fuel wood, 1
Fungi, biological degradation by, 101–103
    resistance to, in chemically modified wood, 403–405

G

Gamma radiation, 425–426
Gas chromatography, for acetyl group determination, 68
Geotextiles, 355–356
Gingko sp., polysaccharide content of wood of, 60
Gleditsia sp., chemical composition of wood of, 56
Glued laminated timber (glulam), 281–282, 290
Gordonia sp., chemical composition of wood of, 69
Grain, raised, in weathering, 170–171
Grain orientation, 13–14
    in weathering, 170
Green volume, of wood, 29
Green wood, 28
    moisture content of, 77–79
Gribbles, 110
Growth characteristics, and wood strength, 307
Growth rings, 15–17

H

Half-bordered pits, 21
Hardboard, 287, 298
Hardeners, in adhesives, 271
Hardness, of wood, 316
Hardwood(s), 11–12
  acetyl content of, 55
  bark of, chemical composition, 47
  carbohydrate, lignin, and ash content of, 36–37
  cell wall component distribution in, 50–52
  cells of, 18
  cellulose composition of, 150
  characteristics of selected, for painting, 143
  chemical composition of, 317–318
  chemical composition of North American, 56–57
  hemicellulose composition of, 41, 150
  methoxyl content of, 54
  microscopic structure of, 24–27
  polysaccharide content of North American, 60
  structure of, 16–17
  ultraviolet degradation of, 160–162
  vessels, and vessel elements, of, 24–25, 146, 147
Heartwood, 10, 11, 12–13, 146
  moisture content of green, 78
  preservative penetration into, comparison of species, 114
Heat(s), of combustion, 127
Helium plasma treatment, 464, 465–466
Hemicellulose(s), 27–28, 39–40, 320–321
  in bark, 48
  determination of, 63–64
  in hardwoods, 41, 150
  in softwoods, 42–43, 150
  sugar monomers of, 40
  in wood cells, 148
Hevea sp., tension wood in, 30
Hexamethyldisiloxane (HMDSO)-RF plasma treatment, 460, 463
Hexanediol diacrylate (HDDA), 427
High density fiberboard, 287, 298
Historical use, of wood, 1–5
Holocellulose, 37
  preparation of, 62–63
Hooke's law, 305–306
Hot melt adhesives, 269
Hydrogen bonding, 223–224
  under shear forces, 323, 324
  and wood strength, 322–325
Hydroperoxides, in ultraviolet degradation of wood, 157–158
Hydroxyethyl methacrylate (HEMA), 427
Hydroxymethylated resorcinol (HMR), 253

I

Identification, of wood, 31–32
Impreg, 441
Inorganic additives
  effect on char yield and aromatic carbon content, 131–132
  effect on oxygen index and levoglucosan yield, 130
  effect on thermogravimetric analysis, 136
  that reduce heat content of volatile gases, 134–135
Inorganics, 50
  in bark, 50
Insects, biological degradation by, 103–105
  beetles, 108–109
  carpenter ants, 107–108
  carpenter bees, 108
  marine borers, 110–111
  termites, 105–107
Insulation board, 286, 296–297
Insulation properties, of wood
  moisture and, 90
Interphase, defined, 219
Intumescent systems, 134
Inverse gas chromatography (IGC), in surface characterization, 206–207
Ion exchange capacity, of wood, 358–359
Ionic bond formation, 224
Irradiance, defined, 151
Isocyanates
  in adhesives, 257–261
  as cross-linking agents, 433
  reaction with wood, 388
Isotherms, sorption, 81–82, 359–360

J

Juglans sp., 17, 27
  tension wood in, 30
Juniperus spp.
  chemical composition of wood of, 58
  pitting in, 24
  polysaccharide content of wood of, 60
Juvenile wood, 30–31, 145–146
  cell wall chemistry in, 52–53
  cross grain checking in, 146

K

Khaya sp., 28
Klason lignin preparation, 65

L

Laguncularia sp., chemical composition of wood of, 56
Laminated strand lumber (LSL), 282–283, 290–291
Laminated timbers, 281–282
Laminated veneer lumber (LVL), 282–283, 290–291
Larix spp., 17
  chemical composition of wood of, 58
  polysaccharide content of wood of, 60
Laser ionization mass spectroscopy (LIMA), 202
Latewood, 16, 144–145
  cell wall chemistry of, 50–51
  ultraviolet degradation of, 167–170
Leach-resistant fire retardants, 136–137
Levoglucosan, from cellulose, 125

Index

Libocedrus sp., chemical composition of wood of, 58
Licaria, 27
Lignin, 19, 27–28, 43–45, 149, 321
  in bark, 49
  chemical isolation of, 45
  classification of, 44–45
  heat of combustion for, 127
  photo-oxidation mechanism in, 161
  precursors of, 43–44
  preparation of Klason, 65
  ultraviolet absorption spectra for, 158
  ultraviolet degradation of, 164–165
  units of, 150
  in wood cells, 148
Lignin adhesives, 267
Lignocellulosic surfaces, properties of, 207–208, 358
Liquidambar sp.
  chemical composition of wood of, 56, 69, 71
  polysaccharide content of wood of, 60
Liriodendron sp., 17
  chemical composition of wood of, 56, 69
Lithocarpus sp., chemical composition of wood of, 57
Load factors, and wood strength, 311–312
  axial, 314–315
  flexural, 312–314
London dispersion force, 222
Long-horned beetles, 109
Low density fiberboard, 286, 296–297
Low polarity plastics, wood adhesion with, 267–268
Lumen modification, 421–422
  monomer polymerization in, 422–423
    cross-linking agents, 433
    methods of, 423–426
    monomers, 426–433
Lyctid powder post beetles, 108

M

Macroscopic properties, of wood, 142
Magnolia sp., chemical composition of wood of, 69
Maleic anhydride (MA), 427
Maleic anhydride-grafted polypropylene (MAPP), 368
Malus, 14
Manilkara, 26
Marine borers, wood damage by, 110–111
  resistance to, in chemically modified wood, 405
Mass spectroscopy, 196, 202–204
Mechanical properties, of wood, 305–307
  average coefficients of variations for, 308
  chemical modification of, 408–410, 439
  factors affecting, 307–316
    axial loading, 314–315
    flexural loading, 312–314
    growth characteristics, 307
    hardness, 316
    load, 311–312
    moisture, 90, 308–309, 310
    shear force, 316

shock resistance, 316
specific gravity, 307, 308
temperature, 309, 311
Medium density fiberboard, 286, 297–298
Melamine formaldehyde adhesives, 255–257
Melamine resins, in monomer polymerization, 432
Melt blown polymer unit, and air forming, 353
Metal alloys, as fire retardants, 134
Methoxyl content, of common hardwoods and softwoods, 54
Methoxyl groups, determination of, 65–67
Methyl methacrylate (MMA), 426–427
Methylation, reaction with wood, 389
Microbial degradation, 100–103
  and wood strength, 332–336 (*see also* Biological degradation)
Microfibril orientation, 319
Micropholis, 27
Microscopic methods, for surface characterization, 188–195
Microscopic structure, of wood, 21–27
Microwave plasma treatment, 466–469
Microwave process, solid wood, 412–413
Middle lamella, ultraviolet degradation of, 164–165
Milalenca sp., chemical composition of wood of, 57
Modulus of elasticity, 306, 313
Modulus of rigidity, 306
Modulus of rupture, 312–313
Moisture content, 28, 77–79
  and cell wall elastic limit, 87
  of chemically modified wood, 397–398, 458–459
  distribution of, 83
  and effects on reactions with wood, 392
  equilibrium, 80–81
  and fiber saturation point, 79–80
  and moisture cycles, 89
  and properties of wood, 90
  and sorption isotherms, 81–82, 359–360
  and swelling, 82–83
  and swelling in liquids other than water, 94–97
  and swelling in wood composites, 94
  and swelling measurement, 84
  and swelling pressure, 87–89
  and water repellency and dimensional stability, 91–94
  and water sorption and activation energy, 84–87
  and wood strength, 90, 308–309, 310, 329
Moisture exclusion efficiency, 438
Moisture retention, cold plasma treatment and effect on, 267–268, 458–459
Mold and stain fungi, biological degradation by, 100–101
Molecular spectroscopy, in surface characterization, 196–199
Monomer formation, from cellulose, 125–126
Monomer polymerization, 422–423
  in adhesive–wood composites, 236–237, 239–240
  cross-linking agents in, 433
  methods of, 423–426
  monomers used in, 426–433
    acrylics, 426–429
    acrylonitrile, 432–433

Monomer polymerization, 422–423 (*Continued*)
   melamine resins, 432
   polyesters, 431–432
   styrene, 429–431
  in wood lumens, 422–434
Mulch mats, fiber, 361

N

Nanocomposites, 287–288, 298–300
Nitrogen, as fire retardant, 135
Nitrogen plasma treatment, 464, 465–466
Nonsubterranean termites, 107
Novolak resins, 251
Nyssa spp., chemical composition of wood of, 57, 69

O

O-quinonoid moiety, 157
Ochroma, 26
Old-house borers, 109
Oriented strand board (OSB), 282–283, 293–295
Oriented strand lumber (OSL), 282–283
Oxydendron sp., chemical composition of wood of, 69
Oxygen plasma treatment, 462–466

P

Paint adhesion, and weathering, 143, 173
Paints, and stains, against weathering, 177–178
Parallel strand lumber (PSL), 282–283, 290–291
Partial discharges, 455
Particleboard, 284, 293
Pectin, in cell wall, 43
Pentachlorophenol, EPA approved consumer information on wood treated with, 116–120
Perforation plate, 25
Performance testing, bonding, 243–245
Peroxides, as polymerization initiators, 424
Persea sp., chemical composition of wood of, 69
pH, of bark, and wood, 50
Phenol formaldehyde adhesives, 250–252
Phenol–resorcinol formaldehyde adhesives, 252–254
Pholads, 110
Phosphorus, as fire retardant, 135, 136
Photochemical reactions, requirements for, 151–152
Photons, of energy, 150
Physical properties, of wood, 27–29
Picea spp., 11, 16
  chemical composition of wood of, 58, 71
  microscopic structure of, 22
  polysaccharide content of wood of, 60
  tension wood in, 30
Pillbugs, 111
Pine, chemically modified
  antishrink efficiency of, 401
  biological decay resistance of, 403, 406
  degree of substitution in, 396
  dimensional stability of, 400

  equilibrium moisture content of, 398
  fiber saturation point of, 397
  mechanical properties of, 409
  reaction rates for, using different chemicals, 391
  stability of acetyl groups in, 394, 395
  thermal properties of, 406, 407, 408
  volume changes in, 392
  weight percent gain in, 396
Pinus spp., 11, 16
  cell wall chemistry of, 51–52
  chemical composition of wood of, 58–59, 71
  compression wood in, 30
  latewood in, 16
  pitting in, 24
  polysaccharide content of wood of, 60
  specific gravity of, 29
  ultraviolet degradation of, 158
Pits, 20–21
Pitting, cross-field, 24
Planes of section, of wood, 13–14
Plasma(s), 449–450
  application of, for deposition and surface modification, 454
  atmospheric pressure cold, 454–456
  classification of, 450–451
  conditions for, 453–454
  energy distribution of electrons in, 451–453
  natural and man-made, 450–451
  parameters of, 451–454
Plasma frequency, 453
Plasma sheath, 454
Plasma treated wood
  dimensional stability of, 459–460
  moisture retention of, 458–460
  plasma comparisons in, 458–470
  weathering degradation of, 460–461
Plastic strength, 325
Platanus sp.
  chemical composition of wood of, 69
  polysaccharide content of wood of, 60
Plywood, 282, 291–293
Podocarpus, 16, 17
  axial parenchyma of, 23
Poly(ethylene vinyl acetate) adhesives, 263–265
Poly(vinyl acetate) adhesives, 263–265
Polydimethylsiloxane (PDMSO) plasma treatment, 460–462
Polyester monomers, 431–432
Polyethylene glycol, 441
Polyethylene resins, 367
Polymer classes, 247–248
Polymer impregnation, 439–441
Polymeric diphenylmethane diisocyanate, 258–260
Polymerization
  of adhesive, 236–237, 246–248
    and bonding strength, 239–240
  of liquid monomers, in wood lumens, 422–434
Polyphenols, 45
Polypropylene
  oxygen plasma treatment of, effects of, 462–465
Polypropylene resin, 367–368

Polysaccharide(s), 149
   hydrogen bonding of, 322–325
   in North American hardwoods and
      softwoods, 60–61
Polyurethane adhesives, 260–261
Polyvinyl chloride resin, 368
Populus spp., 17, 26
   chemical composition of wood of, 57, 69, 71
   polysaccharide content of wood of, 60
Powder post beetles, 108
Precision, of ash content analysis, 62
Preservatives
   in adhesives, 271
   antiswell and antishrink efficiency of, 87
   against biological degradation, 111–113
      handling and use of wood treated with, 115–120
      retention level of some, 112
      study of, results, 113
      timber preparation and conditioning for, 113
      treatment process for, 115
   and effects on wood strength, 337–343
   penetration of heartwood by, in selected
      species, 114
   against ultraviolet degradation, 174–178
Pressure sensitive adhesives, 269
Procedure(s)
   for alpha-cellulose preparation, 64
   for ash content analysis, 62
   for chlorite hollocellulose preparation, 63
   for extraction, 61
   for Klason lignin preparation, 65
   for methoxyl group determination, 67
Proportional limit
   below, 322, 324
   beyond, 325
   fiber stress at, 313
   work to, 314
Protection, of wood, 111–120, 448–449, *see also*
      Preservatives
Protein glues, 265–266
Proteins, in cell wall, 43
Prunus sp., 17
   chemical composition of wood of, 57
Pseudotsuga menziesii, 16, 17, 20
   chemical composition of wood of, 58
   pitting in, 24
   polysaccharide content of wood of, 60
   resin canal complexes of, 23
Pulp molding, 353–354
PVA adhesives, 263–265
Pyrolysis, 121, 123
   and combustion, 122–128
   products of, 124

Q

Quercus spp., 17
   chemical composition of wood of, 57, 69–70, 71
   fibers in, 26
   latewood in, 16
   macerated cells of, 25
   polysaccharide content of wood of, 60

R

Radial system, of tree trunk, 13
Radiant exposure, defined, 151
Radiation, as polymerization initiator, 425–426
Raman band assignment, in softwood cellulose and lignin
      spectra, 198–199
Rays
   hardwood, 26–27
   sections of, 28
   softwood, 23–24
Reaction rates, 391–392
Reaction wood, *see* Compression (reaction) wood
Reagent penetration, in chemically modified wood, 395
Reagents, and materials
   for acetyl group determination, 68
   for alpha–cellulose preparation, 64
   for chlorite hollocellulose preparation, 63
   for extraction, 61
   for Klason lignin preparation, 65
   for methoxyl group determination, 66
Recycling wood products, 2–3
Report(s)
   for acetyl group determination, 68
   for alpha-cellulose preparation, 64
   for ash content analysis, 62
   for methoxyl group determination, 67
Resin canal complexes, of softwood, 21–23
Resins, 45
   amino–, as leach resistant fire retardants, 136–137
   compression in heating and curing with, 440
   epoxy, 261–263, 440
   formaldehyde based, 249–257
   melamine, 255–257, 432
   Novolak, 251
   polyethylene, 263–265, 367, 441
   polypropylene, 367–368
   polyurethane, 260–261
   polyvinyl, 263–265, 368
   Resole, 251–252
   synthetic, 441
   urea–formaldehyde, 236, 254–255
Resole resins, 251–252
Resorcinol formaldehyde adhesives, 252–254
Robinia, 26

S

Salix sp., 26
   chemical composition of wood of, 57
Sample preparation
   for acetyl group determination, 68
   for ash content analysis, 62
   for chlorite hollocellulose preparation, 63
   for extraction, 55
   for methoxyl group determination, 66

Sampling procedure, 53–54
Santalum, 13
Sapwood, 10, 11, 12–13, 146
  moisture content of green, 78
Southern pine, volumetric swelling coefficients for, 88
Sassafras sp., chemical composition of wood of, 70
Scales, of wood, spatial, 229–230
Scanning electron microscopy (SEM), in surface characterization, 190–192
Secondary ion mass spectroscopy (SIMS), 196, 202–204
Secondary layers, of cell wall, 18–19, 147, 149, 320–321
Sequoia sempervirens, 17
  chemical composition of wood of, 59
Setting, of adhesive, 234–236
  defined, 219
Shear forces, 316
Sheath potential, 454
Shipworms, 110
Shock resistance, of wood, 316
Shrinkage, wood, 28–29, 79
  average radial, tangential, and volumetric, 80
Simple pits, 21
Soft rot fungi, 103
Softwood(s), 11–12
  acetyl content of, 55
  bark of, chemical composition, 47
  carbohydrate, lignin, and ash content of, 36–37
  cell wall component distribution in, 50–52
  cells of, 18
  cellulose composition of, 150
  characteristics of selected, for painting, 143
  chemical composition of, 317–319
  chemical composition of North American, 57–59
  hemicelluloses in, 42–43, 150
  methoxyl content of, 54
  microscopic structure of, 21–24
  polysaccharide content of North American, 60–61
  structure of, 16–17
  tracheids of, 21, 146–147, 148
  ultraviolet degradation of, 160–162
Solid wood microwave process, 412–413
Solvent, adhesive, 270–271
Solvent loss, in setting of adhesive, 235–236
Sorbents, fiber, 360–361
Sorption isotherms, 81–82, 359–360
Spaceboard, 354
Spatial scales, of wood, 229–230
Specific gravity, of wood, 29, 142, 144
  and wood strength, 307, 308
Spectral irradiance, defined, 151
Spectral radiant exposure, defined, 151
Spectroscopic methods, for surface characterization, 196–204
Spruce, acetylated, effects of size on acetyl content in, 395
Sputtered neutral mass spectroscopy (SNMS), 202
Stability of bonded chemicals, in chemically modified wood, 393–395

Stabilizers
  in adhesives, 271
  against ultraviolet degradation, 175–176
Starch, in cell wall, 43
Static secondary ion mass spectroscopy (SSIMS), 196, 202–204
Staybwood, 440
Staypak, 440
Strength, of wood, 303–305, *see also* Chemical modification; Weathering
  acidity and, 328
  adsorption of chemicals and, 328–329
  chemical composition of wood and, 316–325
  environmental aspects of, 326–337 (*see also* Weathering)
  and extractive content, 336–337
  extractives and, 336–337
  factors affecting mechanical aspects of, 307–316
    axial loading, 314–315
    flexural loading, 312–314
    growth characteristics, 307
    hardness, 316
    loading, 311–312
    moisture, 308–309
    shear force, 316
    shock resistance, 316
    specific gravity, 307, 308
    temperature, 309, 311
  mechanical aspects of, 305–307
  microbial degradation and, 332–336 (*see also* Biological degradation)
  at microscopic and macroscopic level, 322–325
  moisture content and, 90, 308–309, 310, 329
  structural aspects of, 325–326
  thermal degradation and, 330–332 (*see also* Thermal degradation; Thermal properties)
  treatment methods and effects on, 337–343 (*see also* Preservatives)
  ultraviolet degradation and, 141–142, 329 (*see also* Ultraviolet degradation)
Stress, 305–307
Stress–strain curve, 322
Structural adhesive, defined, 220
Structural composite lumber, 282–283, 290–291
Structure, of wood, 11–21, 146–149
  cellular, 17–21
  chemical composition of, 27–29, 35–71, 149, 316–325
  macroscopic properties of, 2, 9–10, 142, 317
  microscopic structure of, 21–27, 317
  and strength, 325–326
Styrene monomers, 429–431
Substrate failure, defined, 219
Subterranean termites, 105–107
Sugars, in bark, 48, 49
Surface preparation, for adhesives, 225–226, 231–234
Surface properties, 187–188, 447–448
  characterization of
    by atomic force microscopy, 192–195
    by confocal laser scanning microscopy, 188–190
    by contact angle analysis, 204–206

Index

by electron spectroscopy, 199–202
by inverse gas chromatography, 206–207
by mass spectroscopy, 202–204
by molecular spectroscopy, 196–199
by scanning electron microscopy, 190–192
by total surface energy, 207–208
cold plasma modification of, 457–470
Surface treatment, of wood, 448–449
cold plasma, 457–470
Swelling, *see also* Moisture content
of dry wood, 82–83
in liquids other than water, 94–97
measurement of, 84
and moisture content, 82–83
pressure of, 87–89
tangential, in water and other solvents, 95–97
in wood composites, 94
and wood strength, 328–329
Swelling pressure, 87–89
Swietenia, 12, 28

T

Tabebuia, 26
Tack, defined, 219–220
Tannin adhesives, 266–267
Tannins, 45–46, 48
Tar, formation of, 126
Taxodium disticum, 17
chemical composition of wood of, 59
Technology, wood, 27–29
Tectona, 12, 17
Temperature, and wood strength, 309, 311
Tension, perpendicular to grain, 315
Termites, wood damage by, 105–107
resistance of chemically modified wood to, 402–403
Terpenes, in bark, 45, 47
Tetrafluoromethane plasma treatment, 464, 466, 468
Texture, of wood, 145
Thermal conductivity, promoting, 134
Thermal degradation, 121–130
fire retardants and, 128–137
resistance to
in chemically modified wood, 405–406
in wood–polymer composites, 438
and wood strength, 330–332
Thermal properties, 122–128
Thermodynamic methods, in surface characterization, 204–208
Thermogravimetric analysis (TGA), 122–123, 128
Thermoplastic composites
biological degradation of, 374
history of technology of, 366–367
markets for, 375–377
materials used in, 367–368
mechanical properties of, 372
moisture properties of, 373–374
performance improvement of, 374–375
performance of, 372–374
production processes for, 368–371
weathering and fire retardancy of, 374
Thermoplastic matrix materials, 367–368
Thinner, in adhesive, 271
Thuja spp., 16
chemical composition of wood of, 59
polysaccharide content of wood of, 61
Tilia spp., 14, 26
chemical composition of wood of, 57, 71
Total surface energy, in surface characterization, 207–208
Tracheids, 21, 146–147, 148
confocal laser scans of, 189–190
scanning electron micrographs of, 191, 192
Tree(s)
hardwood, 11–12
softwood, 11–12
structure of, 10–11
Tree trunk
axial and radial systems of, 13
cellular structure of, 17–21
growth rings of, 15–17
heartwood of, 12–13
microscopic structure of, 21–27
planes of section of, 13–14
sapwood of, 12–13
vascular cambium of, 14–15
zones of, 10–11
Tropical woods, ultraviolet degradation of, 173
Tsuga spp., 24
chemical composition of wood of, 59, 71
polysaccharide content of wood of, 61
Tunnel flame spread tests, 129, 130

U

Ulmus spp., chemical composition of wood of, 57, 70
Ultraviolet degradation, 141–142, 329
acids in, effects of, 163
and biological degradation, 173
chemicals that retard, 174–178
depth of, 162–163
free radical formation in, 154–157
hydroperoxides in, 157–158
mechanism of, 153
physical aspects of, 164–171, 329
radiation spectra in, 149–152
reaction products of, and chemical analysis, 158–162
resistance to, in chemically modified wood, 406–408
of tropical woods, 173
of wood composites, 171–173
Ultraviolet spectrum, 149–151
Unit stress, 307
Urea, as fire retardant, 134
Urea formaldehyde adhesives, 254–255
Urea–formaldehyde (UF) resins, 236

V

Vascular cambium, 14–15
Vazo catalysts, 424–425
Vessels, and vessel elements, of hardwoods, 24–25, 146, 147
Vibrational properties, of wood, moisture and, 90
Volume changes, in chemically modified wood, 392–393

W

Waferboard, 283, 293–295
Water repellency, 90–94
   in weathering, 170
Water repellent effectiveness (WRE), 92–93
Water repellent preservatives, 177
Water sorption, *see also* Moisture content
   rate of, 84–87
Wavelength interaction, with chemical moieties, 151–152
Weathering, 140–142
   acids in, effects of, 163, 328
   and biological degradation, 100–120, 173
   chemical changes in, 152–154, 328–329
   compared to decay, 140–141
   depth of degradation in, 162–163
   free radical formation in, 154–157
   hydroperoxides in, 157–158
   normal and accelerated, studies of, 165–170
   and paint adhesion, 143, 173
   physical aspects of, 164–171 (*see also* Strength)
   reaction products of, and chemical analysis, 158–162
   resistance to
      in chemically modified wood, 406–408
      in plasma treated wood, 460–462
      in wood–polymer composites, 439
   of tropical woods, 173
   ultraviolet radiation in, 149–152 (*see also* Ultraviolet degradation)
   of wood composites, 171–173
Weathering retardants, 174–178
Weight percent gain, 391, 395–396
Wetting, flow, and penetration, by adhesives, 230–234
White rot fungi, 102–103
Wood
   acidity of, 328
   adhesion of, 215–246 (*see also* Adhesion; Adhesive(s))
   axial and radial systems of, 13
   biological degradation of, 99–120 (*see also* Biological degradation; Preservatives)
   cells and cell walls in, 17–19 (*see also* Cell wall)
   characteristics of selected, for painting, 143
   chemical composition of, 27–29, 35–71, 149, 316–322
      cellulose and hemicelluloses, 37–43
      component distribution, 50–52
      extractives, 45–46
      inorganics, 50
      lignin, 43–45
      and wood strength, 322–325
   chemical modification of, 381–429 (*see also* Chemical modification; Chemical reaction systems; Degradation)
   chemistry of, 35–46, 50, 149
      analytical procedures, 53–55, 61–71
   cold plasma modification of, 457–470
   composites of, 225, 279–281, 300 (*see also* Composites)
   compression (reaction), 30–31, 146
   function of, in living tree, 9–10, 15
   growth rings of, 15–17
   heartwood of, 12–13, 146
   heat of combustion for, 127
   historical use of, 1–5
   ion exchange capacity and sorptive characteristics of, 358–359
   juvenile, 30–31, 145–146
   juvenile wood and reaction wood, chemical composition of, 52–53
   macroscopic properties of, 2, 9–10, 142, 317
   microscopic structure of, 21–27, 317
   moisture content of, 77–98 (*see also* Moisture content)
   physical properties of, 27–29
   pit structure of, 20–21
   planes of section of, 13–14
   plasma treated, 447–473 (*see also* Plasma treated wood)
   sapwood of, 12–13, 146
   self-adherence of, 248–249
   specific gravity of, 29, 142, 144
   strength of, 303–347 (*see also* Strength)
   structure of, 11–21, 146–149, 316–322
   surface characterization of, 187–211 (*see also* Surface properties)
   texture of, 145
   thermal properties of, 121–138 (*see also* Thermal degradation)
   vascular cambium of, 14–15
   weathering of, 139–185 (*see also* Weathering)
   zones of, 10–11
Wood bonding variables, 218
Wood fiber, *see* Fiber(s)
Wood fiber plastics, *see* Thermoplastic composites
Wood flour, 367
Wood identification, 31–32
Wood–polymer composites, 421–422, 434, *see also* Monomer polymerization
   abrasion resistance of, 437
   commercial use of, 439
   decay resistance in, 438–439
   dimensional stability of, 437–438
   fire resistance in, 438
   hardness of, 434
   moisture exclusion efficiency of, 438
   polymer impregnation in production of, 439–440
   properties of, after different modifications, 435–436
   Rockwell hardness of, 429
   toughness of, 434, 437
   volumetric swelling of, in water, 428

Index

water soluble polymers and synthetic resins in production of, 441
weathering resistance in, 439
Wood technology, 27–29
Wood thermoplastic composites, *see* Thermoplastic composites
Work to maximum load, 314
Work to proportional limit, 314

X

X–ray photoelectron spectroscopy (XPS), 196, 199–202
Xanthocyparis, 20

Y

Young equation, 90